Neglected Tropical Diseases and Phytochemicals in Drug Discovery

Neglected Tropical Diseases and Phytochemicals in Drug Discovery

Edited by

Chukwuebuka Egbuna
Chukwuemeka Odumegwu Ojukwu University,
Uli, Nigeria

Muhammad Akram
Government College University,
Faisalabad, Pakistan

Jonathan Chinenye Ifemeje
Chukwuemeka Odumegwu Ojukwu University,
Uli, Nigeria

WILEY

This edition first published 2022

Registered Office
John Wiley & Sons, Inc., 111 River Street, Hoboken, NJ 07030, USA

Editorial Office
111 River Street, Hoboken, NJ 07030, USA

For details of our global editorial offices, customer services, and more information about Wiley products visit us at www.wiley.com.

Wiley also publishes its books in a variety of electronic formats and by print-on-demand. Some content that appears in standard print versions of this book may not be available in other formats.

Library of Congress Cataloging-in-Publication Data

Names: Egbuna, Chukwuebuka, editor. | Akram, Muhammad, editor. |
Ifemeje, Jonathan Chinenye, editor.
Title: Neglected tropical diseases and phytochemicals in drug discovery /
edited by Chukwuebuka Egbuna, Muhammad Akram, Jonathan Chinenye Ifemeje.
Description: Hoboken, NJ : Wiley, 2022. | Includes index.
Identifiers: LCCN 2021032011 (print) | LCCN 2021032012 (ebook) | ISBN
9781119616603 (hardback) | ISBN 9781119617112 (adobe pdf) | ISBN
9781119617150 (epub)
Subjects: LCSH: Tropical medicine. | Phytochemicals–Therapeutic use. |
Materia medica, Vegetable. | Drug development.
Classification: LCC RC961.N4383 2021 (print) | LCC RC961 (ebook) | DDC
616.9/883–dc23
LC record available at https://lccn.loc.gov/2021032011
LC ebook record available at https://lccn.loc.gov/2021032012

Cover Design: Wiley
Cover Images: Courtesy of Egbuna; gailhampshire/Berberis vulgaris berries/Flickr;
Anna Frodesiak; Courtesy of Chukwuebuka Egbuna

Set in 9.5/12.5pt STIXTwoText by Straive, Pondicherry, India

Contents

List of Contributors

Surajudeen Abiola Abdulrahman
Health Education England, Leicester, United Kingdom

Yusuf Abubakar
Department of Biochemistry, Natural Product Research Laboratory, Bauchi State University, Gadau, Nigeria

Charles Oluwaseun Adetunji
Applied Microbiology, Biotechnology and Nanotechnology Laboratory, Department of Microbiology, Edo University Iyamho, Auchi, Nigeria

Juliana Bunmi Adetunji
Nutritional and toxicological Research Laboratory, Department of Biochemistry Sciences, Osun State University, Osogbo, Nigeria

Muhammad Akram
Department of Eastern Medicine and Surgery, Government College University Faisalabad, Faisalabad, Pakistan

Osikemekha Anthony Anani
Laboratory of Ecotoxicology and Forensic Biology, Department of Biological Science, Faculty of Science, Edo University Iyamho, Iyamho, Nigeria

Chibueze Gospel Awuchi
Department of Environmental Technology, Federal University of Technology Owerri, Owerri, Nigeria

Chinaza Godswill Awuchi
Department of Physical Sciences, Kampala International University, Kampala, Uganda

Zarrin Basharat
Jamil-ur-Rahman Center for Genome Research, Dr. Panjwani Center for Molecular Medicine and Drug Research, International Center for Chemical and Biological Sciences, University of Karachi, Karachi, Pakistan

Samrat Bhattacharyya
Department of Botany, Sister Nibedita Government General Degree College for Girls, University of Calcutta, Kolkata, West Bengal, India

D. Chandan
Department of Biotechnology and Bioinformatics, School of Life Sciences, JSS Academy of Higher Education and Research, Mysore, India

Dipak Chetia
Department of Pharmaceutical Sciences, Dibrugarh University, Dibrugarh, Assam, India

Chukwudi Jude Chikwendu
Department of Biochemistry, Faculty of Natural Sciences, Chukwuemeka Odumegwu Ojukwu University, Uli, Nigeria

Tehseen M. Dhorajiwala
Dhorajiwala, Freelance Bioinformatician, Thane, Maharashtra, India

Darcia D'mello
Department of Life Sciences, Garden City University, Bangalore, Karnataka, India

Francis Ushie Ebuara
Department of Microbiology, Faculty of Pure and Applied Sciences, Federal University Wukari, Wukari, Nigeria

Chukwuebuka Egbuna
Department of Biochemistry, Faculty of Natural Sciences, Chukwuemeka Odumegwu Ojukwu University, Uli, Nigeria
Nutritional Biochemistry and Toxicology Unit, World Bank Africa Centre of Excellence, Centre for Public Health and Toxicological Research (PUTOR), University of Port-Harcourt, Port Harcourt, Nigeria

Abeer M. El Sayed
Department of Pharmacognosy, Faculty of Pharmacy, Cairo University, Cairo, Egypt

Goddidit Esiro Enoyoze
Department of Biological Sciences, Plant Biology and Biotechnology Unit, Edo University Iyamho, Iyamho, Nigeria

Nebechi Jane Ezeofor
Department of Food Technology, School of Applied Science and Technology, Federal Polytechnic, Oko, Nigeria

Shahira M. Ezzat
Department of Pharmacognosy, Faculty of Pharmacy, Cairo University, Cairo, Egypt
Department of Pharmacognosy, Faculty of Pharmacy, October University for Modern Sciences and Arts (MSA), Cairo, Egypt

Mihnea-Alexandru Găman
"Carol Davila" University of Medicine and Pharmacy, Bucharest, Romania
Department of Hematology, Center of Hematology and Bone Marrow Transplantation, Fundeni Clinical Institute, Bucharest, Romania

Kurubaran Ganasegeran
Clinical Research Center, Seberang Jaya Hospital, Ministry of Health Malaysia, Penang, Malaysia

Iram Ghaffar
Department of Eastern Medicine and Surgery, Government College University, Faisalabad, Pakistan

S. M. Gopinath
Department of Biotechnology,
Davangere University,
Davangere, India

Sumit T. Haldar
Freelance Bioinformatician,
Thane, Maharashtra,
India

Leena Hameed
Faculty of Eastern Medicine,
Hamdard University Karachi,
Karachi, Pakisthan

Jayashree V. Hanchinalmath
Department of Life Sciences,
School of Sciences, Jain University,
Bangalore, Karnataka, India

Jonathan Chinenye Ifemeje
Department of Biochemistry, Faculty
of Natural Sciences, Chukwuemeka
Odumegwu Ojukwu University,
Uli, Nigeria

Mehwish Iqbal
Institute of Public Health, Dow
University of Health Sciences,
Karachi, Pakistan

Shaista Jabeen
Department of Zoology, DKM
College for Women (Autonomous),
Vellore, India

Goutam Kumar Jena
Department of Pharmaceutics, Roland
Institute of Pharmaceutical Sciences,
Berhampur, Odisha, India

Shyam Sundar Kesh
Department of Veterinary Clinical
Complex, Faculty of Veterinary and
Animal Sciences,
West Bengal University of Animal
and Fishery Sciences,
Kolkata, West Bengal,
India

Johra Khan
Department of Medical Laboratory
Sciences, College of Applied Medical
Sciences, Majmaah University,
Saudi Arabia; Health and Basic
Sciences Research Center,
Majmaah University,
Majmaah, Saudi Arabia

Arisha Khoso
HEJ Research Institute of Chemistry,
International Center for Chemical
and Biological Sciences, University of
Karachi, Karachi, Pakistan

Shiva Prasad Kollur
Department of Sciences, Amrita
School of Arts and Sciences,
Mysore, Karnataka, India

Umme Laila
Department of Eastern Medicine,
Government College University
Faisalabad, Faisalabad,
Pakistan

Olugbenga Samuel Micheal
Cardiometabolic Research Unit,
Department of Physiology, College of
Health Sciences, Bowen University,
Iwo, Nigeria

Ejaz Mohiuddin
Faculty of Eastern Medicine,
Hamdard University Karachi, Karachi,
Pakisthan

Andrew G. Mtewa
Chemistry Section, Malawi Institute
of Technology, Malawi University
of Science and Technology,
Thyolo, Malawi
Pharm-Biotechnology and Traditional
Medicine Center (PHARMBIOTRAC),
Mbarara University of Science and
Technology, Mbarara, Uganda

Chinam Niranjan Patra
Department of Pharmaceutics, Roland
Institute of Pharmaceutical Sciences,
Berhampur, Odisha, India

Uchenna Estella Odoh
Department of Pharmacognosy and
Environmental Medicines, Faculty of
Pharmaceutical Sciences, University
of Nigeria, Nsukka, Nigeria

Alloysius Chibuike Ogodo
Department of Microbiology, Faculty
of Pure and Applied Sciences, Federal
University Wukari, Wukari, Nigeria

Chidera Favour Ogodo
Department of Medical Services,
Federal University Wukari,
Wukari, Nigeria

Tolulope Olawumi Oladosun
Department of Microbiology,
Federal University of Agriculture,
Makurdi, Nigeria

Olugbemi Tope Olaniyan
Laboratory for Reproductive Biology
and Developmental Programming,
Department of Physiology,
Edo University Iyamho,
Iyamho, Nigeria

Ahmed Olatunde
Department of Biochemistry,
Abubakar Tafawa Balewa University,
Bauchi, Nigeria

Michael C. Olisah
Department of Medical Biochemistry,
Faculty of Basic Medical Sciences,
Chukwuemeka Odumegwu Ojukwu
University, Uli, Nigeria

Frances N. Olisaka
Environmental and Public Health
Microbiology Laboratory,
Department of Biological Sciences,
Faculty of Science,
Benson Idahosa University,
Benin City, Nigeria

Phebean Ozolua
Applied Microbiology, Biotechnology
and Nanotechnology Laboratory,
Department of Microbiology,
Edo University Iyamho,
Auchi, Nigeria

Santwana Palai
Department of Veterinary
Pharmacology and Toxicology,
College of Veterinary Science and
Animal Husbandry, Bhubaneswar,
Odisha, India

Jyanaranjan Panda
Department of Chemistry, Roland Institute of Pharmaceutical Sciences, Berhampur, Odisha, India

Seema J. Patel
Department of Biotechnology, Davangere University, Davangere, Karnataka, India
Department of Food Technology, Davangere University, Davangere, Karnataka, India

Sharangouda J. Patil
Department of Zoology, NMKRV College for Women (Autonomous), Bangalore, Karnataka, India

Shashank M. Patil
Department of Biotechnology and Bioinformatics, School of Life Sciences, JSS Academy of Higher Education and Research, Mysore, India

S. S. Patil
Virology Laboratory, ICAR – National Institute of Veterinary Epidemiology and Disease Informatics (NIVEDI), Bangalore, Karnataka, India

Kingsley C. Patrick-Iwuanyanwu
Nutritional Biochemistry and Toxicology Unit, World Bank Africa Centre of Excellence, Centre for Public Health and Toxicological Research (PUTOR), University of Port Harcourt, Port Harcourt, Nigeria
Department of Biochemistry, Faculty of Science, University of Port Harcourt, Port Harcourt, Nigeria

Prabhurajeshwar
Department of Biotechnology, Davangere University, Davangere, India

T. Pramod
Department of Microbiology, The Oxford College of Science, Bangalore, Karnataka, India

Shashanka K. Prasad
Department of Biotechnology and Bioinformatics, School of Life Sciences, JSS Academy of Higher Education and Research, Mysore, India

Ramith Ramu
Department of Biotechnology and Bioinformatics, School of Life Sciences, JSS Academy of Higher Education and Research, Mysore, India

Saher Rahat
Department of Eastern Medicine and Surgery, Government College University, Faisalabad, Pakistan

Mithun Rudrapal
Department of Pharmaceutical Chemistry, Rasiklal M. Dhariwal Institute of Pharmaceutical Education and Research, Chinchwad, Pune, Maharashtra, India

Mohamed A. Salem
Department of Pharmacognosy, Faculty of Pharmacy, Menoufia University, Menoufia, Egypt

Lalit R. Samant
Molecular Genetics Research Laboratory, Bai Jerbai Wadia Hospital for Children, Mumbai, Maharashtra, India

S. R. Santhosh Kumar
Department of Biotechnology, Davangere University, Davangere, Karnataka, India
Department of Food Technology, Davangere University, Davangere, Karnataka, India

Prithvi S. Shirahatti
Department of Biotechnology, Teresian College, Mysore, India

Chandan Shivamallu
Department of Biotechnology and Bioinformatics, School of Life Sciences, JSS Academy of Higher Education and Research, Mysore, Karnataka, India

Kirankumar Shivasharanappa
Department of Life Sciences, Garden City University, Bangalore, Karnataka, India

Amna Siddique
Department of Eastern Medicine and Surgery, Government College University, Faisalabad, Pakistan

Chandrashekar Srinivasa
Department of Biotechnology, Davangere University, Davangere, Karnataka, India
Department of Food Technology, Davangere University, Davangere, Karnataka, India

R. Suhas
Postgraduate Department of Chemistry, JSS College of Arts, Commerce and Science, Mysore, Karnataka, India

K. Sumana
Department of Microbiology, School of Life Sciences, JSS Academy of Higher Education and Research, Mysore, India

P. Sushma
Department of Biotechnology and Bioinformatics, School of Life Sciences, JSS Academy of Higher Education and Research, Mysore, Karnataka, India

Verwiyeh Silas Tatah
Department of Biochemistry, Faculty of Pure and Applied Sciences, Federal University Wukari, Wukari, Nigeria

Habibu Tijjani
Department of Biochemistry, Natural Product Research Laboratory, Bauchi State University, Gadau, Nigeria

Bui Thanh Tung
VNU School of Medicine and Pharmacy, Vietnam National University, Hanoi, Vietnam

Chukwuemelie Zedech Uche
Department of Medical Biochemistry and Molecular Biology, Faculty of Basic Medical Sciences, University of Nigeria, Enugu Campus, Nigeria.

Ahmed Zayed
Department of Pharmacognosy, Faculty of Pharmacy, Tanta University, Tanta, Egypt

Preface

Neglected tropical diseases (NTDs) represent a group of tropical diseases that are most prevalent in the rural and low-income countries where there are poor healthcare facilities, poor sanitary conditions, and a low standard of living. NTDs affect more than one billion people in both tropical and subtropical regions of the world, causing deaths of over 500 000 people annually while leaving survivors with a lifelong disability. NTDs are so named "neglected" because they often does not get the deserved attention like other diseases such as cancer, diabetes, HIV, COVID-19, and so on. There are about twenty (20) different kinds of NTDs. In this book, an effort was made to capture drug discovery progress and opportunities from medicinal plants and phytochemicals for the treatment of them all. In Part I: Introduction to NTDs, three (3) chapters were dedicated to providing a general overview of NTDs, the disease cycles, and potential druggable targets. In Part II: Protozoan Infections, drug discovery opportunities from medicinal plants and phytochemicals for the treatment of giardiasis, amoebiasis, and leishmaniasis were presented. In Part III: Helminth Infections, five (5) chapters on helminthiasis, human echinococcosis, lymphatic filariasis, dracunculiasis (Guinea worm disease), onchocerciasis (River blindness), and foodborne trematodiases were presented. The chapters provide comprehensive information about the potentials of phytochemicals against helminth infections. In Part IV: Bacterial Infections, four (4) chapters on Buruli ulcer, leprosy, trachoma, and yaws (endemic treponematoses) were presented. In Part V: Viral Infections, two (2) chapters on dengue fever and rabies were presented. In Part VI: Fungal and Ectoparasitic Infections, two (2) chapters on eumycetoma and ectoparasites (scabies and myiasis) were presented while in Part VII: Non-classified NTDs, special chapters were dedicated to malaria and human tuberculosis. In this book, the chapter authors (from key institutions) made a frantic effort to discuss the pathogenic mechanisms of each NTD and the potentials of phytochemicals for their treatment. Well-illustrated diagrams were added and one chapter on a computational approach to drug discovery was added. This book is designed to be useful to medicinal chemists, drug discovery scientists, students (graduate and post students), and senior academics. This book will also be useful to pharmaceutical industries (R&D) and health institutions such as the World Health Organization. The book editors are grateful to all chapter contributors and are open to receiving feedback and comments from potential users.

Chukwuebuka Egbuna
Chukwuemeka Odumegwu Ojukwu University, Nigeria
Muhammad Akram
Government College University, Faisalabad, Pakistan
Jonathan Chinenye Ifemeje
Chukwuemeka Odumegwu Ojukwu University, Nigeria

Part I

Introduction to Neglected Tropical Diseases

1

Epidemiology of Neglected Tropical Diseases

Kurubaran Ganasegeran[1] *and Surajudeen Abiola Abdulrahman*[2]

[1] *Clinical Research Center, Seberang Jaya Hospital, Ministry of Health Malaysia, Penang, Malaysia*
[2] *Health Education England, Leicester, UK*

List of Abbreviations

CATT	Card agglutination test for trypanosomiasis
DALYs	Disability-adjusted life years
DSE	Direct smear examination
HAT	Human African Trypanosomiasis
IDM	Intensified disease management
LMICs	Low- and middle-income countries
MTCT	Mother-to-child transmission
NTDs	Neglected tropical diseases
PCR	Polymerase chain reaction
PCT	Preventive chemotherapy with transmission control
SES	Socioeconomic status
STH	Soil-transmitted helminthiasis
USS	Ultrasound scan
WHO	World Health Organization

1.1 Introduction

Neglected tropical diseases (NTDs) encompass parasitic (helminth or protozoan), bacterial, viral, and fungal infections commonly found within the tropical and subtropical regions globally. NTDs are predominantly reported across poverty-stricken

Neglected Tropical Diseases and Phytochemicals in Drug Discovery, First Edition.
Edited by Chukwuebuka Egbuna, Muhammad Akram, and Jonathan Chinenye Ifemeje.

countries due to poor living conditions and sanitation systems, coupled with the inability to access safe water or clean food, and poor self-hygiene behaviors [1]. Although global scale-up interventions and responses to control NTDs are continually being executed, these efforts are often challenged by the re-emergence of cases at an escalated pace in endemic areas due to climate change and global warming [2]. Such phenomenological changes pose substantial challenges to international agencies and voluntary organizations' efforts to eliminate or address potential outbreaks. Vector-borne diseases like dengue, chikungunya, filariasis, onchocerciasis, and trypanosomiasis has been prevalent in recent years among vulnerable populations in endemic areas. Data reported to date have shown that almost a billion people worldwide, particularly from developing countries, were infected with NTDs [1].

NTDs burden are measured in terms of disability-adjusted life years (DALYs). The magnitude of NTDs burden is shown in Table 1.1. In 2010 alone, the total NTDs burden accounted for 16 355 thousand cases, with the bulk of these cases originating from low- (4458 thousand) and lower-middle (8871 thousand) income countries [3]. As for 2020, the total burden of NTDs is projected to be 7362 thousand cases, with low- and lower-middle income countries accounting for 2009 thousand and 4047 thousand cases, respectively [3].

Table 1.1 Burden of disease in 2010, 2015, and 2020 (projected) for selected NTDs.

	Burden of disease in DALYs, in thousands		
Disease	**2010**	**2015**	**2020 (Projected)**
Human African Trypanosomiasis	560	372	0
Chagas disease	550	253	316
Leishmaniasis	3198	1357	305
Soil-transmitted helminthiasis	5134	4443	2200
Schistosomiasis	3308	3514	1646
Lymphatic filariasis	2736	2071	2353
Onchocerciasis	564	1136	224
Foodborne trematodiases	665	1066	NA
Leprosy	215	489	39
Trachoma	334	279	278
Dengue	1243	2610	NA
Rabies	2297	1672	NA

NA, not available.
Source: Adapted from Mitra and Mawson [1] and Stolk et al. [3].

The World Health Organization (WHO) has executed two broad initiatives to control NTDs, namely preventive chemotherapy with transmission control (PCT) and intensified disease management (IDM) programs [4]. Targeted NTDs that use PCT interventions include lymphatic filariasis, onchocerciasis, schistosomiasis, and soil-transmitted helminthiasis (STH). Through PCT, safe, efficacious, and cost-effective drugs are administered periodically to the entire population at risk. In contrast, IDM focuses on NTDs that lack resources for large-scale interventions. Among the diseases include buruli ulcer, Chagas disease, Human African Trypanosomiasis (HAT), and leishmaniasis. While these strategies require continued funding and commitments, there is a need to explore new interventions and novel drug therapies in view of changing landscapes – from climate, healthcare, to geopolitical variations globally. In line with the objectives of the current book, this chapter aims to systematically appraise the global burden and epidemiological determinants of NTDs.

1.2 Protozoan Infections

1.2.1 Human African Trypanosomiasis

HAT, also known as sleeping sickness, is a vector-borne parasitic disease caused by protozoan parasites of the genus, *Trypanosoma,* species, *brucei* [5] and transmitted by insect vectors tsetse flies. With an estimated 65 million people at-risk across 36 countries in Sub-Saharan Africa, rural dwelling, poor access to quality health services (including limited surveillance, diagnosis, and treatment), displacement of populations, war, and poverty are important risk factors that facilitate the transmission of the disease [6]. Whereas over 70% of reported cases in the past 10 years occurred in the Democratic Republic of Congo, a wide variation in incidence has been reported between and within countries, and major epidemics were reported up until the late 1990s.

The chain of infection of HAT and control strategies are summarized in Figure 1.1. With a relatively long incubation period of weeks to months or even years, infection with HAT manifests in two distinct stages that correlates with the pathogenesis of the infection and symptomatology of the disease. In the first stage, also called the hemo-lymphatic stage, the trypanosomes multiply in subcutaneous tissues, blood, and lymph causing symptoms such as fever, myalgia, arthralgia, headaches, itching, and enlarged lymph nodes. The second stage – the neurologic or meningo-encephalitic stage – heralds the infection of the central nervous system as the parasites cross the blood–brain barrier causing symptoms such as change in behavior, confusion, sensory disturbances, problems with coordination, and sleep disturbances which can be rapidly fatal. Chronicity of

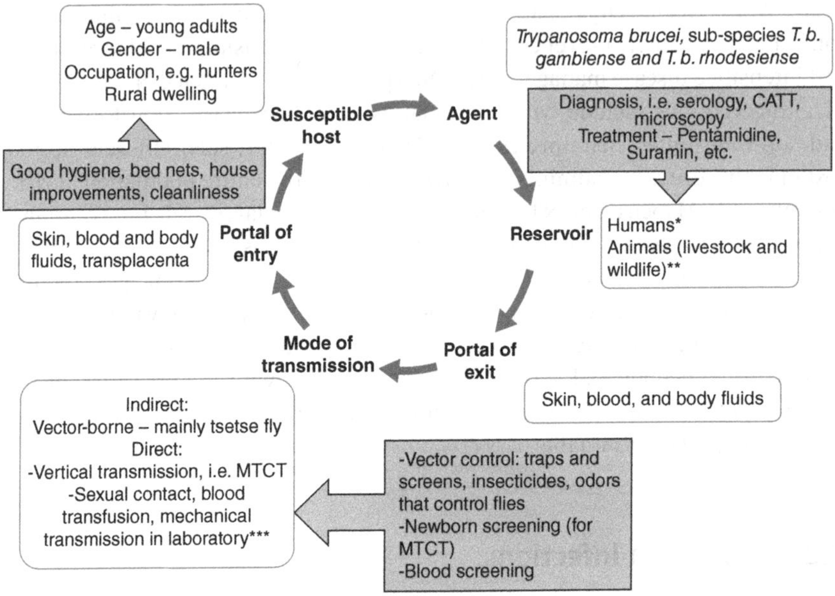

Figure 1.1 Chain of infection and control strategies for Human African Trypanosomiasis (HAT). *Mainly for gambiense HAT which is generally an anthroponotic infection; **Mainly for rhodesiense HAT which is generally a zoonotic disease; ***Rare cases; *Note:* Other medications used for treatment include melarsoprol, eflornithine, nifurtimox, and fexinidazole. *Source:* Based on Franco et al. [5].

infection with prolonged morbidity has been well documented. It is important to highlight that under the WHO leadership, global concerted efforts aimed at strengthening surveillance, prevention, and control measures have resulted in significant decline in 18-year incidence rate by up to 95% (from 2000 to 2018) with a road map targeting its elimination as a public health problem by 2020 and interruption of transmission (zero cases) for 2030 [6, 7].

1.2.2 Chagas Disease (American Trypanosomiasis)

Chagas disease – also called American Trypanosomiasis – is a potentially life-threatening vector-borne parasitic disease caused by protozoan parasite *Trypanosoma cruzi*, and transmitted by insect vectors *Triatoma infestans* (or "kissing bug"). The vector lives in crack walls or roof of rural homes and peridomiciliary structures, and infection is believed to be transmitted through smearing of the bug feces or urine through a skin crack (following a bite/blood meal), the eyes, or mouth. Up to 7 million people, mostly in Latin America, are reported to be infected with the disease as at 2019 [8]. In recent decades, effects of immigration and globalization have led to a change in the epidemiological profile of the disease, with urbanization of the disease and increasing incidence reported in the United States, Canada, many European and some African, East Mediterranean, and Western Pacific Countries [9]. Chagas disease is a complex, multidimensional public health problem with a strong interplay of socioeconomic and environmental determinants necessitating multi-sectorial prevention and control strategies/approaches.

The chain of infection of Chagas disease and control strategies are summarized in Figure 1.2. With an incubation period ranging from 1–2 weeks (for vector-borne acute Chagas disease) up to 120 days (for transfusion-and transplant-associated infections), the disease has two distinct phases: (i) The initial acute phase – typically lasts about two months after infection, and characterized by parasitaemia, absence of symptoms or mild and nonspecific symptoms. Where present, symptoms include fever, headache, enlarged lymph glands, myalgia, difficulty with breathing, abdominal and chest pain; (ii) the chronic phase is characterized by involvement of the cardiac and digestive muscles as well as neurological or mixed presentations.

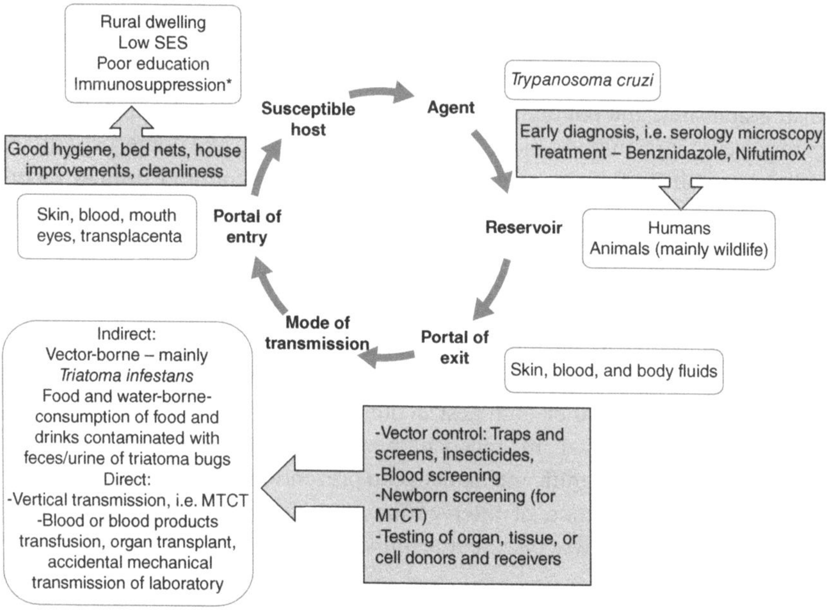

Figure 1.2 Chain of infection and control strategies for Chagas disease. *Immunosuppression can lead to reactivation of Chagas disease; ^Treatment is most effective when given soon after the infection at the onset of acute phase of the disease. The medications are contraindicated in pregnant women, people with liver or kidney failure, and in people with background neurological or psychiatric disorder. *Source:* Based on Nunes et al. [10].

Dilated cardiomyopathy (occurring in up to 30% of cases) is the most important and severe manifestation of human chronic Chagas disease and is characterized by heart failure, ventricular arrhythmias, heart blocks, thromboembolic phenomena, and sudden death [10]. Given that there is no vaccine for Chagas disease, control efforts are targeted at elimination of transmission (through vector control, blood screening) and early healthcare access of the infected and ill population.

1.2.3 Leishmaniasis

Leishmaniasis is a vector-borne parasitic disease caused by protozoan parasite *Leishmania donovani,* and transmitted by infected female sand fly *Phlebotomus argentipes* [11] during a blood meal. An estimated 700 000 to 1 million cases occur annually, with Brazil, India, Ethiopia, Sudan, and South Sudan accounting for a majority of reported cases in 2018 [12]. There are three main forms of leishmaniasis, including visceral (also called Kala azar, the most serious form of the disease and potentially fatal), cutaneous (the most common, and highly endemic in Iran, Sudan, and Somalia), and mucocutaneous (over 90% of cases occurring in Bolivia, Brazil, Ethiopia, and Peru) [12]. The epidemiology of leishmaniasis depends hugely on a dynamic interplay of agent characteristics (including pathogenicity), vector biology, herd immunity, and the local ecological characteristics of the transmission sites, and population dynamics (mainly human behavior). Risk factors for disease transmission include poverty, poor housing and domestic sanitary conditions, malnutrition, population mobility, urbanization and human incursion into forest areas, as well as climate change (temperature, rainfall, and humidity significantly impacts vector biology).

The chain of infection of leishmaniasis and control strategies are summarized in Figure 1.3. With an incubation period of two weeks to several months and years, symptoms may include fever, weight loss, anemia, and enlargement of the spleen and liver in visceral leishmaniasis; skin ulcers and scarring in cutaneous leishmaniasis, and partial or total destruction of the mucous membranes of the nose, mouth, and throat in mucocutaneous leishmaniasis. Asymptomatic infection constitutes a significant challenge to prevention and control efforts [13] necessitating a combination of intervention strategies to address the complex biological system that drives disease transmission.

1.2.4 Amoebiasis

Amoebiasis is a parasitic disease caused by protozoan parasite *Entamoeba histolytica* [14]. It is the third most deadly parasitic infection worldwide [15], and causes non-febrile intestinal disease and febrile extra-intestinal disease mainly affecting the liver. It is prevalent in low- and-middle income countries (LMICs) and mainly transmitted through the fecal–oral route, although a few cases of

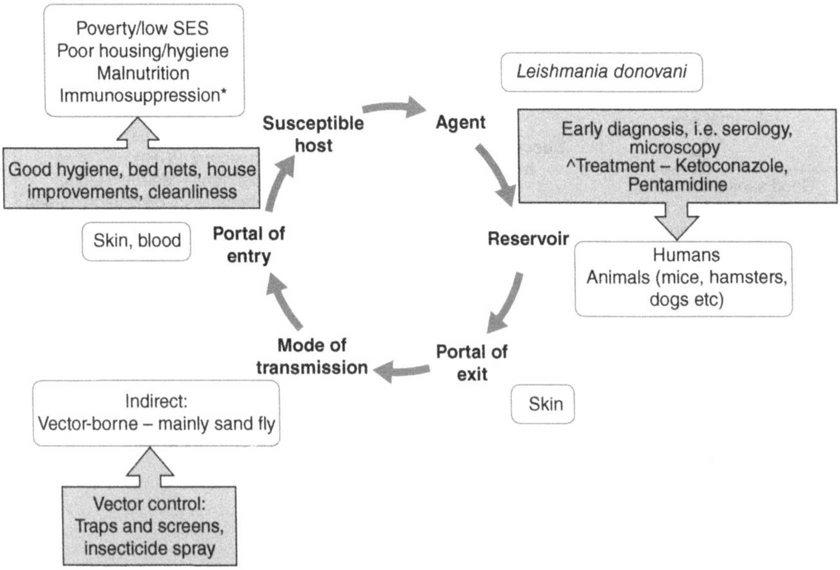

Figure 1.3 Chain of infection and control strategies for leishmaniasis. *Immunosuppression can lead to relapse of leishmaniases; ^Treatment depends on several factors, notably type of disease, concomitant pathologies, parasite species, and geographic location. *Source:* Based on Courtenay et al. [13].

sexual transmission especially among men who have sex with men have been reported [16, 17]. Risk factors include poor sanitary and food hygiene, unprotected sexual intercourse, migration, and/or traveling to endemic areas.

The chain of infection of amoebiasis and control strategies are summarized in Figure 1.4. The average incubation period of amoebiasis is two to four weeks, although this can range from months to years. The clinical spectrum of amoebiasis ranges from asymptomatic infection to fulminant colitis (with diarrhea and/or dysentery, abdominal pain, and weight loss) and peritonitis to extraintestinal amoebiasis, the most common form of which is amoebic liver abscess (presents with right upper quadrant pain with/without diarrhea) [18]. Prevention and control efforts are aimed at early diagnosis and treatment, health education, improvement of sanitation, and personal hygiene [19].

1.3 Helminth Infections

1.3.1 Soil-Transmitted Helminthiasis Infections

STH infections are principally caused by intestinal nematodes, namely the whipworm (*Trichuris trichiura*), the roundworm (*Ascaris lumbricoides*), and two species of hookworms (*Ancylostoma duodenale* and *Necator americanus*) [20].

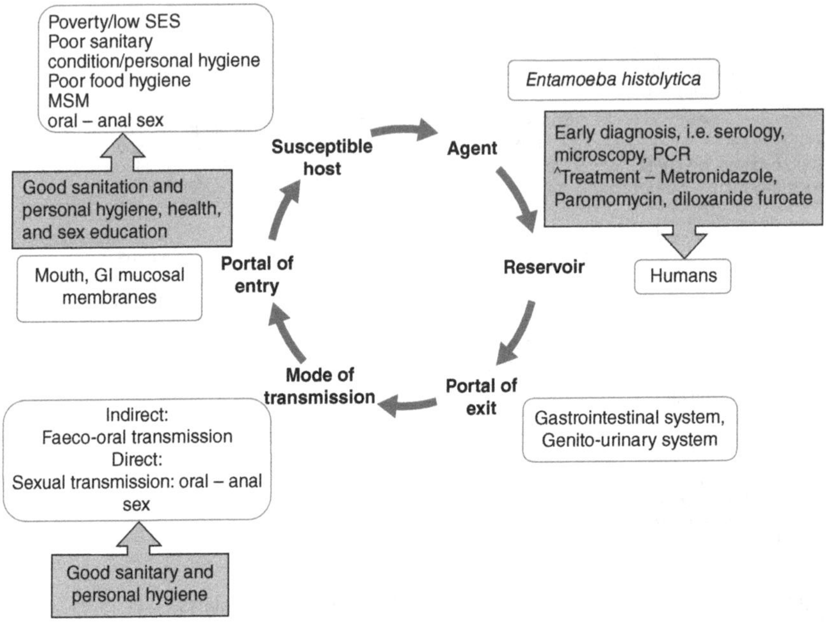

Figure 1.4 Chain of infection and control strategies for amoebiasis. ^Metronidazole or other nitroimidazoles are used for treatment of intestinal amoebiasis while paromomycin or diloxanide furoate are used to achieve luminal cure and prevent relapse. *Source:* Based on Watanabe [18].

Strongyloides stercoralis has much lower prevalence than the former four [21]. In areas of poor sanitation, infections are transmitted by eggs passed in the feces of infected people, which often contaminate the soil [1]. Infection by *Ascaris* or *Trichuris* occurs through consumption of contaminated food (such as ingested vegetables plucked from the soil that were not carefully washed or cooked), water and poor self-hygiene (such as when children play in the contaminated soil and subsequently put their hands in the mouth without washing) [21]. In contrast, hookworms directly penetrate the skin, as a consequence of walking barefooted on contaminated soil [21]. Approximately 24% of the world's populations are infected with STH, mostly from tropical and subtropical areas with poor sanitations, overcrowding, and countries with poor socioeconomic status (SES). Regions mostly affected include the sub-Saharan Africa, the Americas, China, and East Asia with preschool- and school-going children being highly susceptible [21].

The chain of infection of STH and control strategies are summarized in Figure 1.5. Eggs passed in feces typically take three weeks to mature in the soil before they become infective [21]. *Trichuris trichiura*, *Ascaris lumbricoides*,

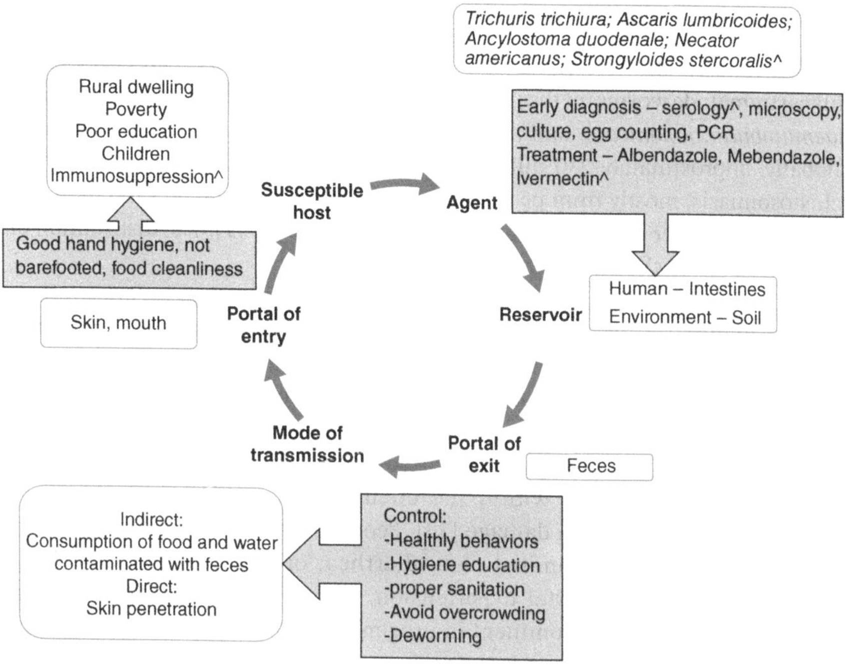

Figure 1.5 Chain of infection and control strategies for soil-transmitted helminthiasis (STH). ^Indicate key facts related to *Strongyloides stercoralis,* a less common intestinal parasite with more peculiar characteristics, being sensitive to only serological test. The parasite has the ability to reproduce within the host and in immunocompromised individuals, and is sensitive to Ivermectin, a drug to be made available from 2021 at an affordable price according to the WHO. *Source:* Based on WHO [21].

Ancylostoma duodenale, and *Necator americanus* typically do not multiply in the human host, and reinfection only occurs in the environment. These infections may cause clinical manifestations such as diarrhea, abdominal pain, malnutrition, malaise and weakness, or impaired growth development. Diagnoses are mainly conventional methods – microscopy, culture, or egg counting – but advancement in research has prompted the applications of polymerase chain reaction [22]. *Strongyloides stercoralis* is capable of multiplying within the human host and in immunocompromised individuals, escalating skin and gastrointestinal complications that could lead to chronic malnutrition. The organism may cause hyper-infection, a condition that could be highly fatal [21]. Diagnosis is only sensitive through serological test, thus making it more difficult to detect as compared with other organisms [22]. The improvement of sanitation systems, and continuation of health behavior education programs, deworming, and proper hygiene are the best measures to control STH infections.

1.3.2 Schistosomiasis

Schistosomiasis, also known as bilharzia, is a parasitic disease caused by blood flukes (trematode worms) of the genus *Schistosoma,* species *mansoni, intercalatum, haematobium, japonicum,* and *mekongi* that were known to infect humans [1]. Globally, approximately 230 million people were estimated to be infected with schistosomiasis, mostly from poor communities without adequate sanitation and safe drinking water across tropical and subtropical regions [1]. The distribution of schistosomiasis is shown in Table 1.2.

The chain of infection of schistosomiasis and control strategies are summarized in Figure 1.6. People are susceptible to schistosomiasis as soon as the larvae (released by aquatic snails) penetrate the skin while being in contact with infested water. The larvae subsequently develop into schistosomes and inhabit within the blood vasculature, allowing the females to release their eggs. While some of the eggs are released out of the body through feces or urine, others are trapped within the circulatory system that could cause immune reactions or organ damage [23]. People with intestinal schistosomiasis may complain of abdominal pain, diarrhea, or blood in the stool [23]. Hepatomegaly, ascites, portal hypertension, and splenomegaly are common signs [23]. Hematuria is prominent in urogenital schistosomiasis. In children, schistosomiasis causes anemia, stunting, and learning disabilities [23]. Diagnostic aims are to detect parasite eggs in stool or urine specimens. Efforts to control schistosomiasis are highly dependent on public health education and behavior programs, improved sanitation systems and better hygiene practices, and elimination of snail habitats [1].

Table 1.2 Global distribution of schistosomiasis.

Type	Species	Geographical distribution
Intestinal schistosomiasis	*Schistosoma mansoni*	Africa, Middle East, Caribbean, Brazil, Venezuela, Suriname
	Schistosoma japonicum	China, Indonesia, Philippines
	Schistosoma mekongi	Several districts of Cambodia, Laos
	Schistosoma intercalatum	Rain forest areas of Central Africa
Urogenital schistosomiasis	*Schistosoma haematobium*	Africa, Middle East, France

Source: Data from World Health Organization [23].

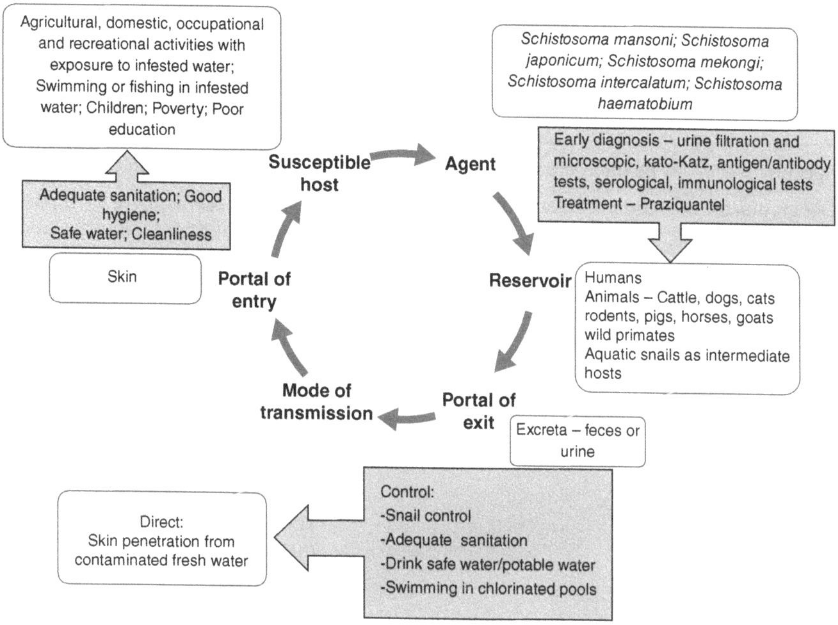

Figure 1.6 Chain of infection and control strategies for schistosomiasis. *Source:* Based on WHO [23].

1.3.3 Echinococcosis

Echinococcosis is a zoonotic disease caused by parasites, particularly tapeworms of the genus *Echinococcus*. The two most distinct forms of human echinococcosis include cystic echinococcosis (hydatidosis) and alveolar echinococcosis [24]. Characteristics and distribution of human echinococcosis are summarized in Table 1.3.

The chain of infection of echinococcosis and control strategies are summarized in Figure 1.7. Deworming of dogs with praziquantel, improved hygiene in slaughtering houses, vaccination of sheep, and public education are crucial measures for the control of human echinococcosis.

1.3.4 Lymphatic Filariasis

Lymphatic filariasis (also known as elephantiasis) is a chronic disabling parasitic disease caused by filarial worms – *Wuchereria bancrofti*, *Brugia malayi*, and *Brugia timori*, transmitted by mosquito vectors, Culex (common in urban or semi-urban settings), Anopheles (found in rural areas), and *Aedes* (prevalent in endemic islands in the Pacific) [1, 26]. The infection causes lymphatic obstruction, resulting

Table 1.3 Human echinococcosis and geographical distribution.

Type	Cystic echinococcosis (also known as hydatidosis)	Alveolar echinococcosis
Species	*Echinococcus granulosus*	*Echinococcus multilocularis*
Geographical distribution	Globally distributed except in the Antarctica. Prevalence rates range between 5 and 10% in parts of Argentina, Peru, East Africa, Central Asia, and China	Northern hemisphere – China, Russia, continental Europe, and North America. Emerging cases are found in Tajikistan, Pakistan, South Korea, Belgium, the Netherlands, Slovakia, Hungary, Lithuania, Latvia, Slovenia, Morocco, Taiwan, Thailand, and Denmark. High prevalence rates reported in France, Germany, Switzerland, Poland, and Lithuania. China and Kyrgyzstan topped the prevalence rates
Incubation period	Many years	5–15 years
Transmission	Herbivores or omnivores are intermediate hosts. They become infected by ingesting parasitic eggs from contaminated food and water. The parasite subsequently develops into larvae in the viscera. Carnivores are definitive host for the parasite, harboring the mature tapeworm in the intestine. They are infected through the consumption of viscera of intermediate hosts that contain the parasite larvae. Humans are regarded as the accidental intermediate hosts, acquiring the infection similarly to the intermediate hosts, but they do not transmit the infection to the definitive host	
Clinical manifestations	Abdominal pain, nausea, vomiting, chronic cough, chest pain, dyspnea, anorexia, weight loss, and weakness	Growth of tumor-like-lesion in the liver causing weight loss, abdominal pain, malaise, and hepatic failure. Larval metastasis to spleen, lungs, and brain could be fatal
Diagnosis	Ultrasonography, complemented by computed tomography or magnetic resonance imaging scans	
Treatment	PAIR (Puncture, Aspiration, Injection, Re-aspiration) procedure, surgery, anti-infective drug treatment	Radical (tumor-like) surgery and anti-infective prophylaxis with albendazole

Sources: Adapted from World Health Organization [24] and Baumann et al. [25].

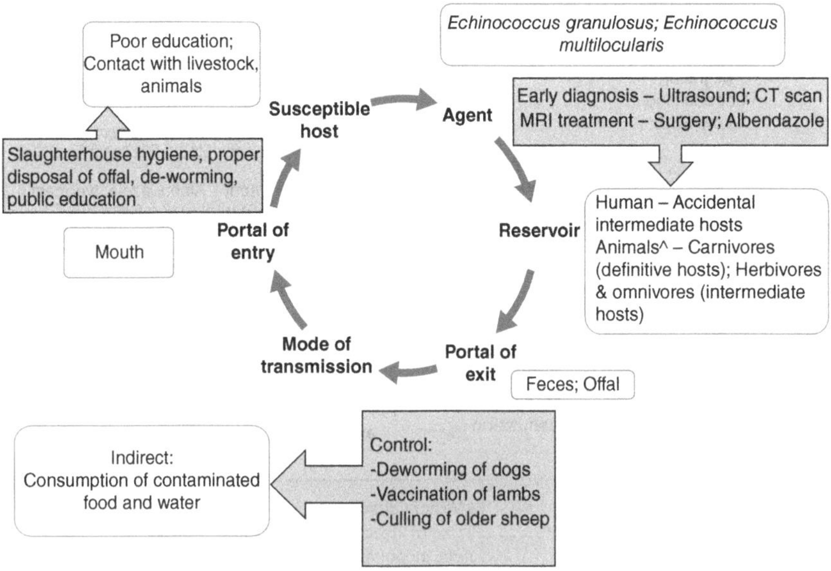

Figure 1.7 Chain of infection and control strategies for echinococcosis. ^Human infection in cystic echinococcosis mostly occurs in dog–sheep–dog cycle, or other domestic animals such as goats, swine, cattle, camels, and yaks. Alveolar echinococcosis attributed to wild carnivores such as foxes and rodents (as intermediate hosts) and domestic dogs or cats (definitive hosts). *Source:* Based on WHO [24].

in swelling of the lower extremities (elephantiasis) or genitals (hydrocele) [1]. It is currently estimated that almost 900 million people across 49 countries globally are at risk to be infected with lymphatic filariasis. Approximately 120 million people were infected throughout the tropics and subtropics, with the bulk coming from the sub-Saharan countries [1, 26, 27].

The chain of infection and control strategies are exhibited in Figure 1.8. Incubation period (time interval between invasion of infective larvae to the presence of clinical manifestations) ranges between 8 and 16 months, or may be more [29]. Adult worms nested in the lymphatic vessels disrupt the function of the lymphatic system. These worms produce millions of microfilariae (immature larvae) that circulate in the blood. Microfilariae assimilated through a blood meal from an infected host mature in mosquitoes into infective larvae, and subsequently reintroduced into the human body when these mosquitoes bite people. This cycle of transmission continues [26]. The disease could be asymptomatic but could also manifest as acute or chronic conditions. At acute stage, the disease could manifest as local inflammation involving the skin, lymph nodes, or lymphatic vessels, while in the chronic stage, lymphedema, elephantiasis, or hydrocele becomes prominent [26]. Thick smear

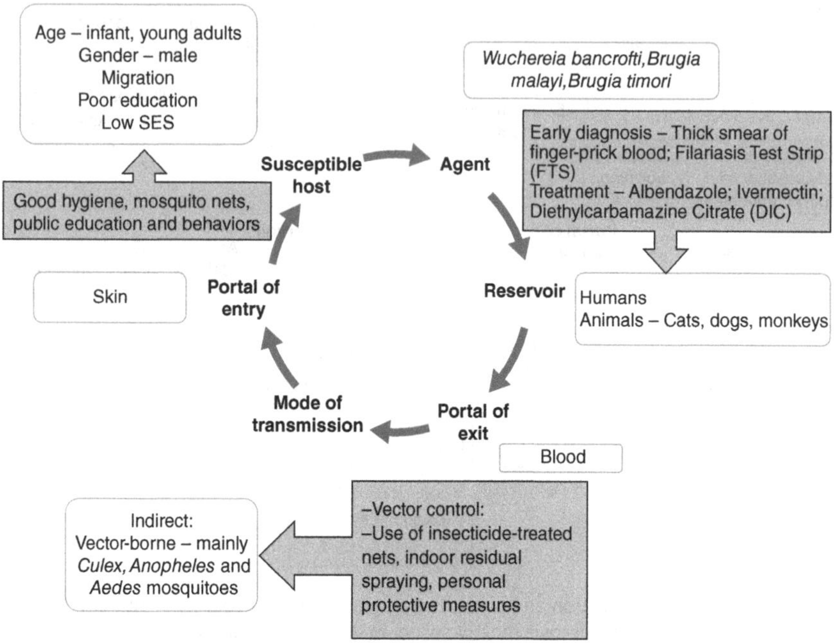

Figure 1.8 Chain of infection and control strategies for lymphatic filariasis. *Source:* Based on WHO [28].

finger prick blood test is the gold standard [28]. Control measures are mainly targeted at elimination of transmission, through vector control.

1.3.5 Onchocerciasis ("River Blindness")

Onchocerciasis is caused by the parasitic worm *Onchocerca volvulus*, and transmitted to humans through the bite of infected blackflies of the genus *Simulium*. With an estimated global burden of over 20 million in 2017, 99% of infected persons live in 31 African countries, although some foci of infection have also been reported in Latin America and Yemen [30]. Risk of infection is generally associated with living or working in close proximity to streams or rivers where there are *Simulium* blackflies, and invariably agricultural practices.

The chain of infection of onchocerciasis and control strategies are summarized in Figure 1.9. With an average incubation period ranging from three months to two years, symptoms include severe itching, disfiguring skin conditions, and visual impairment, including permanent blindness. There is no vaccine for onchocerciasis; therefore, prevention and control efforts have been focused on establishment of community direct treatment with ivermectin (CDTI) and vector control toward elimination in endemic countries.

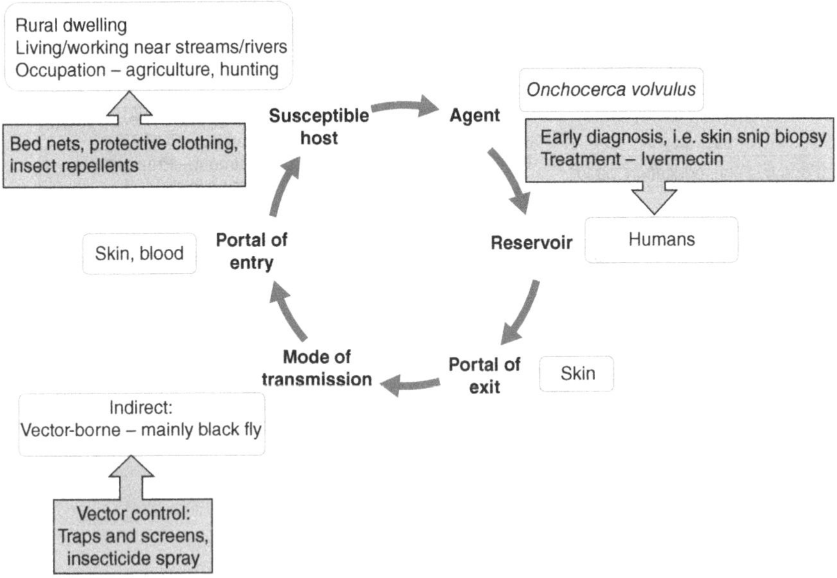

Figure 1.9 Chain of infection and control strategies for onchocerciasis. *Source:* Based on WHO [30].

1.3.6 Foodborne Trematodiases

Foodborne trematodiases are a group of zoonotic parasitic infections caused by flatworms or "flukes" transmitted by consumption of food (particularly raw fish, crustaceans, and aquatic vegetables) contaminated by larval stage of the parasite [31]. The pathogenic parasites include liver flukes (*Clonorchis sinensis, Fasciola gigantica, Fasciola hepatica, Opisthorchis felineus,* and *Opisthorchis viverrini*), lung flukes (*Paragonimus* spp.), and intestinal flukes (*Echinostoma* spp., *Fasciolopsis buski*, and the heterophyids). There are typically two intermediate hosts in the chain of infection, first being a mollusc (freshwater snail) and the second, freshwater fish or crustacean. Humans are the definitive host. Foodborne trematodiases are an important public health and economic problem, most prevalent in East Asia and South America, and cause 2 million life years lost to disability and death worldwide every year. Risk factors of infection include poor sanitation and food hygiene, limited access to safe drinking water, cultural food practices, and agricultural practices.

The chain of infection of foodborne trematodiases and control strategies are summarized in Figure 1.10. With an incubation period of about one to two weeks, the clinical spectrum of infection may range from asymptomatic infection to acute manifestations with variable symptoms depending on the specific parasite (i.e. fever, lung, liver, or intestinal symptoms) and chronic infections [32]. Safe

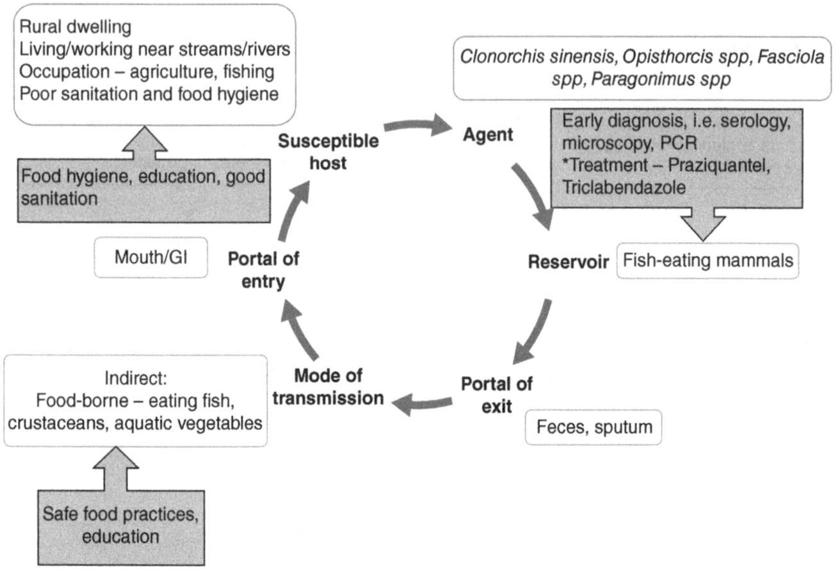

Figure 1.10 Chain of infection and control strategies for foodborne trematodiases. *Praziquantel for Clornichiasis and Opisthorciasis; Triclabendazole for Fascioliasis and Paragonimiasis. *Source:* Based on Keiser and Utzinger [32].

and efficacious medicines are available to prevent and treat foodborne trematodiases, in addition to other interventions such as information, education, and communication on safe food practices, improved sanitation, and veterinary public health measures, within a broad multi-sectoral prevention and management strategy.

1.3.7 Dracunculiasis (Also Called Guinea Worm Disease)

Dracunculiasis is a crippling parasitic disease caused by the nematode *Dracunculus medinensis*, and transmitted through consumption of stagnant water contaminated with parasite-infected copepods or water fleas [33]. On course for global eradication with a grossly reduced annual incidence of double digits in the last nine years, the disease is endemic in rural Africa, particularly Chad, South Sudan, Ethiopia, and Mali [34]. Risk factors include rural dwelling, poor access to safe drinking water, and agricultural practices. Seasonal variation in transmission has also been reported.

The chain of infection of dracunculiasis and control strategies are summarized in Figure 1.11. With a long incubation period of approximately 12–14 months, symptoms include localized redness and tenderness of the skin, usually in the

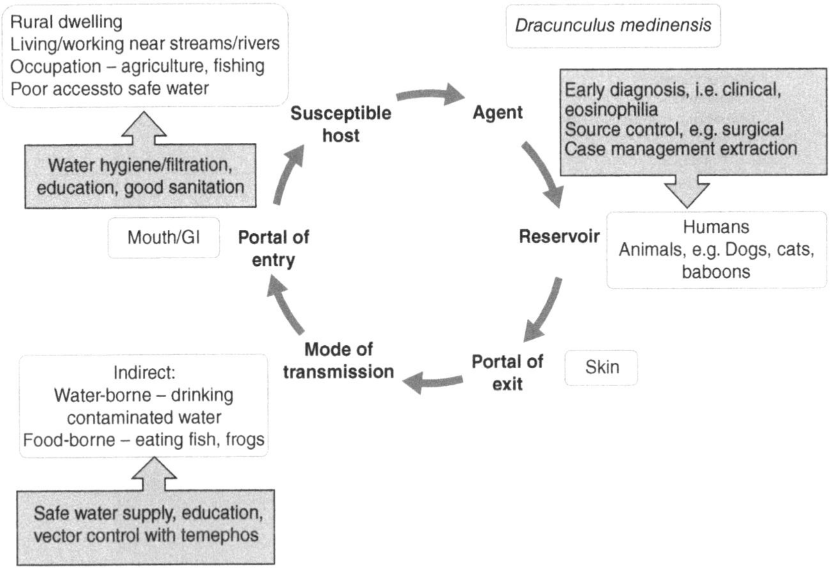

Figure 1.11 Chain of infection and control strategies for dracunculiasis. *Source:* Based on Sankara et al. [35].

lower leg (although any part of body could be involved) where the adult worm emerges [35]. Other symptoms include general allergic symptoms and intense inflammatory reaction. Complications include septic arthritis, cellulitis, sepsis, abscesses, and contractures. As there is no vaccine or specific medication for treatment of dracunculiasis, prevention is the cornerstone of management with strategies including improved surveillance, vector control, provision of safe drinking water, promoting health education, and behavioral change [36].

1.4 Bacterial Infections

1.4.1 Yaws

Yaws is a chronic non-venereal contagious disease caused by *Treponema pallidum* subspecies *pertenue* [37]. It commonly affects the skin, bone, or cartilage [38]. Humans are the only reservoir. Transmission occurs through person to person, mainly affecting children [38]. Yaws is prevalent across poor communities in warm, humid, and tropical areas of Africa, Asia, Latin America, and the Pacific [38]. Around 256 343 yaws cases were reported between 2010 and 2013, all of which were from LMICs. Approximately 84% of these cases originated from Papua New Guinea, Solomon Islands, and Ghana [37].

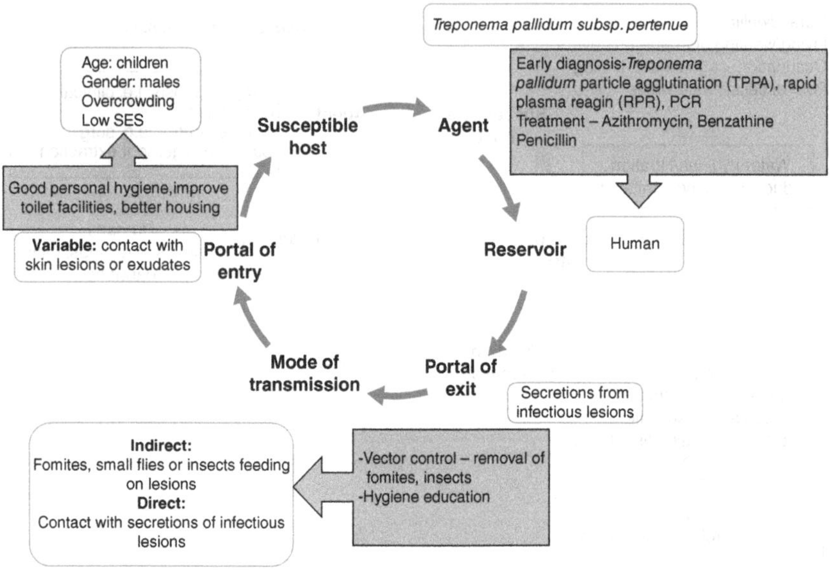

Figure 1.12 Chain of infection and control strategies for yaws. *Source:* Based on WHO [38].

The chain of infection of yaws and control strategies are shown in Figure 1.12. Incubation period is between 9 and 91 days, averaged at 21 days [38]. Yaws presents as skin lesion (papilloma teemed with bacteria), facilitating straightforward diagnosis, but ulcerated lesions require laboratory-based serological confirmation. Relapse may occur weeks to months or even years after the primary infection – mostly presenting as multiple raised yellow lesions, pain and swelling of long bones, and dactylitis [37, 38]. There is no vaccine for yaws. Health education and personal hygiene is the key to control of transmission.

1.4.2 Trachoma

Trachoma is a chronic infection of the conjunctiva or cornea, caused by serotypes A, B, Ba, and C of *Chlamydia trachomatis* [39]. The disease has an acute, self-limiting phase and a chronic advanced phase that occurs mainly in adulthood, corresponding to prior childhood infections. The latter causes scarring of the tarsal mucosa, entropion, trichiasis, corneal opacity, and blindness [40]. The disease is a public health concern across 44 countries globally and has caused irreversible blindness to approximately 1.9 million people till date [41]. As of March 2020, an estimated 137 million people living in trachoma-endemic settings were projected to be at risk of blindness. Hyperendemicity areas of trachoma are prevalent in the poorest-rural settings of Africa, Central and South America, Asia, Australia, and the Middle East [41].

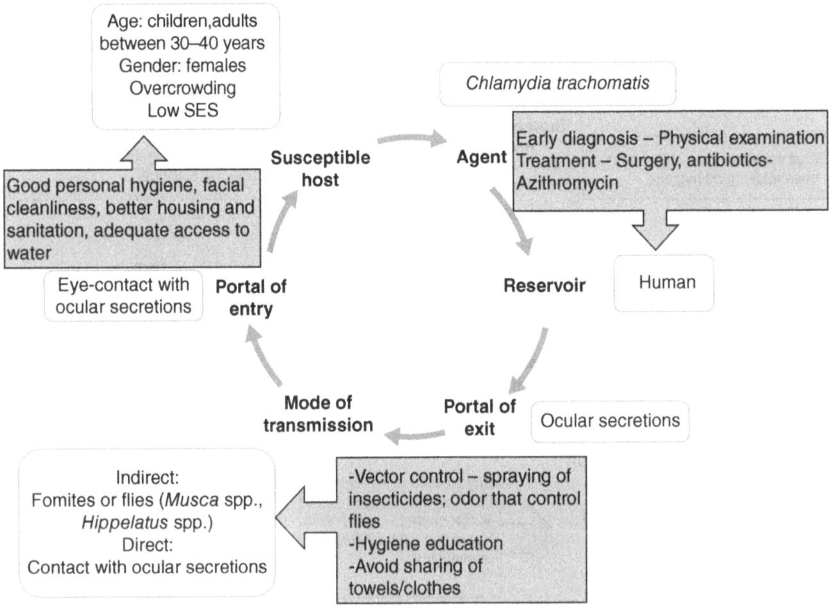

Figure 1.13 Chain of infection and control strategies for trachoma. *Source:* Based on Hu et al. [39].

The chain of infection of trachoma and control strategies are shown in Figure 1.13. Incubation period is between 5 and 12 days [29]. The spread of infection is mainly through personal contacts (hands, clothes, or bedding) and by mechanical transmission via flies. Repeated infection is likely over years. Such recurrence leads to conjunctival scarring and trichiasis, predisposing to constant pain, light intolerance, scarring of cornea, and ultimately visual impairment or blindness [29, 41]. Facial and personal hygiene, adequate clean water, proper sanitation, and control of vector (eye-seeking flies) are crucial measures to control the spread of infection [1].

1.4.3 Leprosy

Leprosy or Hansen's disease is caused by the bacillus, *Mycobacterium leprae* [42]. Leprosy commonly affects the skin, peripheral nerves, mucosa of the upper respiratory tract, and eyes [43]. Globally, over 200 000 new leprosy cases across 159 countries were reported as of 2018 [43]. India, Brazil, and Indonesia accounted for approximately 80% of the total cases worldwide [44]. Around 14% of the population from the Democratic Republic of Congo, Ethiopia, Madagascar, Mozambique, Nigeria, the United Republic of Tanzania, Bangladesh, Nepal, Sri Lanka, Myanmar, and the Philippines were detected to have leprosy as of 2015 [45].

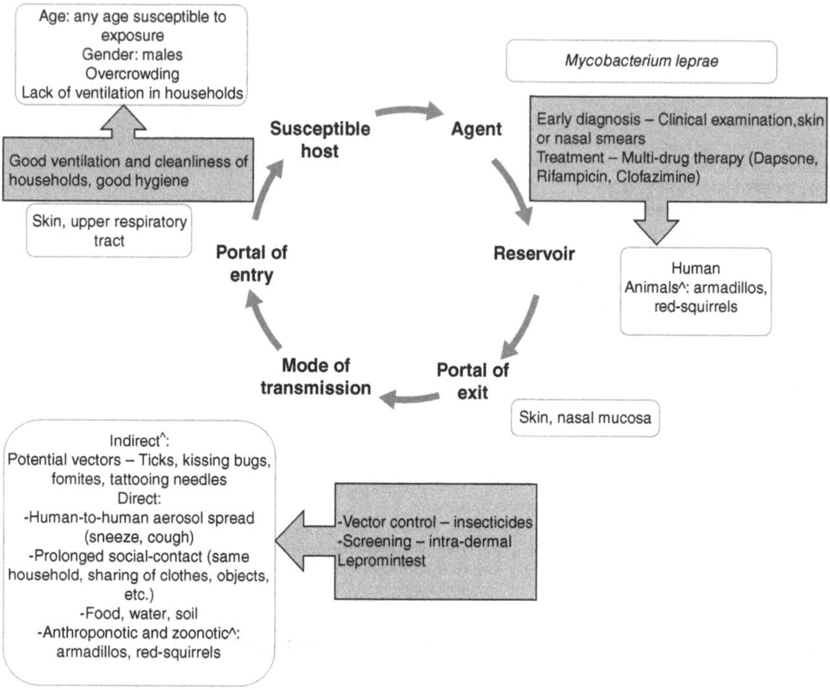

Figure 1.14 Chain of infection and control strategies for leprosy. ^Indicates key facts from the latest evidence. *Source:* Based on WHO [43].

The chain of infection of leprosy and control strategies are shown in Figure 1.14. The mean incubation period of the disease is 5 years but symptoms may occur as early as 1 year or as long as 20 years at post exposure [43]. People can be susceptible to the disease via aerosol spread from sneezing or coughing, or via close contact with infected individuals [46]; however, newer evidences have suggested the possibility of zoonotic or vector-borne transmission [47]. Clinical examination is the choice for diagnosis, as symptoms and signs manifest with skin lesions, or in advanced stages, the presence of nodules or lumps in skin or face, nasal depression, loss of fingers or toes, or other deformities. Thick skin or nasal smears could be useful for the detection of acid-fast bacilli [29]. Multi-drug therapy (MTD) is required to treat leprosy, while proper hygiene, good ventilation and cleanliness, safer housing together with appropriate vector control and avoidance of prolonged social contact with infected individuals eliminate the risk of infection.

1.4.4 Buruli Ulcer

Buruli ulcer, a necrotizing skin infection caused by the pathogen *Mycobacterium ulcerans,* commonly infects the skin and subcutaneous tissues, leading to ulcerations or wounds. Over 32 countries have reported the disease [48], commonly being identified in West Africa – particularly Ivory Coast, Ghana, Benin, and Cameroon [1]; the Americas, Asian, and Western Pacific regions – particularly in the tropics and subtropical regions, except for China and Japan [49].

The chain of infection of buruli ulcer and control strategies are shown in Figure 1.15. The incubation period is between weeks to five months [50]. People can be susceptible by contact with fresh water insects, plants, or fish, or in contact with animals such as possums, dogs, horse, etc. [51, 52]. Clinical diagnosis is highly sensitive as compared with laboratory diagnostics [53]. Symptoms begin with painless nodules and swelling of the arms or legs, and may subsequently develop into large ulcers. The mycobacterium releases mycolactone toxin that has local immunosuppressive properties [49]. Avoiding contact with aquatic environment, wearing long sleeves or pants to prevent insect bites, with regular cleaning of wounds, and covering them could prevent the spread of infection.

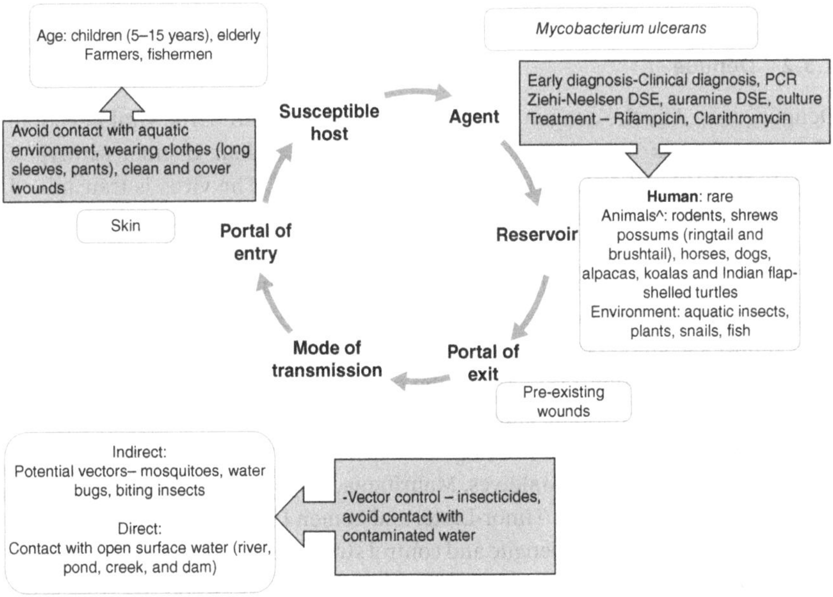

Figure 1.15 Chain of infection and control strategies for buruli ulcer. ^Indicates key facts from the latest evidence. *Source:* Based on Loftus et al. [50].

1.5 Viral Infections

1.5.1 Rabies

Rabies (also known as hydrophobia) is a zoonotic, acute-fatal viral infection caused by the rabid virus (genus *Lyssavirus;* family *Rhabdoviridae*). The disease affects the central nervous system, causing brain and spinal cord inflammation [54]. Rabies is prevalent across all continents worldwide, except the Antarctica. Over 95% of mortalities as a consequence of rabies infection occur in Asia and Africa [55]. India sustains a high global burden of rabies, with annual death rates amounting to approximately 17000–20000 cases [1].

The chain of infection of rabies and control strategies are shown in Figure 1.16. The incubation period is highly variable (usually three to eight weeks), but may be as early as four days and up to many years [29]. Any person around an infected animal with rabies virus is highly at risk of contracting the disease [56]. Diagnosis is mostly clinical, with appropriate history taking and clinical manifestations (usually starts with headache, malaise, sore throat, fever, and subsequently followed by pain or tingling sensation at the bite site before development of widespread excitation of the sensory, motor, and sympathetic nervous system) [29]. The disease is always fatal. The only prevention of rabies is immunization.

1.5.2 Dengue

Dengue, caused by the dengue virus (DENV), is a mosquito-borne viral infection. The virus (genus *Flavivirus;* family *Flaviviridae*) has four distinct serotypes, namely DENV-1, DENV-2, DENV-3, and DENV-4 [57]. The virus is transmitted mainly by female mosquitoes, *Aedes aegypti* and *Aedes albopictus* [57]. The escalated burden of dengue is highly prevalent in the tropics, and highly influenced by the environment – rainfall, urbanization, temperature, and humidity [57]. Approximately 390 million dengue cases are projected to be detected annually, with almost 3.9 billion people being at risk to be infected [57]. From 129 countries at high risk of infection [58], 70% of the cases were predominantly from Asian countries [59], affecting Bangladesh, Malaysia, Philippines, and Vietnam [57]. As of 2020, high infection rates were reported in Bangladesh, Brazil, Cook Islands, Ecuador, India, Indonesia, Maldives, Mauritania, Mayotte (Fr), Nepal, Singapore, Sri Lanka, Sudan, Thailand, Timor-Leste, and Yemen [57].

The chain of infection of dengue and control strategies are shown in Figure 1.17. Incubation period is 4–10 days after the bite of an infected mosquito. High fever (40 °C/104 °F) accompanied by two of the following symptoms is a classical manifestation of dengue: severe headache, retro-orbital pain, muscle or joint pains, nausea, vomiting, swollen glands, and rash. Severe dengue leads to the following warning signs: severe abdominal pain, persistent vomiting, rapid

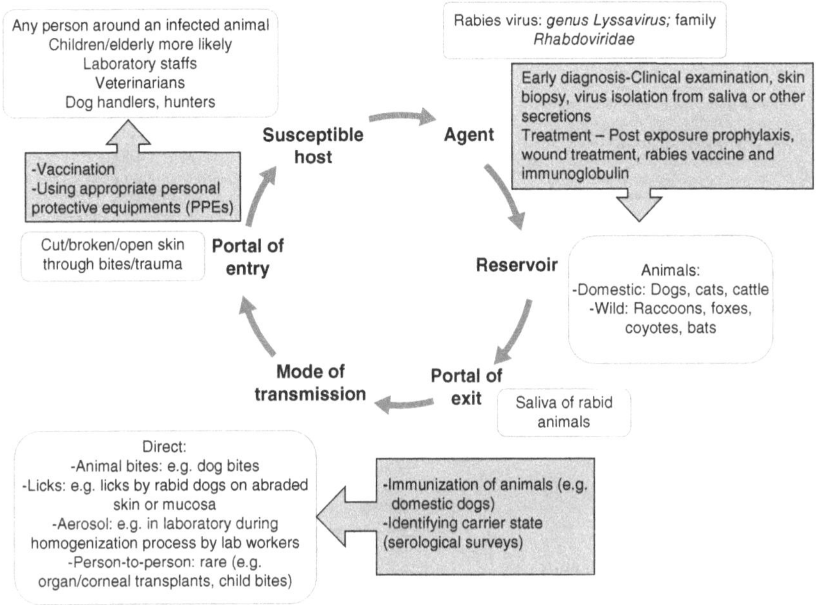

Figure 1.16 Chain of infection and control strategies for rabies. *Source:* Based on Singh et al. [54].

breathing, gum bleeding, fatigue, restlessness, or blood in vomit [57]. Diagnosis is mainly through clinical examination, tourniquet test, blood investigations, and serology. Treatment is supportive – analgesics, fluid replacement, and bed rest [1]. Prevention is highly dependent on vector control measures.

1.5.3 Chikungunya

Chikungunya is caused by the chikungunya virus (CHIKV – genus *Alphavirus;* family *Togaviridae*), and mainly transmitted by the vectors *Aedes aegypti* or *Aedes albopictus* [60]. Vertical transmissions are rare [61]. Chikungunya, once being a NTD, has re-emerged in India during 2006 [29]. The disease has been identified in over 60 countries across Asia, Africa, Europe, and the Americas [60]. Recent outbreaks were reported in the Reunion, Mauritius, India, coastal Italy, and Bangladesh [1]. The incubation period of chikungunya is between four and seven days [29].

The chain of infection of chikungunya and control strategies are shown in Figure 1.18. Clinical manifestations include abrupt onset of fever, accompanied by joint pain, muscle pain, headache, nausea, fatigue, and rash [60]. Serological tests such as enzyme-linked immunosorbent assays (ELISA) and virus isolation through reverse transcriptase–polymerase chain reaction (RT–PCR) are common

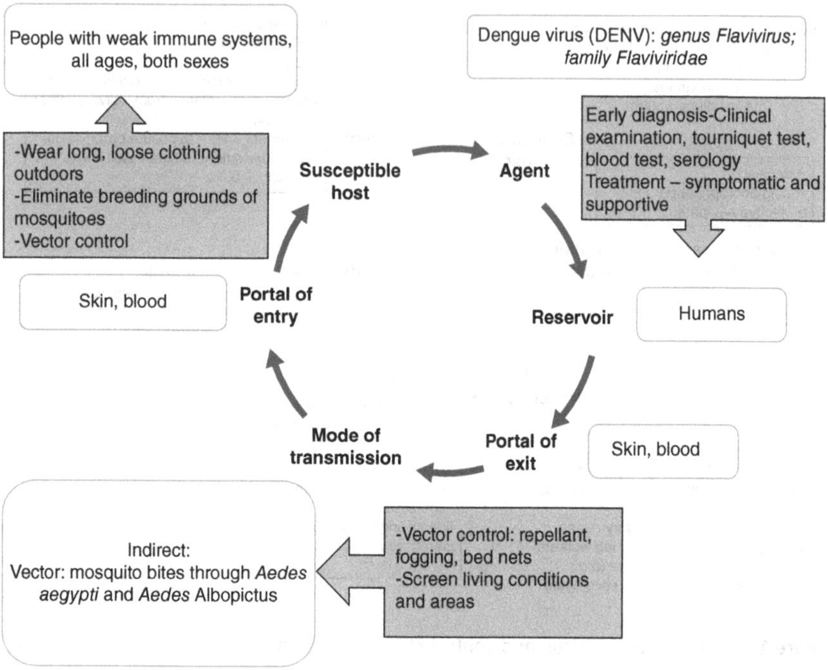

Figure 1.17 Chain of infection and control strategies for dengue. *Source:* Based on WHO [57].

diagnostic methods. Treatment is supportive – antipyretics, analgesics, and fluid administration [60]. Prevention is mainly through avoidance of mosquito bites and vector control measures [1].

1.6 Fungal and Ectoparasitic Infections

1.6.1 Chromoblastomycosis, Mycetoma/Eumycetoma, and Other Deep Mycoses

Chromoblastomycoses, mycetoma, and other deep mycoses are a spectrum of chronic, granulomatous, suppurative mycosis of the skin and subcutaneous tissues, as well as systemic mycosis caused by traumatic inoculation of melanized or brown-pigmented fungi [62]. Chromoblastomycosis being the commonest of the subcutaneous mycoses, is caused by a large number of fungi, mainly *Cladosporium carrionii, Phialophora verrucosa, Fonsecaea pedrosoi, Fonsecaea monophora,* and *Rhinocladiella aquaspersa* [63]. Other subcutaneous mycoses include sporotrichosis (caused by *Sporothrix schenckii*) and mycetoma (caused

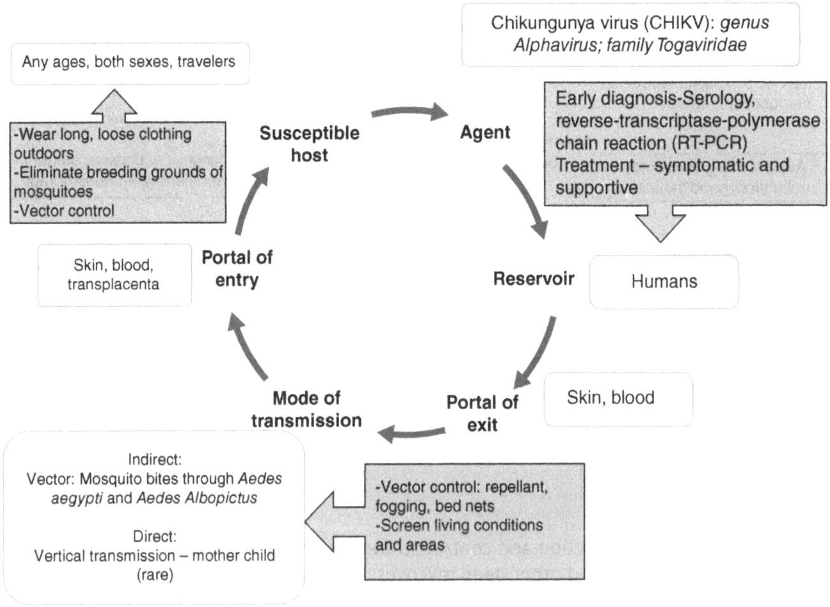

Figure 1.18 Chain of infection and control strategies for chikungunya. *Source:* Based on WHO [60].

by several species of fungi and bacteria) [64]. *Coccidioides immitis*, *Histoplasma capsulatum*, *Blastomyces dermatitidis*, and *Paracoccidioides brasiliensis* are the primary systemic fungal pathogens, whereas the opportunistic ones include *Cryptococcus neoformans*, *Candida* spp., *Aspergillus* spp., *Penicillium marneffei*, the *Zygomycetes*, *Trichosporon beigelii*, and *Fusarium* spp. [65]. The disease occurs in tropical and subtropical environments and endemic in certain parts of Africa, Asia, Europe, and Latin America. Risk factors include rural dwelling, agricultural and manual work, walking barefooted, low SES, and immune suppression.

The chain of infection and control strategies of chromoblastomycosis and other deep mycoses are summarized in Figure 1.19. Incubation period is highly variable and can range from several weeks to years [66]. In general, most cases have a slow, chronic course and patients slowly develop wide-ranging symptoms (depending on causative agent) that may include skin lesions (nodules, verrucas, plaques, scar tissues, and tumors), or symptoms related to lungs, spleen, liver, bone marrow, and brain infections (in the case of deep mycoses). In general, prevention and control can be difficult, as is treatment, although successful treatment options include antifungals, antibiotics, surgery, cryotherapy, thermotherapy, and immunotherapy [67].

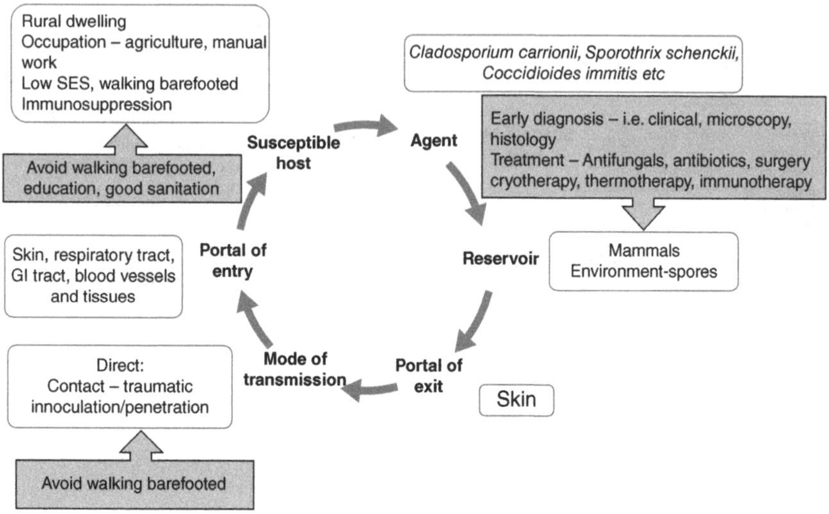

Figure 1.19 Chain of infection and control strategies for Chromoblastomycosis, mycetoma/eumycetoma, and other deep mycoses. *Source:* Based on Bonifaz et al. [66].

1.6.2 Scabies

Scabies is a parasitic disease caused by infestation of the skin by mite *Sarcoptes scabiei* through direct contact, and to a lesser extent, via fomites [68]. WHO estimates that over 200 million people globally are infected with scabies at any point in time [69]. It is a common dermatological problem and a public health concern, endemic in many resource-poor tropical settings, with an estimated prevalence of 5–10% among children. Outbreaks are often reported in close groups and residential and nursing care settings for elderly people [70]. Risk factors include armed conflicts, displacement, homelessness, overcrowding, and communal use of clothes, beds, and pillows which are quite common findings in resource-poor settings [71].

The chain of infection of scabies and control strategies are summarized in Figure 1.20. Following an incubation period of about 4–6 weeks [72], patients develop an allergic reaction to the presence of the mite proteins and feces in the scabies burrows and often present with severe itch, linear burrows, and vesicles around the finger webs, wrists, upper and lower limbs, and belt area, and less commonly penis and scrotum in adult males, breast in adult females, and palms, soles, ankle, and scalp in children. Complications may include secondary bacterial infections [73], impaired renal functions, rheumatic heart disease, and development of crusted scabies and hyperkeratotic skin. Effective drug treatments are available for scabies. These include topical scabicides such as permethrin,

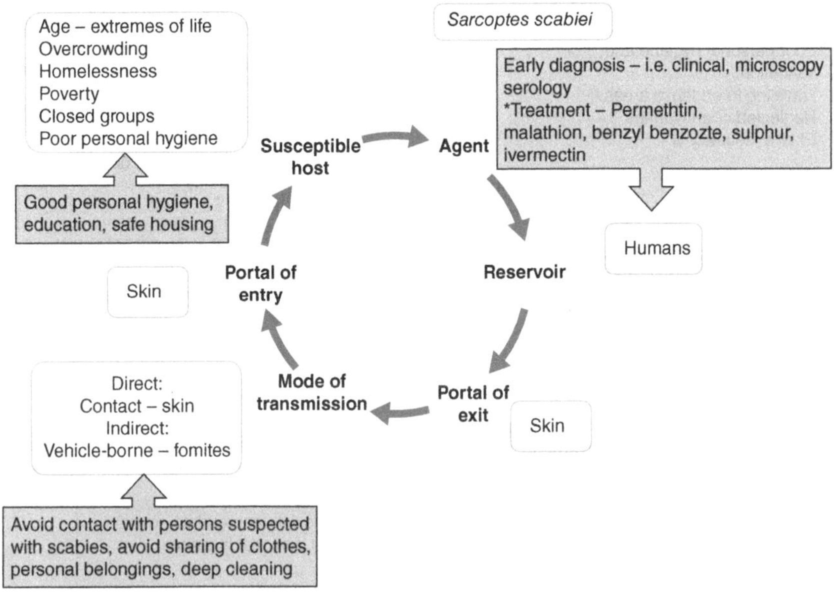

Figure 1.20 Chain of infection and control strategies for scabies. *Treatment should involve close contacts or household members of an index case, regardless of whether they have symptoms or not. *Note:* Safety of ivermectin in pregnant women and children under 15 kg of body weight has not been established. *Source:* Based on Cassell et al. [72].

malathion, benzyl benzoate, or sulphur ointment. Oral ivermectin is also highly effective. Broadly, prevention and control strategies involve multi-sectoral approaches targeted at providing public information and education on personal hygiene, safe housing, treatment of asymptomatic household contacts, and mass drug administration with ivermectin in some high prevalence settings.

1.6.3 Myiasis

Human myiasis is a parasitic disease caused by infestation of dead or living tissue by larvae or maggots of *Dipterous* flies. Human myiasis has a worldwide distribution; however, it is more prevalent in tropical regions [74] and determined by geographic distribution of the different fly species. For example, while *Cordylobia anthropophaga* is responsible for most of the cases of human myiasis in Africa, *Dermatobia hominis* is believed to be responsible for obligate myiasis in Mexico and most of Central and South America. Other species commonly reported include *Cochliomyia hominivorax* which causes facultative myiasis in the New World, *Oestrus ovis* and *Sarcophaga haemorrhoidalis* which have been reported in Ecuador [75]. Risk factors of myiasis include traveling to endemic areas, having neglected open wounds, low SES, poor

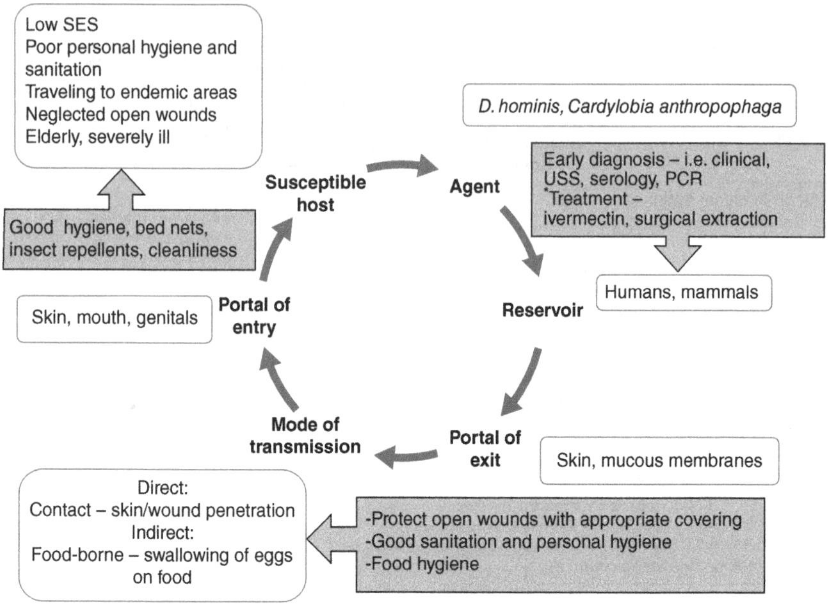

Figure 1.21 Chain of infection and control strategies for myiasis. *Treatment would depend on site of infection, severity, and symptoms. *Source:* Based on Solomon et al. [78].

sanitation and hygiene, and proximity to domestic animals. Certain host characteristics such as being elderly or severely ill, those with intellectual disability, and drug addiction also confers higher risk of myiasis [76, 77].

The chain of infection of myiasis and control strategies are summarized in Figure 1.21. Myiasis can have variable incubation period, presenting symptoms and mode of transmission depending on the part of the body involved, specific species of fly, and geographic location. Although cutaneous myiasis is the common form, many organs may be infested with the larvae of the flies [78]. Prevention and control is centered on improvement of basic sanitation and personal hygiene, education, personal protection, and fly control. Treatment is based on full extraction of the larva (such as maggot debridement therapy) and limited success had been achieved with ivermectin treatment [79].

1.7 Future Direction

NTDs prevention and treatment initiatives will highly benefit from research that advocates novel drug discoveries and innovations to provide better screening and diagnostic options. It should be noted that some of the current interventions, such as preventive chemotherapy, may require futuristic approaches, drugs, and technologies to be more efficacious to treat and combat the emergence of new infections across newer endemic areas, in view of the changing environmental and healthcare landscapes, thus identifying potential risk factors that were never detected before is highly crucial. Governments, volunteer agencies, and pharmaceutical companies should embark on newer drug explorations, for example, the use of phytochemicals in large-scale populations that could be more cost-effective and affordable in the quest to treat NTDs.

In view of changing climate and environment, an integrated system for vector control is required to reduce the breeding sites of mosquitoes, use of repellants, or applications of anti-larval measures, coupled with personal protection to prevent diseases like dengue or chikungunya spread within local communities. Evidences have showed that community awareness programs, health education, and behavior change are tantamount to prevent disease transmission like buruli ulcer or dracunculasis. In view of their effectiveness, these programs and efforts need to be accelerated and amalgamated among countries with high risk of local transmissions toward achieving the elimination roadmap of NTDs.

1.8 Conclusion

With the upcoming review of WHO NTD Roadmap performance against 2020 targets, it is hoped that renewed global commitments to combat the NTDs (as contained in the London Declaration on Neglected Tropical Diseases) and lessons learnt in the past decades will strengthen international partnerships and form the fulcrum upon which renewed global resource mobilization, innovations, and implementation strategies for NTDs are based. Further, it is hoped that multi-sectoral collaborations and partnerships (including poverty eradication, safe housing and water supply, education, and communication) aimed at breaking the chain of transmission of NTDs will be fostered and enshrined in the upcoming 2021–2030 roadmap. More than ever before, the world, as a matter of urgency, now needs effective vaccines and treatments for NTDs in order to slow down and reverse the morbidity and mortality associated with NTDs in a sustainable manner.

Acknowledgments

We thank the Director General of Health, Ministry of Health Malaysia for the support to publish this chapter.

References

1 Mitra, A.K. and Mawson, A.R. (2017). Neglected tropical diseases: epidemiology and global burden. *Trop. Med. Infect. Dis.* 2 (3): 36.

2 Githeko, A.K., Lindsay, S.W., Confalonieri, U.E., and Patz, J.A. (2000). Climate change and vector-borne diseases: a regional analysis. *Bull. World Health Organ.* 78: 1136–1147.

3 Stolk, W., Kulik, M., le Rutte, E. et al. (2016). Between-country inequalities in the neglected tropical disease burden in 1990 and 2010, with projections for 2020. *PLoS Negl. Trop. Dis.* 10 (5): e0004560.

4 World Health Organization (WHO) (2017). Neglected Tropical Diseases. Program. http://www.who.int/neglected_diseases/en (accessed 13 August 2020).

5 Franco, J.R., Simarro, P.P., Diarra, A., and Jannin, J.G. (2014). Epidemiology of human African trypanosomiasis. *Clin. Epidemiol.* 6: 257–275.

6 World Health Organization (WHO) (2020). Fact sheets: trypanosomiasis, Human African (sleeping sickness). https://www.who.int/news-room/fact-sheets/detail/trypanosomiasis-human-african-(sleeping-sickness) (accessed 3 August 2020).

7 World Health Organization (WHO) (2012). Accelerating work to overcome the global impact of neglected tropical diseases: a roadmap for implementation. 1–22. http://whqlibdoc.who.int/hq/2012/WHO_HTM_NTD_2012.1_eng.pdf (accessed 3 August 2020).

8 World Health Organization (WHO) (2020). Fact sheets: Chagas disease (also known as American trypanosomiasis). https://www.who.int/news-room/fact-sheets/detail/Chagas-disease-(american-trypanosomiasis) (accessed 3 August 2020).

9 Moncayo, Á. and Silveira, A.C. (2017). Current epidemiological trends of Chagas disease in Latin America and future challenges: epidemiology, surveillance, and health policies. In: *American Trypanosomiasis Chagas Disease*, 59–88. Elsevier.

10 Nunes, M.C., Dones, W., Morillo, C.A. et al. (2013). Chagas disease: an overview of clinical and epidemiological aspects. *J. Am. Coll. Cardiol.* 62 (9): 767–776.

11 Chowdhury, R., Mondal, D., Chowdhury, V. et al. (2014). How far are we from visceral leishmaniasis elimination in Bangladesh? An assessment of epidemiological surveillance data. *PLoS Negl. Trop. Dis.* 8 (8): e3020.

12 World Health Organization (WHO) (2020). Fact sheets: leishmaniasis. https://www.who.int/news-room/fact-sheets/detail/leishmaniasis (accessed 4 August 2020).

13 Courtenay, O., Peters, N.C., Rogers, M.E., and Bern, C. (2017). Combining epidemiology with basic biology of sand flies, parasites, and hosts to inform leishmaniasis transmission dynamics and control. *PLoS Pathog.* 13 (10): e1006571.

14 Haque, R., Huston, C.D., Hughes, M. et al. (2003). Amebiasis. *N. Engl. J. Med.* 348 (16): 1565–1573.

15 Shamsuzzaman, S.M. and Hashiguchi, Y. (2002). Thoracic amebiasis. *Clin. Chest Med.* 23 (2): 479–492.

16 Allason-Jones, E., Mindel, A., Sargeaunt, P. et al. (1986). Entamoeba histolytica as a commensal intestinal parasite in homosexual men. *N. Engl. J. Med.* 315 (6): 353–356.

17 Billet, A.C., Rousseau, A.S., Piroth, L., and Martins, C. (2019). An underestimated sexually transmitted infection: amoebiasis. *BMJ Case Rep.* 12 (5): e228942.

18 Watanabe K. (2020). Amoebiasis. BMJ Best Practice. https://bestpractice.bmj.com/topics/en-gb/553 (accessed 4 August 2020).

19 Erko, B.E., Birrie, H.A., and Tedla, S.H. (1995). Amoebiasis in Ethiopia. *Trop. Geogr. Med.* 47 (1): 30.

20 Parija, S.C., Chidambaram, M., and Mandal, J. (2017). Epidemiology and clinical features of soil-transmitted helminths. *Trop. Parasitol.* 7 (2): 81–85.

21 World Health Organization (WHO) (2020). Fact sheets: soil-transmitted helminth infections. https://www.who.int/news-room/fact-sheets/detail/soil-transmitted-helminth-infections (accessed 8 August 2020).

22 Khurana, S. and Sethi, S. (2017). Laboratory diagnosis of soil transmitted helminthiasis. *Trop. Parasitol.* 7 (2): 86–91.

23 World Health Organization (WHO) (2020). Fact sheets: schistosomiasis. https://www.who.int/news-room/fact-sheets/detail/schistosomiasis (accessed 8 August 2020).

24 World Health Organization (WHO) (2020). Fact sheets: echinococcosis. https://www.who.int/news-room/fact-sheets/detail/echinococcosis (accessed 8 August 2020).

25 Baumann, S., Shi, R., Liu, W. et al. (2019). Worldwide literature on epidemiology of human alveolar echinococcosis: a systematic review of research published in the twenty-first century. *Infection* 47: 703–727.

26 World Health Organization (WHO) (2020). Fact sheets: lymphatic filariasis. https://www.who.int/news-room/fact-sheets/detail/lymphatic-filariasis (accessed 8 August 2020).

27 Joshi, P.L. (2018). Epidemiology of lymphatic filariasis. In: *Lymphatic filariasis* (ed. B. Tyagi). Singapore: Springer.

28 World Health Organization (WHO) (2020). Health topics – diagnosis of lymphatic filariasis. https://www.who.int/lymphatic_filariasis/epidemiology/epidemiology_diagnosis/en (accessed 8 August 2020)

29 Park, K. (2009). *Parks Textbook of Preventive and Social Medicine*, 20e. India: Bhanot.

30 World Health Organization (WHO) (2020). Fact sheets: onchocerciasis. https://www.who.int/news-room/fact-sheets/detail/onchocerciasis (accessed 10 August 2020).

31 World Health Organization (WHO) (2020). Fact sheets: foodborne trematode infections. https://www.who.int/news-room/fact-sheets/detail/foodborne-trematode-infections (accessed 20 August 2020).

32 Keiser, J. and Utzinger, J. (2009). Food-borne trematodiases. *Clin. Microbiol. Rev.* 22 (3): 466–483.

33 Rawla, P. and Jan, A. (2020). Dracunculiasis. In: *StatPearls* [Internet]. StatPearls Publishing.

34 World Health Organization (WHO) (2020). Fact sheets: dracunculiasis (guinea worm disease). https://www.who.int/news-room/fact-sheets/detail/dracunculiasis-(guinea-worm-disease) (accessed 20 August 2020).

35 Sankara, D.P., Korkor, A.S., Agua-Agum, J., and Biswas, G. (2016). Dracunculiasis (Guinea worm disease). In: *Neglected Tropical Diseases-Sub-Saharan Africa*, 45–61. Cham: Springer.

36 Hopkins, D.R. (2020). Dracunculiasis. In: *Hunter's Tropical Medicine and Emerging Infectious Diseases*, 10e (eds. E.T. Ryan, D.R. Hill, T. Solomon, et al.), 878–881. Canada: Elsevier.

37 Mitja, O., Marks, M., Konan, D.J.P. et al. (2015). Global epidemiology of yaws: a systematic review. *Lancet Glob. Health* 3: e324–e331.

38 World Health Organization (WHO) (2019). Fact sheets: yaws. https://www.who.int/news-room/fact-sheets/detail/yaws (accessed 8 August 2020).

39 Hu, V.H., Harding-Esch, E.M., Burton, M.J. et al. (2010). Epidemiology and control of trachoma: systematic review. *Trop. Med. Int. Health* 15: 673–691.

40 Ramadhani, A.M., Derrick, T., Holland, M.J., and Burton, M.J. (2016). Blinding trachoma: systematic review of rates and risk factors for progressive disease. *PLoS Negl. Trop. Dis.* 10: e0004859.

41 World Health Organization (WHO) (2020). Fact sheets: trachoma. https://www.who.int/news-room/fact-sheets/detail/trachoma (accessed 9 August 2020).

42 Lastoria, J.C. (2014). Leprosy: review of the epidemiological, clinical, and etiopathogenic aspects – Part 1. *An. Bras. Dermatol.* 89 (2): 205–218.

43 World Health Organization (WHO) (2020). Fact sheets: leprosy. https://www.who.int/news-room/fact-sheets/detail/leprosy (accessed 9 August 2020).

44 Blok, D.J., De Vlas, S.J., and Richardus, J.H. (2015). Global elimination of leprosy by 2020: are we on track? *Parasit. Vectors* 8: 548.

45 Akpolat, N.D., Akkus, A., and Kaynak, E. (2018). An Update on the epidemiology, diagnosis and treatment of Leprosy. In: *Hansen's Disease – The Forgotten and Neglected Disease* (ed. W. Ribon). United Kingdom: IntechOpen https://doi.org/10.5772/intechopen.80557.

46 Job, C.K. (1981). Leprosy – the source of infection and its mode of transmission. *Lepr. Rev.* 52 (Suppl 1): 69–76.

47 Ploemacher, T., Faber, W.R., Menke, H. et al. (2020). Reservoirs and transmission routes of leprosy: a systematic review. *PLoS Negl. Trop. Dis.* 14 (4): e0008276.

48 Simpson, H., Deribe, K., Tabah, E.N. et al. (2019). Mapping the global distribution of Buruli ulcer: a systematic review with evidence consensus. *Lancet Glob. Health* 7: e912–e922.

49 World Health Organization (WHO) (2020). Fact sheets: buruli ulcer. https://www.who.int/health-topics/buruli-ulcer#tab=tab_2 (accessed 12 August 2020).

50 Loftus, M.J., Trubiano, J.A., Tay, E.L. et al. (2018). The incubation period of Buruli ulcer (*Mycobacterium ulcerans* infection) in Victoria, Australia – remains similar despite changing geographic distribution of disease. *PLoS Negl. Trop. Dis.* 12 (3): e0006323.

51 Roltgen, K. and Pluschke, G. (2015). *Mycobacterium ulcerans* disease (buruli ulcer): potential reservoirs and vectors. *Curr. Clin. Microbiol Rep.* https://doi.org/10.1007/s40588-015-0013-3.

52 Singh, A., McBride, W.J.H., Govan, B., and Pearson, M. (2018). Potential animal reservoir of *Mycobacterium ulcerans*: a systematic review. *Trop. Med. Infect. Dis.* 3: 56.

53 Eddyani, M., Sopoh, G.E., Ayelo, G. et al. (2018). Diagnostic accuracy of clinical and microbiological signs in patients with skin lesions resembling buruli ulcer in an endemic region. *Clin. Infect. Dis.* 67: 827–834.

54 Singh, R., Singh, K.P., Cherian, S. et al. (2017). Rabies – epidemiology, pathogenesis, public health concerns and advances in diagnosis and control: a comprehensive review. *Vet. Q.* 37 (1): 212–251.

55 World Health Organization (WHO) (2020). Fact sheets: rabies. https://www.who.int/news-room/fact-sheets/detail/rabies (accessed 13 August 2020).

56 World Health Organization (WHO). Health topics: rabies. https://www.who.int/rabies/about/en (accessed 13 August 2020).

57 World Health Organization (WHO) (2020). Fact sheets: dengue. https://www.who.int/news-room/fact-sheets/detail/dengue-and-severe-dengue (accessed 13 August 2020).

58 Brady, O.J., Gething, P.W., Bhatt, S. et al. (2012). Refining the global spatial limits of dengue virus transmission by evidence-based consensus. *PLoS Negl. Trop. Dis.* 6 (8): e1760.

59 Bhatt, S., Gething, P.W., Brady, O.J. et al. (2013). The global distribution and burden of dengue. *Nature* 496 (7446): 504–507.

60 World Health Organization (WHO) (2020). Fact sheets: chikungunya. https://www.who.int/news-room/fact-sheets/detail/chikungunya (accessed 13 August 2020).

61 Contopoulos-Ioannidis, D., Newman-Lindsay, S., Chow, C., and LaBeaud, A.D. (2018). Mother-to-child transmission of Chikungunya virus: a systematic review and meta-analysis. *PLoS Negl. Trop. Dis.* 12 (6): e0006510.

62 World Health Organization (WHO) (2019). Fact sheets: mycetoma. https://www.who.int/news-room/fact-sheets/detail/mycetoma (accessed 20 August 2020).

63 Kurien, G., Sugumar, K., and Chandran, V. (2017). Chromoblastomycosis (Chromomycosis). In: *StatPearls* [Intenet]. Treasure Island, FL: StatPearls Publishing.

64 Carrasco-Zuber, J.E., Navarrete-Dechent, C., Bonifaz, A. et al. (2016). Cutaneous involvement in the deep mycoses: a literature review. Part I – subcutaneous mycoses. *Actas Dermosifiliogr.* 107 (10): 806–815.

65 Walsh, T.J. and Dixon, D.M. (1996. Chapter 75). Spectrum of mycoses. In: *Medical Microbiology*, 4e (ed. S. Baron), 919–925. Galveston, TX: University of Texas Medical Branch at Galveston.

66 Bonifaz, A., Carrasco-Gerard, E., and Saul, A. (2001). Chromoblastomycosis: clinical and mycologic experience of 51 cases. *Mycoses* 44 (1–2): 1–7. http://www.ncbi.nlm.nih.gov/pubmed/11398635. (accessed 20 August 2020).

67 Brito, A.C. and Bittencourt, M.D. (2018). Chromoblastomycosis: an etiological, epidemiological, clinical, diagnostic, and treatment update. *An. Bras. Dermatol.* 93 (4): 495–506.

68 Walton, S.F. (2010). The immunology of susceptibility and resistance to scabies. *Parasite Immunol.* 32: 532–540.

69 World Health Organization (WHO) (2020). Fact sheets: scabies. https://www.who.int/news-room/fact-sheets/detail/scabies (accessed 21 August 2020).

70 Hewitt, K.A., Nalabanda, A., and Cassell, J.A. (2015). Scabies outbreaks in residential care homes: factors associated with late recognition, burden and impact. A mixed methods study in England. *Epidemiol. Infect.* 143 (7): 1542–1551.

71 Ugbomoiko, U.S., Oyedeji, S.A., Babamale, O.A., and Heukelbach, J. (2018). Scabies in resource-poor communities in Nasarawa state, Nigeria: epidemiology, clinical features and factors associated with infestation. *Trop. Med Infect. Dis.* 3 (2): 59.

72 Cassell, J.A., Middleton, J., Nalabanda, A. et al. (2018). Scabies outbreaks in ten care homes for elderly people: a prospective study of clinical features, epidemiology, and treatment outcomes. *Lancet Infect. Dis.* 18 (8): 894–902.

73 Swe, P.M., Reynolds, S.L., and Fischer, K. (2014). Parasitic scabies mites and associated bacteria joining forces against host complement defence. *Parasite Immunol.* 36 (11): 585–593.

74 Calvopina, M., Ortiz-Prado, E., Castañeda, B. et al. (2020). Human myiasis in Ecuador. *PLoS Negl. Trop. Dis.* 14 (2): e0007858.

75 Ortega-Munoz, G., Luzuriaga-Neira, N., Salazar-Silva, R., and Rodriguez-Hidalgo, R. (2019). Oestrus ovis in Ecuador: importance in the Andean sheep farming. *Vet. World* 12: 522–526.

76 Singh, A. and Singh, Z. (2015). Incidence of myiasis among humans – a review. *Parasitol. Res.* 114: 3183–3199.

77 Dominguez, J., Cueva, J., Cusco, C. et al. (2015). Miasis orbital severa causada por Cochliomyia hominivorax en la region andina de Ecuador. *Rev. Mex. Oftalmol.* 90: 43–47.

78 Solomon, M., Lachish, T., and Schwartz, E. (2016). Cutaneous myiasis. *Curr. Infect. Dis. Rep.* 18 (9): 28.

79 Francesconi, F. and Lupi, O. (2012). Myiasis. *Clin. Microbiol. Rev.* 25 (1): 79–105.

2

Neglected Tropical Diseases, Phytochemicals, Protein Targets, and Mechanisms in Drug Discovery

Lalit R. Samant[1], *Tehseen M. Dhorajiwala*[2], *and Sumit T. Haldar*[3]

[1] *Molecular Genetics Research Laboratory, Bai Jerbai Wadia Hospital for Children, Mumbai, Maharashtra, India*
[2] *Dhorajiwala, Freelance Bioinformatician, Thane, Maharashtra, India*
[3] *Freelance Bioinformatician, Thane, Maharashtra, India*

List of Abbreviations

ATP	Adenosine tri-phosphate
HAT	Human African Trypanosomiasis
HPLC	High-performance liquid chromatography
NMR	Nuclear Magnetic Resonance
NS	Nonstructural protein
NTDs	Neglected Tropical Diseases
PDB	Protein Data Bank
PZQ	Praziquantel
rRNA	Ribosomal ribonucleic acid
Tb	*Trypanosoma brucei*
TBG	*Trypanosoma brucei gambiense*
TBR	*Trypanosoma brucei rhodesiense*
TR	Trypanothione reductase
tRNA	Transfer ribonucleic acid
WHO	World Health Organization

2.1 Introduction

Neglected tropical diseases (NTDs) collectively consist of 20 diverse infectious diseases that affect more than one billion people in developing countries of tropical

Neglected Tropical Diseases and Phytochemicals in Drug Discovery, First Edition.
Edited by Chukwuebuka Egbuna, Muhammad Akram, and Jonathan Chinenye Ifemeje.

and subtropical regions [1]. NTDs mostly affect the poor population, majorly those who lack access to potable water, proper sanitation, and cannot afford basic health services to protect themselves. NTDs can cause chronic damage which would become irreversible if left untreated and can lead to lifelong disability. Commonly, bacteria, helminth parasites, protozoans, and viruses are the causative agents of NTDs and they are transmitted by various vectors like blackflies, mosquitoes, sandflies, snails, tsetse flies, and various other organisms which infest the disease-causing agent when the host comes in contact. The most common NTDs include lymphatic filariasis, onchocerciasis, trachoma, schistosomiasis, and soil-transmitted helminth like ascariasis, hookworm infection, and trichuriasis (Figure 2.1).

NTDs are responsible for 500 000 deaths annually [2]. NTDs such as African sleeping sickness, visceral leishmaniasis, and rabies are lethal and NTDs filariasis, foodborne trematodiases, leishmaniasis, hookworm infection, and schistosomiasis are responsible for causing lifelong disability. According to the estimated statistics of 2015 by WHO, Disability-adjusted life years for soil-transmitted helminthiasis was 4443.47 thousand years followed by Schistosomiasis with 3513.85 thousand years and then Dengue fever with 2610.08 thousand years. In a 2010 study report by the Global Burden of Disease, the estimated Disability-adjusted life years of the NTDs was 26.06 million years [3].

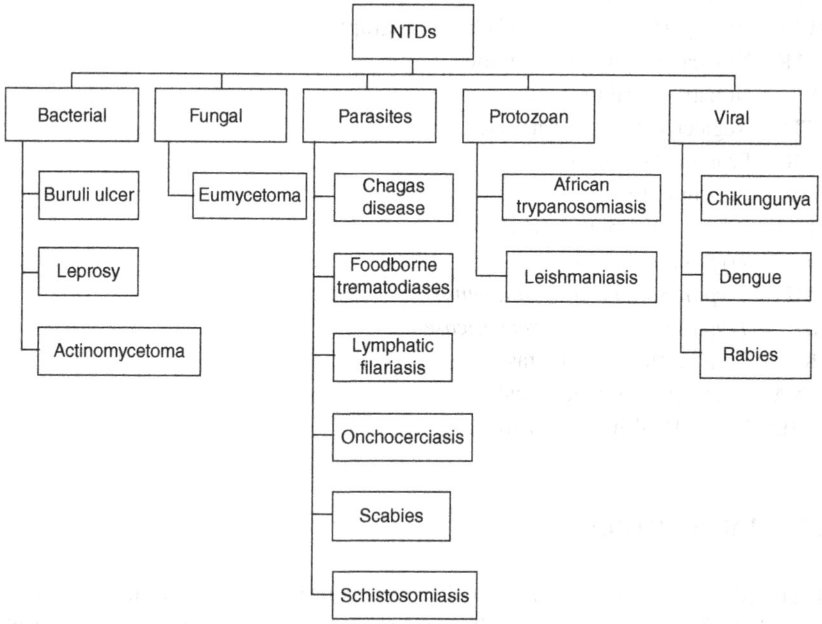

Figure 2.1 List of NTDs discussed in this chapter.

NTDs target the poorest population causing chronic damage and leading to further poverty amongst the people living such communities. With NTDs causing lifelong disability many are unable to support their families and communities economically. Children lose basic educational opportunities and also have an impact on their childhood and social life.

With another one billion people at risk of getting NTD, various preventive measures are taken like mass drug administration, control, or eradication of vectors. An effective approach for mass drug administration is where communities take responsibility for collecting and delivering the drug from mass drug administration programs. The 66th World Health Assembly held in May 2013, intended to form strong measures against NTDs and improving the health condition of affected people [1].

The majority of the affected population is still lacking access to proper medication and cannot afford basic preventive amenities. Poverty has led such people to seek another inappropriate medication option. Phytochemicals having biological activity against these NTDs can serve as a cheaper and effective option and can also lead to the production of cheap naturally derived drugs.

2.2 African Trypanosomiasis

Human African Trypanosomiasis (HAT) also known as sleeping sickness is one of the most neglected tropical diseases owing to the recently growing number of cases specified in the sub-Saharan African region. It is caused by a parasite *Trypanosoma brucei* which is transmitted to humans through the bite of the tsetse fly [4]. According to WHO, there are around 70 million people that are at risk of developing the disease in Africa.

The parasite for African trypanosomiasis belongs to the *Trypanosoma brucei* family having three subspecies: *Tb* gambiense, *Tb rhodesiense,* and *Tb brucei* [4]. Most of the HAT cases (around 97%) are caused by *Trypanosoma bruceia gambiense* which is prevalent in the sub-Saharan African region, while a small percentage (3%) of the disease is caused by *Tb rhodesiense* but with a more severe form and outbreaks in Eastern and Southern Africa. *Tb Brucei* is the causative agent of African trypanosomiasis in animals like cattle but not in humans [5].

A tsetse fly will bite an infected animal or an infected person with African trypanosomiasis this bite will cause free unicellular parasitic flagellate protozoa to be ingested in the gut of the fly. While the parasite is inside the gut, the trypanosome will transform into a more severe and infective type due to certain biological changes and reach the salivary glands of the fly so it can easily be transferred to a host through the bite of the fly [6].

After a human is infected, the disease is manifested in two stages: stage 1 and stage 2. The trypanosomes spread through the bloodstream of the host to different vital organs like the spleen, lymph nodes, heart, liver, pancreas, ovaries, testes, and eyes which is part of the first stage of the disease also known as the hemo-lymphatic stage followed by a lesion on the skin which occurs within a few days after the bite. This stage will also be accompanied by symptoms like recurring episodes of fever. The disease progression period into stage 2 is different in the case of *Tb rhodesiense* HAT and *Tb gambiense* HAT, the progression to stage 2 in the case of *Tb rhodesiense* HAT is rapid, i.e. few weeks the reason why it is more severe and fatal, while in case of *Tb gambiense* HAT the disease takes around few months to progress to stage 2. The primary process that happens while in stage 2, is that the parasite crosses the blood–brain barrier of the human host and enters the central nervous system which will cause more severe symptoms like headache, difficulty in concentration, sleepiness giving it the name of the sleeping sickness disease. The last stage occurs when a patient is diagnosed by meningoencephalitis [6, 7].

The treatment for the disease includes only four drugs recognized to date and there is no serious research on the disease done by the relevant agencies as there is less funding given by the government despite the large number of cases leading it to be classified as a neglected disease by WHO. For the first stage of the disease, suramin is prescribed against Tbr-HAT, and pentamidine in cases of Tbg-HAT. During the second stage of HAT, melarsoprol is prescribed against both Tbr and Tbg HAT. Lastly, eflornithine is proven effective against Tbg-HAT during the second stage [8].

A study conducted to determine the structure and function of target protein L-threonine3-dehydrogenase from the protozoa *Trypanosoma brucei* by Adjogatse et al. [9] revealed insights into the role of the protein in the metabolic pathway of the protozoa. The structure for the protein was solved using X-ray crystallography and is deposited in the PDB database. L-threonine 3-dehydrogenase metabolized L-threonine in the gut of *Trypanosoma brucei* and to yield acetyl Co-A, which is crucial for the procyclic forms of the parasite. L-Threonine 3-dehydrogenase along with another enzyme catalyzes the degradation of L-threonine to yield glycine and acetyl-CoA. This pathway is an important requirement for fatty acid synthesis and energy production by the parasite. The inhibition of the L-threonine 3-dehydrogenase thus will be lethal to the parasite hence the protein can be used in target-based drug therapy for efficient treatment of the disease [9].

New medications and better management of the disease is needed to eradicate it from the rural regions of Africa but requires active funding and participation by the government into setting up research and healthcare facilities against human trypanosomiasis as this disease had been plaguing the continent since a century but new cases continue to be reported exponentially every year.

2.2.1 Phytochemicals Against African Trypanosomiasis

Many phytochemicals have been identified against African trypanosomiasis, Schmidt et al. [10] conducted *in vitro* study to identify monoterpenes against African trypanosomiasis. The most active monoterpene identified was terpinen-4-ol identified against the disease. Other phytochemicals tested *in vitro* are α-Pinene, C6-epoxycarboxylic acid (3,4-epoxy-3-methyl hexanoic acid) extracted from *Inula montbretiana*, Lipiferolide, extracted from *Liriodendron tulipifera* (Magnoliaceae), Aromadendrene, Alloaromadendrene, Cupacinosides extracted from *Cupania cinerea* (Sapindaceae), pseudoguaianolide helenalin extracted from *Arnica* and *Helenium* species showing good activity against African trypanosomiasis. [10]

Schmidt et al. [11] assayed quinones isolated from plants, phytochemicals identified having activity against African trypanosomiasis. The phytochemical of interest is 2-(1-hydroxyethyl)-naphtho[2,3-b]furan-4,9-quinone extracted from *Kigelia pinnata* (Bignoniaceae) [11].

The antitrypanosomal activity of alkaloids was studied and out of them, the most active were berberine, emetine, and quinidine which showed high anti-trypanosomal activity equal to the standard drugs suramin and diminazene aceturate [11].

Hoet et al. [12] mentioned natural products having antitrypanosomal activity. Three oxoaporphines were isolated from the bark of the plant *Unonopsis buchtienii* (Annonaceae), and the most potent phytochemical was O-methylmoschatoline showing activity against *Tb brucei*. Alkaloid named Pancracine having a 5,11-methano-morphanthridine ring was isolated from the bulb of *Narcissus angustifolius* subsp. Transcarpathicus shows antitrypanosomal activity [12].

Camptothecin is a pentacyclic alkaloid isolated from *Camptotheca acuminata* (Nyssaceae), this alkaloid can inhibit DNA topoisomerase I, and *in vitro* studies proved it can block DNA replication of trypomastigotes, causing the death of the parasite. Quinone class of phytochemicals like diospyrin obtained from the bark of *Diospyros montana* (Ebenaceae) has been shown to have antitrypanosomal activity *in vitro* on *Tb brucei* [12].

In vitro studies have proven that inactivation of the enzyme trypanothione or its biosynthesis could lead to the death of the parasite [13]. Phytochemicals that can bind to SH- group present in the enzyme trypanothione can be used as a potential lead for the treatment of African trypanosomiasis, polyacetylene Carlina oxide from *Carlina acaulis* (Asteraceae) can bind to the SH group showing antitrypanosomal activity. Intercalating DNA agents can also be used as antitrypanosomal agents, berberine, and sanguinarine have shown significant activity against African trypanosomiasis. Sanguinarine can be used as antitrypanosomal as it inhibits microtubule formation. Camptothecin which inhibits the DNA topoisomerase I has also shown *in vitro* activity against *T. brucei*. Vinblastine has shown antitrypanosomal activity through the inhibition of farnesyl transferase and tubulin formation [13].

2.3 Buruli Ulcer

Buruli ulcer is a necrotic skin disease caused by the mycobacterium species *Mycobacterium ulcerans* primarily affecting tropical regions like West Africa, sub-Saharan regions and also reported in around 33 countries but the disease is poorly understood and there is no proper treatment available because of negligence by health sectors leading it to be labeled as a neglected tropical disease. It causes skin lesions causing necrosis of the subcutaneous muscles and even the bones if not treated at an early stage leading to disabilities and amputation of the sufferer. The main source of transmission of *M. ulcerans* to humans is bruised skin coming in direct contact to stagnant contaminated water with *M. ulcerans* including pond, swamps, and even aerosols but exact sources remain unknown [14].

The disease progresses very slowly and the early stages maybe non-ulcerative with patients exhibiting small nodules or papules. The disease later is manifested as skin ulcer that has weakened edges occurring mostly on the limbs of the patient caused due to a bacterial exotoxin released by the bacteria known as Mycolactone. This toxin further causes the necrosis of the subcutaneous tissues and surrounding regions of the ulcer through the necrosis of different cells in the host like leucocytes, adipocytes, and fibroblasts [15, 16].

The progression of the infection inside the body after entering through the skin is studied and it is considered that the *M. ulcerans* parasitizes the host macrophages multiplying inside them intracellularly thereby inhibiting phagocytosis of the bacterium lowering the host immunity and gradually invading the healthy tissues [15, 16].

M. ulcerans produces virulence factors like mycolactone, mycobactin, and a plasmid-encoded polyketide which helps in the pathogenicity of the disease. The cell wall is made up of these virulence factors, fatty acids, and glycolipids. Synthesis of this cell wall components and virulence factors is carried out by certain pathways in the bacteria which are activated by the protein Sfp-type 4′-phosphopantetheinyl transferase (PPTases). There are two types of PPTases in all bacteria which include the acyl carrier protein synthase (AcpS) and PPTase involved in the surfactin biosynthesis (Sfp), the type of PPTase present in *M. ulcerans*. The 3D structure of Sfp-type PPTases and MuPPT from *M. ulcerans*. MuPPT (PDB id 4QJL) was solved using X-ray crystallography technique and is available in the PDB database. These SFP type PPTases are majorly involved in the production of secondary metabolites in the bacterium helping it to multiply and survive inside the host. As this protein has many important mechanisms associated with the virulence and metabolism of the bacterium, inhibition of this enzyme with a suitable lead compound will cause the death of the bacterium inside the host. Targeted drug therapy against this enzyme will lead to new potential leads against the disease. Due to its important role in multiple aspects of mycobacterial secondary metabolism and concomitant bacterial viability, the discovery of specific inhibitors of this enzyme will enable new therapeutic leads for the treatment of mycobacterial infections. Vickery et al. [17] solved the 3D X-ray crystallographic structures of two Sfp-type PPTases from *Mycobacteria*, PptT from *M. tuberculosis* [17].

Till today there is no vaccination available and treatment usually requires surgical removal of the ulcer if the ulcer is large in size. Recently, a combination of antimycobacterium drugs used for the treatment of tuberculosis, rifampicin, and either streptomycin or amikacin are able to inhibit *M. ulcerans* in human lesions [18].

2.3.1 Phytochemicals Against Buruli Ulcer

Bamberger et al. [19] found the following phytochemicals to be effective in the treatment of Buruli ulcer: Ferulenol extracted from *Ferula communis*; norditerpenoid 12-demethylmulticauline extracted from *S. multicaulis*; Z-phytol, e-phytol, and phytol from Lucas volkensii; tryptanthin from *Strobilanthus cusia*; 7-methyljuglone from *Euclea natalensis*; diospyrin from *Euclea natalensis*; dehydrocostus-lactone from *Saussurea lappa*; and cryptolepine from *Cryptolepsis sanguinolenta* [19].

Fokou et al. [20] mentioned phytochemicals against Buruli ulcer; some of them are Genin acemannan from *Aloe vera* (apigenin, vitexin, isovitexin, sterol, stigmasterol, β-D-sitosterol, β-D-glucoside, sapogenins, alkaloids, triterpenae alcohol, and 1-triacontanol) from *J. curcas* (capsanthin, capsorubin, zeaxanthin, cryptoxanthin, and lutein) from *C. annum* (holamine, holaphyllamine and holaphylline) from *H. floribunda* (alkaloid spiganthine, volatile alkaloids, isoquinoline and actidine isomer, and three quaternary alkaloids, choline, benzoylcholine, and 2,3-dimethylacrolyl choline, phenolcarboxylic acids) from *S. anthelmia* (eugenol [70–85%], eugenyl acetate [15%] and β-caryophyllene [5–12%]) from *S. aromaticum* [20].

2.4 Chagas Disease

Chagas disease is a parasitic disease caused by *Trypanosoma cruzi* through the bite of a blood-sucking bug also known as the kissing bug of the subfamily Triatominae. The disease is mostly occurring in tropical America labeled by WHO as a neglected tropical disease [21]. The outbreaks of the disease have been occurring outside tropical America becoming a global endemic which leads to improvements in understanding and control of the disease, but the disease remains a burden on poor Latin countries where 10 000 people die every year and more than 25 million people are at high risk of contracting the disease [22].

The transmission of disease to humans is insect vector-based, and three potential vectors from the family Triatominae including *Triatoma infestans*, *Rhodnius prolixus*, and *Triatoma dimidiate* is considered to be the main source of *T. cruzi* transmission to humans, the vector remaining unaffected by the parasite. The disease can also be transferred through blood transfusion, orally through the ingestion of contaminated food, and from mother to fetus. Aside from humans the disease can also occur in domestic and wild animals further hampering the socio-economic status of the poor countries [23, 24].

Triatominae bug infected with *T. cruzi* will transfer the parasite to the host during a blood meal, ejecting trypomastigotes in its feces near the site of the bite which will enter the host through the wound or intact mucosal membrane like the conjunctiva which lead to swelling of the eye known as lymphadenitis of the preauricular ganglia giving a characteristic (Romaña sign) shape. If the bite is on other parts of the skin it will lead to swelling of the subcutaneous tissue with local edema. Once the trypomastigotes penetrate the cells near the site of the bite they will be differentiated into amastigotes which will multiply inside the host by binary fission and differentiate into trypomastigotes which will be released into the bloodstream and infect a variety of cells and tissues of the host leading to the manifestation of different symptoms. When a triatominae bug bites a person infected with Chagas disease, the trypomastigotes will be differentiated into epimastigotes in the vector's midgut where they will multiply and lastly differentiates into infective trypomastigotes in the hindgut [25].

After a person is infected with the disease, the disease is manifested into two stages, the acute and the chronic phase. The acute phase which includes symptoms like fever, inflammation at the site of the bite also known as the inoculation chancre, enlarged lymph nodes, swelling of the liver and spleen, unilateral palpebral edema lasting around four to eight weeks, the symptoms of the acute phase usually resolves on its own but in 90% of people and 60% of people will never develop any chronic symptoms as they have antibodies against *T. cruzi* these people are known to suffer from intermediate from Chagas disease, but around 30–40% people can develop the chronic phase of Chagas disease which can affect the cardiac, digestive, or cardio digestive system. These developments are usually observed after 10–30 years of being diagnosed with chronic Chagas disease. The most frequently involved organ in the chronic phase of most of the patients is the heart (14–45%) which causes progressive heart failure ultimately leading to the death of the patient [23, 24].

In the chronic stage, the gastrointestinal problem is observed in (10–21%) of patients affecting the GIT organs like the esophagus and the colon of the patient [24].

The protein which is primarily involved in metabolism and other important processes in the parasite *T. cruzi* is Trypanothione reductase (TR). TR has been extensively studied for decades as an important target for Chagas disease and its 3D structure has been determined using X-ray crystallography deposited in PDB of *T. cruzi* TR (PDB code 1BZL). Parasites like *T. cruzi* are intracellular inside the host, thus becoming sensitive to reactive oxygen species ROS, which is produced by *T. cruzi* under severe oxidative stress. TR plays an important role in the survival of the parasite inside the host by protecting it against the oxidative damage produced by the infected host immunity thus helping it to survive and grow inside the host. Besides protection against oxidative stress, TR plays an important role in redox metabolism, deoxyribonucleotide synthesis, metabolism, and other important cellular processes. Owing to the importance of this enzyme in the survival of the host's body, potential leads against this target can help in curing Chagas disease [26].

There are only two drugs marketed to date which are studied to be effective against Chagas disease, which is benznidazole and nifurtimox. Benznidazole is safer and efficient against *T. cruzi*. Newer drugs with increased efficacy and safety need to be developed to completely eradicate the disease [21].

2.4.1 Phytochemicals Against Chagas Disease

A study conducted revealed that active phytochemicals like ent-kaur-16-en-19-oic acid exhibited activity against Chagas disease and clerodane trans-dehydrocrotonin isolated from the stem of *Croton cajucara* (Euphorbiaceae) showed inhibition of trypomastigote forms of *T. cruzi*. [10]

Schmidt et al. [11] conducted a study using natural quinones to test against *T. cruzi*. These included plumbagin from *Pera benensis*, embelin from *Oxalis erythrorhiza*, diospyrin from *Diospyros Montana* with all showing prospects against *T. cruzi*. Also, atisinium chloride and 13-oxocardiopetamine showed considerable *in vivo* activity against *T. cruzi* similar to the standard drug [11].

Hoet et al. [12] mentioned natural products having antitrypanosomal activity, which includes thiol-containing enzymes from garlic extract. Ajoene extracted from garlic shows activity against *T. cruzi* by inhibiting the enzyme TR.

Flavonol similar to quercetagetin but has an absence of hydroxyl on C_6 terminal, has been shown to inhibit *Trypanosoma cruzi* causative agent of Chagas disease [12].

2.5 Chikungunya

Chikungunya is a viral disease caused by the chikungunya virus of the Togovirus family. The virus is classified in the Alphavirus family and is transferred to humans through the bite of an infected mosquito that is *Aedes albopictus* and *Aedes aegypti* and, the disease has affected people all around the world in major countries of Africa, Asia, and the Indian subcontinent and remains a reemerging type of disease and as of 2019 there is no approved drug or vaccine available against the virus, medication to treat local symptoms and long term symptoms are prescribed for the management of the disease. The eradication of the disease can be bought by mosquito population control and through the use of mosquito repellant in high-risk regions [27].

Chikungunya virus is a single-stranded positive-sense RNA molecule enclosed by a lipid envelop about 12 kb in weight. The virus consists of two envelop glycoprotein on its surface that are the E1 and E2 glycoprotein embedded in a heterodimer form on the surface responsible attachment of the virus to host cell membrane fusion [28].

It consists of structural proteins which include C, E1, E2, E3, and nonstructural protein which include nsP1, nsP2, nsP3, and nsP4. Out of the four nonstructural proteins the most important role played in virus replication and proteolytic cleavage of nonstructural polyprotein into individual protein after replication is carried by nsp2 protein. It has several other enzymatic roles and plays a critical role in the replication of the virus inside the host by inducing oxidative stress in the host cells. Oxidative stress-induced inside the host will change the functional role of the proteins through posttranslational modification. Thus, targeting the nsp2 protein of the chikungunya virus can directly help in the inhibition of the virus cause of its important function in viral replication inside the host [29, 30].

The infected mosquito takes a blood meal of the host, the virus spreads inside the host through the bloodstream spreading to organs like the spleen, liver, and the lymph nodes affecting the joints, and the muscles through the destruction of fibroblasts, skeletal muscle progenitor cells, and myofibers. Infection with the chikungunya virus confers lifelong immunity to the person infected [27].

The infection causes chikungunya fever in the host which is characterized by fever, rash, headache, vomiting, and extreme joint pains leading to arthritis. The clinical manifestation of the disease is differentiated into two phases: the acute and chronic phases. The acute phase lasts for one to two weeks with symptoms like fever, body ache, and persistent symptoms like pain and swelling of the joints, ankles which can last up to several months to years often causing immobility due to extreme pain and stiffness of joints. Chikungunya virus damages the cartilage, skeletal muscle cells, and the fibroblast where the virus replicates primarily before spreading to other parts of the body causing extensive inflammation and pain in the joints. In the chronic phase, the patient will suffer from recurring episodes of joint pain which would be migratory occurring in different joint or nonmigratory occurring in the same joint as the previous phase. Generally, the age group of extremely young and old are affected the worst by chikungunya due to a lower level of immunity [27].

As there is no medication against the disease, usually paracetamol and anti-inflammatory drugs are prescribed to control the systematic symptoms of fever, body ache, and antirheumatic drugs for management of joint pains might be prescribed [31].

2.5.1 Phytochemicals Against Chikungunya

Extraction of phytochemicals was carried out using Reverse phase-HPLC and GC-MS on ethanolic extracts of *C. dactylon* (Murali et al. [32]) and phytochemicals like luteolin and apigenin having antiviral activity against chikungunya virus was obtained from the study [32].

In vitro studies were carried out to obtain a potent antiviral agent against the chikungunya virus by Gómez-Calderón et al. [33] using open column chromatography from the ethanolic extract of *M. americana* seeds. Two potent phytochemicals of coumarinic type were obtained that were named coumarin A and coumarin B showing full inhibition of the virus *in vitro* [33].

Cytopathic assay was adopted to study the inhibition activity of three phytochemicals by Lani ct al. [34] performed, out of which the most potent phytochemical identified was Silymarin, extracted from milk seeds of *Silybum marianum* [34].

Phytochemical studies were conducted of the plant croton Mauritius by Nina Corlay et al. [35] two phytochemicals 12-O-decanoylphorbol-13-acetate and the new 12-O-decanoyl-7-hydroperoxy-phorbol-5-ene-13-acetate from *Croton mauritianus* were isolated and viral cell assay was performed using the compounds. Both the compounds showed inhibition of the chikungunya virus [35].

2.6 Dengue

Dengue fever is a viral disease caused due to the dengue virus belonging to the family Flaviviridae. The virus is classified into four different serotypes and is transmitted to humans through the bite of a mosquito of species Aedes most specifically *A. aegypti* is the vector [36, 37]. The disease has acquired an endemic status affecting around 50–100 million people per year from over 100 countries specifically in the tropical regions of the world particularly Africa, Southeast Asia, Americas, and is recognized as a neglected tropical disease by WHO [38]. The disease causes a wide range of symptoms like fever, rash, and can also progress to serious effects like hemorrhagic fever and dengue shock effect [36, 37].

The virus is a single-stranded positive-sense RNA, which is common for all the 4 subtypes of the virus differing only serologically. The virus is made of structural proteins and nonstructural proteins. The structural proteins include the capsid protein which is encapsulated in the viral genome, membrane proteins, and envelop protein that is present on the virus surface and involved in the attachment of the virus to the host cell and thus can be targeted to inhibit the virus. The nonstructural proteins include NS1, NS2A, NS2B, NS3, NS4A, NS4B, and NS5 expressed in host cells after dengue infection [39–41]. The NS5 protein is the largest protein out of all the nonstructural proteins and is conserved in all the four types of dengue virus (1–4) with around 70% sequence identity thus can be used as the potential target in drug design for all type of dengue infections. The NS5 proteins play a crucial role in the virus replication inside the host body through its two domain that is a methyltransferase domain (MTase) at its N-terminus and an RNA-dependent RNA polymerase (RdRp) at its C-terminus while also lowering the host immunity by interacting with different host proteins. The MTase performs the capping of viral RNA and performs methylation activity while the RdRp domain initiates viral replication. Because of its conserved nature across all virus serotypes and its importance in viral replication, NS5 protein is the most suitable drug target for the treatment of dengue [40, 41].

The dengue virus is transferred to the mosquito *A. aegypti* where it multiplies inside the midgut of the insect. The infected mosquito transfers the virus to a human host during a blood meal, the virus is inoculated in the dermis and epidermis where they will infect the Langerhans cells of the epidermis and the keratinocytes. The infected cells then circulate through the bloodstream to the lymph nodes where the virus will further multiply by infecting cells like the macrophages and the monocytes. The virus can also migrate and replicate in hepatocytes and kupffer cells of the liver causing necrosis of the liver in later stages of the disease [42, 43].

The clinical manifestation of dengue is divided into three phases:

- Phase 1 also known as the febrile phase is a phase accompanied by symptoms like a high-grade fever which will last a few days to a week, nausea, vomiting, headache, muscle ache, and arthralgia. Typically, children can show severe symptoms due to compromised immunity like a febrile seizure.
- Phase 2 also known as the critical phase is characterized by severe symptoms like a decrease in blood platelets, increase in capillary permeability causing plasma leakage causing severe plasma loss which can last for 1–2 days. The extent of plasma leak can also progress to dengue shock and hemorrhagic in some patients from where the chance of death of the patients increases due to organ failure and blood loss.
- Phase 3 is the recovery phase where the patient will show recovery symptoms like reabsorption of the plasma back into the system showing improved platelet count and lowered hematocrit level after which proper treatment can be continued [36, 37].

There is no specific treatment for dengue, supportive care should be taken by monitoring the fluid intake during phase 2 of the infection to prevent dehydration of the patient. Intravenous fluids are injected if the patient cannot take fluids orally to avoid dehydration. Blood transfusions will be carried out if the platelet count is significantly low or if a patient is suffering from extreme dehydration [36, 44].

2.6.1 Phytochemicals Against Dengue

Isolation of flavonoid Baicalin from the roots of the plant *Scutellaria baicalensis* by Moghaddam et al. [45] was conducted and after carrying out several *in vitro* antiviral studies to test the potency of baicalin against dengue virus, it was observed that baicalin shows anti-dengue activity through the use of the concentration of 50 mg/ml [45].

In vitro studies carried on two plants to obtain potent antiviral agent against dengue virus by Gómez-Calderón et al. [33] revealed the antiviral activity of four compounds derived from plants *M. americana* (coumarin A and coumarin B) and *T. cyamosa* (lupeol acetate and voacangine) were studied using mass spectroscopy and nuclear magnetic resonance (NMR) showed *in vitro* inhibition of the dengue virus [33].

HPLC and phytochemical analysis carried on the leaves of *S. mombin* and *S. tuberosa* plants found in brazil by Silva et al. [46] lead to the isolation of phenolic compounds quercetin, rutin, and ellagic acid. These compounds were evaluated for their cytotoxicity and antiviral activity against dengue virus, after carrying out the antiviral assays it was determined that rutin and quercetin have potential anti-dengue activity [46].

In vitro assays carried on the ethanolic extracts from fruits, stems, leaves of the plant *Distictella elongate* by Simões et al. [47] revealed the presence of phytochemicals pectolinari and acacetin-7-O-ruti-noside (2). The mixture of the two phytochemicals extracted from the plant showed eight times more antiviral activity against the dengue virus than pectolinari alone after preforming appropriate antiviral assays [47].

2.7 Leishmaniasis

Leishmaniasis is caused by protozoan parasites of over 20 *Leishmania* species which are transmitted by 90 species of sand-fly. WHO estimates show that 20 000–30 000 deaths and 70 000 to 1 million new cases would occur annually in 88 different countries with 350 million people under threat [48]. Leishmaniasis targets mostly the people living in poor communities, who are malnourished, have a weak immune system, poor housing, and is also associated with population displacement, environmental changes, and lack of financial resources.

Female phlebotomine sand flies transmit Leishmaniasis when they bite a human host. The infective stage of the parasite, i.e. promastigotes is injected from the proboscis of the sand fly during their blood meal. Macrophages and mononuclear phagocytic cells phagocytized those promastigotes which reach the puncture wound. Within these cells, promastigotes transform into amastigotes also known as tissue stage. Amastigotes multiply themselves by simple division and infect more mononuclear phagocytic cells. These amastigotes are then ingested by sandflies during their blood meals. In the gut of these sand flies' amastigotes transform into promastigotes, once developed these promastigotes migrate to the proboscis, ready to infect another human host when the sand fly takes its next blood meal [49].

There are three forms of Leishmaniasis:

a) Visceral leishmaniasis: It is also known as kala-azar which is the main form of leishmaniasis and may cause death if not treated. It usually affects the spleen, liver, and bone marrow causing anemia, enlargement of the spleen and liver, irregular bouts of fever, and weight loss.
b) Cutaneous leishmaniasis: This is the most common form of leishmaniasis skin lesions and can worsen into an ulcer-causing serious disability and leaving behind a permanent scar.
c) Mucocutaneous leishmaniasis: It is one of the least common forms of leishmaniasis but can cause partial or total damage to mucous membranes of the nose, mouth, and throat [48].

Out of 20 disease-causing *Leishmania* species *Leishmania donovani, L. infantum*, or *L. chagasi* can cause visceral leishmaniasis in humans. More than 15 species such as *Leishmania major, Leishmania tropica, Leishmania aethiopica, Leishmania mexicana, Leishmania amazonensis, Leishmania braziliensis, Leishmania panamensis,* and *Leishmania guyanensis* can cause cutaneous leishmaniasis in humans [50].

Paromomycin is used for treating both visceral and mucocutaneous leishmaniasis. Amphotericin B is used against visceral leishmaniasis while liposomal amphotericin B is used in the treatment of mucocutaneous leishmaniasis. Other drugs for the treatment of visceral leishmaniasis include meglumine antimoniate, sodium stibogluconate, and miltefosine [51]. Amphotericin B causes cell death by irreversibly binding to sterols (ergosterol) causing disruption of membrane integrity, leaking out intracellular components [52, 53]. Mode of the action of Paromomycin in *Leishmania* species is not understood yet, although it is known that it inhibits protein synthesis by interacting with ribosomal RNA subunits [54, 55].

Ogungbe et al. [56] performed molecular docking of antileishmanial phytochemical phenolic compounds against potential protein targets in Leishmaniasis causing agents. After docking, the potential protein targets found are listed in Table 2.1 [56].

2.7.1 Phytochemicals Against Leishmaniasis

Mishra et al. [57] in their review article reported that alkaloids such as quinoline, isoquinoline, pyrimidine–carboline, steroidal, and diterpene have leishmanicidal activity [57].

Table 2.1 List of protein targets with respective PDB ID for various species of *Leishmania*.

Organism	Enzyme/protein target	PDB ID
L. major	Dihydroorotate dehydrogenase	3GYE, 3MHU, and 3MJY
L. major	Methionyl-tRNA synthetase	3KFL
L. major	Nucleoside diphosphate kinase B	3NGS, 3NGT, and 3NGU
L. major	Nucleoside hydrolase	1EZR
L. major	N-myristoyl transferase	2WSA, 3H5Z, and 4A30
L. major	Oligopeptidase B	2XE4
L. major	Phosphodiesterase 1	2R8Q
L. major	Pteridine reductase 1	1E7W, 1W0C, 2BF7, and 3H4V
L. major	Tyrosyl-tRNA synthetase	3P0H and 3P0J
L. major	Uridine diphosphate-glucose pyrophosphorylase	2OEF and 2OEG
L. major	Deoxyuri-dine triphosphate nucleotidohydrolase	2YAY and 2YB0
L. donovani	Cyclophilin	2HAQ and 3EOV
L. donovani	Dihydroorotate dehydrogenase	3C61
L. donovani	N-myristoyltransferase	2WUU
L. mexicana	Glycerol-3-phosphate dehydrogenase	1EVZ, 1M66, 1N1E, and 1N1G
L. mexicana	Pyruvate kinase	1PK1, 3HQP, and 3PP7
L. infantum	Sterol 14α-demethylase	3L4D
L. infantum	Nicotinamidase1	3R2J

Source: Based on Ogungbe et al. [56].

2.8 Leprosy

Leprosy is an infectious skin disease caused by the bacteria *Mycobacterium leprae*. The disease mainly affects the tropical countries with more than 211 009 cases in the year 2017 around the world especially in underdeveloped or developing countries like Brazil, Africa, and Southeast Asia [58]. Early detection of the disease can lead to effective treatment but in many cases due to unawareness about the disease, it is diagnosed at a later stage causing disability and disfiguration of the skin in most severe cases. The disease is spread through inhalation of the bacilli via the nasal route and skin [59].

The infecting agent is *M. Leprae,* a gram-positive bacillus having a rod-like structure when observed under a microscope and the optimal temperature required for it to grow is 27–30 °C making it easier to grow on the skin surface, nerves, upper airways [60]. The primary mode of transmission of the bacteria is through inhalation of the bacteria present in the nasal secretions of an infected person. The bacteria can also be transmitted through blood transfusion, skin erosion, and breast milk of an infected mother to the fetus. Some people are genetically more susceptible to the infection than others due to certain genetic variations in their genome making them less immune to the bacteria [59, 60].

Till date, only two protein structures of *M. leprae* are deposited in the PDB that does not have any homology to humans, which includes only Holliday junction DNA helicase RuvA and riboflavin synthase thus making them potential drug target against leprosy. Riboflavin synthase plays an important role in Riboflavin synthesis inside the bacteria which is required for cell growth and survival. RuvA processes Holliday junctions which in turn play an important role in recombination repair of the bacterial genome. Due to their critical roles in bacterial survival and growth, these two proteins can be considered as potential drug targets [61].

After a person is infected with the bacteria, the clinical manifestation of the diseases occurs in five forms that are two polar forms of tuberculoid (paucibacillary) and lepromatous (multibacillary) leprosy) and three subtypes of intermediate stage leprosy-borderline tuberculoid (BT), mid-borderline (BB), borderline lepromatous (BL), and depending upon the bacterial load, host immunity and a number of lesions on the screen this classification is done [32, 62, 63].

2.8.1 Indeterminate Leprosy

This is an asymptomatic stage that is rarely identified due to the absence of any prominent symptoms other than hypopigmented macules occurring on the body. This stage can last up to years and is often misdiagnosed to other skin ailments, depending upon the host immunity, the disease will progress to other forms of leprosy.

This stage is further classified into three groups: BT leprosy, borderline BB leprosy, and BL leprosy [32, 62, 63].

Tuberculoid form occurs when the host has a good immune response against the disease, causing a milder and localized form of leprosy. There will be a single or few skin lesions in the form of macules or papules [32, 62, 63].

Lepromatous leprosy occurs when the host immunity response is poor against the disease. This causes generalized or widespread disease, causing numerous red-brown lesions on the skin, nerves. The lesions will be symmetrical in nature and this stage can also cause edema of the feet [32, 62, 63].

The treatment of the disease requires a combination of antibiotics inhibiting the bacteria like dapsone with rifampicin, and clofazimine is typically prescribed for a duration of one to two years. The proteins Inactive dihydropteroate synthase 2 and dihydropteroate synthase 1 are targets of the drug dapsone [64].

Early detection of the disease is the primary requirement to prevent any long-term nerve damage or serious disability like blindness, paralysis of the hands or the feet, and disfiguration [64].

2.8.2 Phytochemicals Against Leprosy

Ghosh [65] mentioned phytochemicals like hydnocarpic acid from the Hydnocarpus species which has been used for the treatment of leprosy since the early 1920s. The plant has also other active molecules like taraktophyllin, 3,4-dihydroxybenzyl alcohol, 3-hydroxy-4 methoxybenzoic acid, and chaumoogric acid which has exhibited antibacterial activity [65].

The plant *Psoralea corylifolia* has been studied and registered in pharmaceutical index for effective treatment of leprosy having active phytochemicals like corylifols a–c (prenylfoavanoids) in seeds; psoralen, isopsoralen, and neobavaisflavones in dried fruits. neoba-vaislfloavone, borachin, bavaislfavooz, bavachalcone, bavachromene psoralidin, corylifolinin, barachini psoralenoides, isopsoralenoside, and coumarins, Diadzein (4 : 7 dyhydroxuisoflavon) and genistein (4′6′7 trihydroxyisoflavon) from the plant. Brombexine from leaf extract of *Adhatoda vasica*, Allicin from *Allium sativum*, asiaticoside from *Centella asiatica* have been studied for the treatment of leprosy [65].

Studies about the antileprotic activity of the Flavonolignans from the hydrocarpus species by Sahoo et al. [66] revealed phytochemicals like hydnocarpic, chaulmoogric, and garlic acid. These acids are usually used in oils that are applied for the treatment of leprosy. Both the compounds show activity against the *M. leprae* bacteria by activating host immunity and destroying the cell wall of the bacterium or through chemotaxis [66].

2.9 Lymphatic Filariasis

Lymphatic filariasis often called Elephantiasis is caused by parasitic filarial worms that are transmitted by mosquitoes and cause a part of a person's body to swell in massive proportions. These worms attack the lymphatic system of the host and cause pain, disfiguration of the body due to enlargement of body parts, and severe disability. Globally, 856 million people across 52 countries are under threat of this parasitic infection. WHO estimates that 120 million people have been affected by lymphatic filariasis, 15 million people suffer from lymphedema, 25 million men suffer from hydrocele [67].

Only three types of nematodal worms (roundworms) cause lymphatic filariasis of which, 90% of the cases are caused by *Wuchereria bancrofti*, remaining cases are caused by *Brugia malayi* and *Brugia timori*. Mosquitoes act as a vector and are responsible for the transmission of these parasitic worms. Third stage infective filarial larvae are introduced onto human skin by a mosquito during its blood meal. These larvae then penetrate the bite wound and reside in the lymphatic vessel of the host. Here, the larvae develop into the adult worms and produce sheathed microfilariae. Microfilariae then migrate from lymph to the bloodstream where they get ingested by a mosquito during its blood meal. Once ingested by the mosquito, microfilariae lose their sheath and migrate to the midgut of the cardiac portion where they travel into thoracic muscles. Microfilariae complete its three larval stages here in the thoracic muscles and after development migrate to the mosquito's proboscis through the hemocoel, ready to infect another human host [68].

Current treatment for lymphatic filariasis includes the administration of a two-drug combination either albendazole with diethylcarbamazine or albendazole with ivermectin. Albendazole with ivermectin is believed to be an optimal treatment but a recent study by Thomsen et al. [69] suggests that a triple-drug therapy, i.e. albendazole + ivermectin + diethylcarbamazine is more efficient than double drug treatment. Albendazole binds to the colchicine-sensitive site of tubulin (tubulin β-2 chain) and blocks the polymerization of tubulin proteins into microtubule [70]. This leads to weak uptake of glucose and depletes glycogen stores of adult and larval stages of the parasite. The energy production of the parasite is reduced due to a decrease in the production of ATP which modeled the β-tubulin protein of *Burgia malayi* for *in silico* docking analysis of benzimidazole analog for treating lymphatic filariasis. Since the 3D structure of the β-tubulin protein of *B. malayi* is not available, the structure was modeled using its amino acid sequence [71]. In search of novel drug targets against filarial nematodes, Choi et al. [72] suggest that Calumenin protein found in *B. malayi* can be considered as a novel drug target. Functional loss of Calumenin protein will make the worms more susceptible to drugs, infertility, and can also cause abnormal locomotion and reduced body size. Choi et al. [72] modeled the 3D structure of Calumenin protein from *B. malayi* as only the amino acid sequence is available and not 3D structure.

2.9.1 Phytochemicals Against Lymphatic Filariasis

In search of a new class of antifilarial agents against *Burgia malayi*, Sashidhara et al. [73] found that Galactolipids from *Bauhinia racemose* are effective against the worm. From the n-hexane fraction of ethanolic extract of *B. racemose* leaves, (2S)-1,2-di-O-linolenoyl-3-O-a-galactopyranosyl-(1/6)-O-b-galactopyranosyl glycerol is active against adult worm and microfilariae whereas (2S)-1-O-linolenoyl-2-O-palmitoyl-3-O-a-galactopyranosyl-(1/6)-O-b-galactopyranosyl glycerol and

(2S)-1-O-oleoyl-2-O-palmitoyl-3-O-a-galactopyranosyl-(1/6)-O-b-galactopyranosyl glycerol were effective only against microfilariae [73]. A drug discovery research carried by Kalani et al. [74]; Glycyrrhetinic acid isolated from the roots of *Glycyrrhiza glabra* showed activity against microfilariae of *B. malayi* [74].

Ranjini et al. [75] reported a list of phytoconstituents that are isolated from plants extracted which have been screened for *in vivo* and *in vitro* antifilarial activity. The potential phytoconstituents which require further molecular studies are (−) Epicatechin, (−) Laminaribiitol, 12-hydroxy-9-octadecenoic acid; 12-methoxy-*N*(4)-methylakuammicine; 17-Acetoxy-nor-echitamine; 19-Epischolaricine; 2-(4-methoxy phenyl) imino 4-methyl-1, 3-thiozolan-4yl); 2′-(*R*)-O-Acetylglaucarubinone; 4-Hydroxy-α-tetralone; 4-O-(b-D-glucopyranosyl-6-sulphate) gallic acid; 7,4′-dihydroxyflavon, Akuammicine, Akuammidine, Aloe emodina, Andrographiside, Andrographolide, Apigenin, Apocynin, Archidic acid, Azadirachtin, β-sitosterol, Boeravinone A, Boeravinone B, Calctin, Calotropin, Calotropises quiterpenol, Carapolide A, Cardenolide (200-oxovoruscharin), Catechin, Chebulagic acid, Chebulanin, Chebulinic acid, Chlorogenic acid, Chlorogenin, Columbin, Corilagin, Courmaric acid, Cycloartanone, Demethylwedelolactone, Dihydroagnosterol, Dihydrobenzofuran neolignan rel-(2α, 3β)-7-O-methylcedrusin, Diosgenin, Dodecanoicacid, Echitamidine, Echitamine, Echitamine, Epitetraphyllin B (volkenin), Euphol, Ferulicacid, Flavonoids, Gummiferol, Hecogenin, Hypaphorine, Iridoids, Isocoumarins, Isoricinoleic acid, Isoshinanolone, Isoshinanolone, Isovanillic acid, Krisofanat acid, Luteonin, Mansonin, Mexicanolide-methylangolensate, Monoterpenoidnaphthaquinones, Myricetin, Naphthaquinones, Neogitogenin, Neohecogenin, Nimbidin, Nimbin, Nimbolin B, Nor-echitamine, Nα-formylechitamidine, Oligorhamnosides, Palmatine, Palmitate, Palmitic acid, *p*-hydroxybenzoic acid, Physcion, Proanthocyanidins, Protocatechuic acid, Pseudoephedrine, Punicalagin, Quassinoid, Quercetin, Repenol, Repenone, Rhein, Ricinoleic acid, Robustadial A, Robustadial B, Ruscogenin, Saponin, Scopolamine, Scopoline, Stearic acid, Stearic, Steroids, Stigmasterol, Strebloside, Tannins, Tembeterine, Tetracosanoic acid, Tetralone-4-O-β-D-glucopyranoside, Tetraphyllin B (barterin), Tinosporoside, Tirucallol, Tubotaiwine, Vellesamine, Wedelolactone, Zingerone, ά-calotropeols, α-2-sitosterol, α-amyrin, α-zingiberol, β-calotropeols, β-sitosterol, and β-sitosterolglucoside [75].

2.10 Mycetoma

Mycetoma, also known as Madura's foot, is a subcutaneous infection caused by certain bacteria and fungi [76]. Cases of mycetoma are mostly reported in tropical and subtropical regions. People living in poor communities, manual workers, and males are commonly affected.

Mycetoma can be distinguished into two types based on the causative agents: Actinomycetoma caused by bacteria and Eumycetoma which is caused by fungi. Several species of fungi and bacteria cause mycetoma in which the most common of them are *Actinomadura madurae, Actinomadura pelletieri, Nocardia brasiliensis,* and *Streptomyces somaliensis* which are aerobic actinomycetes and the common causative agent of eumycetoma is *Madurella mycetomatis* [77]. Both forms of mycetoma show similar symptoms; they enter into the body through a minor break or trauma to the skin. They then develop painless masses under the skin; the skin usually hardens near these masses. Over time, these masses grow larger, and then the skin ruptures and forms sinus tracts that discharge pus and grains containing the disease-causing organism.

Despite the similarity in symptoms and appearance, both the forms are treated differently, actinomycetoma is treated with antibiotics like amikacin, amoxicillin-clavulanate, cotrimoxazole, dapsone, gentamicin, isepamicin, meropenem, netilmicin, rifampicin, streptomycin, tobramycin, and trimethoprim-sulfamethoxazole whereas eumycetoma is treated with surgery and prolonged administration of imidazoles such as itraconazole or itraconazole with terbinafine [78].

Itraconazole which is used in the treatment of various fungal infections binds to Lanosterol 14-α demethylase (Uniprot ID: P50859) which is an enzyme that converts lanosterol to ergosterol. By inhibiting the synthesis of ergosterols, itraconazole increases cellular permeability of fungal cell membranes leading to the leakage of cellular components. It is also known to inhibit the transformation of yeasts into its mycelial form and disrupt endogenous respiration [79].

Amikacin interrupts the protein synthesis of bacteria by binding to the 30S ribosomal subunit, interfering with mRNA binding sites and tRNA acceptor sites thereby disrupting cell growth. The protein target of amikacin in *Escherichia coli* (strain K12) is 30S ribosomal protein S12 (Uniprot ID: P0A7S3) [80].

Gentamicin, netilmicin, streptomycin, and tobramycin belong to the aminoglycoside class of drugs and have a similar action. They bind to 30S ribosomal protein S12 (Uniprot ID: P0A7S3) of *Escherichia coli* (strain K12) and 16S rRNA of enteric bacteria and other eubacteria, interfering with the initiation complex and causing misreading of mRNA due to which wrong amino acids are inserted into polypeptide chain-forming nonfunctional peptides and nonfunctional monosomes [81–84].

Rifampicin causes cell death of various Gram-positive and Gram-negative bacteria by inhibiting DNA-dependent RNA polymerase stopping RNA synthesis. In *Escherichia coli* (strain K12) rifampicin binds to DNA-directed RNA polymerase subunit beta (Uniprot ID: P0A8V2) [85].

Meropenem inhibits cell wall synthesis in Gram-positive and Gram-negative bacteria by binding to penicillin-binding-protein targets. In *Escherichia coli* and

Pseudomonas aeruginosa, it binds to penicillin-binding-protein 2, 3, and 4 while in *Staphylococcus aureus*, it binds to penicillin-binding-protein 1, 2, and 4 [86].

2.10.1 Phytochemicals Against Mycetoma

Elfadil et al. [87] investigated the *in vitro* antifungal activity of medicinal plants against *Madurella mycetomatism* which is a major causative agent of eumycetoma. They found that crude methanol extract of *Boswellia papyrifera* having phytochemicals β-amyrin, β-amyrone, β-sitosterol, and Stigmatriene were most active against the *M. mycetomatis* of which the only Stigmatriene showed inhibitory actions at lower concentrations [87].

2.11 Onchocerciasis

Onchocerciasis (subcutaneous filariasis), also called river blindness is an eye and skin disease caused by a parasitic infection with the filarial worm *Onchocerca volvulus*. More than 99% of the infected population lives in 31 African countries and it was estimated by the Global Burden of Disease Study that out of 20.9 million *Onchocerca volvulus* infections, 1.15 million infected people had vision loss, and 14.6 million infected people had a skin disease [88].

Onchocerciasis is acquired when a third-stage filarial larva is introduced onto the human skin by an infected blackfly of *Simulium* species during its blood meal. The larvae then penetrate the bite wound and reach subcutaneous tissues and reside in nodules of the subcutaneous connective tissues where it develops itself into adult filariae. Some of these nodules may contain numerous male and female adult worms. The female worms produce unsheathed microfilariae which are found in the skin and the lymphatics of connective tissues and can also be occasionally found in peripheral blood, urine, and sputum. The next stage of this microfilariae starts when it is ingested by a blackfly during its blood meal. The microfilariae penetrate the blackfly's midgut and migrate to the thoracic muscles through hemocoel. The microfilariae then develop into the first larval stage and subsequently grows into third-stage larvae. These third-stage infective larvae migrate to the blackfly's proboscis which is ready to get introduced onto human hosts when the blackfly takes its next blood meal [89].

Patients suffering from skin disease suffer from terrible itching while those suffering from eye disease develop eye manifestations as the death of microfilariae is very toxic for the skin and eye. This leads to disfiguration skin diseases and irreversible blindness after repeated years of exposure. To date, there is no medication or vaccine available for the prevention of Onchocerciasis. Current treatment includes the administration of Ivermectin which kills larvae and prevents them from causing damage but it does not kill the adult worms [88]. However, antibiotic Doxycycline is also recommended as it kills the Wolbachia bacteria on which the adult worm depends for its survival. Hence Doxycycline weakens the worm by killing the Wolbachia bacteria [90]. Ivermectin binds to glutamate-gated chloride ion channels in nerve cells of the microfilaria and causes increased uptake of chloride ions through cell membranes resulting in hyperpolarization of the cell which in turn leads to paralysis and then the death of the microfilaria [91]. Doxycycline binds reversibly to the A site on the ribosome-RNA complex of the 30S ribosomal subunit and 50S ribosomal subunit (to a lesser extent) of bacteria, blocking the binding of aminoacyl tRNA. This leads to inhibition of bacterial protein synthesis as a result of inhibiting cell growth [92, 93].

An *in vitro* study conducted by Gloeckner et al. [94] on enzyme chitinase of *O. volvulus* states that it can be considered as a potential protein target as biosynthesis of chitin is important for development in the larval stage and competitive binding of the drug to this enzyme target can control the growth of *O. volvulus* [94].

Choi et al. [72] suggest that calcium-binding-protein calumenin in *O. volvulus* can serve as a novel protein target for filarial worms. Loss of calumenin function enhances drug acceptability and causes infertility [72].

2.11.1 Phytochemicals Against Onchocerciasis

Metuge et al. [95] carried out an investigative study to find the efficacy of metabolites isolated from roots of *Cyperus articulates* on *Onchocerca ochengi,* which is a close relative of *Onchocerca volvulus.* Two phytochemicals; Mustakone and Linoleic acid (9*Z*,12*Z*)-octadeca-9,12-dienoic acid were proved to have anti-filarial activity as they were able to kill both the microfilariae and adult worms of *O. ochengi* [95].

2.12 Rabies

Rabies is a viral infection caused due to rabies virus of the Rhabdoviridae family causing acute neurological symptoms. The disease is classified as zoonotic, typically transferred from an infected animal to humans, in most cases through the bite of infected dogs [96]. Rabies mostly affects developing countries like Asia and Africa causing 55 000 deaths worldwide every year and is classified as a neglected tropical disease by WHO. Since it is a fatal disease if not detected at an early stage, proper prevention like animal control and vaccination of people in high-risk regions should be undertaken by the government and health departments [97].

Rabies virus is an RNA virus of the Lyssavirus genus. It is enveloped single-stranded virus and has a negative-sense RNA genome. It encodes structural proteins which include nucleoprotein (N), phosphoprotein (P), matrix protein (M), glycoprotein (G), and RNA-dependent RNA polymerase (L) [96, 98].

The glycoprotein g of the virus is present in the form of protrusion covering the viral envelop, it is the primary protein that helps in the entry of the virus into the target host cell through attachment to the receptor present on the host's cell. It induces endocytosis of the virus undergoing conformational changes during endocytosis due to the pH causing membrane fusion between the virus and host cell membrane. It is also able to induce viral neutralizing antibodies thereby allowing replication of the virus. Due to its importance in the virus entry and replication, it is one of the most promising targets for drug designing and vaccine development [96, 99].

The primary reservoirs of rabies are dogs and bat. Rabies transmission through dogs is most common in Asia and Africa while transmission through bats is common in America. The virus is transmitted through the saliva of an infected animal through open wounds and bite on the skin of the host. Once the virus enters the host it enters the central nervous system and primarily replicates in the muscular cells and the brain [97].

Clinical manifestation of the disease occurs in five stages: incubation period, prodrome, acute neurologic phase, coma, and death. The incubation period of the virus inside the host is one to two months followed by the prodrome and neurological phase. Symptoms like fever, headache, neuropathic pain at the site of the bite occur as the virus multiplies through the peripheral nerves to the central nervous system after attaching to the axons of nerves causing encephalomyelitis. During the prodrome stage, other symptoms like fever, headache, weakness is accompanied as the virus travels to the dorsal root ganglia and causes the destruction of CNS neuron to neuron, invading all other organs and constantly secreted through saliva. This is the final stage which will be followed by coma and death after multiple organ failure of the patient [100, 101].

Treatment of rabies requires cleansing of the wound using diluted povidone-iodine solution followed by the combination of vaccine which includes Imovax Rabies and RabAvert vaccines, human rabies immune globulin (HRIG), and antiviral drugs which include ribavirin, amantadine, interferon-alpha, and acyclovir could help in the clearance of the virus but since it is a fatal disease, the chances of survival are very low, till date only a few people have survived the virus. As there is high mortality, constant preventive measures should be taken to avoid contact with unfamiliar animals, animal vaccination, educating people to vaccinate them as a preventive measure, and healthcare professionals coming in contact with such patients should use gloves and masks [97].

2.12.1 Phytochemicals Against Rabies

The medicinal plants used to treat rabies in Ethiopia were validated through *in vivo* studies. Meresa et al. [102] claim that the leaf extracts of the plant *Salix Subserrata* along with chloroform and methanol has antirabies activity in infected mice. The phytochemicals isolated from this plant include flavonoids like rutin, luteolin-7-glucoside, quercetin and other phytochemicals among which are (+) catechin, 1,2-benzenedicarboxylic acid, bis (2-ethylhexyl) ester, saligenin, methyl 1-hydroxy-6- oxocyclohex-2-enecarboxylate, catechol, propyl acetate, β-sitosterol, and β-sitosterol glucopyranoside [102].

Phytochemicals isolated from onion-like kaempferol and quercetin have antiviral activity and have shown to inhibit various viruses including rabies virus [103]. They inhibit the proteins of the virus which is required for it to replicate inside the host [103].

2.13 Scabies

Sarcoptes scabiei var hominins is a mite that causes parasitic infestation called Human Scabies. These microscopic mites burrow into the epidermis layer of the human host and lay eggs which trigger the host's immune response causing severe itching and skin rash. More than 130 million people are affected by scabies according to WHO. Scabies is distributed in every country mostly affecting young children, elderly people living in poor communities, and the people living in hot, tropical climates [104].

The life cycle of *Sarcoptes scabiei* can be distinguished into four stages:

Stage 1: Egg – As adult female mites burrowing under the skin, they lay two to three eggs per day which hatch in three to four days.

Stage 2: Larva – Once the egg hatches larvae come out to the skin surface and then burrows to construct molting pouches into the outermost layer of skin. This stage lasts for three to four days.

Stage 3: Nymph – Larva molts into Nymph which looks similar to the adults but is smaller in size. Before molting into an adult, the nymph grows larger in size inside the molting pouch.

Stage 4: Adult – Nymph molts into an adult, active males penetrate female's molting pouch and mate. Impregnated females are fertile for the rest of their life; they leave their molting pouch to find a permanent burrow site and lay eggs until her death [105].

Oral medication against Scabies includes the administration of broad-spectrum anti-parasitic drug Ivermectin. Permethrin, 5% Malathion in aqueous base,

10–25% benzyl benzoate emulsion or 5–10% sulfur ointment is topical scabicide applied all over the body [104]. Ivermectin kills the parasite when it binds selectively to glutamate-gated chloride ion channels in the invertebrate muscle causing hyperpolarization of cells which leads to paralysis [91]. Thus, glutamate-gated chloride ion channels in microfilariae can be considered as a potential drug target [106]. Similar to Ivermectin, Permethrin causes delayed repolarization and paralysis of the parasite by disrupting the sodium channel current of the nerve cell membrane [107]. A wide-spectrum organophosphate insecticide Malathion is a cholinesterase inhibitor that acts as a neurotoxin [108]. Benzyl benzoate also has a toxic effect on the nervous system and has been known to kill the mite within five minutes [109]. Octasulfur commonly known as Sulphur is lethal to scabies mites but its mechanism of action is still not clear [110].

2.13.1 Phytochemicals Against Scabies

In Rwanda, calves suffering from scabies were treated with powder of the roots of *Neorautanenia mitis* A. Rich (Fabaceae). Puyvelde et al. [111] discovered that the principal compound responsible for the anti-scabies activity is 12a-hydroxyrotenone known as Rotenolone. Alcoholic extract of the roots was then used to prepare a solution and ointment to study the anti-scabies effect on human patients. In an extensive study carried out on 400 patients suffering from infected and noninfected scabies, it was concluded that the treatment with solution took three to four days whereas with ointment it was seven days [111].

A comparative study of Ivermectin and Permethrin with Tea tree oil for anti-scabies activity conducted by Walton et al. [112] showed that 1,8-cineole, α-terpineol, and Terpinen-4-ol have anti-scabies activity with Terpinen-4-ol being the primary active component [112].

Fang et al. [113] studied the efficacy of ten essential oils against *Sarcoptes scabiei*. Clove oil, Palmarosa, and Tea tree oil were found to be most efficient in killing the Scabies causing mites. The major phytochemical components determined to have anti-scabies activity were α-terpinene, γ-terpinene, terpinen-4-ol, α-terpineol from tea tree oil, acetyl eugenol, eugenol from clove oil, geraniol from palmarosa oil, 1,8-cineole, α-terpineol from eucalyptus oil, and citronellol, citronellyl formate, geraniol, linalool, linalyl acetate from geranium oil [113].

2.14 Schistosomiasis

Schistosomiasis, also known as snail fever is a parasitic disease acquired when a person's skin comes in contact with contaminated freshwater which is infested with parasitic blood flukes known as schistosomes. According to WHO, 240 million people worldwide are affected with Schistosomiasis and about 700 million people living in subtropical and tropical regions are at high risk; especially people living in poverty that lack access to potable water and proper sanitation facilities [114].

The life cycle of *Schistosoma* starts when a person suffering from schistosomiasis releases *Schistosoma* eggs into the freshwater bodies through excretion. The eggs then hatch into miracidia and infect freshwater snail. These infestations are very precise, a particular species of *Schistosoma* genera will infect snails of the genus that is specific for that species of *Schistosoma*. The miracidia then undergoes the next phase of their life cycle in the snail, the successive generation of miracidia are called sporocysts. After the development of sporocysts, the host snail releases cercariae into the water column. These cercariae swim in search of a human host. Once cercariae penetrate the human skin, they shed their fork tail to become schistosomula and enter blood vessels. Schistosomulae then migrates through several tissues to reach its residence veins and develop into an adult worm. The paired adult worms reproduce and the female worm lays eggs into the perivesical systems. The eggs are then eliminated out of the body with feces or urine [115].

Different species of *Schistosoma* genera cause different types of Schistosomiasis. *S. haematobium* causes urogenital schistosomiasis which is transmitted by Bulinus snails. *S. mansoni, S. japonicum, S. mekongi, S. guineensis,* and *S. intercalatum* all cause gastrointestinal schistosomiasis. Also, *S. haematobium, S. mansoni,* and *S. japonicum* can cause central nervous system diseases [114, 116].

Urogenital schistosomiasis causes hematuria and fibrosis of the bladder ureter. Other symptoms of urogenital schistosomiasis include genital lesions, nodules in the vulva, and vaginal bleeding in females whereas it can cause infertility and can induce infection of the prostate, seminal vesicles, and other organs in males. Symptoms of gastrointestinal schistosomiasis include abdominal pain, blood in the stool, diarrhea, liver enlargement, and sometimes enlargement of the spleen.

Praziquantel (PZQ) is the main drug that has a direct inhibitory action on schistosomes. The exact mechanism of action is still unknown but it is established that PZQ increases calcium influx into the cell which induces muscle contraction and alterations in parasites integument [117, 118].

2.14.1 Phytochemicals Against Schistosomiasis

In a research conducted by Ali et al. [119] to investigate the anti-schistosomal activity of two active constituents isolated from leaves of Egyptian medicinal plants; it was found that D-mannitol isolated from leaves of *Ixora undulata* Roxb. and pectin isolated from *Linum grandiflorum* Desf. both were found effective against Schistosomiasis. A patient suffering from Schistosomiasis would have a reduction in plasma glucose, liver glucose, and glycogen levels due to liver injury caused by Schistosomes. When mice infected with *S. mansoni* were given D-mannitol or pectin, a significant upregulation of hepatic glucose and glycogen levels was observed compared to the infected and Praziquantel-treated mice groups [119].

Ghareeb et al. [120] extracted phytochemicals from the leaves of *Eucalyptus camaldulensis* and tested it against *S. mansoni*. The 5 *n*-hexane/EtOAc fraction isolated from ethyl extract of the leaves showed 100% mortality in miracidia and cercariae of *S. mansoni* at a concentration of 200 mg/l. The phytochemicals identified in these fractions were: 3,4,5-trihydroxybenzoic acid (gallic acid), 3,5,7,3′,4′-pentah ydroxy-flavanone (dihydroquercetin or taxifolin), methyl gallate, 5,7,3′,4′-flavon-3-ol (quercetin), 5,7,3′,4′-tetrahydroxy-flavone (luteolin), and 3′,5,7-trihydroxy-4′-metho xyflavanone-7-O-rutinoside (hesperidin or hesperetin 7-O-rutinoside) [120].

Moraes [121] in his review article presented a list of naturally occurring phytochemicals and compounds derived from natural products having anti-schistosomal activities. Aspidin, Betulin, Budlein-A, Curcumin, Kalata B1, Kalata B2, Dermaseptin 01, Desaspidin, Epiisopiloturine, Flavaspidic acid, Hesperidin, Kaempferol, Karavilagenin C, (+)-limonene epoxide, Methylene-bis-aspidinol, Nerolidol, Phytol, Piperamide 1, Piplartine, Plumbagin, Rotundifolone, Sativan, Sanguinarine, Solamargine, Solasonine, and Quercetin 3-O-β-drhamnoside are the naturally occurring phytochemicals which are reported to have anti-schistosomal activity [121].

Table 2.2 presents a list of active phytochemicals against respective NTDs.

Table 2.2 List of phytochemicals against NTDs.

Diseases	Phytochemicals
African trypanosomiasis	2-(1-Hydroxyethyl)-naphtho[2,3-b]furan-4,9-quinone; 6α,9α-dihydroxy-4(13),7-coloratadiene-11,12-dial; Alloaromadendrene, Aromadendrene, Berberine, C6-epoxycarboxylic acid (3,4-epoxy-3-methyl hexanoic acid), Camptothecin, Cupacinosides, Diospyrin, Emetine, Lipiferolide, Mukaadial, Muzigadial, O-methylmoschatoline, Pancracine, Polyacetylene Carlina oxide, Pseudoguaianolide helenalin, Quinidine, Sanguinarine, Terpinen-4-ol, Vinblastine, and α-Pinene
Buruli ulcer	1-Triacontanol; 2,3-dimethylacrolyl choline; 7-methyljuglone, Acemannan, Actidine isomer, Alkaloid spiganthine, Apigenin, Benzoylcholine, Capsanthin, Capsorubin, Choline, Cryptolepine hcl, Cryptoxanthin, Dehydrocostus-lactone, Diospyrin, E-Phytol, Eugenol, Eugenyl acetate, Ferulenol, Genin, Holamine, Holaphyllamine, Holaphylline, Isoquinoline, Isovitexin, Lutein, Norditerpenoid 12-demethylmulticauline, Phenolcarboxylic acids, Phytanol, Sapogenins, Sterol, Stigmasterol, Triterpenae alcohol, Tryptanthin, Vitexin, Z-phytol, Zeaxanthin, β-caryophyllene, β-D-glucoside, and β-D-sitosterol

(*Continued*)

Table 2.2 (Continued)

Diseases	Phytochemicals
Chagas disease	13-Oxocardiopetamine, Ajoene, Atisinium chloride, Clerodane trans-dehydrocrotonin, Diospyrin, Embelin, Ent-kaur-16-en-19-oic acid, Plumbagin, and Quercetin
Chikungunya	12-o-Decanoyl-7-hydroperoxy-phorbol-5-ene-13-acetate, 12-o-decanoylphorbol-13-acetate, Apigenin, Coumarin A, Courmarin B, Decanoylphorbol-13 acetate, Luteolin, and Silymarin
Dengue	Acacetin-7-O-rutinoside, Baicalein, Baicalin, Coumarin A, Courmarin B, Ellagic acide, Lupeol acetate, Pectolinarin, Quercetin, Rutin, and Voacangine
Leishmaniasis	Quinoline, Isoquinoline, Pyrimidine-carboline, Steroidal, and Diterpene
Leprosy	3-Hydroxy-4-methoxybenzoic acid; 3,4-dihydroxybenzyl alcohol; 3,4-dihydroxybenzoic acid, Allicin, Asiaticoside, Barachini psoralenoides, Bavachalcone, Bavachromene psoralidin, Bavaislfavooz, Borachin, Brombexine, Chaumoogric acid, Corylifolinin, Corylifols A, Corylifols B, Corylifols C, Coumarins, Diadzein (4:7 dyhydroxuisoflavon), Genistein (4′6′7 trihydroxyisoflavon), Gorlic acid, Hydnocarpic acid, Isopsoralen, Isopsoralenoside, Neobavaisflavones, Neoba-vaislfloavone, Psoralen, and Taraktophyllin
Lymphatic filariasis	(−)-Epicatechin, (−)Laminaribiitol, 12-hydroxy-9-octadecenoic acid; (2*S*)-1, 2-di-O-linolenoyl-3-O-a-galactopyranosyl-(1/6)-O-b-galactopyranosyl glycerol; (2*S*)-1-O-linolenoyl-2-O-palmitoyl-3-O-a-galactopyranosyl-(1/6)-O-b-galactopyranosyl glycerol; (2*S*)-1-O-oleoyl-2-O-palmitoyl-3-O-a-galactopyranosyl-(1/6)-O-b-galactopyranosyl glycerol;12-methoxy-N(4)-methylakuammicine; 17-Acetoxy-nor-echitamine; 19-Epischolaricine; 2-(4-methoxy phenyl) imino) 4-methyl-1, 3-thiozolan-4yl); 2′-(*R*)-O-Acetylglaucarubinone; 4-Hydroxy-α-tetralone; 4-o-(b-D-glucopyranosyl-6-sulphate) gallic acid; 7,4′-dihydroxyflavon, Akuammicine, Akuammidine, Aloe emodina, Andrographiside, Andrographolide, Apigenin, Apocynin, Archidic acid, Azadirachtin, β-sitosterol, Boeravinone A, Boeravinone B, Calctin, Calotropin, Calotropises quiterpenol, Carapolide A, Cardenolide (200-oxovoruscharin), Catechin, Chebulagic acid, Chebulanin, Chebulinic acid, Chlorogenic acid, Chlorogenin, Columbin, Corilagin, Courmaric acid, Cycloartanone, Demethylwedelolactone, Dihydroagnosterol, Dihydrobenzofuran neolignan rel-(2α, 3β)-7-O-methylcedrusin, Diosgenin, Dodecanoicacid (RT = 9.92), Echitamidine, Echitamine, Echitamine, Epitetraphyllin B (volkenin), Euphol, Ferulicacid, Flavonoids, Glycyrrhetinic acid, Gummiferol, Hecogenin, Hypaphorine, Iridoids, Isocoumarins, Isoricinoleic acid, Isoshinanolone, Isoshinanolone, Isovanillic acid, Krisofanat acid, Luteonin, Mansonin, Mexicanolide-methylangolensate, Monoterpenoidnaphthaquinones,

Table 2.2 (Continued)

Diseases	Phytochemicals
	Myricetin, Naphthaquinones, Neogitogenin, Neohecogenin, Nimbidin, Nimbin, Nimbolin B, Nor-echitamine, Nα-formylechitamidine, Oligorhamnosides, Palmatine, Palmitate, Palmitic acid, *p*-hydroxybenzoic acid, Physcion, Proanthocyanidins, Protocatechuic acid, Pseudoephedrine, Punicalagin, Quassinoid, Quercetin, Repenol, Repenone, Rhein, Ricinoleic acid, Robustadial A, Robustadial B, Ruscogenin, Saponin, Scopolamine, Scopoline, Stearic acid, Stearic, Steroids, Stigmasterol, Strebloside, Tannins, Tembeterine, Tetracosanoic acid, Tetralone-4-O-β-D-glucopyranoside, Tetraphyllin B (barterin), Tinosporoside, Tirucallol, Tubotaiwine, Vellesamine, Wedelolactone, Zingerone, ά- calotropeols, α-2-sitosterol, α-amyrin, α-zingiberol, β-calotropeols, β-sitosterol, and β-sitosterolglucoside
Mycetoma	β-amyrin, β-amyrone, β-sitosterol, and Stigmatriene
Onchocerciasis	Mustakone and Linoleic acid or (9Z,12Z)-octadeca-9,12-dienoic acid
Rabies	1,2-Benzenedicarboxylic acid, bis (2-ethylhexyl) ester, Catechin, Catechol, Kaemferol, Luteolin-7-glucoside, Methyl 1-hydroxy-6-oxocyclohex-2-enecarboxylate, Propyl acetate, Quercetrin, Rutin, Saligenin, β-sitosterol, and β-sitosterol glucopyranoside
Scabies	1,8-Cineole or Eucalyptol, 12a-hydroxyrotenone or Rotenolone, Acetyleugenol, Citronellol, Citronellyl formate, Eugenol, Geraniol, Linalool, Linalyl acetate, Terpinen-4-ol, α-terpinene, α-terpineol, and γ-terpinene
Schistosomiasis	(+)-Limonene epoxide, Aspidin, Betulin, Budlein-A, Curcumin, Dermaseptin 01, Desaspidin, D-mannitol, Epiisopiloturine, Flavaspidic acid, Hesperidin, Kaempferol, Kalata B1, Kalata B2, Karavilagenin C, Luteolin, Methylene-bis-aspidinol, Nerolidol, Pectin (a linear chain homogalacturonan polysaccharide), Phytol, Piperamide 1, Piplartine, Plumbagin, Quercetin 3-O-β-drhamnoside, Rotundifolone, Sanguinarine, Sativan, Solamargine, and Solasonine

2.15 Conclusion

NTDs affect a billion people and claim thousands of lives annually. They mostly affect people with low economic income, who lack access to proper healthcare. Such people opt for cheaper treatment methods that mostly include plant-based or plant-derived remedies. In this plant-based remedies, phytochemicals are responsible for showing activity against a particular disease. These phytochemicals can be used directly in the treatment of NTDs, or they can also serve as a lead molecule for synthesizing an effective yet low-cost drug. This study focused on summarizing potential druggable targets of various NTDs and potential phytochemicals that show activity against specific NTDs. The phytochemicals, mentioned in this study, can be used for further *in vitro* and *in vivo* experiments and the drug targets can be used for *in silico* studies.

References

1 WHO (2019). Neglected tropical diseases. https://www.who.int/neglected_diseases/diseases/en (accessed 20 May 2021).

2 Molyneux, D.H., Hotez, P.J., and Fenwick, A. (2005). "Rapid-impact interventions": how a policy of integrated control for Africa's neglected tropical diseases could benefit the poor. *PLoS Medicine* 2 (11): 1064–1070. https://doi.org/10.1371/journal.pmed.0020336.

3 Mitra, A.K. and Mawson, A.R. (2020). Neglected tropical diseases: epidemiology and global burden. *Tropical Medicine and Infectious Disease* 2 (3): 36. https://doi.org/10.3390/tropicalmed2030036.

4 Brun, R., Blum, J., Chappuis, F., and Burri, C. (2010). Reading L6- human African trypanosomiasis. *Lancet* ***375*** (9709): 148–159. https://doi.org/10.1016/S0140-6736(09)60829-1.

5 WHO (2018). Trypanosomiasis, human African (sleeping sickness). https://www.who.int/news-room/fact-sheets/detail/trypanosomiasis-human-african-(sleeping-sickness) (accessed 20 May 2021).

6 Kennedy, P.G.E. (2013). Clinical features, diagnosis, and treatment of human African trypanosomiasis (sleeping sickness). *The Lancet Neurology* 12 (2): P186–P194. https://doi.org/10.1016/S1474-4422(12)70296-X.

7 Simarro, P.P., Jannin, J., and Cattand, P. (2008). Eliminating human African trypanosomiasis: where do we stand and what comes next? *PLoS Medicine* 5 (2): e55. https://doi.org/10.1371/journal.pmed.0050055.

8 Stich, A., Abel, P.M., and Krishna, S. (2002). Human African trypanosomiasis. *BMJ (Clinical Research ed.)* 325 (7357): 203–206.

9 Adjogatse, E., Erskine, P., Wells, S.A. et al. (2018). Structure and function of l-threonine-3-dehydrogenase from the parasitic protozoan *Trypanosoma brucei* revealed by X-ray crystallography and geometric simulations. *Acta Crystallographica Section D: Structural Biology* 74 (9): 861–876. https://doi.org/10.1107/S2059798318009208.

10 Schmidt, T.J., Khalid, S.A., Romanha, A.J. et al. (2012). The potential of secondary metabolites from plants as drugs or leads against protozoan neglected diseases – Part I. *Current Medicinal Chemistry* 19 (14): 2128–2175.

11 Schmidt, T.J., Khalid, S.A., Romanha, A.J. et al. (2012). The potential of secondary metabolites from plants as drugs or leads against protozoan neglected diseases – Part II. *Current Medicinal Chemistry* 19 (14): 2176–2228.
12 Hoet, S., Opperdoes, F., Brun, R., and Quetin-Leclercq, J. (2004). Natural products active against African trypanosomes: a step towards new drugs. *Natural Product Reports* 21: 353–364. https://doi.org/10.1039/b311021b.
13 Wink, M. (2012). Medicinal plants: a source of anti-parasitic secondary metabolites. *Molecules* 17 (11): 12771–12791. https://doi.org/10.3390/molecules171112771.
14 Portaels, F., Silva, M.T., and Meyers, W.M. (2009). Buruli ulcer. *Clinics in Dermatology* 27 (3): 291–305. https://doi.org/10.1016/j.clindermatol.2008.09.021.
15 Wansbrough-Jones, M. and Phillips, R. (2006). Buruli ulcer: emerging from obscurity. *Lancet* 367: P1849–P1858. https://doi.org/10.1016/S0140-6736(06)68807-7.
16 Silva, M.T., Portaels, F., and Pedrosa, J. (2009). Pathogenetic mechanisms of the intracellular parasite *Mycobacterium ulcerans* leading to Buruli ulcer. *The Lancet Infectious Diseases* 9 (11): P699–P710. https://doi.org/10.1016/S1473-3099(09)70234-8.
17 Vickery, C.R., Kosa, N.M., Casavant, E.P. et al. (2014). Structure, biochemistry, and inhibition of essential 4′-phosphopantetheinyl transferases from two species of Mycobacteria. *ACS Chemical Biology* 9 (9): 1939–1944. https://doi.org/10.1021/cb500263p.
18 Johnson, P.D.R., Stinear, T., Small, P.L.C. et al. (2005). Buruli ulcer (*M. ulcerans* infection): new insights, new hope for disease control. *PLoS Medicine* 2: 0282–0286. https://doi.org/10.1371/journal.pmed.0020108.
19 Bamberger, D., Jantzer, N., Leidner, K. et al. (2011). Fighting mycobacterial infections by antibiotics, phytochemicals and vaccines. *Microbes and Infection* 13 (7): 613–623. https://doi.org/10.1016/j.micinf.2010.09.002.
20 Tsouh Fokou, P.V., Nyarko, A.K., Appiah-Opong, R. et al. (2015). Ethnopharmacological reports on anti-Buruli ulcer medicinal plants in three West African countries. *Journal of Ethnopharmacology* 172: 297–311. https://doi.org/10.1016/j.jep.2015.06.024.
21 Pérez-Molina, J.A., Perez, A.M., Norman, F.F. et al. (2015). Old and new challenges in Chagas disease. *The Lancet Infectious Diseases* 15 (11): P1347–P1356. https://doi.org/10.1016/S1473-3099(15)00243-1.
22 WHO (2018). Chagas disease (American trypanosomiasis). https://www.who.int/chagas/home_more/en (accessed 20 May 2021).
23 Rassi, A., Rassi, A., and Marin-Neto, J.A. (2010). Chagas disease. *The Lancet* **375** (9723): 1388–1402. https://doi.org/10.1016/s0140-6736(10)60061-x.
24 Pérez-Molina, J.A. and Molina, I. (2018). Seminar Chagas disease. *The Lancet* **391**: 82–94. https://doi.org/10.1016/S0140-6736(17)31612-4.

25 CDC (2015). Parasites – American trypanosomiasis (also known as Chagas Disease). https://www.cdc.gov/parasites/chagas/biology.html (accessed 20 May 2021).

26 Beltran-Hortelano, I., Perez-Silanes, S., and Galiano, S. (2017). Trypanothione reductase and superoxide dismutase as current drug targets for trypanosoma cruzi: an overview of compounds with activity against Chagas disease. *Current Medicinal Chemistry* **24** (11): 1066–1138. https://doi.org/10.2174/0929867323666161227094049.

27 Amdekar, S., Parashar, D., and Alagarasu, K. (2017). Chikungunya virus-induced arthritis: role of host and viral factors in the pathogenesis. *Viral Immunology* **30** (10): 691–702. https://doi.org/10.1089/vim.2017.0052.

28 Thiberville, S.D., Moyen, N., Dupuis-Maguiraga, L. et al. (2013). Chikungunya fever: epidemiology, clinical syndrome, pathogenesis, and therapy. *Antiviral Research* 99 (3): 345–370. https://doi.org/10.1016/j.antiviral.2013.06.009.

29 Saisawang, C., Kuadkitkan, A., Smith, D.R. et al. (2017). Glutathionylation of chikungunya nsP2 protein affects protease activity. *Biochimica et Biophysica Acta, General Subjects* 1861 (2): 106–111. https://doi.org/10.1016/j.bbagen.2016.10.024.

30 Rashad, A.A., Mahalingam, S., and Keller, P.A. (2014). Chikungunya virus: emerging targets and new opportunities for medicinal chemistry. *Journal of Medicinal Chemistry* **57** (4): 1147–1166. https://doi.org/10.1021/jm400460d.

31 Caglioti, C., Lalle, E., Castilletti, C. et al. (2013). Chikungunya virus infection: an overview. *New Microbiologica* 36 (3): 211–227.

32 Murali, K.S., Sivasubramanian, S., Vincent, S. et al. (2015). Anti-chikungunya activity of luteolin and apigenin rich fraction from *Cynodon dactylon*. *Asian Pacific Journal of Tropical Medicine* **8** (5): 352–358. https://doi.org/10.1016/S1995-7645(14)60343-6.

33 Gómez-Calderón, C., Mesa-Castro, C., Robledo, S. et al. (2017). Antiviral effect of compounds derived from the seeds of *Mammea americana* and *Tabernaemontana cymosa* on Dengue and Chikungunya virus infections. *BMC Complementary and Alternative Medicine 17* (1) https://doi.org/10.1186/s12906-017-1562-1.

34 Lani, R., Hassandarvish, P., Chiam, C.W. et al. (2015). Antiviral activity of silymarin against chikungunya virus. *Scientific Reports* 5 https://doi.org/10.1038/srep11421.

35 Corlay, N., Delang, L., Girard-Valenciennes, E. et al. (2014). Tigliane diterpenes from *Croton mauritianus* as inhibitors of chikungunya virus replication. *Fitoterapia* 97: 87–91. https://doi.org/10.1016/j.fitote.2014.05.015.

36 Ranjit, S. and Kissoon, N. (2011). Dengue hemorrhagic fever and shock syndromes. *Pediatric Critical Care Medicine* 12 (1): 90–100. https://doi.org/10.1097/PCC.0b013e3181e911a7.

37 Rajapakse, S., Rodrigo, C., and Rajapakse, A. (2012). Treatment of dengue fever. *Infection and Drug Resistance* 5: 103–112.

38 Kularatne, S.A.M. (2015). Dengue fever. *BMJ*: h4661. https://doi.org/10.1136/bmj.h4661.

39 Rodenhuis-Zybert, I.A., Wilschut, J., and Smit, J.M. (2010). Dengue virus life cycle: viral and host factors modulating infectivity. *Cellular and Molecular Life Sciences* 67: 2773–2786. https://doi.org/10.1007/s00018-010-0357-z.

40 El Sahili, A. and Lescar, J. (2017). Dengue virus nonstructural protein 5. *Viruses* 9 (4): 91. https://doi.org/10.3390/v9040091.

41 Dwivedi, V.D., Tripathi, I.P., Tripathi, R.C. et al. (2017). Genomics, proteomics, and evolution of dengue virus. *Briefings in Functional Genomics* **16** (4): 217–227. https://doi.org/10.1093/bfgp/clw040.

42 Martina, B.E.E., Koraka, P., and Osterhaus, A.D.M.E. (2009). Dengue virus pathogenesis: an integrated view. *Clinical Microbiology Reviews* 22 (4): 564–581. https://doi.org/10.1128/CMR.00035-09.

43 Carrington, L.B. and Simmons, C.P. (2014). Human to mosquito transmission of dengue viruses. *Frontiers in Immunology* https://doi.org/10.3389/fimmu.2014.00290.

44 Dengue Virus Net (2019). Treatment of dengue. http://www.denguevirusnet.com/treatment/8-treatment-of-dengue.html (accessed 20 May 2021).

45 Moghaddam, E., Teoh, B.T., Sam, S.S. et al. (2014). Baicalin, a metabolite of baicalein with antiviral activity against dengue virus. *Scientific Reports* 4 https://doi.org/10.1038/srep05452.

46 Silva, A.R.A., Morais, S.M., Marques, M.M.M. et al. (2011). Antiviral activities of extracts and phenolic components of two Spondias species against dengue virus. *Journal of Venomous Animals and Toxins Including Tropical Diseases* **17** (4): 406–413.

47 Simões, L.R., Maciel, G.M., Brandão, G.C. et al. (2011). Antiviral activity of Distictella elongata (Vahl) Urb. (Bignoniaceae), a potentially useful source of anti-dengue drugs from the state of Minas Gerais, Brazil. *Letters in Applied Microbiology* **53** (6): 602–607. https://doi.org/10.1111/j.1472-765X.2011.03146.x.

48 WHO (2018). Leishmaniasis. https://www.who.int/leishmaniasis/en (accessed 20 May 2021).

49 CDC (2018). Parasites – leishmaniasis. https://www.cdc.gov/parasites/leishmaniasis/biology.html (accessed 20 May 2021).

50 Barrett, M.P. and Croft, S.L. (2012). Management of trypanosomiasis and leishmaniasis. *British Medical Bulletin* 104 (1): 175–196. https://doi.org/10.1093/bmb/lds031.

51 Aronson, N., Herwaldt, B.L., Libman, M. et al. (2017). Diagnosis and treatment of leishmaniasis: clinical practice guidelines by the Infectious Diseases Society of America (IDSA) and the American Society of Tropical Medicine and Hygiene

(ASTMH)*. *American Society of Tropical Medicine and Hygiene* 96 (1): 24–45. https://doi.org/10.4269/ajtmh.16-84256.

52 DrugBank (2018). Amphotericin B. https://www.drugbank.ca/drugs/DB00681 (accessed 20 May 2021).

53 Saha, A.K., Mukherjee, T., and Bhaduri, A. (1986). Mechanism of action of amphotericin B on *Leishmania donovani* promastigotes. *Molecular and Biochemical Parasitology 19*: 195–200.

54 DrugBank (2019). Paromomycin. https://www.drugbank.ca/drugs/DB01421 (accessed 20 May 2021).

55 Chawla, B., Jhingran, A., Panigrahi, A. et al. (2011). Paromomycin affects translation and vesicle-mediated trafficking as revealed by proteomics of paromomycin – susceptible – resistant *Leishmania donovani*. *PLoS One* 6 (10) https://doi.org/10.1371/journal.pone.0026660.

56 Ogungbe, I.V., Erwin, W.R., and Setzer, W.N. (2014). Antileishmanial phytochemical phenolics: molecular docking to potential protein targets. *Journal of Molecular Graphics and Modelling 48*: 105–117. https://doi.org/10.1016/j.jmgm.2013.12.010.

57 Mishra, B.B., Singh, R.K., Srivastava, A. et al. (2009). Fighting against leishmaniasis: search of alkaloids as future true potential anti-leishmanial agents. *Mini-Reviews in Medicinal Chemistry* 9 (1): 107–123.

58 WHO (2019). Leprosy. https://www.who.int/news-room/fact-sheets/detail/leprosy (accessed 20 May 2021).

59 Bhat, R.M. and Prakash, C. (2012). Leprosy: an overview of pathophysiology. *Interdisciplinary Perspectives on Infectious Diseases 2012*: 1–6. https://doi.org/10.1155/2012/181089.

60 Lastória, J.C. and de Abreu, M.A.M.M. (2014). Leprosy: review of the epidemiological, clinical, and etiopathogenic aspects – Part 1. *Anais Brasileiros de Dermatologia* 89 (2): 205–218. https://doi.org/10.1590/abd1806-4841.20142450.

61 Anusuya, S. and Natarajan, J. (2013). The eradication of leprosy: molecular modeling techniques for novel drug discovery. *Expert Opinion on Drug Discovery* 8 (10): 1239–1251. https://doi.org/10.1517/17460441.2013.826188.

62 Talhari, C., Talhari, S., and Penna, G.O. (2015). Clinical aspects of leprosy. *Clinics in Dermatology* 33 (1): 26–37. https://doi.org/10.1016/j.clindermatol.2014.07.002.

63 Leprosy, Chapter: 5. Pathogenesis, 10–22. http://nlep.nic.in/pdf/Ch%205%20Pathogenesis.pdf

64 CDC (n.d.). Hansen's disease (leprosy) diagnosis and treatment. https://www.cdc.gov/leprosy/treatment/index.html (accessed 20 May 2021).

65 Ghosh, G. (2017). Traditional use of plants against leprosy in India: a review of the recent literature. *Journal of Innovations in Pharmaceutical and Biological Sciences* 4 (4): 55–64.

66 Sahoo, M.R., Dhanabal, S.P., Jadhav, A.N. et al. (2014). Hydnocarpus: an ethnopharmacological, phytochemical and pharmacological review. *Journal of Ethnopharmacology* 154 (1): 17–25. https://doi.org/10.1016/j.jep.2014.03.029.

67 WHO (2018). Lymphatic filariasis. https://www.who.int/lymphatic_filariasis/en (accessed 20 May 2021).

68 CDC (2018). Parasites – lymphatic filariasis. https://www.cdc.gov/parasites/lymphaticfilariasis/biology_w_bancrofti.html (accessed 20 May 2021).

69 Thomsen, E.K., Sanuku, N., Baea, M. et al. (2016). Efficacy, safety, and pharmacokinetics of coadministered diethylcarbamazine, albendazole, and ivermectin for treatment of bancroftian filariasis. *Clinical Infectious Diseases* 62 (3): 334–341. https://doi.org/10.1093/cid/civ882.

70 DrugBank (2018). Albendazole. https://www.drugbank.ca/drugs/DB00518 (accessed 20 May 2021).

71 Sharma, O.P., Pan, A., Hoti, S.L. et al. (2012). Modeling, docking, simulation, and inhibitory activity of the benzimidazole analogue against β-tubulin protein from *Brugia malayi* for treating lymphatic filariasis. *Medicinal Chemistry Research* 21: 2415–2427. https://doi.org/10.1007/s00044-011-9763-5.

72 Choi, T., Cho, J.H., Ahnn, J., and Song, H. (2018). Novel findings of anti-filarial drug target and structure-based virtual screening for drug discovery. *International Journal of Molecular Sciences* 19 (11): 3579. https://doi.org/10.3390/ijms19113579.

73 Sashidhara, K.V., Singh, S.P., Misra, S. et al. (2012). Galactolipids from *Bauhinia racemosa* as a new class of anti filarial agents against the human lymphatic filarial parasite, *Brugia malayi*. *European Journal of Medicinal Chemistry* **50**: 230–235. https://doi.org/10.1016/j.ejmech.2012.01.057.

74 Kalani, K., Kushwaha, V., Verma, R. et al. (2013). Glycyrrhetinic acid and its analogs: a new class of antifilarial agents. *Bioorganic & Medicinal Chemistry Letters* **23** (9): 2566–2570. https://doi.org/10.1016/j.bmcl.2013.02.115.

75 Ranjini, G., Selvakumari, E., and Gopal, V. (2013). Perception of natural antifilarial drugs. *International Research Journal of Pharmacy* **4** (1): 27–30.

76 WHO (2018). Mycetoma. https://www.who.int/buruli/mycetoma/en (accessed 20 May 2021).

77 Relhan, V., Mahajan, K., Agarwal, P., and Garg, V.K. (2017). Mycetoma: an update. *Indian Journal of Dermatology* **62** (4): 332–340. https://doi.org/10.4103/ijd.IJD_476_16.

78 Welsh, O., Al-Abdely, H.M., Salinas-Carmona, M.C., and Fahal, A.H. (2014). Mycetoma medical therapy. *PLoS Neglected Tropical Diseases 8* (10) https://doi.org/10.1371/journal.pntd.0003218.

79 DrugBank (2018). Itraconazole. https://www.drugbank.ca/drugs/DB01167 (accessed 20 May 2021).

80 DrugBank (2019). Amikacin. https://www.drugbank.ca/drugs/DB00479 (accessed 20 May 2021).

81 DrugBank (2019). Gentamicin. https://www.drugbank.ca/drugs/DB00798 (accessed 20 May 2021).

82 DrugBank (2018). Netilmicin. https://www.drugbank.ca/drugs/DB00955 (accessed 20 May 2021).

83 DrugBank (2019). Streptomycin. https://www.drugbank.ca/drugs/DB01082 (accessed 20 May 2021).

84 DrugBank (2019). Tobramycin. https://www.drugbank.ca/drugs/DB00684 (accessed 20 May 2021).

85 DrugBank (2019). Rifampicin. https://www.drugbank.ca/drugs/DB01045 (accessed 20 May 2021).

86 DrugBank (2019). Meropenem. https://www.drugbank.ca/drugs/DB00760 (accessed 20 May 2021).

87 Elfadil, H., Fahal, A., Kloezen, W. et al. (2015). The in vitro antifungal activity of Sudanese medicinal plants against *Madurella mycetomatis*, the eumycetoma major causative agent. *PLoS Neglected Tropical Diseases* 9 (3) https://doi.org/10.1371/journal.pntd.0003488.

88 WHO (2018). Onchocerciasis. https://www.who.int/onchocerciasis/en (accessed 20 May 2021).

89 CDC (2013). Parasites – Onchocerciasis (also known as river blindness). https://www.cdc.gov/parasites/onchocerciasis/biology.html (accessed 20 May 2021).

90 Hoerauf, A., Mand, S., Volkmann, L. et al. (2003). Doxycycline in the treatment of human onchocerciasis: kinetics of Wolbachia endobacteria reduction and of inhibition of embryogenesis in female Onchocerca worms. *Microbes and Infection* 5: 261–273.

91 DrugBank (2019). Ivermectin. https://www.drugbank.ca/drugs/DB00602 (accessed 20 May 2021).

92 SMPDB (2018). Doxycycline action pathway. http://smpdb.ca/view/SMP0000291?highlight[compounds][]=DB00254&highlight[proteins][]=DB00254 (accessed 20 May 2021).

93 DrugBank (2019). Doxycycline. https://www.drugbank.ca/drugs/DB00254 (accessed 20 May 2021).

94 Gloeckner, C., Garner, A.L., Mersha, F. et al. (2009). Repositioning of an existing drug for the neglected tropical disease Onchocerciasis. *PNAS* 107 (8): 3424–3429. https://doi.org/10.1073/pnas.0915125107.

95 Metuge, J.A., Smith, B.B., James, A.M. et al. (2014). Anti-onchocerca metabolites from *Cyperus articulatus*: isolation, in vitro activity, and in silico "drug-likeness.". *Natural Products and Bioprospecting* 4: 243–249. https://doi.org/10.1007/s13659-014-0023-5.

96 Nigg, A.J. and Walker, P.L. (2009). Overview, prevention, and treatment of rabies. *Pharmacotherapy* 29 (10): 1182–1195. https://doi.org/10.1592/phco.29.10.1182.

97 Yousaf, M.Z., Qasim, M., Zia, S. et al. (2012). Rabies molecular virology, diagnosis, prevention, and treatment. *Virology Journal* 9: 50. https://doi.org/10.1186/1743-422X-9-5022348291.

98 Zhu, S. and Guo, C. (2016). Rabies control and treatment: from prophylaxis to strategies with curative potential. *Viruses* 8 (11): 279. https://doi.org/10.3390/v8110279.

99 UniProt (n.d.). UniProtKB – P03524 (GLYCO_RABVE) Glycoprotein. https://www.uniprot.org/uniprot/P03524#function (accessed 20 May 2021).

100 Rupprecht, C.E. (1996. Chapter 61. https://www.ncbi.nlm.nih.gov/books/NBK8618). Rhabdoviruses: rabies virus. In: *Medical Microbiology*, 4e (ed. S. Baron). Galveston, TX: University of Texas Medical Branch at Galveston (accessed 20 May 2021).

101 Koury, R. and Warrington, S.J. (2019). Rabies [Updated 18 November 2018]. In: *StatPearls* [Internet]. Treasure Island, FL: StatPearls Publishing https://www.ncbi.nlm.nih.gov/books/NBK448076 (accessed 20 May 2021).

102 Meresa, A., Moges, H., Tadele, A. et al. (2017). Medicinal plants used for the management of rabies in Ethiopia – a review. *Medicinal Chemistry* 7 (2): 795–806. https://doi.org/10.4172/2161-0444.1000431.

103 Yahia, E.M. (2018). *Onion (page 1152), Fruit and Vegetable Phytochemicals*, 2e. Mexico: Wiley Blackwell.

104 WHO (2018). Scabies. https://www.who.int/lymphatic_filariasis/epidemiology/scabies/en (accessed 20 May 2021).

105 CDC. (2010). Parasites-Scabies. https://www.cdc.gov/parasites/scabies/biology.html (accessed 20 May 2021).

106 Laing, R., Gillan, V., and Devaney, E. (2017). Ivermectin – old drug, new tricks? *Trends in Parasitology* 33: P463–P472. https://doi.org/10.1016/j.pt.2017.02.004.

107 DrugBank (2019). Permethrin. https://www.drugbank.ca/drugs/DB04930 (accessed 20 May 2021).

108 DrugBank (2018). Malathion. https://www.drugbank.ca/drugs/DB00772 (accessed 20 May 2021).

109 DrugBank (2019). Benzyl benzoate. https://www.drugbank.ca/drugs/DB00676 (accessed 20 May 2021).

110 DrugBank (2019). Sulfur. https://www.drugbank.ca/drugs/DB09353 (accessed 20 May 2021).

111 Van Puyvelde, L., Heyndrickx, G., Brioen, P. et al. (1990). Development of an anti-scabies drug from the roots of neorautanenia mitis. *Planta Medica* **56** (6): 643. https://doi.org/10.1055/s-2006-961289.

112 Walton, S.F., Mckinnon, M., Pizzutto, S. et al. (2004). Acaricidal activity of *Melaleuca alternifolia* (tea tree) oil: In vitro sensitivity of *Sarcoptes scabiei* var *hominis* to terpinen-4-ol. *Archives of Dermatology* 140: 563–566.

113 Fang, F., Candy, K., Melloul, E. et al. (2016). In vitro activity of ten essential oils against *Sarcoptes scabiei*. *Parasites & Vectors* 9: 594. https://doi.org/10.1186/s13071-016-1889-3.

114 WHO (2018). Schistosomiasis. https://www.who.int/en/news-room/fact-sheets/detail/schistosomiasis (accessed 20 May 2021)

115 CDC (2018). Parasites – Schistosomiasis. https://www.cdc.gov/parasites/schistosomiasis/biology.html (accessed 20 May 2021)

116 Gryseels, B., Polman, K., Clerinx, J., and Kestens, L. (2006). Human schistosomiasis. *Lancet* 368 (9541): 1106–1118. https://doi.org/10.1016/S0140-6736(06)69440-3.

117 Vale, N., Gouveia, M.J., Rinaldi, G. et al. (2017). Praziquantel for schistosomiasis: single-drug metabolism revisited, mode of action, and resistance. *Antimicrobial Agents and Chemotherapy* 61 (5): e02582-16. https://doi.org/10.1128/AAC.02582-16.

118 Doenhoff, M.J., Cioli, D., and Utzinger, J. (2008). Praziquantel: mechanisms of action, resistance, and new derivatives for schistosomiasis. *Current Opinion in Infectious Diseases* 21 (6): 659–667. https://doi.org/10.1097/QCO.0b013e328318978f.

119 Ali, S.A., El-regal, N.S., and Saeed, S.M. (2015). Antischistosomal activity of two active constituents isolated from the leaves of Egyptian medicinal plants. *Infectious Diseases: Research and Treatment*: 5–16. https://doi.org/10.4137/IDRT.S24342.

120 Ghareeb, M.A., Habib, M.R., Mossalem, H.S., and Abdel-Aziz, M.S. (2018). Phytochemical analysis of *Eucalyptus camaldulensis* leaves extracts and testing its antimicrobial and schistosomicidal activities. *Bulletin of the National Research Centre 42* (1): 16. https://doi.org/10.1186/s42269-018-0017-2.

121 De Moraes, J. (2008). Antischistosomal natural compounds: present challenges for new drug screens. In: *Current Topics in Tropical Medicine* (ed. A.J. Rodriguez-Morales), 333–358. Intech Open.

3

Novel Bioactive Lead Compounds for Drug Discovery Against Neglected Tropical Diseases, Leishmaniasis, Lymphatic Filariasis, Trypanosomiasis (African Sleeping Sickness and Chagas Disease), and Schistosomiasis

Abeer M. El Sayed[1] *and Chukwuebuka Egbuna*[2,3]

[1] *Department of Pharmacognosy, Faculty of Pharmacy, Cairo University, Cairo, Egypt*
[2] *Department of Biochemistry, Faculty of Natural Sciences, Chukwuemeka Odumegwu Ojukwu University, Uli, Nigeria*
[3] *Nutritional Biochemistry and Toxicology Unit, World Bank Africa Centre of Excellence, Centre for Public Health and Toxicological Research (PUTOR), University of Port Harcourt, Port Harcourt, Nigeria*

List of Abbreviations/Definitions

AIDS	Acquired immunodeficiency syndrome
BALB/c	An albino, laboratory-bred strain of the house mouse after which a number of common substrains are derived
CC_{50}	Cytotoxicity concentration of drug is that results in 50% of mortality of macrophage
CL	Cutaneous leishmaniasis
clone D6	Human anti-CTLA4 Recombinant Antibody
CYP51	Cytochrome P450 [CYP] subfamily 51
Ex vivo	Ex vivo refers to experimentation or measurements done in or on tissue from an organism in an external environment with minimal alteration of natural conditions
GAPDH	Glyceraldehyde-3-phosphate dehydrogenase
HAT	Human African trypanosomiasis
IC_{50}	The concentration of drug that causes 50% growth inhibition
K11777	A potent, irreversible cysteine protease inhibitor (*N*-methyl-piperazine-Phe-homoPhe-vinylsulfone-phenyl)
LF	Lymphatic filariasis
MIC	Minimum inhibitory concentration

Neglected Tropical Diseases and Phytochemicals in Drug Discovery, First Edition.
Edited by Chukwuebuka Egbuna, Muhammad Akram, and Jonathan Chinenye Ifemeje.

ML	Mucocutaneous leishmaniasis
NPs	Natural products
NTDs	Neglected tropical diseases
PFK	Phosphofructo kinase
RNAi i	RNA interference (RNAi) is the process by which the translation of a protein is prevented by selective degradation of its encoded mRNA.
ROS	Reactive oxygen species
SI	The selectivity index
Tbb BSF	*T. b. brucei* bloodstream forms
Tbb PF	*T. b. brucei* procyclic forms
TbCatB	*Trypanosoma brucei* cathepsin B-like protease
VL	Visceral leishmaniasis
WHO	World Health Organization

3.1 Introduction

A well-known approach in medicine is the usage of native or distant popular knowledge to classify new drugs against diseases or infections. There are approximately one billion people who are at risk of suffering from one of the 17 serious diseases classified by the World Health Organization (WHO) as neglected tropical diseases (NTDs) [1]. These diseases signify a chief cause of morbidity, infirmity, and mortality in tropical regions of the world. Owing to the absence of financial investment into research and development of new drugs and almost nonexistent public awareness in high-income countries, these diseases are named "neglected." Being related with poor hygienic and socioeconomic circumstances, they could also be named diseases of neglected populations [1].

NTDs include infections by pathogens of viral (e.g. Dengue fever) bacterial (e.g. Leprosy, Trachoma), helminth (e.g. Schistosomiasis, Filariasis), as well as those of the protozoan origin (e.g. African sleeping sickness, Chagas' disease, Leishmaniasis). In certain surroundings where NTDs prevail, malaria, the most prevalent "protozoan" infection – though not presently treated as such by WHO can also be measured a neglected disease. Nevertheless, the current fractional achievements in the fight to eradicate or even eliminate some of these diseases, which have been attained by WHO's consequential policies of disease nursing, vector control, preventive chemotherapy, and others, the development of new, safe, and affordable drugs remains a crucial need [2].

The existing pharmacotherapies have some shortcomings such as unwanted side effects, high degree of toxicity, lack of availability, and/or problematic application under the life conditions of affected populace. Also, disease resistance and emergence or reemergence of unaffected pathogens together has triggered a new wave of research in NTDs for the need for new chemical entities with potent activities against the NTD-pathogens with no or less side effects [3].

It is on record that protozoan parasitic diseases among others are still an unsolved public health problem in the present millennium, causing morbidity, and mortality worldwide. Trypanosomatid infections amongst the various infections that afflict mankind and livestock, in tropical and subtropical areas of the globe are responsible for more than 20 million cases primarily; and it is assessed that half a billion people are at risk of contracting these diseases [4]. In particular, *Trypanosoma* is responsible for Chagas disease in South America and sleeping sickness (or human African trypanosomiasis, HAT) in Africa [5], whereas *Leishmania* is responsible for cutaneous and visceral infections, endemic in 88 countries primarily in the Horn of Africa, South Asia, and Latin America, with cases also registered in Spain and in the south of Italy [6]. Trypanosomiasis and leishmaniasis lack visibility in the developed countries because of its main prevalence in the poorest areas of the planet. The WHO for this reason, characterizes them as the most challenging among NTDs.

Aimed at the treatment of various serious ailments, nature has provided various sources of drugs. Secondary metabolites from an immense diversity of living organisms consequently represent an enormous source of chemical entities which has been a source of new drugs, directly in their innate form or after optimization. Amongst the usual bases for novel bioactive chemicals, bacteria and fungi, and plants have conventionally played the main part; though, gradually over the decades, many stimulating new active particles are found in marine life forms.

3.2 Prevalence of Neglected Tropical Diseases

NTDs have a higher prevalence in tropical and subtropical regions, and affects approximately one billion people worldwide [7]. The WHO recognized 17 NTDs [7], in which the protozoan-borne diseases, leishmaniasis, and HAT are the main focus of this chapter.

In the quest for new drugs for NTDs, it has been observed that there are some tricky dishonesties in the part of pharmaceutical companies because they may not be able to recuperate the budget of developing and generating treatments for NTDs. NTDs affect the humblest persons around the world and are often ignored by drug designers or agencies responsible for drug admission, such as administration officials, public health databases, and even the news media [8–10]. Between 1975 and 2004, of the 1556 new drugs approved, only 1.3% was exactly developed for the treatment of NTDs, even though these diseases causes 11.4% of the worldwide disease burden [11].

3.2.1 Leishmaniasis

Leishmaniasis is a NTD found across the tropics, subtropics, and southern Europe (Figure 3.1). It emerges as complex and clinically diverse illness which are transmitted through the bite of sandflies and caused by protozoan *Leishmania* species. The process progresses in the posh life cycle in sandflies from promastigotes stage to amastigotes in humans (Figure 3.2). The *Leishmania* species causing the infection is predictable by the severity of the disease and whether the parasitic infection progresses to a systemic disease or not. Meanwhile, the sensitivity toward diverse medications highly differs among the *Leishmania* species. Leishmaniasis is measured by the WHO to be one of the world's most neglected diseases. The present drugs used to treat leishmaniasis were developed more than 40 years ago yet with some side effects and resistance to its treatment.

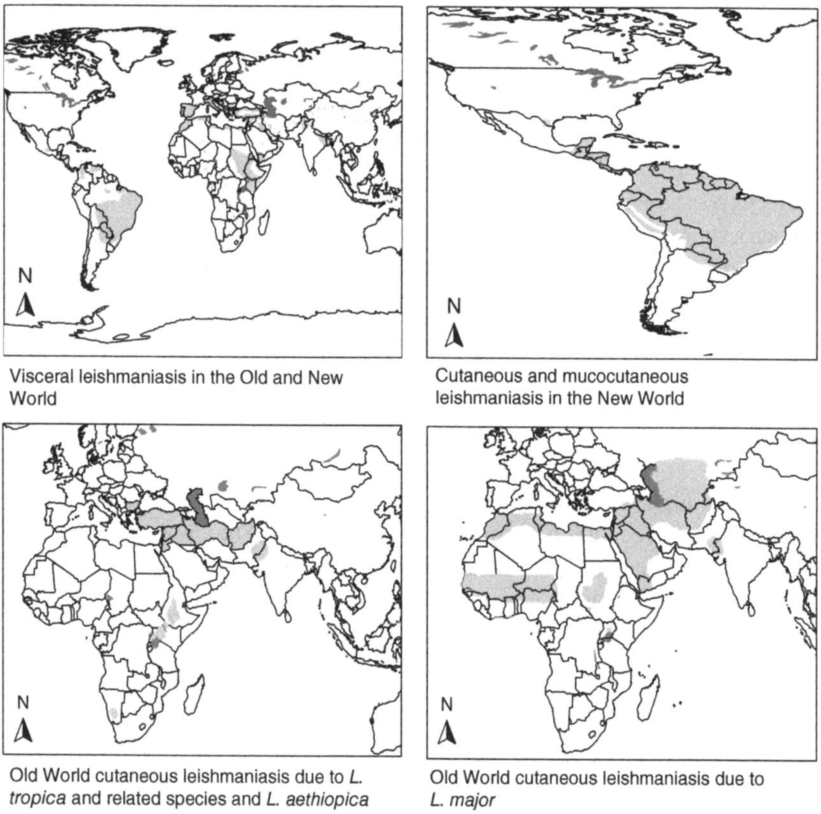

Figure 3.1 Geographical distribution of leishmaniasis. *Source*: Essential leishmaniasis maps, World Health Organization, 2020. Freely taken from https://www.who.int/leishmaniasis/leishmaniasis_maps/en (accessed 20 September 2020).

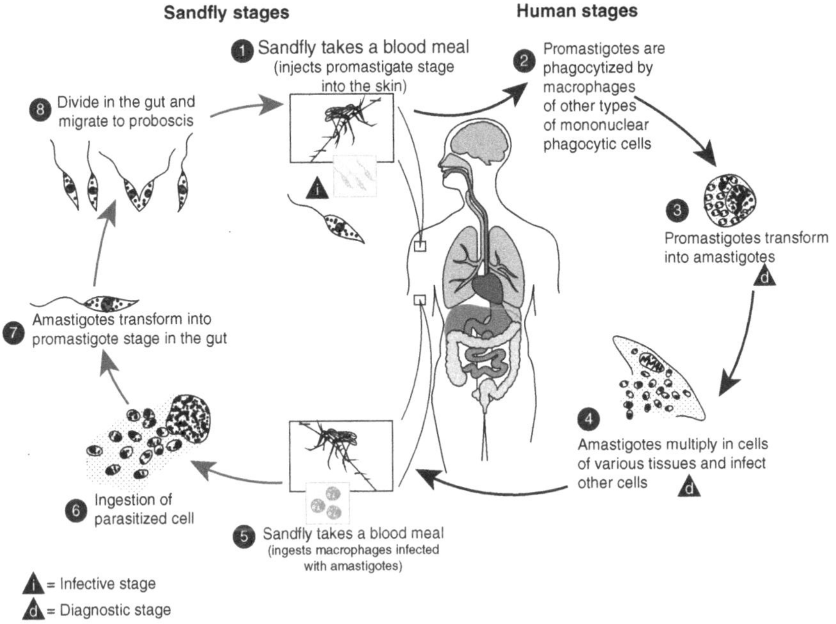

Figure 3.2 Leishmaniasis life cycle. *Source:* Parasites – Leishmaniasis, Centers for Disease Control and Prevention, 4 September 2020. Freely taken from https://www.cdc.gov/parasites/leishmaniasis/biology.html (accessed 4 September 2020).

There are over 20 species of *Leishmania* implicated to be the causative agent of Leishmaniasis in which its transmission to humans is by the bite of infected female phlebotomine sandflies [12]. This disease presents a varied range of clinical symptoms which include manifestations of mucocutaneous, visceral, or cutaneous leishmaniasis (CL). CL (Figure 3.3) is caused by *Leishmania major*, *L. tropica*, *L. aethiopica*, and certain zymodemes from *L. infantum*. Also, other etiologic species implicated are *Leishmania braziliensis,* the most common species, followed by *L. amazonensis*, *L. guyanensis*, and *L. panamensis*. Though, other species such as *Leishmania mexicana*, *L. pifanoi*, *L. venezuelensis*, *L. peruviana*, *L. shawi*, and *L. lainsoni*, which chiefly appear in the Amazon region and Central parts of America can also be related to CL. *Leishmania donovani*, a viscerotropic species, can be implicated in CL throughout or next visceral leishmaniasis (VL). Mucocutaneous leishmaniasis (ML) affects the nasal and oral mucosa and is caused by *L. braziliensis*, *L. panamensis*, *L. guyanensis*, and *L. amazonensis*, *L. majora*, and *L. infantum* [13, 14]. VL is caused by *L. Donovani* and by *L. infantum* [15]. Although CL tends to spontaneously resolve, ML causes severe facial disfigurement while VL is fatal if untreated, causing a global annual mortality estimated at 59 000 cases [16].

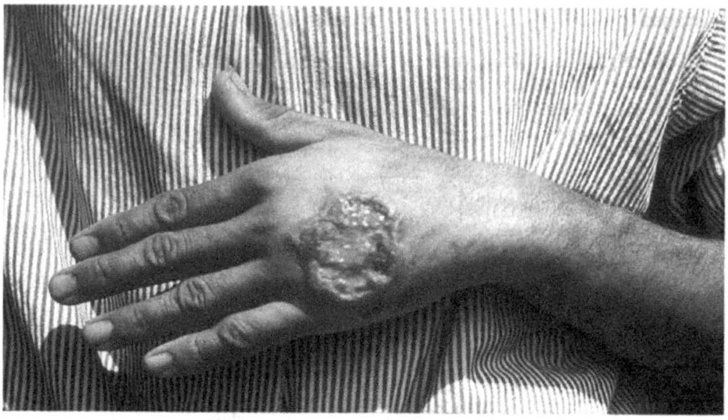

Figure 3.3 Skin ulcer due to leishmaniasis, hand of Central American adult. *Source:* CDC/ Dr. D.S. Martin. Freely taken from https://en.wikipedia.org/wiki/Leishmaniasis#/media/File:Skin_ulcer_due_to_leishmaniasis,_hand_of_Central_American_adult_3MG0037_lores.jpg (accessed 4 September 2020).

The current treatment for leishmaniasis is chemotherapy; though it poses limitations such as difficult route of administration, toxicity, and lack of efficacy in endemic areas. The use of pentavalent antimonials (sodium stibogluconate (Figure 3.4) and meglumine antimoniate) has continued despite the efforts to find new drugs for the treatment of leishmaniasis which was developed more than 60 years back with notable side effects, including abdominalcolic, nausea, diarrhea, skin rashes, hepatotoxicity besides cardiotoxicity. Moreover, side effects of antimonials has been a rising problem around four decades [20, 21]. The polyene antibiotic amphotericin B (Figure 3.4) and its liposomal formulation are amongst the chemotherapeutic agents used as second-line treatment for leishmaniasis, particularly VL [15]. Although it is highly effective, even in antimony-unresponsive patients, amphotericin B has limits due to its renal toxicity and inopportune slow intravenous administration [22]. Liposomal amphotericin B was favored, due to its milder toxicity profile over conventional amphotericin B, but its use remains very limited as a result of its high cost [17]. In South America, an aromatic diamine (pentamidine, Figure 3.4), has been used in the treatment of CL only [23] but harsh adverse effects, including diabetes mellitus, hypoglycemia, shock, myocarditis, and renal toxicity has limited its use [24]. Parmomycin discovered in 1950 is an aminoglycoside antibiotic with labeled leishmanicidal activity; though, this drug has been recognized, it also has some side effects and is costly to acquire [25, 26].

In 2002, Miltefosine (Figure 3.4), is registered as the first, and remains the lone, oral mediator used for the treatment of all types of leishmaniasis [17–19] even though gastrointestinal side effects (anorexia, nausea, vomiting, and diarrhea), hepatotoxicity and insufficiency were reported [27]. Despite efforts to fight this disease for the past 10 years and to permit the utilization of lipid formulations of

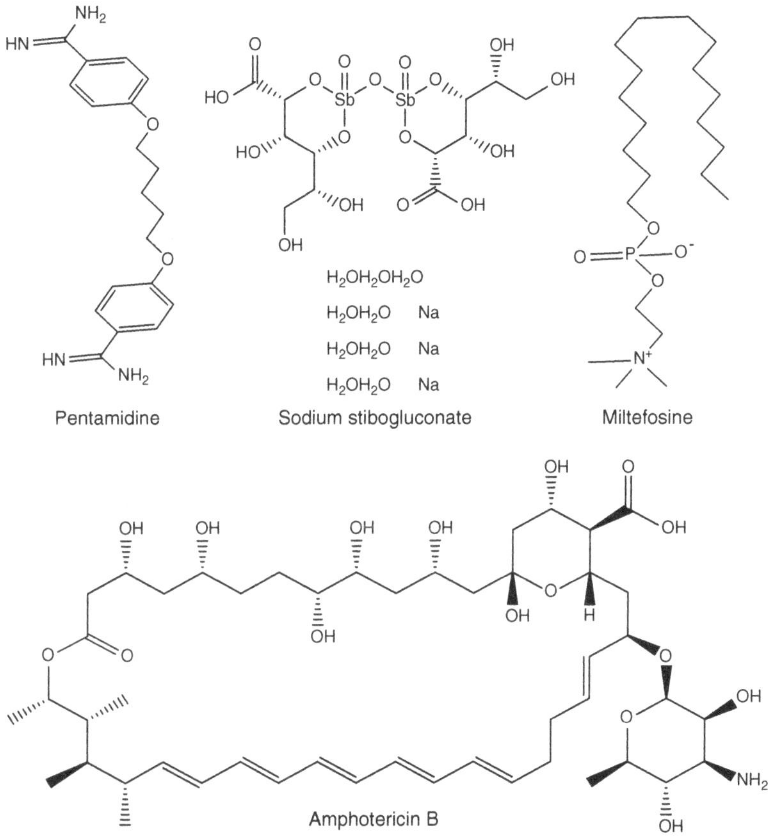

Figure 3.4 Chemical structures of some drugs used against leishmaniasis. *Sources*: Based on Dorlo et al. [17], Sundar and Rai [18], and Singh and Sivakumar [19].

amphotericin B, miltefosine, and paromomycin has been used for the treatment of leishmaniasis in several endemic countries including Brazil. Again, the preference for the use of pentavalent antimonials or conventional amphotericin B exists, despite their inherent toxicities and sophisticated route of administration [27].

In general, heterocyclic compounds are most highly represented among the lead compounds, comprising 55% of the total (Table 3.1) [28]. Eleven percent (11%) of the principal compounds are single-ring heterocyclic structures, 21% had 2-ring structures, 14% were classified as quinolines, and 7% are 3-ring phenothiazines. Ten percent (10%) of the leads are alkaloids and 20% are hydrocarbon structures composed of aromatics and terpenes (Table 3.1). The chemical libraries that have been screened included large numbers of known bioactives and drugs so it was not surprising that 27 out of 84 (32%) leads had been formerly used clinically.

Table 3.1 Chemical class of lead compounds newly identified or previously known to have anti-leishmanial activity [28].

Compound[a]		
Heterocyclic compounds		Class
	1-ring	
		Furans
		Pyridines
		Piperidines
		Pyrans
	2-ring	
		Isoquinolines
		Quinolines
		Purines
		Benzopyrans
		Bicyclocomp.
	3-ring	
		Phenothiazines
		Acridines
		Xanthenes
		Phenanthridines
Alkaloids		
Hydrocarbons		
	Aromatics	
	Terpenes	
	Acyclic	
Polycyclic compounds		
	Macrocyclic	
	Steroids	
Amines		
	Ethylamines	

Table 3.1 (Continued)

Compound[a]		
Heterocyclic compounds		Class
	Q. ammonium	
	Polyamines	
Lactones		
Onium comp.		
Sulfur comp.		
Others*b*	Amides	
	Phenols	
	Carboxili acids	
	Organometallic compounds	

[a] Chemical name according the Chemical Class National Library of Medicine, National Institutes of Health, United States 2010.
[b] Others: Amidines, Phenols, Carboxilic Acids, Organometallic compounds [28].
Source: Modified from Osorio et al. [28].

Accordingly, review of the published literature about the mechanisms of action of the 84 lead compounds revealed that the likely mode of action or parasite targets are the cell wall (21%), cell metabolism (17%), the host immune reaction (13%), apoptosis (7%), and DNA interaction (6%).

The screening of compounds for anti-leishmania activity within the ex vivo model, including screening for toxicity using the HepG2 hepatocyte cell line enabled the choice of 84 lead compounds, 69 of which had not been identified previously to possess anti-leishmania activity. While the utilization of cell lines to predict in vivo toxicity has some limitations [29], the $HepG_2$ cell line is considered a good predictor of human toxicity [30, 31]. Although the measurement of ATP in viable cells has some limitations, this method is one of the most reliable methods to estimate cell toxicity [32], multiparametric toxicity (based on features such as apoptosis markers and membrane integrity) and other in vitro models may be desirable to help in the assessment of hepatic cytotoxicity [33] and further lead optimization [32]. Ultimately, the final selection of lead compounds will require in vivo studies to identify dose limiting toxicities and to evaluate whether the compound's pharmacokinetic and ADME (absorption/distribution/metabolism/excretion) properties make it suitable for use in VL. Already, it is clear that certain of the lead compounds (e.g. topical and antiseptic agents) are unlikely to be useful in treating systemic infection.

3.2.2 Lymphatic Filariasis

Lymphatic filariasis (LF) is deemed to be a major health problem in tropical and subtropical countries (Figure 3.5). Adult worms may live for many years within the infected individual, producing microfilariae and thereby facilitating transmission of the disease through the vector mosquitoes. Filarial nematodes are liable for several diseases such as elephantiasis, onchocerciasis, and tropical pulmonary eosinophilia. The drug of choice for controlling filariasis exhibits numerous side effects. Existing strategies to regulate filariasis are not thought to be completely safe and victorious.

LF is a vector-borne, long-standing chronic disease which is prevalent in many parts of the tropics and subtropics of the planet affecting many people as the second leading cause for long-term and permanent disability in the world [27]. LF is caused by the lymph-dwelling nematode parasites, *Wuchereria bancrofti*, *Brugiamalayi*, and *Brugia timori*. Ninety-one percent (91%) of LF infections are caused by *W. bancrofti* while *B. malayi* and *B. timori* are responsible for the remaining 9% in South and South East Asia [34]. Forty percent (40%) of the world's 120 million cases of LF occur in Africa (approximately 46–51 million cases) [35–40], with an estimated 382–394 million people at risk of infection, including 176 million children [39, 41]. In Africa, all of the LF cases are caused by a spread of *W. bancrofti* by culicine and anopheline mosquitoes [40]. *W. bancrofti* infection produces series of clinical manifestations, with lymphedema and hydrocele, the most clinical features due to the associated disability, disfigurement, and stigma [42]; with global estimates that 12.5% of LF infections may end in lymphedema and 20.8% in hydrocele [41]. There are approximately five million cases of lymphedema and eight million cases of hydrocele in Africa. LF is also associated with huge economic losses, impairing economic activity of up to 88% [43]. Quite often, poor people are infected by the disease and suffer from physical, mental, and socioeconomic hardships [44]. In 72 countries, about 1.4 billion people are at risk of infection where filaria is endemic [44, 45].

The filarial worms need two hosts (Figure 3.6) for their development, i.e. (1) human or other mammalian hosts like primate or cat, and (2) mosquito vector from numerous species of the genera *Culex*, *Aedes*, and *Anopheles* during which the larval development takes place. Symptoms of LF ranges from subclinical lymphangiectasia with severe edema and elephantiasis. These lymphatic dwelling parasites cause damage/blockage to the lymphatic system [46], resulting in the clinical symptoms like lymphedema, elephantiasis, and hydrocele (Figure 3.7). Scrotal, hydrocele elephantiasis and chyluria are only seen with bancroftian filariasis.

The treatment includes a single-dose of mass drug administration yearly for 4–6 years for all people in the endemic region except pregnant women and

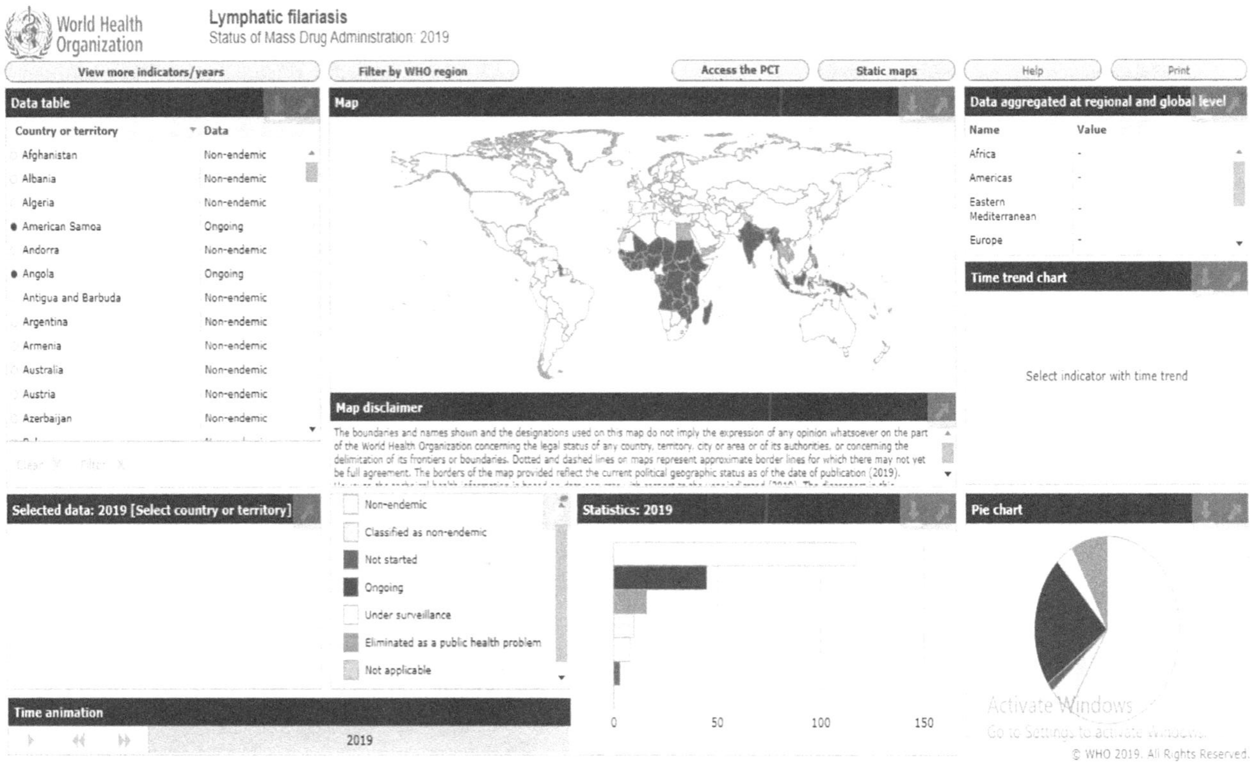

Figure 3.5 Global distribution of lymphatic filariasis. *Source:* World Health Organization (WHO). Freely taken from https://apps.who.int/neglected_diseases/ntddata/lf/lf.html (accessed 4 September 2020).

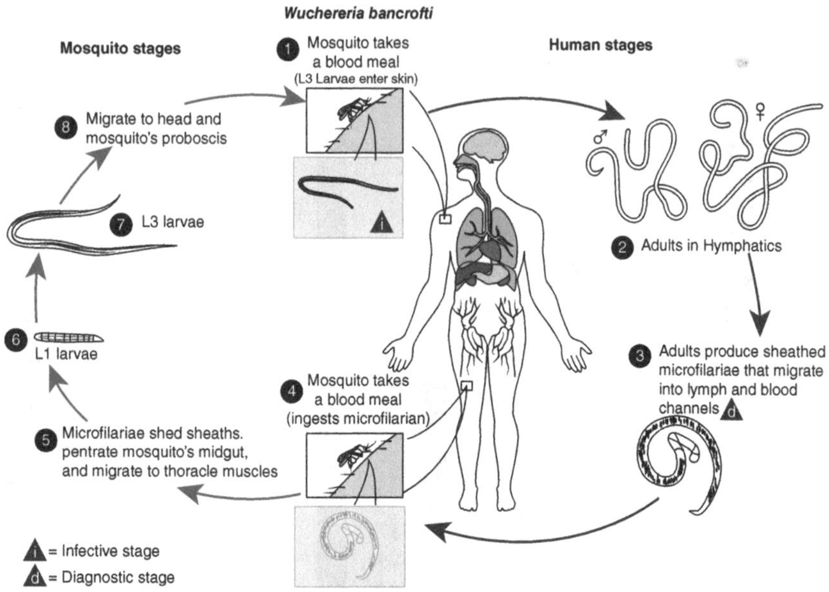

Figure 3.6 Life cycle of *Wuchereria bancrofti*. *Source:* Lymphatic Filariasis-Biology. Centers for Disease Control and Prevention, 4 September 2020. Freely taken from https://www.cdc.gov/parasites/lymphaticfilariasis/biology_w_bancrofti.html (accessed 4 September 2020).

Figure 3.7 Lymphatic filariasis sufferers. *Source:* World Health Organization (WHO). Freely taken from https://www.who.int/lymphatic_filariasis/disease/en (accessed 4 September 2020).

children up to 2 years of age. Diethylcarbamazine and albendazole combination (Figure 3.8) is given in most endemic zones, except in some areas of Africa where bancroftian filariasis is also present with onchocerciasis, therefore a combination of ivermectin and albendazole is recommended [47, 48].

Current drugs used for mass drug administration implemented by national programs are ivermectin, diethylcarbamazine, and albendazole (Figure 3.8) which have been the drugs of choice for filariasis control [49]. These drugs are real in reducing microfilariae totals but not effective in killing adult worms [50]. Later, these drugs deliver only incomplete benefit to infected patients, and very often are associated with adverse reactions. Diethylcarbamazine has been stated to cause side effects such as gastrointestinal disturbance, fever, headache, malaise, and a skin rash that reduces patient's adherence to regimen [51].

The unavailability of vaccines as well as the pressure of augmented danger of expansion of drug resistant worms calls for an urgent need of a cheap, nontoxic, and novel antifilarial drug with long-term antimicrofilarial or macrofilaricidal activity [52, 53].

A few medicinal agents have been derived from plants and utilized in conventional therapeutics. Natural products of plant origin with insecticidal properties

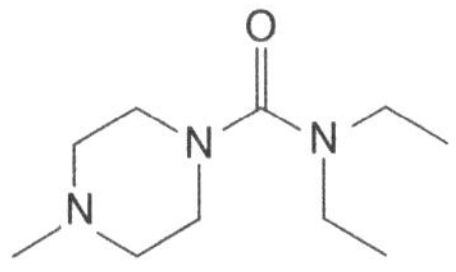

Diethylcarbamazine

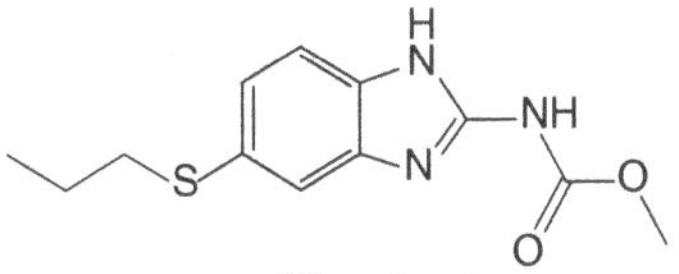

Albendazole

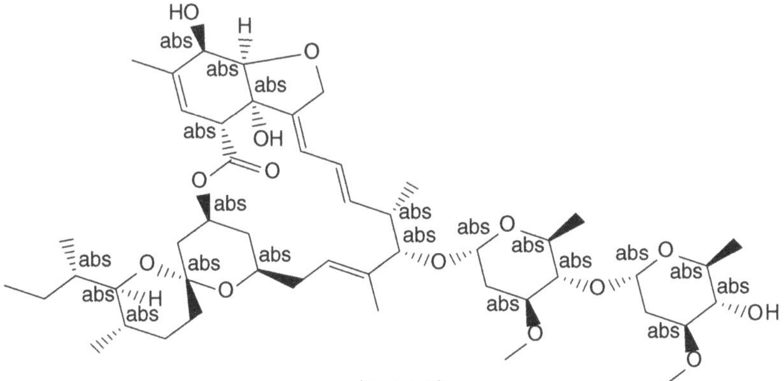

Ivermectin

Figure 3.8 Drugs of choice against lymphatic filariasis. *Sources*: Based on Molyneux [47], Ichimori and Ottesen [48], and Gayen et al. [49].

have been used for the control of a diversity of insect pests and vectors [54]. Various medicinal plants containing pentacyclic triterpene and oleanolic acid have antifilarial activities [55]. Several antifilarial herbs have also been discovered through research on medicinal plants used by local healers.

3.2.3 Trypanosomiasis

3.2.3.1 African Trypanosomiasis (African Sleeping Sickness)

Another NTD with great crash in Africa (Figure 3.9) is HAT or sleeping sickness. HAT is caused by the protozoan *T. brucei* and is transmitted by insects of the genre *Glossina*, recognized as tsetse flies (Figure 3.10). Annually, these parasites infect nearly 30 000 people consistent with official data of reported cases [56] and another 60 million live at-risk areas [57]. The WHO reports that *Trypanosoma brucei gambiense* causes 98% of HAT cases.

Two subspecies of *T. brucei* (Figure 3.11a) are associated with the event of HAT: *Trypanosoma brucei* gambiense, which is endemic in Western and Central African Republic and features a chronic course of infection; and *Trypanosoma brucei rhodesiense*, which is endemic in Eastern and Southern Africa and exhibits a more acute pattern of progression compared to *T. b. gambiense* [59, 60]. The medical appearance of HAT (Figure 3.11b) consists of two recognizable stages: the first is

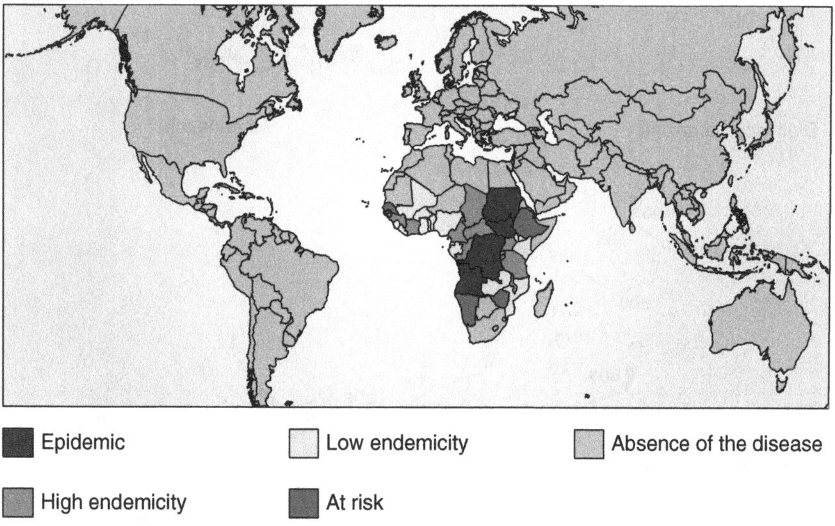

Figure 3.9 Global distribution of human African trypanosomiasis (sleeping sickness). *Source:* TDR-Disease Watch Focus: Human African trypanosomiasis, World Health Organization, 9 September 2020. Freely taken from https://www.who.int/tdr/publications/disease_watch/afrtryp/en (accessed 20 September 2020).

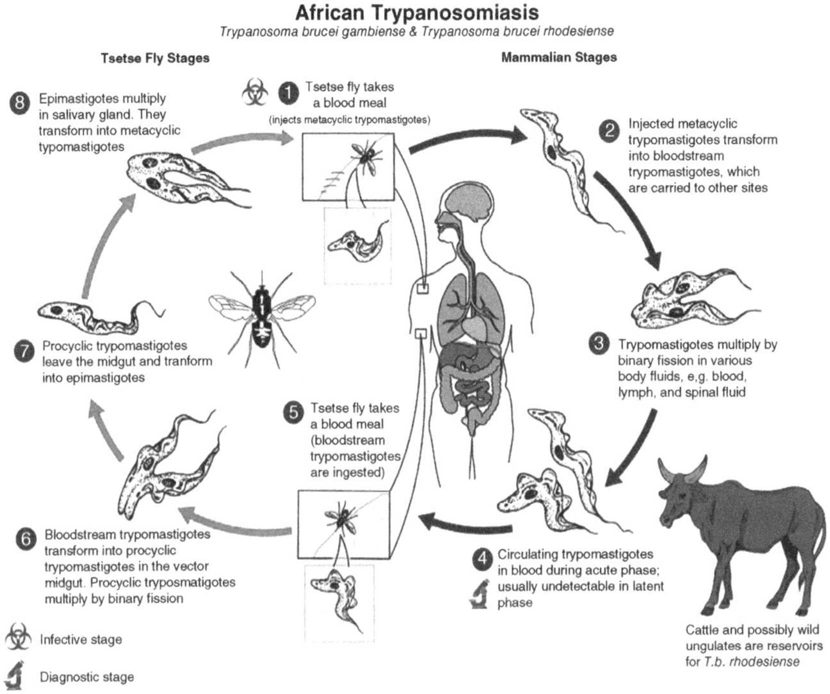

Figure 3.10 African trypanosomiasis life cycle. *Source:* DPDx; Global Health, Division of Parasitic Disease, Centers for Disease Control and Prevention, 4 September 2020. Freely taken from https://www.cdc.gov/dpdx/trypanosomiasisafrican/modules/SleepingSick_LifeCycle_lg.jpg (accessed 4 September 2020).

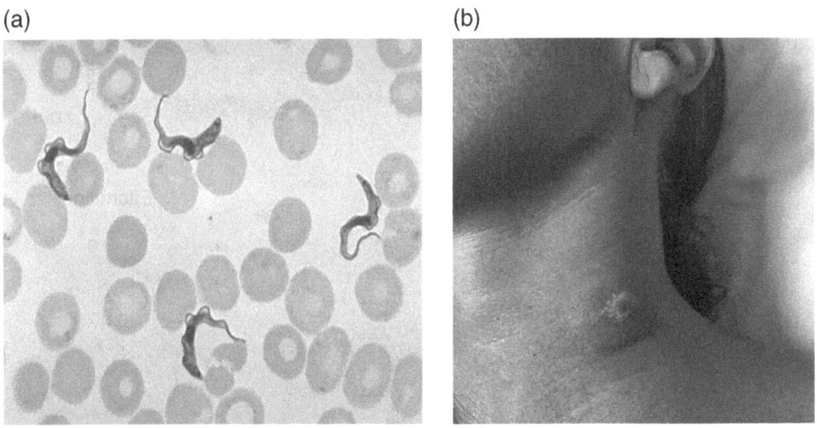

Figure 3.11 (a) Blood smear from patient with African trypanosomiasis. *Source:* CDC/Dr. Myron G. Schultz. Freely taken from https://en.wikipedia.org/wiki/African_trypanosomiasis#/media/File:Trypanosoma_sp._PHIL_613_lores.jpg (accessed 4 September 2020). (b) Chancre of Human African Trypanosomiasis in a Spanish traveler returning from Tanzania. *Source:* Gómez-Junyent et al. [58]. https://doi.org/10.1371/journal.pntd.0005324 (accessed 4 September 2020).

hemo-lymphatic stage, or phase I; and the second is the late encephalitic stage involving the central systema nervosum, or phase II. In phase I, the patient experiences episodes of fever lasting 1–7 days that occur with generalized lymphadenopathy alongside other nonspecific symptoms including malaise, headache, arthralgia, generalized weakness, and weight loss. In phase II, the vermin undergo the blood brain barrier and proliferate within the central systema nervosum, causing an encephalitic response that results in death if the infection is untreated or insufficiently treated.

For the treatment of phase I HAT, pentamidine is used against *T. b. gambiense*, whereas suramin (Figure 3.12) is preferred against *T. b. rhodesiense*. For both treatments, side effects have been observed. In at least half of the patients, pentamidine causes important toxicity with hypoglycemia being the most serious. A variety of side effects, comprising vomiting, nausea, fatigue, and shock followed by neurological complications and renal toxicity such as headache and peripheral neuropathy, have been reported for suramin [61].

For phase II HAT, melarsoprol is active against both *T. b. rhodesiense* and *T. b. gambiense*, whereas eflornitine and nifurtimox (Figure 3.12) are effective only against *T. b. gambiense*. Eflornithine has replaced melarsoprol for *T. b. gambiense* in many endemic countries, and its use is suggested together with nifurtimox [62]. Melarsoprol is very toxic and could cause death. The side effects for eflornithine alone include, fever, seizures, and diarrhea; all capable of causing death. Diarrhea, infections, fever, or hypertension are reported for nifurtimox–eflornithine mixture therapy. Therefore, the toxicity of all currently available

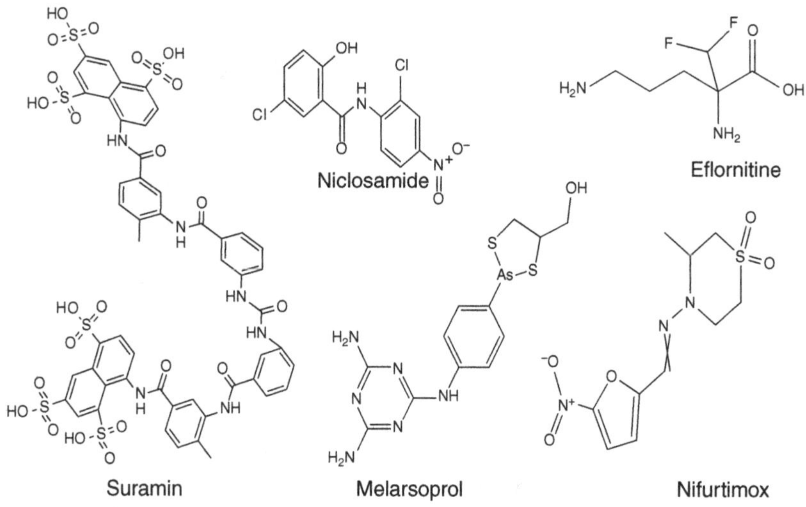

Figure 3.12 Drugs against human African trypanosomiasis.

drugs to treat HAT, the inconvenience of parenteral management, the need of a certain drug and the increasing incidence of treatment failures, make the development of new therapeutic agents against HAT vital.

3.2.3.2 American Trypanosomiasis (Chagas Disease)

American trypanosomiasis, also known as Chagas disease is a tropical parasitic disease that is endemic in 18 countries on the American continent especially in endemic areas of the Southern Cone region, Central America, and Mexico where the main vector lives and cohabited (Figure 3.13). The disease was named "Chagas" after the Brazilian physician who discovered it in 1909. Chagas disease is caused by *Trypanosoma cruzi* (Figures 3.14 and 3.15a), a parasitic euglenoid that is transmitted to humans by *Triatominae,* a blood-sucking insect that is also known as the kissing bug. Besides the kissing bug, *T. cruzi* can also be transmitted from human-to-human through blood transfusion, organ transplant, or from mother-to-child. Also, the parasite can be transmitted by laboratory accident or eating food contaminated by the parasite. However, some preventable measures can be put in place to reduce the spread of the disease. One such measure is through the eradication of the vector by the use of insecticides or treated nets. Again, blood and organ screening for the parasite before transfusion or transplant and diagnosing women who are pregnant could be some viable means to prevent transmission.

The symptoms of Chagas disease could be grouped into acute and chronic and if not treated could be lifelong. In the acute phase, a mild fever, headache, or

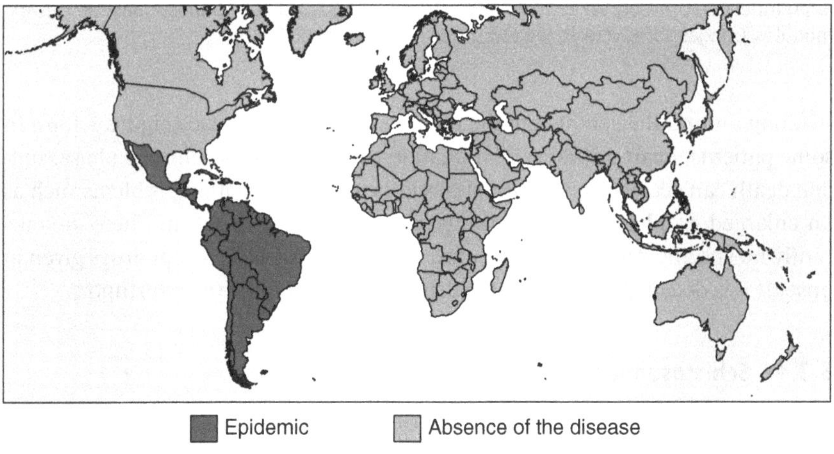

Figure 3.13 Global distribution of American trypanosomiasis (Chagas disease). *Source:* TDR-Disease Watch Focus: Chagas disease, World Health Organization, 20 September 2020. Freely taken from https://www.who.int/tdr/publications/disease_watch/chagas/en (accessed 20 September 2020).

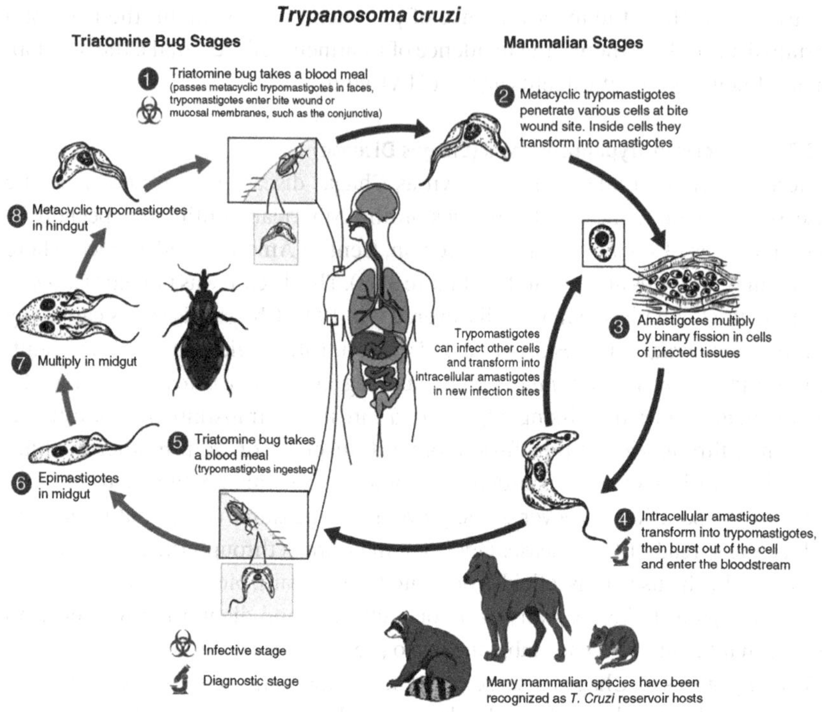

Figure 3.14 American trypanosomiasis life cycle (*Trypanosoma cruzi* life cycle). *Source:* DPDx – American Trypanosomiasis, Centers for Disease Control and Prevention, 13 September 2020. Freely taken from https://www.cdc.gov/dpdx/trypanosomiasisamerican/modules/Chagas_LifeCycle_lg.jpg (accessed 13 September 2020).

swelling around the spot where the parasite entered can ensue (Figure 3.15b). In some patients, heart or brain inflammation can occur. In the chronic phase, sudden death can occur from abnormalities in heart rhythm. Other problems such as an enlarged esophagus and other digestive problems can set in. There are currently no specific drugs for the treatment of Chagas disease except drugs given at onset if it is detected on time. These drugs are benznidazole or nifurtimox.

3.2.4 Schistosomiasis

Schistosomiasis is a known example of a tropical parasitic disease with major incidence found across Africa (Figure 3.16). In the past decade, an increasing effort has been made toward its control. There are two main chemical interventions employed to chop off the schistosomiasis life cycle (Figure 3.17). The primary one consists of therapy, to treat infected individuals. The second uses molluscicidal

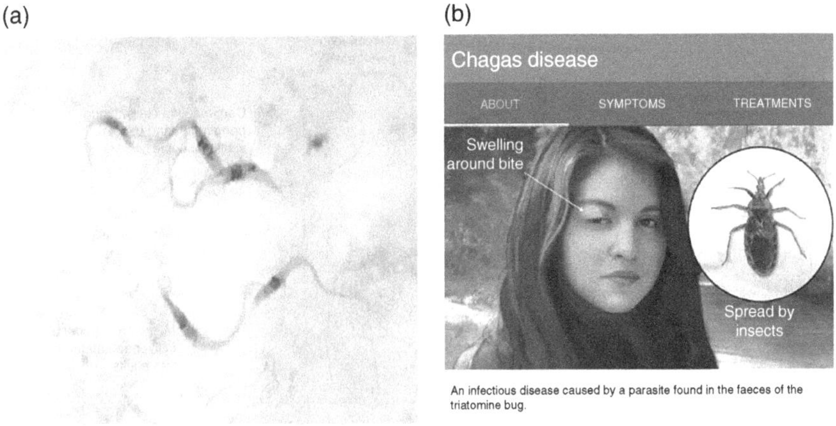

Figure 3.15 (a) Photomicrograph of *Trypanosoma cruzi*, the causative agent of Chagas disease. *Source:* Dr. Myron G. Schultz/Centers for Disease Control and Prevention (CDC). Freely taken from https://www.britannica.com/science/Chagas-disease (accessed 4 September 2020). (b) Bruised eyelid swelling of a lady suffering from Chagas disease. *Source: The Yucatan Times*: Over 20 cases of "Chagas" Disease confirmed during February in Yucatán. Freely taken from https://www.theyucatantimes.com/2019/02/over-20-cases-of-chagas-disease-confirmed-during-february-in-yucatan (accessed 4 September 2020).

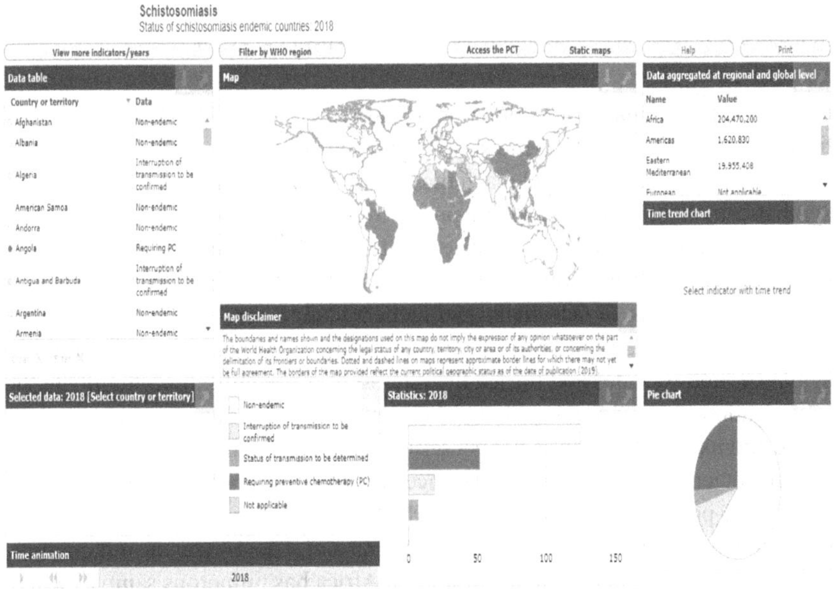

Figure 3.16 Global distribution of schistosomiasis. *Source:* World Health Organization (WHO). Freely taken from https://apps.who.int/neglected_diseases/ntddata/sch/sch.html (accessed 20 September 2020).

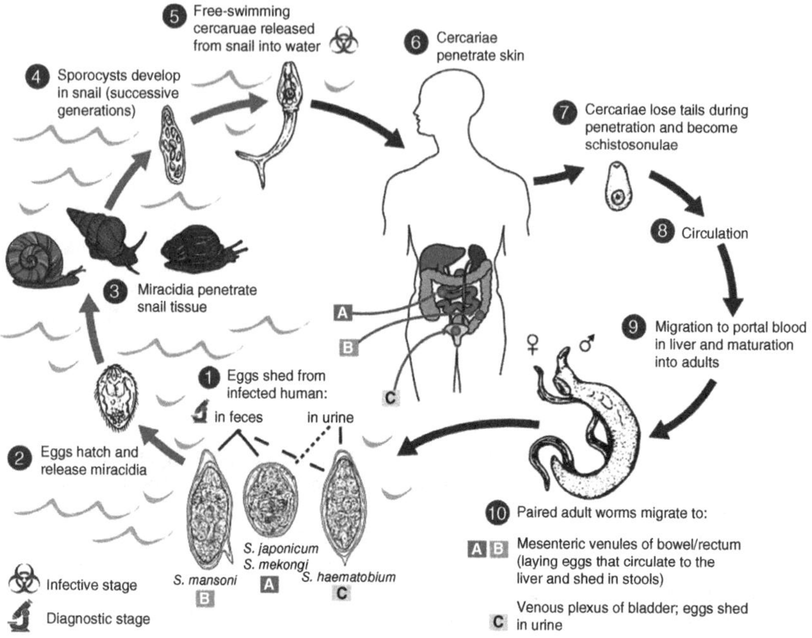

Figure 3.17 Schistosome life cycle. *Source:* DPDx – Schistosomiasis Infection, Centers for Disease Control and Prevention, 13 September 2020. Freely taken from https://www.cdc.gov/dpdx/schistosomiasis/index.html (accessed 13 September 2020).

compounds to regulate the population of the vector snails. This last strategy has been considered together as the simplest preventive actions, since it eliminates the parasites before they could reach the human hosts. To this end, niclosamide, a synthetic molluscicide has shown promising features against schistosomiasis. Nevertheless, undesirable side effects related to this compound limits its use. While synthetic drugs often display higher molluscicidal activity than natural products, their lack of selectivity and biodegradation makes them undesirable. This problem has caused criticism on the utilization of the currently available synthetic molluscicides and turned the interest of the scientific community to the investigation of compounds of natural origin. In fact, the dilemma between synthetic and natural products is not yet settled, and still remains a challenge to seek out an efficient and eco-friendly substance to combat the snails selectively. Around 76% of the populace in Africa lives nearby lakes, rivers, and other water bodies contaminated with snail intermediate hosts [63–66]. Those living near dam reservoirs are at major risk [63], and subsaharan Africa has numerous samples where the infection has emerged or where there has been a dramatic rise within the prevalence of schistosomiasis as a results of irrigation project construction

[63, 67]. Global climate change and heating can also be a factor [68]. The highest occurrence and intensities of human schistosomiasis occur in school-aged children, adolescents, and young adults who also suffer from morbidity and mortality severity of the disease. There are two major forms of schistosomiasis found in Africa. Approximately two-thirds of the schistosomiasis cases are caused by *Schistosoma haematobium* (Figure 3.18), which represents a crucial form of severe tract disease [69]. In 2003, van der Werf et al. [69] found that in Africa, 70 million and 32 million individuals out of 682 million people had experienced hematuria and dysuria, respectively [69]. Kidney failure accounts for an outsized proportion of the predictable 150 000 deaths from tract schistosomiasis in Africa, and there's also a big association between major bladder wall pathology and epithelial cell carcinoma [70]. A big percentage of girls and men with urinary schistosomiasis acquire genital ulcers and other lesions [71]. Within the previous, urogenital schistosomiasis may be a significant explanation for poor reproductive health, including sexual dysfunction and infertility [72]. Genital schistosomiasis also promotes the horizontal transmission of HIV/AIDS in Africa [73]. Intestinal schistosomiasis from *S. mansoni* causes most of the remaining cases in Africa. *S. intercalatum* causes a second sort of intestinal schistosomiasis, but with a limited distribution in Central and West African Republic [74].

As well as the organ-specific pathology designated for both *S. haematobium* and *S. mansoni* infections, there is increasing sign for more generalized morbidity resulting from the chronic inflammation of long-standing infections [71,

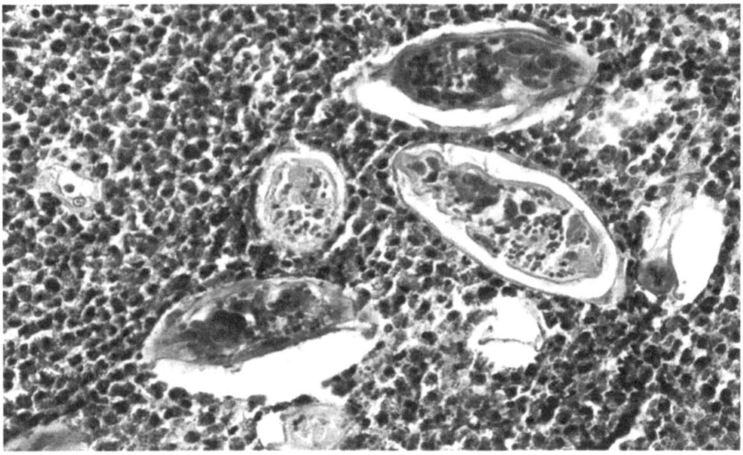

Figure 3.18 Bladder histopathology shows eggs of *Schistosoma haematobium* surrounded by intense infiltrates of eosinophils. *Source:* CDC/Dr. Edwin P. Ewing, Jr. Freely taken from https://en.wikipedia.org/wiki/Schistosoma_haematobium#/media/File:Schistosomiasis_haematobia.jpg (accessed 4 September 2020).

75]. Among the foremost important manifestations are iron deficiency anemia and anemia of chronic inflammation, growth stunting and malnutrition, fatigue and diminished physical fitness, and impaired cognitive development [71, 75]. It has been optional that the true disease load for schistososomiasis may be numerous fold higher than preceding estimates [71] possibly making this infection the most important NTD in Africa, in addition to *S. haematobium* and HIV coinfections [76].

3.3 Novel, Economical, and Efficacious Therapeutics from Medicinal Plants Against NTDs

3.3.1 Phytochemicals Effective Against Leishmania Parasite

Compounds containing the heterocyclic quinoline ring system were frequently identified as active inhibitors, representing 14% of the lead compounds. The quinoline leads might be further divided into several distinct classes. Variety of the leads bear obvious structural relation to antimalarial quinolines, including several 4-aminoquinolines with basic side chains, as well as 8-aminoquinolines and a 2-arylquinoline derivative containing the core structure of the antimalarial drug mefloquine. Also represented are 8-hydroxyquinoline antifungals like clioquinol, in addition to more novel quinoline diones and dimeric 2-aminoquinolines. Several of the quinoline-containing compounds exhibited low toxicity and activity at micromolar concentrations to the HepG2 cell line, signifying that they are noble leads for optimization. Earlier studies have described other quinoline-containing compounds with anti-leishmania activity. For example, a 2-substitued quinoline alkaloid decreased by 79.6% the parasite load in the liver of BALB/c mice infected with *L. donovani* [77]. For the treatment of VL, sitamaquine (Figure 3.19), an 8-aminoquinoline, has undergone a phase II study [78]. Quinoline (Figure 3.19) ring structure is common in known drugs and other bioactive molecules and such compounds can exhibit distinct bioactivities depending on their specific structures. Previously, quinoline byproducts have been stated to affect electron transport and generate lethal oxidative radicals against *Leishmania*, and also to inhibit cysteine proteases [79], a gene family important for *Leishmania* virulence [80]. Other quinoline derivatives have also been shown to alter the vesicle and endocytosis in *Plamodium falciparum* [81]. Consequently, together with these prior observations, suggests the likely possibility that multiple distinct bioactivities are represented among the quinoline leads identified and that multiple pharmacologically orthogonal candidates might be selected for further lead optimization studies.

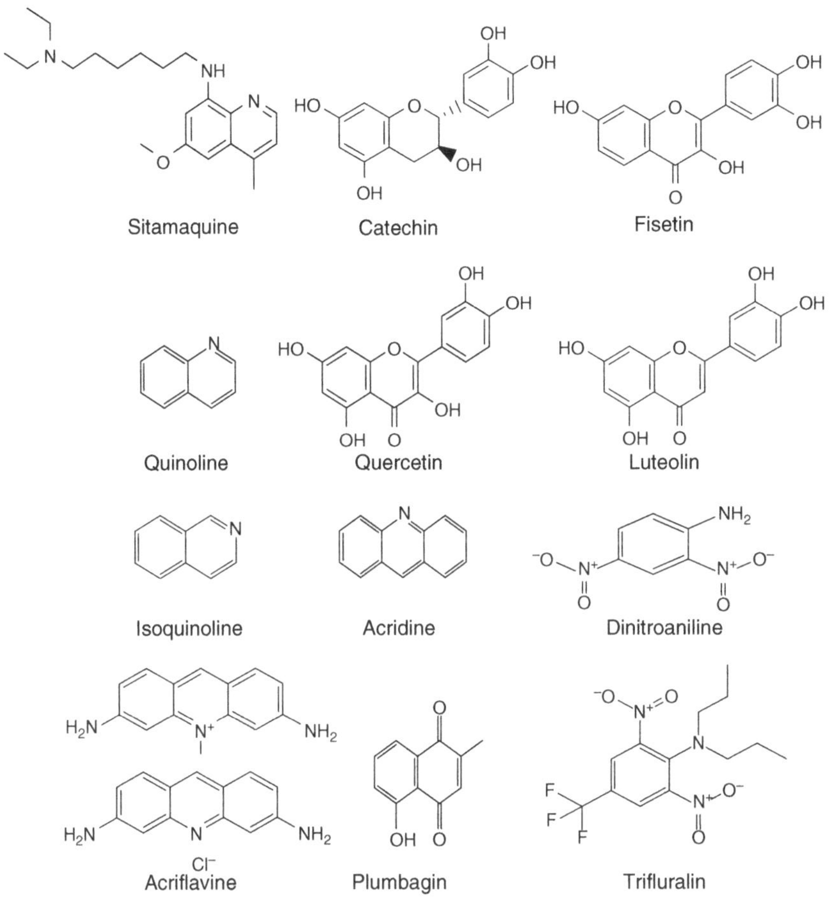

Figure 3.19 Some compounds active against leishamniasis. *Source:* Based on Dietze et al. [78].

In an *ex vivo* screening, alkaloids also represented an important fraction of the lead combinations (10%) identified. This parallels the observation that alkaloids derived from natural yields have been active against *Leishmania* species [82].

As mentioned previously, there is a need of a competent drug because there is resistance to drugs administered aimed at handling leishmaniasis, coupled with their high cost, toxicity and parenteral route of administration; have been regarded as a great concern especially in endemic areas of developing countries. Hence, there could be no doubt that the search for novel agents having anti-leishmanial or leishmanicidal potency is one of the critical challenges in the field of the current drug detection program and is of universal apprehension [83–85].

Designed for this point, some researchers have been looking for new alternatives in nature as an important source of drugs used in medicine [86–88]. It has been assessed that there are about 250 000 medicinal plant species in the world. Though, the biological actions of only about 6% of them have been screened. Moreover, only about 0.75% medicinal herbal compounds have been considered in clinical trials [89, 90]. The main qualities of herbal medicine include their low cost, low incidence of serious adverse effects and good efficacy [83, 91]. Unluckily, the worth of most of these phytochemicals has been exposed only in *in vitro* studies and therefore are still in their initial stages of preclinical trials to be appropriate in clinical practice and conversion of their results into clinical practice is a neglected field. It is assessed that herbal products are implicated directly or indirectly in the production of about 25% of medicines [92].

Several studies have revealed that a diversity of plant species exhibits inhibitory activity against certain types of parasites such as *L. major* [84], *L. amazonensis* [93], *L. tropica* [94], *L. aethiopica* [95], *L. braziliensis* [96], *L. Mexicana* [87], *L. infantum* [97, 98], *L. chagasi* [99], and *L. donovani* [100].

Some recent *in vitro* and *in vivo* animal studies regarding the effectiveness of medicinal plants and plant-derived compounds such as the crude extracts and essential oils against CL and VL caused by different *Leishmania* species have been summarized in Table 3.2. In order to evaluate the anti-leishmanial or leishmanicidal activities of a compound, several criteria are considered including minimum inhibitory concentration (MIC), IC_{50} and also the selectivity index [85]. The MIC of a compound or drug against the parasite has to be defined. Also, because the IC_{50} is described, the concentration of a medicine which causes 50% growth inhibition of amastigote or promastigote sorts of *Leishmania* could be desirable [106]. IC_{50} is employed for macrophages of the host, also cytotoxicity concentration (CC_{50}) of drug could be described as that which leads to 50% of mortality of macrophage from usually BALB/c mice model [108] or as the case may be.

When selectivity index is greater than 10 the anti-leishmanial drug or natural compound is safe [85]. Rondon et al. [99] investigated the leishmanicidal activities of the greeneries of *Aloe vera* (aloe), and *Ricinus communis* (castor) and the seeds of *Coriandrum sativum* (coriander) against amastigotes and promastigotes of *L. infantum*. These extracts were compared at different concentrations of 6.25, 12.5, 25, 50, and 100 μg/ml with 40 μg/ml pentamidine and amphotericin B as positive controls respectively. All fractions were effective against *L. infantum* promastigotes with no significant difference compared to pentamidine. The observation was linked to iron-dependent enzymes and membrane lysis of the parasite thanks to the presence of flavonoids, alkaloids, phytoestrols, and saponins.

The necessity of undertaking in vivo studies with *R. communis* ester fraction is based on its low cytotoxicity yet with high efficacy against the parasite. As well, 2-substituted quinolone, a medicinal plant derivative, is effective against

Table 3.2 *In vitro* and *in vivo* anti-leishmanial activity of plant compounds.

Plant	Part used	References
Croton pullei var. *glabrior*	Stem bark	[101]
Lippia sidoides Cham. (*Alecrim pimenta*)	Aerial parts	[93]
Valeriana wallichii DC	Root	[102]
Aloe vera vs. *Tamarix aphylla*	Leaf vs. bark	[94]
Cymbopogon citratus and citral	Aerial parts	[103]
Galipea longiflora Krause	Bark	[101]
Calophyllum brasiliense	Leaf	[104]
Moringa stenopetala	Root	[95]
Vanillosmopsis arborea and α-bisabolol	Stem	[92]
Curcuma longa Linn.	Spice turmeric	[101]
Piper regnellii var. *pallescens*	Leaf	[105]
Strychnos pseudoquina	Stem bark	[101]
Vitis vinifera L.	Leaf	[97]
Physalis angulate	Stem	[101]
Aloe secundiflora	Leaf	[84]
Annona muricata acetogenins, annonacinone and corossolone, and *Platymiscium floribundum* coumarin scoparone	Stem of *P. floribundum*and leaves and seeds of *A. muricata*	[101]
Senna Spectabilis	Flower	[101]
Thymus capitellatus Hoffmanns. and Link	Aerial parts	[98]
Berberis vulgaris and *B. barbering* (its major principl, an isoquinoline alkaloid)	Root (rich in berberine)	[85]
Chenopodium ambrosioides	Not available	[106]
Arrabidaea chica	Leaf	[107]
Olea europeaea, five olive tree varieties (Limouni, Zarrazi, Dhokkar, Toffehi, Chemlali Tataouine)	Leaf	[88]
Bixa orellana	Seed	[106]
Tanacetum parthenium (L.) Schultz-Bip	Aerial part	[101]
Tridax procumbens alone and in combination with *Allium sativum* (garlic)	Whole plant and bulbs, respectively	[87]
Berberis vulgaris L.	Root bark	[101]
Baccharis uncinella	Leaf	[101]

(*Continued*)

Table 3.2 (Continued)

Plant	Part used	References
Lippia sidoides Cham.	Leaf	[107]
Mangifera indica L.	Leaf	[116]
[105]*Artemisia annua* (*artemisinin*)	Leaf and seed with floral part	[101]
[108]*Trichosanthes dioica* Roxb. (pointed gourd)	Root	[107]
Withanolides (*Withania somnifera*, Ashwagandha)	Leaf	[107]
[101]*Croton Cajucara* (red sacaca)	Leaf	[106]
Eryngium foetidum L.	Aerial part	[107]
[101]*Pleumeria pudica*	Leaf	[90]
Fumaria parviflora Lam.	Whole plant	[91]
Millettia species	Stem bark	[107]
[101]*Withania somnifera* (winter cherry)	Whole plant	[90]
Tinospora cordifolia in combination with cisplatin	Whole plant	[99]
Combination of *Asparagus racemosus* (shatavari) with cisplatin	Whole plant	[99]

promastigotes and amastigotes of *L. infantum*, and *L. amaznensis*, [109] and simpler than miltefosin against *L. donovani* [110].

Ogungbe et al. [111] indicated 352 phytochemicals of phenolic nucleus displaying antiparasitic activity; comprising 92 flavonoids, 52 coumarins, 52 lignans, 41 isoflavonoids, 34 chalcones, 25 quinones, 20 chromenes, 10 aurones, 9 xanthones, 8 stilbenoids, 6 cannabinoids, and 3 miscellaneous phenolic compounds against 24 *Leishmania* enzymes. It was demonstrated that two aurones (bracterin triacetate and 4,6-dibenzoyl-2-[phenylhydroxymethyl]-3(2*H*)-benzofuranone), one chalcone (crotaorixin), five coumarins (mammeaA/AA, mammeaB/BA, mammeaB/BA cyclo F, umckalin, and scoparone), six flavonoids (diplacone, quercetin-3′,4′, 7-trimethyl ether, cannflavin A, 3′-*O*-methyldiplacone, 3′-*O*-methyldiplacol, and 4′-*O*-methyldiplacone), one isoflavonoid (sophoronol E), three lignans (aristolignin, 3,3′,4,5-tetramethoxy-4′,5′-methylenedioxy-7,7′-epoxy-lignan and 4,5,4′,5′-dimethy lenedioxy-3,3′-dimethoxy-7,7′-epoxylignan), and one stilbenoid (machaeriol B) can be measured as promising drugs and are worthy of further investigation for the management of leishmaniasis [111].

Essential oils prepared by hydrodistillation are composed of various small hydrophobic molecules like sesquiterpenes and terpenes, especially monoterpenes diffusing easily across cell membranes and consequently gaining access to intracellular targets [98]. Terpenes are hydrocarbons formed from isoprene units which can easily penetrate the lipid bilayer of cell wall and make changes within the integrity of cell structures and mitochondrial membrane [92]. There are several information on the action of volatile oils on endemic *Leishmania* species liable for CL and VL [103]. Pharmacological application of volatile oils against human parasitic infections is unclear because of the potential risk for toxicity [98]. Though, numerous studies have established that the crude volatile oils and their major compounds at the effective concentrations present low or no toxicity to host cells [107]. It has lately been reported that volatile oils from *Artemisina annua* [101], *Croton argyropylloides* [101], *Ligustim chuanxiong* [101], and *Menta villosa* [112] are effective and nontoxic in animal models with leishmaniasis. Furthermore, certain essential oils resulting from plants such as *Croton cajucara* have immunomodulatory effects and this makes them useful in the management of leishmaniasis [107]

Some plant extracts which exerts anti-leishmanial activity has been attributed to the flavonoids contents [113]. Flavonoids are polyphenolic phytochemicals which naturally exists in fruits and vegetables and are referred to as antioxidants and anticancers agents with a big protective effect against membrane damage [88]. Flavonoids such as catechins (Figure 3.19) are able to form complexes with the parasite cell wall to influence processes requiring cell linking, and hence restrain the parasite growth [84]. In a study by Manjolin et al. [86], it was shown that dietary flavonoids such as fisetin (the most potent alkaloid), luteolin, quercetin (Figure 3.19), and 7,8-hydroxyflavone with low cytotoxicity characteristics are able to inhibit arginase enzyme from *L. amazonensis*. Arginase shows a fundamental role in polyamine biosynthesis which is very important and essential for protecting the parasite against oxidative stress and ROS produced by the host's defense system.

Mushrooms and their metabolites have been tested for their probable anti-leishmanial potentials [114]. Mallick et al. [114] evaluated the anti-leishmanial effect of mushrooms against promastigote and amastigote forms of *L. donovani*. They used 18 extracts resultant from six mushrooms. Among these, *Astraeus hygrometricus* and *Tricholoma giganteum*-induced apoptosis and repressed the growth of *L. donovani* promastigotes. Instead, fractions soluble in water of, *Russula laurocerasi*, *Russula albonigra*, *Russula delica*, *Termitomyces eurhizus*, *A. hygrometricus*, and polyphenolic fractions of *R. laurocerasi* obliged the proliferation of intracellular amastigotes in macrophages. Concerning intracellular amastigotes, the compounds were nontoxic and prompted release of NO and in murine cells IL-12 [101]. They showed that such mushrooms can unlock the way for further phytochemical and pharmacological investigations toward the search of novel anti-leishmanial compounds [114]. A study presented by Mishra et al. [115] has reported leishmanicidal activity of numerous alkaloids including indole, quinoline, isoquinoline (Figure 3.19), pyrimidine-carboline, steroidal, and diterpene derived from various plants as well as those obtained from marine sources (Table 3.3).

Anti-leishmanial activity of acridine and dinitroaniline (Figure 3.19) herbicides in combination with triterpenoid saponin extract and extracts prepared from roots of *Plumbago capensis* was investigated against *L. major* in BALB/c mice [116]. The combination therapy of saponin, acriflavine, plumbagin, and trifluralin (Figure 3.19) is a promising treatment for infection by *L. major*. The activity of extracts of stem wood, root, and bark from *Connarus suberosus* Planch (Connaraceae), *Myrcia linearifolia* Cambess. (Myrtaceae), and *Neea theifera* Oerst. (Nyctaginaceae) against promastigotes of *L. amazonensis* at a concentration of 100 μg/ml were evaluated [117]. The bisnortriterpenes from *Salacia madagascariensis* (Celastraceae); isoiguesterin and 20-*epi*-isoiguesterinol showed potent activities against *Leishmania* sp. [118]. Meanwhile, isoiguesterin and 20-*epi*-isoiguesterinol exhibited similar activities with artemisinin and chloroquine against the D6 clone, being extra selective and potent against *L. donovani* (recognized to cause VL). When compared to amphotericin B used currently in the cure of Leishmaniasis, isoiguesterin and 20-*epi*-isoiguesterinol showed great potential for future selective drug development against *Leishmania*.

Several researchers have evaluated anti-leishmanial properties of the most common plants available in their countries. For example, Ribeiro et al. [119] investigated the *in vitro* anti-leishmanial activity of 44 extracts and fractions derived from 16 Brazilian plant species against *L. amazonensis*. They used plants with minimum inhibitory concentrations (IC_{50}) in *Leishmania* and their cytotoxic effects on murine macrophages (CC_{50}) for defining the SI of every extract or fraction. The products that offered higher SI values were utilized in the treatment of macrophages infected with *L. amazonensis*. These plants comprise *Campomanesia lineatifolia* Ruiz, *Bowdichia virgiloides* Kunth, *Cecropia pachystachya* Trecul, *Diospyros hispida*, D.C., *Chrysobalanus icaco* L, *Dipteryx alata* Vog., *Eugenia uniflora* L., *Syzygium cumini* L, *Hymenaea courbaril* L., MS Hymenaea stignocarpa Mart. ex. Hayne, *Jacaranda caroba* Vell., *Jacaranda cuspidifolia* Mart., *Jacaranda ulei* Bureau and K. Schum, and *Licania tomentosa* Benth.

Garcia et al. [105] examined the anti-leishmanial action of 48 extracts obtained from 46 Cuban plants mainly from leaves, seeds, and roots against promastigotes and mastigotes of *L. amazonensis*. Only four extracts including *Bambusa vulgaris*,

Table 3.3 Plants and marine sources of alkaloids with leishmanicidal activity [115].

Plant	Family	Compound
Kopsia griffithii	Apocynaceae	Indole alkaloids
Peschirea australis	Apocynaceae	
Bark of *Corynanthe pachyceras*	Rubiaceae	Indole alkaloids
Aspidosperma ramiflorum	Apocynaceae	Indole alkaloids
Bark of *Peschiera van heurkii*	Apocynaceae	Indole alkaloids
Annona foetida	Annonaceae	Isoguinoline alkaloids
Bark of *Guatteria foliosa*	Annonaceae	Isoguinoline alkaloids
Young leaves *Guatteria dumetorum*	Annonaceae	Isoguinoline alkaloids
Bark of *Rollinia emarginata*	Annonaceae	Isoguinoline alkaloids
Unonopsis buchtienii	Annonaceae	Isoguinoline alkaloids
Twigs and leaves of *Ancistrocladus griffithii*	Ancistrocladaceae	Naphthylisoquinoline alkaloids
Roots of *Ancistrocladus likoko* J. Leonard	Ancistrocladaceae	Naphthylisoquinoline alkaloids
Ancistrocladus tanzaniensis	Ancistrocladaceae	Naphthylisoquinoline alkaloids
Albertisia papuana	Menispermaceae	Bisbenzylisoquinoline alkaloids
Pseudoxandra sclerocarpa	Annonaceae	Bisbenzylisoquinoline alkaloids
Leaves of the Bolivian plant *Saracha punctata*	Solanaceae	Steroidal alkaloids
Leaves of *Holarrhena curtisii*	Apocynaceae	Steroidal alkaloids
Psychotria klugii	Rubiaceae	Benzoquinolizidine alkaloids
Species of genera *Aconitum*, *Delphinium* and *Consolida*	Ranunculaceae	Diterpene alkaloids
Bark extract of *Annona foetida*	Annonaceae	Pyrimidine-β-carboline alkaloids
leaves of *Acanthus illicifolius*	Acanthaceae	Pyrimidine-β-carboline alkaloids
Marine sponges (*Amphimedon viridis*, *Acantho strongylophora* sp., *Neopetrosia* sp., *Plakortis angulospiculatus* and *Pachymatisma johnstonii*)	—	Marine alkaloids

(Continued)

Table 3.3 (Continued)

Plant	Family	Compound
Leaves of *Acanthus illicifolius*	Acanthaceae	Miscellaneous alkaloids
Stem bark of *Zanthoxylum chiloperone* var. *angustifolium*	Rutaceae	Miscellaneous alkaloids

Source: Based on Mishra et al. [115].

Hura crepitans, *Mangifera indica*, and *Simarouba glauca* exhibited selective activity against the parasite intracellular amastigotes of *L. amazonensis*. Collectively, among the 46 evaluated plant species, the highest inhibition of promastigotes growth (IC_{50} = 16.4 μg/ml) was caused by *H. crepitans* which showed minor toxicity against host cells (IC_{50} = 390.5 μg/ml), with an excellent selectivity index equal to 24. In general, the extracts prepared from *H. crepitans*, *B. vulgaris*, and *S. glauca* showed more promising anti-leishmanial activity against both stages of the parasite.

A study conducted in Saudi Arabia evaluated the *in vitro* antiprotozoal potential of the methanol extracts of 51 plants and some of their petroleum ether, chloroform, ethyl acetate, and aqueous fractions against *L. infantum* amastigotes and also *Trypanosoma brucei*, *Plasmodium falciparum*, and *T. cruzi* [120]. Among these, only nine crude extracts (18%) showed pronounced anti-leishmanial activity with IC_{50} from 0.25 to 2.03 μg/ml. Three plants including *Euphorbia schimperiana*, *Tribulus macropterus*, and *Hypoestes forsskalii* had IC_{50} less than 0.25 μg/ml. Maximum activities were associated with the extracts of *Solanum villosum*, *Verbesina encelioides*, and *T. macropterus* with IC_{50} of 0.51, 0.35, and less than 0.25 μg/ml and SI of 125, 99.3, and 87.5, respectively.

To stem the high occurrence of CL by *L. major* in Iran, Manjili et al. [121] assessed the anti-leishmanial potential of ethanol extracts of seventeen medicinal plants against *L. major* promastigotes. Four plants, *Thymus migricus*, *Carum copticum heirm*, *Satureia hortensis*, and *Caesalpinia gilliesii* showed high anti-leishmanial activities with IC_{50} of 31.25 ± 15.44, 15.625 ± 5.46, 15.625 ± 3.76, and 9.76 ± 1.27 μM, respectively. These plants were lethal against macrophages at concentrations higher than those required to hinder the parasite cell growth (IC_{50}, 45.13 ± 3.17, 100.44 ± 17.48, 43.76 ± 0.78, and 39.67 ± 3.29 μM, respectively). Furthermore, glucantime repressed the growth of *L. major* promastigotes by IC_{50} = 254 μg/ml deprived of moving the growth of macrophages. The effectiveness of *T. migricus*, *C. gilliesii*, *C. copticum* heirm, and *S. hortensis* extracts owing to the existence of the active compounds might help as substitute mediators in the control of CL.

Although herbal medicines have been used for a long time for the cure of leishmaniasis, a systematic scientific evaluation of the effectiveness of herbal extracts has rarely been performed. Several methods as double-blind, randomized, clinical experimental were passed by Sattar et al. [122] to investigate the anti-leishmanial potency of *Morinda citrifolia* (Noni) stem extract against CL. Among 40 patients, 30% indicated good improvement and 50% exhibited excellent response. In addition, morindicone and morinthone, two major constituents derived from the extract, displayed good anti-leishmanial activity. Zerehsaz et al. [123] compared the efficacy of intramuscular administration of meglumine antimonial (glucantime) (85 patients, for 20 days, 15–20 mg/kg/day) through current herbal extract containing of a mixture of the pure extracts of *Althaea rosa*, *Althaea officinalis*, and associates of the families Faliaceae, Leguminosae Lythraceae, and Malvaceae (86 patients, as a crude paste covered by a dressing, for five days) on patients with CL through double-blind randomized clinical trial. Moreover, the adverse effects due to treatment including urticarial and generalized pruritus only were observed in the glucantime group [123]. A thorough investigation on anti-leishamnial activity of the plants present along with their toxicity, mechanism of action and physiochemical properties for development of an optimum formulation is urgently needed to confirm their efficacy in the management of leishamniasis [124].

From *Aloe* species quinones have also shown anti-leishmanial activities. These comprise aloin, aloinoside, and microdontin from the leaf latex of *Aloe calidophila* [125]. It is noteworthy that, the activities of the most potent compounds, with IC_{50} values ranging from 3.12 to 10.92 μM against *Leishmania aethiopica* and from 3.70 to 15.26 μM against *Leishmania major*, were similar to the control drug amphotericin B (IC_{50} = 0.12 and 0.07 μM against *L. aethiopica* and *L. major*, respectively). Selectivity indices of aloinoside against *L. aethiopica* and *L. major*, 813.35 and 694.90, respectively were considerably improved than those of, amphotericin B (control) (423.49 and 688.96, respectively). This proposes that the isolated phytochemicals could serve as possible support for the progress of specific, safe, and cost-effective anti-leishmanial drugs [125]. Moreover, the dioncoquinones A and B isolated from *Triphyophyllum peltatum* (Dioncophyllaceae) presented worthy and specific activity against *L. major* by hindering at very low concentrations the growth of the parasite [126]. Klaivanolide, derived from the stems of *Uvaria klaineana* (Annonaceae), was likewise described as a potent molecule (*in vitro* IC_{50} values of 1.75 and 3.12 μM, respectively) against sensitive and amphotericin B-resistant promastigote forms of *L. donovani* [127].

3.3.2 Plants and Phytochemicals Effective Against Filarial Parasite

There is a limited focus on medicinal plants used for LF. Limited data are available on important plants useful in filariasis. It is needful to indicate the importance of thorough study of herbal drugs for fortification of the antifilarial therapeutic range. This old-style therapeutic substitute may very well evidence in terms of cost-effectiveness and patient fulfillment in combating the filariasis.

From the funicles of *Acacia auriculiformis* A. Cunn. ex Benth. (Fabaceae) triterpenoid saponins acaciaside-A and acaciaside-B derived are effective against both microfilaria and adult worm of *Setaria cervi* [102]. Excluding cestocidal effect, these saponins at the application of 4 mg/ml killed 97% of microfilaria of *S. cervi* and 100% of adults in 100 minutes [128]. It has been reported that these saponins enhance the cell membrane lipid peroxidation [129], while the unsaturated conjugated system of these saponins is involved in the creations of free radicals which persuade membrane damage over peroxidation of membranes in helminthes [130].

Aegle marmelos Corr. (Rutaceae) is well-known aimed at its ethanobotanical use for the treatment of Filariasis [131]. Methanol soluble extract of leaves at 100 ng/ml showed total loss of motility of *B. malayi* microfilariae after 48 hours, indicating the inhibition of the vital physiological process in larvae [132]. The leaves were stated to have polyphenolic and coumarins phytochemicals which display dose-dependent increase in lipid peroxidation levels and protein carbonylation. High degree of correlation coefficients was observed between the increase in the different oxidative parameters and the resultant loss of microfilarial motility [133].

Alnus nepalensis D. Don (Betulaceae) crude extract of the leaves showed antifilarial activity *in vitro* on adult worm [134]. *In vivo* assay revealed that chloroform, methanolic, and *n*-butanol soluble fraction produced >50%, 38–40% macrofilaricidal activity, respectively against *B. malayi* along with some sterilizing effect. The isolated compound diarylheptanoid exhibited antifilarial activity [135].

Water-based decoction of the leaves of *Andrographis paniculata* Burm. f. (Acanthaceae) was reported to exhibit antifilarial activity against *Dipetalonema reconditum* microfilariae in dogs [136] and mature worms of subperiodic *B. malayi* [137].

Plant extracts of *Asparagus adscendens* Roxb. (Liliaceae) presented antifilarial activity against *S. cervi*. Both aqueous and alcohol extracts triggered death of microfilariae *in vitro* [138].

Antifilarial activity was also reported for *Azadirachta indica* A. Juss. (Meliaceae) [139]. Flowers aqueous and alcohol extracts of *A. indica* were assessed *in vitro* for its antifilarial activity alongside whole worm, nerve muscle preparation and microfilariae of *S. cervi*. The response on the adult worm was characterized by an initial increase in rate, tone, and largeness of reductions along with reversible paralysis. There is an inhibition of impulsive movements tracked by reversible paralysis in nerve muscle preparation. Flower extract was likewise effective against filarial worm *Setaria degitata* [140]. Antifilarial effect of polyphenol rich ethanol soluble extract obtained from the leaves of *A. indica* was demonstrated against adults and microfilariae of *S. cervi*. A dose-dependent reduction in microfilariae in addition to adult worm feasibility was experimental. The extract caused alteration in morphological characters like thickening of epithelium, cuticle and muscle layers of the treated adult worms. Major apoptotic signs such as a higher quantity of chromatin condensation, presence of fragmented chromatin and brighter fluorescence were experiential in both microfilariae and adult of *S. cervi*. Where, ROS up rule also contributes in antifilarial efficacy [141].

Phytochemicals of galactolipid and catechin class were isolated from the *n*-butanol fraction of ethanol extract of the leaves of *Bauhinia racemosa* Lam. (Caesalpinaeceae). Among the active galactolipids, (2S)-1,2-di-*O*-linolenoyl-3-*O*-α-galactopyranosyl-(1/6)-*O*-β-galactopyranosyl glycerol) arose as the lead molecule which was active on the *B. malayi*, lymphatic filarial parasite [50].

Butea monosperma L. (Fabaceae) is a habitually used medicinal plant in various Ayurvedic formulations in India which exhibits promising adulticidal activity against intestinal worms [142]. Its leaves exhibit major antifilarial activity against adult *S. cervi* with inhibitory concentrations (IC_{50}) of 1.25 and 3.6 mg/ml methanol, hexane–ethanol extracts, respectively [143]. One more study against *B. malayi*, aqueous soluble extracts of roots and leaves showed significant activity [132]. Leaf extract in contrast to the roots displayed dose-dependent inhibition of motility of microfilariae [132]. Polyphenolic phytochemicals existing in leaves increase the levels of lipid peroxidation and protein carbonylation which is responsible for their mortality [133].

Leaves juice and seed powder of *Caesalpinia bonducella* L. (Caesalpiniaceae) is used for the worm infestations. The leaves have antifilarial effect [144]. *C. bonducella* seed kernel extract (2 g/kg) reveal microfilaricidal, macrofilaricidal, and female sterilizing effectiveness against *L. sigmodontis* and *B. malayi* in animal models [145].

Aqueous and ethanol extracts of *Cardiospermum halicacabum* Linn. (Sapindaceae) were reported for its antifilarial activity against *B. pahangi*. There was time and a concentration dependent reduction in motility of adult worms besides the form of release of microfilariae from the female worms [146]. *Cassia alata* Linn (Caesalpiniaceae) extracts were found to possess activity against the model bovine filarial parasite *S. digitata*. Wood extracts of *Cedrus deodara* Roxb. (Pinaceae) were found to possess an important macrofilaricidal activity against *S. digiata* [147].

Centratherum anthelminticum (Willd.) Kuntz (Asteraceae) is an annual herb with anthelmintic properties especially against threadworms. The principal constituent responsible for anthelmintic activity primarily originates in the fruits of the plant. The methanolic and aqueous extract of seed triggered inhibition of spontaneous motility of the adult worm and the nerve–muscle preparation of *S. cervi*, characterized by reduced tone, amplitude, and rate of contractions. Both the methanolic as well as the aqueous extracts cause the death of microfilariae *in vitro*, LC_{50} and LC_{90} being 75 and 32.5 mg/ml, respectively [148].

Antifilarial activity of the leaves methanol extract of the *Excoecaria agallocha* L. (Euphorbiaceae) at a very low dose of (10 μg/ml) revealed a dose-dependent relationship with induction of death of filarial parasite *S. digitata* in the developmental stages. Induction of death is due to decline of motility of microfilarae and loss of membrane integrity of both microfilariae and eggs. It is seen as a potential agent for blocking embryogenesis in filarial parasites in addition to reducing the oxidative stress generated during chronic manifestation of LF in human beings [149].

Fruit extracts of *Ficus racemosa* Linn. (Moraceae) significantly inhibit the impulsive movements of the worm. Both extracts caused death of microfilare *in vitro* at LC_{50} and LC_{90} of 21 and 35 ng/ml for alcoholic extract and 27 and 42 ng/ml for aqueous extract respectively [150].

Liquorice roots contain a unique triterpenic acid "glycyrrhetinic acid" as a major constituent in the form of saponin "glycyrrhizic acid." Glycyrrhetinic acid was inactive against adult worms but is effective against microfilariae of *B. malayi* *in vitro* (IC_{50}: 1.20 μM; LC_{100}: 12.5 μM). Synthetic amide analogs of glycyrrhetinic acid may exert *in vitro* and *in vivo* antifilarial activity [151].

Hibiscus mutabilis (Malvaceae) extract and its isolated compounds have antiparasitic activity [152]. Leaves methanolic extract besides ferulic acid isolated from *H. mutabilis* exhibited major antifilarial activity against together microfilariae and macrofilariae of *S. cervi*. Antifilarial effect of ferulic acid mediates through induction of apoptosis by generating oxidative stress and by downregulation and modification of the extent of some antioxidants like glutathione, glutathione-S-transferase, and superoxide dismutase of the filarial nematode *S. cervi* [153].

The crude extracts of the leaves of *Hibiscus sabdariffa* Linn. (Malvaceae) was tested against the human filarial parasite, *B. malayi in vitro*. The *n*-butanol fraction at 250 μg/ml killed 100% microfilariae. Leaf extract at 500 mg/kg (administered for five days) produced macrofilaricidal (about 30%) activity *in vivo* against *B. malayi* [152].

Crude extract prepared from the stem of *Lantana camara* Linn. (Verbenaceae) exhibits considerable *in vivo* antifilarial activity against *B. malayi*. *L. camara* stem extract administered at the dose of 1 g/kg for five days killed 43% of adults of the surviving female worms. All *B. malayi* worms were killed *in vitro* with extracts at concentration of 0.031 mg/ml [55]. The antiparasitic activity of *Leucas aspera* (Willd.) Linn. (Lamiaceae) was tested against the model bovine filarial parasite *S. digitata* [139]. Flower and stem extracts of *L. cephalotes* Spreng (Labiatae) hinder the impulsive mobility of filarial parasite *S. cervi*. The alcoholic extracts of flower and stem produced initial stimulation of the movements followed by paralysis of the entire worm. Alcoholic extract of flower produced reversible paralysis while that of stem produced irreversible paralysis [154].

In traditional medicine *Mallotus philippensis* (Lam.) Muell. Arg (Euphorbiaceae) is commonly used for the elimination of intestinal worms and, ringworm as well as freckles. Leaf aqueous and alcoholic extracts have antifilarial effect against *S. cervi* probably due to alteration of membrane permeability [154].

The gum extract (125 mg/ml) of *Moringa oleifera* Lam. (Moringaceae) causes irreversible loss of motility of microfilariae. *In vivo* study shows that the extract (500 mg/kg, p. o. for five days) causes 69% adulticidal and 83% female worm's sterilization in primary broadcast [155].

Neurolaena lobata Linn. (Asteraceae) extract has antifilarial effect on adult worm's motility of *B. pahangi* in a concentration and time-dependent manner. The concentrations of 10 μg/ml prevent microfilarial release by females [156].

Piper betle Linn. (Piperaceae) is reported for antiparasitic activity [157]. Adulticidal and female sterilizing efficacy was observed for crude methanol (100 mg/kg) alongside with its *n*-hexane fraction (30 mg/kg). Methanol extract (100 mg/kg) along with its *n*-hexane fraction significantly raises the antibody producing cells count and hemagglutinating antibody in addition to cell-mediated immune responses (lymphoproliferation, macrophage activation, and delayed type hypersensitivity) in mice [158].

The crude methano extract of the root of *Plumbago indica/rosea* Linn. (Plumbaginaceae) was screened for antifilarial activity against adults of the cattle filarial worm *S. digitata*. The lowest concentrations (0.01 mg/ml) of extract cause 83.3% of worms immobilization after six hours. The isolated compound (plumbagin) exhibited immobilization of worms at 0.0006 mg/ml [159].

The aqueous (250 μg/ml) and alcoholic (120 μg/ml) extracts of fruits and the leaves alcohol extract (270 μg/ml) of *Pongamia pinnata* Linn. (Fabaceae) cause inhibition of impulsive movements of the worm *S. cervi* [160].

Antifilarial effectiveness of alcohol extract of leaves and seeds of *Psoralea corylifolia* Linn. (Fabaceae) was confirmed on spontaneous movements of the entire worm and therefore the nerve muscle of *S. cervi* is characterized by initial, short lasting, small increase in tone of contractions followed by paralysis. Alcohol extracts of both leaves and seeds caused the death of microfilariae *in vitro*, LC_{50} and LC_{90} being 15 and 25 ng/ml for alcohol extract of leaves and 12 and 18 ng/ml for alcohol extract of seeds, respectively [154].

Ricinus communis Linn. (Euphorbiaceae) methanolic extract revealed antifilarial activity in a dose-dependent manner as apparent from initiation of death in the embryogenesis of filarial parasite *B. malayi*. The extract also shows dose-dependent inhibition of microfilariae motility [161].

Solamargine, a steroidal alkaloid glycoside from ripe berries of *Solanum khastanum*, has Antifilarial effect. Solamargine at 4 mg/ml concentration killed 100% microfilariae and adults of *S. cervi* in 88 and 60 minutes, respectively [128].

Bark aqueous extract of *Streblus asper* Lour. (Moraceae) has been described to exhibit considerable macrofilaricidal activity against *Litomosoides carinii* and *Brugia malayi* in rodents. *In vitro* and *in vivo* activity of two cardiac glycosides asperoside and strebloside at the dose of 50 mg/kg against *B. malayi* has been reported [162]. An effective and safe clinical efficacy filaricidal "Filacid" has been reported from the bark of *S. asper* [163]. Asperoside and strebloside have been stated for *in vitro* effects on *Setaria cervi* females. Both cause death of the worms, inhibition of motility and glucose uptake of the parasites. These C-glycosides also inhibit the incorporation of glucose into macromolecules of *Setaria cervi* females which results in lowered profiles of certain enzymes activities like glucokinase, malate dehydrogenase, and succinate dehydrogenase, which suggest that the lethal effects were because of significant effects on glucose metabolism [164]. The interference of asperoside and strebloside with glutathione metabolism of the adult *S. cervi* was also reported which consequences in the disruption in various vital activities of the parasites that ultimately result in the death of the parasites [165].

Dried stem of *Tinospora crispa* L. Hook.f. & Thomson (Menispermiaceae) exhibited strong antifilarial activity against adult worms of subperiodic *B. malayi* [166].

An *in vitro* study on motility in addition to reduction in formazan formation in MTT reduction assay using adult *S. digitata* worms displays that the extract and a phenolic monoterpene (2-isopropyl-5-methyl phenol with a position isomer 5-isopropyl-2-methyl phenol) isolated from *Trachyspermum ammi* Linn. (Apiaceae) have adulticidal property. Isolated phytochemicals were further investigated *in vivo* against the *B. malayi* in *M. coucha* rodent and shows significant macrofilaricidal and female sterility [167].

Root extract of *Vitex negundo* Linn. (Verbenaceae) at 100 ng/ml concentration showed complete loss of motility of *B. malayi* microfilariae *in vitro* after 48 hours of incubation [132, 168]. Leaf extract revealed in the *in vitro* system a promising adulticidal activity against adult filarial worm *S. cervi*. The treated worms got totally immobilized due to the lethal effects of the plant extracts at lower concentrations in a dose-dependent manner. MTT reduction assay on the worms treated with the drug shows significant inhibition of motility which is evident from the low levels of mitochondrial enzyme which reduces the MTT to formazan [169].

From *Withania somnifera* Dunal. (Solanaceae), Withaferin A, has been isolated and reported for its *in vivo* larvicidal effects at thc lowest concentration 7.8 μg/ml against *B. malayi*. It was also demonstrated that there was a reduction of 63.6% microfilariae and 66.2% defective embryogenesis in female worms [170].

Xylocarpus granatum Koenig (Meliaceae), dried seeds extract displayed antifilarial activity against adult worms of subperiodic *B. malayi* [166]. Gedunin and photogedunin, two compounds isolated from *X. granatum* fruit exhibited *in vitro* and *in vivo* antifilarial activity against *B. malayi*, human lymphatic filarial parasite. Gedunin (IC_{50} 0.239 μg/ml, CC_{50} 212.5 μg/ml) and photogedunin (IC_{50} 0.213 μg/ml, CC_{50} 262.3 μg/ml) at five daily doses of 100 mg/kg subcutaneously revealed excellent adulticidal efficacy resulting in 80 and 70% death of transplanted adult *B. malayi* [137], respectively.

Alcoholic extracts of *Zingiber officinale* Rosc. (Zingiberaceae) rhizomes at 100 mg/kg condensed concentration of microfilariae in blood by extreme of 98% in dogs infected with *D. immitis* [167].

3.3.3 Plants and Phytochemicals Effective Against Trypanosomasis

The assessment of antitrypanosomal activity of traditionally used plants has been a major area of contemporary research focus. Moreover, NPs are supposed to have major advantages as lead molecules over synthetic molecules [171]. Natural phytochemicals, for example, quinones, alkaloids, terpenes, and polyphenols originating from these extracts have been revealed to be strong growth inhibitors of *T. cruzi* [172]. Triterpenoids and sterols from the plants are described to have antitrypanosomal activity [173]. Antitrypanosomal activities of alkaloids like actinodaphine, dicentine, cassythine isolated from *Cassytha filiformis* [173]. The trypanocidal activity of numerous flavonoids such as quercetagetin [174] has been beforehand reported. An azaanthraquinone early reported in *A. nilotica* was associated with the observed antitrypanosomal effects [175]. The criteria for selecting a specific natural product for studies are either supported by preexisting knowledge on the normal use of the source species in therapy (ethnobotanical knowledge) or the look for structurally related molecules with known pharmacologically active agents from chemical databases [171, 176–179]. Alkaloids are categorized by nitrogen-containing phytochemicals that are naturally occurring. Diversity of fungi, plants, and animal species have produced numerous bioactive alkaloids against a wide variety of diseases [180–184].

Leaves, stem bark, and roots of *Ancistrocladus* sp. (Ancistrocladaceae) are recognized to be bases of naphthylisoquinoline alkaloids [185, 186]. Ancistrocladidine, ancistrotanzanines B and C, ancistrotectoriline A, *O*-methylancistrocladidine, ancistroealaines A and B, ancistrocongolines A-D, ancistrocladiniums A and B (13), *N*-6′-epi-ancistrocladinium B, ancistrolikokine D, and dioncophylline E are examples of naphthylisoquinoline antiparasitic alkaloids from *Ancistrocladus* sp. and *Dioncophyllum thollonii* (Dioncophyllaceae). The assessment of the biological activities of these compounds presented them to be an occasional set and hopeful class of antiprotozoal and antiviral agents that are only found in plants of the Dioncophyllaceae and Ancistrocladaceae, mostly from Africa. They possess antitrypanosomal activities against *Trypanosoma brucei rhodesiense*, together with good to reasonable activities against *Trypanosoma cruzi* and *Leishmania donovani*. It might be worth mentioning that the isoquinoline gibbet has also been explored unnaturally for the detection of novel antiprotozoals and antimicrobials [187, 188]. Other bioactive alkaloids containing the aporphines; actinodaphnine and cassythine from *Cassytha filiformis* (Lauraceae), a plant whose alkaloidal extract presented activity against *T. b. brucei* (with an IC_{50} value of 2.2 μg/ml). This confirms the use of this plant in African folkloric medicine to treat African Trypanosomiasis and other diseases [174, 189]. The compounds displayed antitrypanosomal activities, with IC_{50} values of 10.29–17.60 μM. They represent good preliminary support that could be optimized so as to improve the efficiency and discrimination in the search for new bioactive molecules with trypanocidal effects. Other trypanocidal alkaloids include the quinolines; waltheriones E–L, 8-deoxoantidesmone and antidesmone from *Waltheria indica* (Malvaceae) [187]. The root extract showed activities against *T. cruzi* (IC_{50} = 0.74 μg/ml), *T. b. brucei* (2.3% survival at 20 μg/ml), and *T. b. rhodesiense* (IC_{50} = 17.4 μg/ml) [187]. Through the isolation of waltherione L, with a slightly higher IC_{50} (3.1 μM), showed potent growth inhibition to the amastigote form of *T. cruzi* , with IC_{50} values lower than that of the reference drug benznidazole (IC_{50} = 2.9 μM). Structure–activity relationships provide suggestions that, a methoxy group, bound to the nitrogen atom is important for activity. This group at this position increased the lethality of *T. cruzi*. Lastly, an assessment of the IC_{50} values of the isolated combinations against *T. brucei* sp. and *T. cruzi* revealed selective toxicity toward the latter. This recommends that the waltherione scaffold is a possible initial peak for new safe anti-trypanocidal drug progress, though antidesmone has already been original for its potential as an antiprotozoal drug since 2003 [187, 190].

An indolosesquiterpene alkaloid polysin, from *Polyalthia suaveolens* (Annonaceae), was isolated with the recognized alkaloids; greenwayodendrin-3-one, 3-*O*-acetyl greenwayodendrin, *N*-acetyl polyveoline, and polyveoline. These alkaloids have exhibited motivating activities on selected glycolytic enzymes, e.g. glyceraldehyde-3-phosphate dehydrogenase (GAPDH) phosphofructo kinase (PFK), and aldolase [188]. Of specific notice are polysin and 3-*O*-acetyl greenwayodendrin. Polysin deeds as a competitive reversible inhibitor against *T. brucei* PFK (K_i = 10 μM), while acetyl greenwayodendrin acts as a discriminating inhibitor of *T. brucei* aldolase (with IC_{50} ~ 0.5 μM). Meanwhile, polyveoline proceeded as a selective inhibitor of *T. brucei* PFK and is a diverse rescindable inhibitor of *T. brucei* GAPDH. These phytochemicals, consequently, characterize a good preliminary point for the proposal of new selective and potent antitrypanosomal drugs.

Phytochemical terpenoids constitute a great and varied class of certainly occurring secondary metabolites, with attractive physiological and pharmacological functions [191, 192]. The sesquiterpenoids, muzigadiolide, muzigadial, 6α, 9α-dihydroxy-4(13),7-coloratadiene-11,12-dial, mukaadial, and ugandensidial from *Warburgia ugandensis* (Canellaceae), the East African medicinal plant have shown anti-*Trypanosoma* activities [193]. The compounds showed *in vitro* activities (with IC_{50} values reaching from 0.6 to 6.4 μM) against *T. b. rhodesiense*, the parasite liable for African sleeping sickness. Mukaadial had been isolated before, likewise viewing antitrypanocidal activity [194]. An extra dialdehyde functional group to the sesquiterpene lactone backbone, together through a hydroxyl attached to C-9 is suggested to donate to the activity of the compounds.

Sphaeranthus bullatus (Asteraceae) has been the origin of several compounds [195–197], including the carvotacetone derivatives; 3-acetoxy-7-hydroxy-5-tigloyloxycarvotacetone, 3,7-dihydroxy-5-tigloyloxycarvotacetone, and 3-aceto xy-5,7-dihydroxycarvotacetone,which demonstrate anti-leishmanial activities, with IC_{50} values of 2.5, 10.6, and 2.9 μM, respectively, against the parasite *L. donovanii* promastigotes. Additional terpenoids contain the abietane diterpenoids, uncinatone and taxodione, from the roots of *Clerodendrum eriophyllum* (Verbenaceae) [195], which displayed potent anti-leishmanial activities (with IC_{50} values of 0.25 and 0.61 μM, respectively) against *L. donovanii*. The actions of the crude extracts, e.g. the ethyl acetate extracts of *Newbouldia laevis* (Bignoniaceae) and *Eucalyptus maculata* (Myrtaceae) and the hexane extract of *Polyalthia longifolia* (Annonaceae) as well as their isolated active compounds; 16-α-hydroxy-cleroda-3-13(-14)z-dien-15,16-olide, kolavenic acid, polyalthiadioic acid and the triterpenoid 3*β*,13*β*-dihydroxy-urs-11-en-28-oic acid were observed against different trypanosomes strains (s427 WT, B48, and AQP2/3KO [198]. The kolavic acid derivative; monomethyl ester-15-kolavic acid demonstrated interesting selective inhibitory activity (IC_{50} value of 12 μM) against *T. brucei* which is isolated from *Entada abyssinica* (Fabaceae) [199]. Other bioactive diterpenoids include 15-oxo-*ent*-kaur-16-en-19-oic acid, 15α-acetoxy-*ent*-kaur-16-en-19-oic acid or xylopic acid, and 16,17-epoxy-15-oxo-*ent*-k auran-19-oic acid or xylopioxyde, from the fruits of *Xylopia aethiopica* (Annonaceae) [200].

The genus *Aframomum* (Zingiberaceae), has become a source of antitryposonomals. Sceptrumlabdalactone B, derived from the rhizomes of *A. sceptrum*, was used to treat infectious diseases including Trypanosomiasis of the African people (sleeping sickness), and sceptrumlabdalactone A [201–203]. The activity of Sceptrumlabdalactone B (with IC_{50} value of 5.7 μM) against *L. donovani* was similar to that of reference drugs (IC_{50} of 2.5 and 3.0 μM expected at pentamidine and miltefosine, individually). Additionally, this molecule demonstrated selective activity for *L. donovani*, when compared with the activity against *T. b. brucei*.

The antischistosomal activity of Asparagalin A derived from *Asparagus stipularis* (Asparagaceae) has been tested [204]. It was found that that it reduced the potency of older female worms that lay eggs. Also, it affects the egg-laying capacity in the form of volume dependence [205]. *Elaeodendron schlechteranum* (Celastraceae) contains tingenin B or 22β-hydroxytingenone [206]. This compound has displayed an extensive diversity of activities against *T. cruzi* (IC_{50} < 0.57 μM), *T. brucei* (<0.57 μM), *L. infantum* (1.67 μM), and *P. falciparum* (0.83 μM), confirming the claim of the applicability of the plant in traditional medicine to treat various noninfectious diseases [206].

To treat a variety of infections, including parasitic infections *Plectronia leucantha* Krause (Rubiaceae) was used [206–208]. Ursolic acid and oleanolic acid, all along with other constituents were predominant from the leaves of this plant. An examination of the antitrypanosomal activities of essential oil, the dichloromethane extract, and isolated compounds on *T. b. brucei* bloodstream forms (*Tbb* BSF) and procyclic forms (*Tbb* PF) [209] showed that ursolic acid and oleanolic acid were the greatest bioactive verified compounds [209]. Ursolic acid displayed IC_{50} values of 5.48 and 14.25 μM, respectively, on *Tbb* BSF and *Tbb* PF, while oleanolic acid displayed an IC_{50} value of 16.00 μM on *Tbb* BSF. This could explain why the plant is effective in the traditional treatment of related parasitic ailments. Another identified triterpenoid was polycarpol or lanosta-7,9(11),24-trien-3β,15α-diol from *Piptostigma preussi* (Annonaceae) [210]. The compound presented antitrypanosomal activity by an ED_{50} value of 5.11 μM on *T. brucei* cells. An examination of its manner of action displayed that the compound was replaced by inhibiting *T. brucei* glycolytic enzymes GAPDH and PFK (glycolytic pathway enzymes validated by WHO as good boards for the expansion of drugs against trypanosomiasis), with IC_{50} values of 650 and 180 μM, respectively. Glycolytic enzymes GAPDH are accountable for ATP creation and have been stated to be vital for the survival of Trypanosomatids [207]. From the stem bark of *Vernonia guineensis* (Asteraceae), vernoguinosterol, and vernoguinoside, showed stimulating trypanocidal activity with IC_{50} values in the range 4.60–7.67 μM [211].

Plants from the genus *Zapoteca* (Fabaceae) have been the origin of diverse compounds with antiprotozoal activities as aurantiamide acetate or saropeptide from *Z. portoricensis* [212]. The IC_{50} values of saropeptide were 3.6, 41.7, and 92.0 μM against *T. b. rhodesiense*, *T. cruzi*, and rat skeletal myoblast cell line (L6 cells) respectively. Since irritation poses major difficulties in the progressive stages of Trypanosomiasis, saropeptide signifies a promising natural hit with sensitive effects against *T. b. rhodesiense*.

Diarylheptanoid; letestuianin C and (4*Z*,6*E*)-5-hydroxy-1,7-bis(4-hydroxyphenyl) hepta-4,6-dien-3-one are additional effective antitrypanosomal compounds from the species of *Aframomum letestuianum* [204].

Guttiferone A, garcinol, cambogin, and guttiferone F from *Allanblackia monticola* (Clusiaceae or Guttiferae) fruits and xanthone V1 from *Symphonia globulifera* leaves have shown anti-leishmanial activities [213]. These compounds showed very powerful *in vitro* anti-leishmanial activities, mainly guttiferone A, garcinol, and cambogin, with lower IC_{50} values 0.2, 0.3, and 0.2 μM respectively compared to that of the reference drug, miltefosine (0.46 μM).

A fairly powerful class of antitrypanosomal compounds (taccalonolide A 12-propanoate, taccalonolide T, and taccalonolide S) isolated from *Tacca leontopetaloides* (Taccaceae) exhibits activities against *T. b. brucei* s427 lister strain [214, 215].

Artemetin, from *Vitex simplicifolia* (Verbenaceae) leaves, exhibits promising trypanocidal activity with an IC_{50} value of 4.7 μg/mL and an SI of 9.8 against L6 cells [216]. This activity confirms the effect of *V. simplicifolia* in the traditional treatment of ailments including parasitic diseases. An investigation of *Ageratum conyzoides* (Asteraceae), a plant known for its importance in the management of sleeping sickness traditionally [217], found that it contains numerous flavonoids; eupalestin, 5′-methoxynobiletine, ageconyflavone C, which exhibit antiprotozoal activities, some in the lower micromolar range. Among the tested NPs, ageconyflavone C showed the highest activity against *T. b. rhodesiense.*

From the stem bark of *Allexis cauliflora* (Violaceae), 22-hydroxyclerosterol, and clerosterol were isolated [190] and evaluated for trypanocidal activities, and the activity of 22-Hydroxyclerosterol ($ED_{50} = 1.12\,\mu M$) was found better than that of clerosterol ($ED_{50} = 134.34\,\mu M$). This difference in activity could be attributed to the presence of hydroxyl group at C-22 in the side chain of 22-Hydroxyclerosterol which is absent in clerosterol. Additionally, it was detected that 22-Hydroxyclerosterol when compared to clerosterol was more active and selective on the parasite enzyme glycolytic enzymes (PGI and GAPDH).

3.3.4 Plants and Phytochemicals Effective Against Schistosomiasis

Praziquantel is a heterocyclic pyrazine–isoquinoline which is very active against a good range of trematodes, including all species of human pathogenic schistosomes. Praziquantel although generally well known, may induce abdominal discomfort, bloody diarrhea, nausea, headache, dizziness, urticaria, and rectal bleeding in patients with heavy worm loads [218]. Resistance to praziquantel used for the treatment of schistosomiasis has been reported in Egypt where the drug has been used aggressively for more than 10 years [219]. Again, metrifonate, a well-tolerated and extremely effective drug against *S. haematobium* is used mostly in mass chemotherapy programs [220], probably due to its low cost and low side effects. However, the side effects of this drug when used for mass chemotherapy in communities recently exposed to insecticides or other agricultural chemicals with an anticholinesterase action includes abdominal pain, nausea, vomiting, diarrhea, headache, and vertigo. Similarly, Oxamniquine, a tetrahydroquinoline derivative is employed for the treatment of schistosomiasis. The major limitation of this drug is its side effects which includes dizziness, drowsiness, headache, seizures, and increase in the level of serum transaminases in some patients [221].

Again, the use of artemisinin derivative against schistosomiasis is in question. The reason being that its large-scale use might provoke resistance in malaria parasites. Although this risk appears to be low, mainly due to the very short elimination half-life of artemisinins. It is recommended that it should not be used in areas where both schistosome and malaria parasites coexist [222].

As earlier noted, natural products especially plant extracts and their metabolites have been used traditionally for the treatment of parasitic infections and other diseases. Some herbs used in the treatment of parasitic infections includes Dryopteris filix-mas (*Dryopteris filix mas*) root, tansy (*Tanacetum vulgare*) leaf, wormwood, sweet annie, black walnut (*Juglans nigra*) fruit, and cloves (*Syzygium aromaticum*) [223]. Some plants with antischistosomal activities were presented in Table 3.4.

Table 3.4 Some plants with antischistosomal activities.

Plant	Activities	References
Allium sativum (garlic)	Antiparasitic, treatment of roundworm, pinworm, and hookworm	[224, 225]
Chenopodium ambrosioides now known as *Dysphania ambrosioides* (Wormseed)	Treatment of worm infections in Mexico. A toxic plant	[226]
Cucurbita pepo (Pumpkin seed)	Active against tapeworms	[227]
Olive leaf	Antibacterial, antioxidant, and hypoglycemic. Exhibits activities against malaria, lymphatic disorders, and Schistosomiasis	[228]
Vernonia amygdalina Del (family Compositae)	Schistosomiasis	[229]
Wood and bark of the pau d'arco	Antimalarial, antibacterial, anti-inflammatory, antifungal, and antiparasitic activities	[230]
Mentha crispa (Curled mint)	Relieves the symptoms of giardia and amoeba infections	[231]
Commiphora molmol extract (Mirazid)	Anti-fasciolicidal drug, antischistosomal activity	[232–237]
Ailanthus altissima stem bark	Improves organs (liver–kidney–spleen) damage caused by parasitic infections	[238]
Ziziphus spina-christi root	Antischistosomal activities	[239, 240]

3.4 Future Prospects

Appropriate diagnosis to monitor the incidence of disease and assess the impact of the intervention is important in directing NTD treatment strategies in various control areas, transmission disorders, elimination, and post-elimination surveillance. The contraction of the worldwide pharmaceutical industry in recent years has arguably been a positive factor for work on NTDs, as experienced drug-discovery scientists have migrated from the private sector into academia as a result. Quite unimaginable five years ago, there exists now a large pool of experienced and brilliant medicinal chemists who can be willingly recruited to work on drug leads for NTDs. Most important, however, has been the support of charity for both the funding and improvement of drugs for the treatment of NTDs. Given the long timelines inherent in drug development, it is perhaps too soon to take a full account of the ultimate success of this philanthropic support. The next five years are sure to see more new chemical entities targeting the management of NTDs progressing through human clinical trials. Important drivers of early-stage discovery are going to be continued in basic research and in parasite biology, including the character of host-parasite interactions. The number of people being treated for untreated tropical diseases is on the rise, with one billion people being provided with antiretroviral drugs by 2015, and some countries being able to announce the elimination of certain diseases. With the enormous number of phytochemicals in existence [241–247], it is recommended that researchers explore these natural chemical entities which has served in one way or the other as lead compounds for the development of different kinds drugs against diseases caused by virus, bacteria, fungi, among others.

3.5 Conclusion

This chapter highlights the importance of in-depth research into herbal remedies to strengthen the range of antiparasitic treatments. Medicinal plants represent a potential source of new drugs. Natural products from animals such as mammals, fungi, and high-grade plants have long been known as good sources of chemotherapy to fight a number of diseases, including infectious diseases. The identification of drug targets and/with wealth of information available from the pharmaceutical industry will have a significant impact on the speed of drug development that will be used to treat NTDs. In addition, new biological and host-parasite–drug interaction studies will be a powerful discovery that will provide greater improvements in drug performance and safety. Incentives that result in the involvement of the pharmaceutical industry within the development of medicines against these new goals should be the main goal of the government and donor support agencies to help promote the rapid and effective development of well-designed therapies against NTDs.

References

1 WHO (2013). *Why Are Some Tropical Diseases Called "Neglected"?* World Health Organization http://www.who.int/features/qa/58/en (

2 Weng, H., Chen, H.B., Chen, H.X., and Wang, M.W. (2018). Innovation in neglected tropical disease drug discovery and development. *Infect. Dis. Poverty* 7: 67. https://doi.org/10.1186/s40249-018-0444-1.

3 Nwaka, S., Besson, D., Ramirez, B. et al. (2011). Integrated dataset of screening hits against multiple neglected disease pathogens. *PLoS Negl. Trop. Dis.* 5 (12): e1412.

4 Stuart, K., Brun, R., Croft, S. et al. (2008). Kinetoplastids: related protozoan pathogens, different diseases. *J. Clin. Invest.* 118: 1301–1310.

5 Fevre, E.M., Picozzi, K., Jannin, J. et al. (2006). Human African trypanosomiasis: epidemiology and control. *Adv. Parasitol.* 61: 167–221. https://doi.org/10.1016/S0065-308X(05)61005-6.

6 Bern, C., Maguire, J.H., and Alvar, J. (2008). Complexities of assessing the disease burden attributable to leishmaniasis. *PLoS Negl. Trop. Dis.* 2 (10): e313. https://doi.org/10.1371/journal.pntd.0000313.

7 WHO (2013). *Trypanosomiasis, Human African (Sleeping Sickness).* World Health Organization http://www.who.int/mediacentre/factsheets/fs259/en (

8 Yamey, G. and Torreele, E. (2002). The world's most neglected diseases. *BMJ* 325: 176–177.

9 Trouiller, P., Olliaro, P., Torreele, E. et al. (2002). Drug development for neglected diseases: a deficient market and a public-health policy failure. *Lancet* 359: 2188–2194.

10 Werneck, G.L., Hasselmann, M.H., and Gouvêa, T.G. (2011). An overview of studies on nutrition and neglected diseases in Brazil. *Cien. Saude Colet.* 16: 39–62.

11 DNDi (2013). Diseases and projects. Drugs for Neglected Diseases initiative. http://www.dndi.org/diseases-projects/diseases.html (accessed August 2013).

12 Pinto, M.C., Barbieri, K., Silva, M.C.E. et al. (2011). Octenol as attractant to *Nyssomyia neivai* (Diptera: Psychodidae: Phlebotominae) in the field. *J. Med. Entomol.* 48: 39–44.

13 Goto, H. and Lindoso, J.A. (2010). Current diagnosis and treatment of cutaneous and mucocutaneous leishmaniasis. *Expert Rev. Anti Infect. Ther.* 8: 419–433.

14 Goto, H. and Lindoso, J.A. (2012). Cutaneous and mucocutaneous leishmaniasis. *Infect. Dis. Clin. North Am.* 26: 293–307.

15 Balasegaram, M., Ritmeijer, K., Lima, M.A. et al. (2012). Liposomal amphotericin B as a treatment for human leishmaniasis. *Expert Opin. Emerg. Drugs* 17: 493–510.

16 Den Boer, M., Argaw, D., Jannin, J., and Alvar, J. (2011). Leishmaniasis impact and treatment access. *Clin. Microbiol. Infect.* 17: 1471–1477.

17 Dorlo, T.P.C., Balasegaram, M., Beijnen, J.H., and de Vries, P.J. (2012). Miltefosine: a review of its pharmacology and therapeutic efficacy in the treatment of leishmaniasis. *J. Antimicrob. Chemother.* 67: 2576–2597.

18 Sundar, S. and Rai, M. (2002). Advances in the treatment of leishmaniasis. *Curr. Opin. Infect. Dis.* 15: 593–598.

19 Singh, S. and Sivakumar, R. (2004). Challenges and new discoveries in the treatment of leishmaniasis. *J. Infect. Chemother.* 10: 307–315.

20 Sundar, S. (2001). Drug resistance in Indian visceral leishmaniasis. *Trop. Med. Int. Health* 6: 849–854.

21 Maltezou, H.C. (2010). Drug resistance in visceral leishmaniasis. *J. Biomed. Biotechnol.* https://doi.org/10.1155/2010/617521.

22 Bhandari, V., Kulshrestha, A., Deep, D.K. et al. (2012). Drug susceptibility in leishmania isolates following miltefosine treatment in cases of visceral leishmaniasis and post kala-azar dermal leishmaniasis. *PLoS Negl. Trop. Dis.* 6: e1657.

23 Croft, S.L. and Olliaro, P. (2011). Leishmaniasis chemotherapy – challenges and opportunities. *Clin. Microbiol. Infect.* 17: 1478–1483.

24 WHO (2010). *Control of the Leishmaniases. Report of a Meeting of the WHO Expert Committee on the Control of Leishmaniases.* Geneva, Swiss: World Health Organization.

25 van Griensven, J. and Diro, E. (2012). Visceral leishmaniasis. *Infect. Dis. Clin. North Am.* 26: 309–322.

26 Singh, N., Kumar, M., and Singh, R.K. (2012). Leishmaniasis: current status of available drugs and new potential drug targets. *Asian Pac. J. Trop. Med.* 5: 485–497.

27 WHO (2010). *Working to Overcome the Global Impact of Neglected Tropical Diseases. First WHO Report on Neglected Tropical Diseases.* World Health Organization http://whqlibdoc.who.int/publications/2010/9789241564090_eng.pdf (accessed August 2013).

28 Osorio, Y., Travi, B.L., Renslo, A.R. et al. (2011). Identification of small molecule lead compounds for visceral leishmaniasis using a novel ex vivo splenic explant model system. *PLoS Negl. Trop. Dis.* 5 (2): e962.

29 Donato, M.T., Lahoz, A., Castell, J.V., and Gomez-Lechon, M.J. (2008). Cell lines: a tool for in vitro drug metabolism studies. *Curr. Drug Metab.* 9: 1–11.

30 Scheers, E.M., Ekwall, B., and Dierickx, P.J. (2001). In vitro long-term cytotoxicity testing of 27 MEIC chemicals on Hep G2 cells and comparison with acute human toxicity data. *Toxicol. In vitro* 15: 153–161.

31 Schoonen, W.G., de Roos, J.A., Westerink, W.M., and Debiton, E. (2005). Cytotoxic effects of 110 reference compounds on HepG2 cells and for 60 compounds on HeLa, ECC-1 and CHO cells. II mechanistic assays on NAD(P)H, ATP and DNA contents. *Toxicol. In vitro* 19: 491–503.

32 Gerets, H.H., Hanon, E., Cornet, M. et al. (2009). Selection of cytotoxicity markers for the screening of new chemical entities in a pharmaceutical context: a preliminary study using a multiplexing approach. *Toxicol. In vitro* 23: 319–332.

33 Dambach, D.M., Andrews, B.A., and Moulin, F. (2005). New technologies and screening strategies for hepatotoxicity: use of in vitro models. *Toxicol. Pathol.* 33: 17–26.

34 Ndjonka, D., Rapado, L.N., Silber, A.M. et al. (2013). Natural products as a source for treating neglected parasitic diseases. *Int. J. Mol. Sci.* 14 (2): 3395–3439.

35 Molyneux, D.H., Hotez, P.J., and Fenwick, A. (2005). "Rapid-impact interventions": how a policy of integrated control for Africa's neglected tropical diseases could benefit the poor. *PLoS Med.* 2 (11): e336.

36 Fenwick, A., Molyneux, D., and Nantulya, V. (2005). Achieving the millennium development goals. *Lancet* 365 (9464): 1029–1030. https://doi.org/10.1016/0140-6736(05)71134-X.

37 Michael, E. and Bundy, D.A. (1997). Global mapping of lymphatic filariasis. *Parasitol. Today* 13 (12): 472–476.

38 Zagaria, N. and Savioli, L. (2002). Elimination of lymphatic filariasis: a public-health challenge. *Ann. Trop. Med. Parasitol.* 96 (Suppl 2): S3–S13.

39 WHO (2008). Global programme to eliminate lymphatic filariasis. *Wkly. Epidemiol. Rec.* 83 (37): 333–341.

40 Ottesen, E.A. (2006). Lymphatic filariasis: treatment, control and elimination. *Adv. Parasitol.* 61: 395–441.

41 Ottesen, E.A., Hooper, P.J., Bradley, M., and Biswas, G. (2008). The global programme to eliminate lymphatic filariasis: health impact after 8 years. *PLoS Negl. Trop. Dis.* 2 (10): e317. https://doi.org/10.1371/journal.pntd.0000317.

42 Njenga, S.M., Wamae, C.N., Njomo, D.W. et al. (2007). Chronic clinical manifestations related to wuchereria bancrofti infection in a highly endemic area in Kenya. *Trans. R. Soc. Trop. Med. Hyg.* 101 (5): 439–444. https://doi.org/10.1016/j.trstmh.2006.09.006.

43 Gyapong, J.O., Gyapong, M., Evans, D.B. et al. (1996). The economic burden of lymphatic filariasis in northern Ghana. *Ann. Trop. Med. Parasitol.* 90 (1): 39–48.

44 Evans, D.B., Gelband, H., and Vlassoff, C. (1993). Social and economic factors and the control of lymphatic filariasis: a review. *Acta Trop.* 53 (1): 1–26.

45 Hooper, P.J., Chu, B.K., Mikhailov, A. et al. (2014). Assessing progress in reducing the at-risk population after 13 years of the global programme to eliminate lymphatic filariasis. *PLoS Neglected Tropical Diseases 8* (11): e3333.

46 Babu, S. and Thomas, B.N. (2012). Immunopathogenesis of lymphatic filarial disease. *Semin. Immunopathol.* 34 (6): 847–861.

47 Molyneux, D. (2003). Lymphatic filariasis (elephantiasis) elimination: a public health success and development opportunity. *Filaria J.* 2: 13.

48 Ichimori, K. and Ottesen, E.A. (2011). Eliminating lymphatic filariasis. *Bol. Med. Hosp. Infant. Mex.* 68 (2): 134–137.

49 Gayen, P., Nayak, A., Saini, P. et al. (2013). A double-blind controlled field trial of doxycycline and albendazole in combination for the treatment of bancroftian filariasis in India. *Acta Trop.* 125: 150–156.

50 Sashidhara, K.V., Singh, S.P., Misra, S. et al. (2012). Galactolipids from Bauhinia racemosa as a new class of antifilarial agents against human lymphatic filarial parasite, Brugia malayi. *Eur. J. Med. Chem.* 50: 230–235.

51 Babu, B.V., Rath, K., Kerketta, A.S. et al. (2006). Adverse reactions following mass drug administration during the programme to eliminate lymphatic filariasis in Orissa State, India. *Trans. R. Soc. Trop. Med. Hyg.* 100 (5): 464–469.

52 Dhananjeyan, M.R., Milev, Y.P., Kron, M.A., and Nair, M.G. (2005). Synthesis and activity of substituted anthraquinones against a human filarial parasite, Brugia malayi. *J. Med. Chem.* 48: 2822–2830.

53 Srivastava, S.K., Chauhan, P.M.S., Bhaduri, A.P. et al. (2000). Quinolones: novel probes in antifilarial chemotherapy. *J. Med. Chem.* 43: 2275–2279.

54 Das, N.G., Goswami, D., and Rabha, B. (2007). Preliminary evaluation of mosquito larvicidal efficacy of plant extracts. *J. Vector Borne Dis.* 44 (2): 145–148.

55 Misra, N., Sharma, M., Raj, K., and Bhattacharya, S.M. (2007). Chemical constituents and antifilarial activity of Lantana camara against human lymphatic filariid Brugia malayi and rodent filariid *Acanthocheilonema viteae* maintained in rodent hosts. *Parasitol. Res.* 100 (3): 439–448.

56 Simarro, P.P., Diarra, A., Postigo, J.A.R. et al. (2011). The human African trypanosomiasis control and surveillance programme of the World Health Organization 2000–2009: the way forward. *PLoS Negl. Trop. Dis.* 5: 1007. https://doi.org/10.1371/journal.pntd.0001007.

57 MSF (2008). Relatório anual 2008. Medicina Sem Fronteiras. https://www.msf.org.br/arquivos/Doc/Publicacoes/83.pdf (accessed November 2013).

58 Gómez-Junyent, J., Pinazo, M.J., Castro, P. et al. (2017). Human African trypanosomiasis in a Spanish traveler returning from Tanzania. *PLoS Negl. Trop. Dis.* 11 (3): e0005324. https://doi.org/10.1371/journal.pntd.0005324.

59 Dumas, M. and Girard, P.L. (1979). Sleeping sickness. *Trends Neurosci.* 2: 256–258.

60 Kennedy, P.G.E. (2008). The continuing problem of human African trypanosomiasis (sleeping sickness). *Ann. Neurol.* 64: 116–126.

61 Jacobs, R.T., Nare, B., and Phillips, M.A. (2011). State of the art in African trypanosome drug discovery. *Curr. Top. Med. Chem.* 11: 1255–1274.

62 Brun, R., Don, R., Jacobs, R.T. et al. (2011). Development of novel drugs for human African trypanosomiasis. *Future Microbiol.* 6: 677–691.

63 Steinmann, P., Keiser, J., Bos, R. et al. (2006). Schistosomiasis and water resources development: systematic review, meta-analysis, and estimates of people at risk. *Lancet Infect. Dis.* 6 (7): 411–425. https://doi.org/10.1016/S1473-3099(06)70521-7.

64 Beck-Worner, C., Raso, G., Vounatsou, P. et al. (2007). Bayesian spatial risk prediction of schistosoma mansoni infection in western Cote d'Ivoire using a remotely-sensed digital elevation model. *Am. J. Trop. Med. Hyg.* 76 (5): 956–963.

65 Kabatereine, N.B., Brooker, S., Tukahebwa, E.M. et al. (2004). Epidemiology and geography of schistosoma mansoni in Uganda: implications for planning control. *Trop. Med. Int. Health* 9 (3): 372–380.

66 Ekpo, U.F., Mafiana, C.F., Adeofun, C.O. et al. (2008). Geographical information system and predictive risk maps of urinary schistosomiasis in Ogun state, Nigeria. *BMC Infect. Dis.* 8: 74. https://doi.org/10.1186/1471-2334-8-74.

67 Fenwick, A., Rollinson, D., and Southgate, V. (2006). Implementation of human schistosomiasis control: challenges and prospects. *Adv. Parasitol.* 61: 567–622. https://doi.org/10.1016/S0065-308X(05)61013-5.

68 Mangal, T.D., Paterson, S., and Fenton, A. (2008). Predicting the impact of long-term temperature changes on the epidemiology and control of schistosomiasis: a mechanistic model. *PLoS One* 3 (1): e1438.

69 van der Werf, M.J., de Vlas, S.J., Brooker, S. et al. (2003). Quantification of clinical morbidity associated with schistosome infection in sub-Saharan Africa. *Acta Trop.* 86 (2–3): 125–139.

70 Maxwell, P.D. (2008). The global burden of urinary bladder cancer. *Scand. J. Urol. Nephrol.*: 1–9. https://doi.org/10.1080/03008880802285032.

71 King, C.H. and Dangerfield-Cha, M. (2008). The unacknowledged impact of chronic schistosomiasis. *Chronic Illn.* 4 (1): 65–79. https://doi.org/10.1177/1742395307084407.

72 Swai, B., Poggensee, G., Mtweve, S., and Krantz, I. (2006). Female genital schistosomiasis as an evidence of a neglected cause for reproductive ill-health: a retrospective histopathological study from Tanzania. *BMC Infect. Dis.* 6: 134. https://doi.org/10.1186/1471-2334-6-134.

73 Bradley, M. (1990). Rate of expulsion of necator americanus and the false hookworm ternidens deminutus railliet and henry 1909 (nematoda) from humans following albendazole treatment. *Trans. R. Soc. Trop. Med. Hyg.* 84 (5): 720.

74 Tchuem Tchuente, L.A., Southgate, V.R., Jourdane, J. et al. (2003). Schistosoma intercalatum: an endangered species in Cameroon? *Trends Parasitol.* 19 (9): 389–393.

75 King, C.H., Dickman, K., and Tisch, D.J. (2005). Reassessment of the cost of chronic helmintic infection: a meta-analysis of disability-related outcomes in endemic schistosomiasis. *Lancet* 365 (9470): 1561–1569. https://doi.org/10.1016/S0140-6736(05)66457-4.

76 Kjetland, E.F., Ndhlovu, P.D., Gorno, E. et al. (2006). Association between genital schistosomiasis and HIV in rural Zimbabwean women. *AIDS* 20: 593–600.

77 Fournet, A., Gantier, J.C., Gautheret, A. et al. (1994). The activity of 2-substituted quinoline alkaloids in BALB/c mice infected with *Leishmania donovani*. *J. Antimicrob. Chemother.* 33: 537–544.

78 Dietze, R., Carvalho, S.F., Valli, L.C. et al. (2001). Phase 2 trial of WR6026, an orally administered 8-aminoquinoline, in the treatment of visceral leishmaniasis caused by Leishmania chagasi. *Am. J. Trop. Med. Hyg.* 65: 685–689.

79 Koyama, J. (2006). Anti-infective quinone derivatives of recent patents. *Recent Pat. Antiinfect. Drug Discov.* 1: 113–125.

80 Saravia, N.G., Escorcia, B., Osorio, Y. et al. (2006). Pathogenicity and protective immunogenicity of cysteine proteinase-deficient mutants of *Leishmania mexicana* in non-murine models. *Vaccine* 24: 4247–4259.

81 Roberts, L., Egan, T.J., Joiner, K.A., and Hoppe, H.C. (2008). Differential effects of quinoline antimalarials on endocytosis in *Plasmodium falciparum*. *Antimicrob. Agents Chemother.* 52: 1840–1842.

82 Rocha, L.G., Almeida, J.R., Macedo, R.O., and Barbosa-Filho, J.M. (2005). A review of natural products with antileishmanial activity. *Phytomedicine* 12: 514–535.

83 Wink, M. (2012). Medicinal plants: a source of anti-parasitic secondary metabolites. *Molecules* 17: 12771–12791.

84 Ogeto, T.K., Odhiambo, R.A., Shivairo, R.S. et al. (2013). Antileishmanial activity of *Aloe secundiflora* plant extracts against Leishmania major. *Adv. Life Sci. Technol.* 13: 9–18.

85 Mahmoudvand, H., Ayatollahi-Mousavi, S.A., Sepahvand, A. et al. (2014). Antifungal antileishmanial, and cytotoxicity activities of various extracts of *Berberis vulgaris* (Berberidaceae) and its active principle berberine. *ISRN Pharmacol.* 2014: 602436. https://doi.org/10.1155/2014/602436.

86 Manjolin, L.C., dos Reis, M.B., Maquiaveli Cdo, C. et al. (2013). Dietary flavonoids fisetin, luteolin and their derived compounds inhibit arginase, a central enzyme in Leishmania (Leishmania) amazonensis infection. *Food Chem.* 141: 2253–2262.

87 Gamboa-Leon, R., Vera-Ku, M., Peraza-Sanchez, S.R. et al. (2014). Antileishmanial activity of a mixture of *Tridax procumbens* and *Allium sativum* in mice. *Parasite* 21: 15. https://doi.org/10.1051/parasite/2014016.

88 Sifaoui, I., Lopez-Arencibia, A., Martin-Navarro, C.M. et al. (2014). Activity of olive leaf extracts against the promastigote stage of Leishmania species and their correlation with the antioxidant activity. *Exp. Parasitol.* 141: 106–111.

89 Sen, R. and Chatterjee, M. (2011). Plant derived therapeutics for the treatment of leishmaniasis. *Phytomedicine* 18: 1056–1069.

90 Jameel, M., Islamuddin, M., Ali, A. et al. (2014). Isolation, characterization and antimicrobial evaluation of a novel compound N-octacosan 7β ol, from *Fumaria parviflora* Lam. *BMC Complement. Altern. Med.* 14: 98. https://doi.org/10.1186/1472-6882-14-98.

91 Sarkar, J., Pal, S., Bhattacharya, S., and Biswas, M. (2013). In vitro antileishmanial activity of *Pleumeria pudica* leaf extracts on *Leishmania donovani* promastigotes. *Am. Euras. J. Sci. Res.* 8: 68–71.

92 Colares, A.V., Almeida-Souza, F., Taniwaki, N.N. et al. (2013). In vitro antileishmanial activity of essential oil of *Vanillosmopsis arborea* (Asteraceae) baker. *Evid. Based Complement. Alternat. Med.* 2013: 727042.

93 de Medeiros, M., da Silva, A.C., Cito, A.M. et al. (2011). In vitro antileishmanial activity and cytotoxicity of essential oil from Lippia sidoides Cham. *Parasitol. Int.* 60: 237–241.

94 Iqbal, H., Khattak, B., Ayaz, S. et al. (2012). Comparative efficacy of *Aloe vera* and *Tamarix aphylla* against cutaneous leishmaniasis. *Int. J. Basic Med. Sci. Pharm.* 2: 42–45.

95 Bekele, B., Adane, L., Tariku, Y., and Hailu, A. (2013). Evaluation of antileishmanial activities of triglycerides isolated from roots of *Moringa stenopetala*. *Med. Chem. Res.* 22: 4592–4599.

96 Yamamoto, E.S., Campos, B.L., Laurenti, M.D. et al. (2014). Treatment with triterpenic fraction purified from *Baccharis uncinella* leaves inhibits Leishmania (Leishmania) amazonensis spreading and improves Th1 immune response in infected mice. *Parasitol. Res.* 113: 333–339.

97 Mansour, R., Haouas, N., Ben Kahla-Nakbi, A. et al. (2013). The effect of *Vitis vinifera* L leaves extract on *Leishmania infantum*. *Iran. J. Pharm. Res.* 12: 349–355.

98 Machado, M., Dinis, A.M., Santos-Rosa, M. et al. (2014). Activity of *Thymus capitellatus* volatile extract, 1,8-cineole and borneol against Leishmania species. *Vet. Parasitol.* 200: 39–49.

99 Rondon, F.C., Bevilaqua, C.M., Accioly, M.P. et al. (2012). In vitro efficacy of *Coriandrum sativum*, *Lippia sidoides* and *Copaifera reticulata* against *Leishmania chagasi*. *Rev. Bras. Parasitol. Vet.* 21: 185–191.

100 Sachdeva, H., Sehgal, R., and Kaur, S. (2014). Tinospora cordifolia as a protective and immunomodulatory agent in combination with cisplatin against murine visceral leishmaniasis. *Exp. Parasitol.* 137: 53–65.

101 Oryan, A. (2015). Plant-derived compounds in treatment of leishmaniasis. *Iran J. Vet. Res.* 16 (1): 1–19.

102 Ghosh, M., Babu, S.P., Sukul, N.C., and Mahato, S.B. (1993). Antifilarial effect of two triterpenoid saponins isolated from *Acacia auriculiformis*. *Indian J. Exp. Biol.* 31: 604–606.

103 Machado, M., Pires, P., Dinis, A.M. et al. (2012). Monoterpenic aldehydes as potential anti-Leishmania agents: activity of *Cymbopogon citratus* and citral on *L. infantum*, *L. tropica* and *L. major*. *Exp. Parasitol.* 130: 223–231.

104 Cardoso, B.M., De Mello, T.F., Lera, D.S. et al. (2017). Antileishmanial Activity of a Calophyllum brasiliense Leaf Extract. *Planta Med.* 83: 57–62. https://doi.org/10.1055/s-0042-107673.

105 Garcia, M., Monzote, L., Scull, R., and Herrera, P. (2012). Activity of Cuban plants extracts against *Leishmania amazonensis*. *ISRN Pharmacol.* 2012: 104540. https://doi.org/10.5402/2012/104540.

106 Monzote, L., Garcia, M., Pastor, J. et al. (2014). Essential oil from *Chenopodium ambrosioides* and main components: activity against Leishmania, their mitochondria and other microorganisms. *Exp. Parasitol.* 136: 20–26.

107 Rodrigues, I.A., Azevedo, M.M., Chaves, F.C. et al. (2013). In vitro cytocidal effects of the essential oil from *Croton cajucara* (red sacaca) and its major constituent 7-hydroxycalamenene against *Leishmania chagasi*. *BMC Complement. Altern. Med.* 13: 249. https://doi.org/10.1186/1472-6882-13-249.

108 Islamuddin, M., Farooque, A., Dwarakanath, B.S. et al. (2012). Extracts of *Artemisia annua* leaves and seeds mediate programmed cell death in *Leishmania donovani*. *J. Med. Microbiol.* 61: 1709–1718.

109 Seifert, K. (2011). Structures, targets and recent approaches in anti-leishmanial drug discovery and development. *Open Med. Chem. J.* 5: 31–39.

110 Loiseau, P.M., Gupta, S., Verma, A. et al. (2011). In vitro activities of new 2-substituted quinolines against *Leishmania donovani*. *Antimicrob. Agents Chemother.* 55: 1777–1780.

111 Ogungbe, I.V., Erwin, W.R., and Setzer, W.N. (2014). Antileishmanial phytochemical phenolics: molecular docking to potential protein targets. *J. Mol. Graph. Model.* 48: 105–117.

112 da Silva Machado, F.L., Pacienza-Lima, W., Rossi-Bergmann, B. et al. (2011). Antileishmanial sesquiterpenes from the Brazilian red alga *Laurencia dendroidea*. *Planta Med.* 77: 733–735.

113 Wong, I.L., Chan, K.F., Chen, Y.F. et al. (2014). In vitro and in vivo efficacy of novel flavonoid dimers against cutaneous leishmaniasis. *Antimicrob. Agents Chemother.* 58: 3379–3388.

114 Mallick, S., Dutta, A., Dey, S. et al. (2014). Selective inhibition of *Leishmania donovani* by active extracts of wild mushrooms used by the tribal population of India: an in vitro exploration for new leads against parasitic protozoans. *Exp. Parasitol.* 138: 9–17.

115 Mishra, B.B., Singh, R.K., Srivastava, A. et al. (2009). Fighting against leishmaniasis: search of alkaloids as future true potential anti-Leishmanial agents. *Mini Rev. Med. Chem.* 9: 107–123.

116 Makwali, J.A., Wanjala, F.M., Kaburi, J.C. et al. (2012). Combination and monotherapy of Leishmania major infection in BALB/c mice using plant extracts and herbicides. *J. Vector Borne Dis.* 49: 123–130.

117 da Costa, R.C., Santana, D.B., Araujo, R.M. et al. (2014). Discovery of the rapanone and suberonone mixture as a motif for leishmanicidal and antifungal applications. *Bioorg. Med. Chem.* 22: 135–140.

118 Thiem, D.A., Sneden, A.T., Khan, S.I., and Tekwani, B.L. (2005). Bisnortripterpenes from *Salacia madagascariensis*. *J. Nat. Prod.* 68: 251–254.

119 Ribeiro, T.G., Chavez-Fumagalli, M.A., Valadares, D.G. et al. (2014). Antileishmanial activity and cytotoxicity of Brazilian plants. *Exp. Parasitol.* 143: 60–68.

120 Abdel-Sattar, E., Maes, L., and Salama, M.M. (2010). In vitro activities of plant extracts from Saudi Arabia against malaria, leishmaniasis, sleeping sickness and Chagas disease. *Phytother. Res.* 24: 1322–1328.

121 Kheiri Manjili, H., Jafari, H., Ramazani, A., and Davoudi, N. (2012). Anti-leishmanial and toxicity activities of some selected Iranian medicinal plants. *Parasitol. Res.* 111: 2115–2121.

122 Sattar, F.A., Ahmed, F., Ahmed, N. et al. (2012). A double-blind, randomized, clinical trial on the antileishmanial activity of a *Morinda citrifolia* (Noni) stem extract and its major constituents. *Nat. Prod. Commun.* 7: 195–196.

123 Zerehsaz, F., Salmanpour, R., Handjani, F. et al. (1999). A double-blind randomized clinical trial of a topical herbal extract (Z-HE) vs systemic meglumine antimoniate for the treatment of cutaneous leishmaniasis in Iran. *Int. J. Dermatol.* 38: 610–612.

124 Jain, K. and Jain, N.K. (2013). Novel therapeutic strategies for treatment of visceral leishmaniasis. *Drug Discov. Today* 18: 1272–1281.

125 Abeje, F., Bisrat, D., Hailu, A., and Asres, K. (2014). Phytochemistry and antileishmanial activity of the leaf latex of *Aloe calidophila* Reynolds. *Phytother. Res.* 28: 1801–1805.

126 Bringmann, G., Rudenauer, S., Irmer, A. et al. (2008). Antitumoral and antileishmanial dioncoquinones and ancistroquinones from cell cultures of *Triphyophyllum peltatum* (Dioncophyllaceae) and *Ancistrocladus abbreviatus* (Ancistrocladaceae). *Phytochemistry* 69: 2501–2509.

127 Akendengue, B., Roblot, F., Loiseau, P.M. et al. (2002). Klaivanolide, an antiprotozoal lactone from *Uvaria klaineana*. *Phytochemistry* 59: 885–888.

128 Ghosh, M., Shinhababu, S.P., Sukul, N.C. et al. (1994). Antifilarial effect of solamargine isolated from *Solanum khastanum*. *Int. J. Pharm.* 32: 1–7.

129 Babu, S.P., Sarkar, D., Ghosh, N.K. et al. (1997). Enhancement of membrane damage by saponins isolated from *Acacia auriculiformis*. *Jpn. J. Pharmacol.* 75: 451–454.

130 Nandi, B., Roy, S., Bhattacharya, S., and Babu, S.P.S. (2004). Free radicals mediated membrane damage by the saponins acaciaside A and acaciaside B. *Phytother. Res.* 18 (3): 191–194.

131 Joshi, S.G. (2000). *Medicinal Plants*, 3. Calcutta: Oxford and IBH Publishing Co Pvt Ltd.

132 Sahare, K.N., Anandhraman, V., Meshram, V.G. et al. (2008). Anti-microfilarial activity of methanolic extract of *Vitex negundo* and *Aegle marmelos* and their phytochemical analysis. *Indian J. Exp. Biol.* 46: 128–131.

133 Sharma, R.D., Veerpathran, A.R., Dakshinamoorthy, G. et al. (2010). Possible implication of oxidative stress in antifilarial effect of certain traditionally used medicinal plants in vitro against *Brugia malayi* microfilariae. *Pharm. Res.* 2 (6): 350–354.

134 Chauhan, V.S. and Misra, A.K. (2002). Development of molecular markers for screening of *Alnus nepalensis* (D. Don) genotypes for the nitrogenase activity of actinorhizal root nodules. *Mol. Genet. Genomics* 267: 303–312.

135 Yadav, D., Singh, S.C., Verma, R.K. et al. (2013). Antifilarial diarylheptanoids from *Alnus nepalensis* leaves growing in high altitude areas of Uttarakhand, India. *Phytomedicine* 20: 124–132.

136 Dutta, A. and Sukul, N.C. (1982). Filaricidal properties of a wild herb, *Andrographis paniculata*. *J. Helminthol.* 56: 81–84.

137 Misra, S., Verma, M., Mishra, S.K. et al. (2011). Gedunin and photogedunin of *Xylocarpus granatum* possess antifilarial activity against human lymphatic filarial parasite *Brugia malayi* in experimental rodent host. *Parasitol. Res.* 109 (5): 1351–1360.

138 Singh, R., Khan, N.U., and Singhal, K.C. (1997). Potential antifilarial activity of roots of *Asparagus adscendens* Roxb, against *Setaria cervi* in vitro. *Indian J. Exp. Biol.* 35: 168–172.

139 Mathai, A. and Devi, K.S. (1992). Anti-parasitic activity of certain indigenous plants. *Anc. Sci. Life* 12 (1–2): 271–273.

140 Mishra, V., Parveen, N., Singhal, K.C., and Khan, N.U. (2005). Antifilarial activity of *Azadirachta indica* on cattle filarial parasite *Setaria cervi*. *Fitoterapia* 76: 54–61.

141 Mukherjee, N., Mukherjee, S., Saini, P. et al. (2014). Antifilarial effects of polyphenol rich ethanolic extract from the leaves of *Azadirachta indica* through molecular and biochemical approaches describing reactive oxygen species (ROS) mediated apoptosis of *Setaria cervi*. *Exp. Parasitol.* 136: 41–58.

142 Mukerji, A.K. and Bhaduri, N.V. (1947). The treatment of intestinal worms with the indigenous drugs Butea, Embelia and kamala. *Ind. Med. Gaz.* 82 (2): 66–69.

143 Deshmukh, M., Sahare, K.N., Patidar, R.K. et al. (2014). Antifilarial activity of *Butea monosperma* L. leaves extracts against *Setaria cervi*. *Trends Vector Res. Parasitol.* 1: 1.

144 Datté, J.Y., Traoré, A., Offoumou, A.M., and Ziegler, A. (1998). Effects of leaf extract of *Caesalpinia bonduc* (Caesalpiniaceae) on the contractile activity of uterine smooth muscle of pregnant rats. *J. Ethnopharmacol.* 60: 149–155.

145 Gaur, R.L., Sahoo, M.K., Dixit, S. et al. (2008). Antifilarial activity of *Caesalpinia bonducella* against experimental filarial infections. *Indian J. Med. Res.* 128: 65–70.

146 Khunkitti, W., Fujimaki, Y., and Aoki, Y. (2000). In vitro antifilarial activity of extracts of the medicinal plant *Cardiospermum halicacabum* against *Brugia pahangi*. *J. Helminthol.* 74 (3): 241–267.

147 Nisha, M., Kalyanasundaram, M., Paily, K.P. et al. (2006). In vitro screening of medicinal plant extracts form acrofilaricidal activity. *Parasitol. Res.* 100 (3): 575–579.

148 Singhal, K.C., Sharma, S., and Mehta, B.K. (1992). Antifilarial activity of *Centratherum anthelminticum* seed extracts on *Setaria cervi*. *Indian J. Exp. Biol.* 30 (6): 546–548.

149 Patra, J.K., Mohapatra, A.D., Rath, S.K. et al. (2009). Screening of antioxidant and antifilarial activity of leaf extracts of *Excoecaria agallocha* L. *Int. J. Integr. Biol.* 7 (1): 9–15.

150 Mishra, V., Khan, N.U., and Singhal, K.C. (2005). Potential antifilarial activity of fruit extracts of *Ficus racemosa* Linn. against *Setaria cervi* in vitro. *Indian J. Exp. Biol.* 43 (4): 346–350.

151 Kalani, K., Kushwaha, V., Verma, R. et al. (2013). Glycyrrhetinic acid and its analogs: a new class of antifilarial agents. *Bioorg. Med. Chem. Lett.* 23: 2566–2570.

152 Saxena, K., Dube, V., and Kushwaha, V. (2010). Antifilarial efficacy of *Hibiscus sabdariffa* on lymphatic filarial parasite *Brugia malayi*. *Med. Chem. Res.* 20: 1594–1602.

153 Saini, P., Gayen, P., Nayak, A. et al. (2012). Effect of ferulic acid from *Hibiscus mutabilis* on filarial parasite *Setaria cervi*: molecular and biochemical approaches. *Parasitol. Int.* 61: 520–531.

154 Qamaruddin, Parveen, N., Khan, N.U., and Singhal, K.C. (2002). In vitro antifilarial potential of the flower and stem extracts of *Leucas cephalotes* on cattle filarial parasite *Setaria cervi*. *J. Nat. Rem.* 2 (2): 155–163.

155 Kushwaha, V., Saxena, K., Verma, S.K. et al. (2011). Antifilarial activity of gum from *Moringa oleifera* Lam. on human lymphatic filaria *Brugia malayi*. *Chron. Young Sci.* 2: 201–206.

156 Fujimaki, Y., Kamachi, T., Yanagi, T. et al. (2005). Macrofilaricidal and microfilaricidal effects of *Neurolaena lobata*, a Guatemalan medicinal plant, on *Brugia pahangi*. *J. Helminthol.* 79 (1): 23–28.

157 Tripathi, S., Singh, N., and Shakya, S. (2008). Gender based differences in phenol and thiocyanate contents and biological activity in *Piper betle* L. *Curr. Sci.* 91 (6): 746–749.

158 Singh, M., Shakya, S., Soni, V.K. et al. (2009). The *n*-hexane and chloroform fractions of *Piper betle* L. trigger different arms of immune responses in BALB/c mice and exhibit antifilarial activity against human lymphatic filarid *Brugia malayi*. *Int. Immunopharmacol.* 9: 716–728.

159 Mathew, N., Paily Abidha, K.P., Vanamail, P. et al. (2002). Macrofilaricidal activity of plant *Plumbago Indica*/rosea in vitro. *Drug Dev. Res.* 56 (1): 33–39.

160 Uddin, Q., Parveen, N., Khan, N.U., and Singhal, K.C. (2003). Antifilarial potential of the fruits and leaves extracts of *Pongamia pinnata* on cattle filarial parasite *Setaria cervi*. *Phytother. Res.* 17: 1104–1107.

161 Shanmugapriya, R. and Ramanathan, T. (2012). Antifilarial activity of seed extracts of *Ricinus communis* against *Brugia malayi*. *J. Pharm. Res.* 5 (3): 1448–1450.

162 Chatterjee, R.K., Fatma, N., Murthy, P.K. et al. (1992). Macrofilaricidal activity of the stem bark of *Streblus asper* and its major active constituents. *Drug Dev. Res.* 26: 67–78.
163 Hashmi, S. and Singh, V.K. (2002). *Streblus asper* Lour. – an indigenous drug for the treatment of filariasis. In: *Recent Progress in Medicinal Plants: Ethnomedicine and Pharmacognosy*, vol. 1 (eds. D.K. Majumdar, J.N. Govil and V.K. Singh), 259–319. Houston, TX, USA: SCI Tech Publishing LLC.
164 Singh, S.N., Chatterjee, R.K., and Srivastava, A.K. (1994). Effect of glycosides of *Streblus asper* on motility, glucose uptake, and certain enzymes of carbohydrate metabolism of *Setaria cervi*. *Drug Dev. Res.* 32: 191–195.
165 Singh, S.N., Raina, D., Chatterjee, R.K., and Srivastava, A.K. (1998). Antifilarial glycosides of *Streblus asper*: effect on metabolism of adult *Setaria cervifemales*. *Helminthologia* 35: 173–177.
166 Zaridaha, M.Z., Idid, S.Z., Omar, A.W., and Khozira, S. (2001). In vitro antifilarial effects of three plant species against adult worms of subperiodic *Brugia malayi*. *J. Ethnopharmacol.* 78: 79–84.
167 Datta, A. and Sukul, N.C. (1987). Antifilarial effect of *Zinger officinale* on dirofilaria immitis. *J. Helminthol.* 61: 268–270.
168 Parveen, N. (1991). Antifilarial activity of *Vitex negundo* L. against *Setaria cervi*. *Fitoterapia* 62: 163.
169 Sahare, K.N. and Singh, V. (2013). Antifilarial activity of ethyl acetate extract of *Vitex negundo* leaves in vitro. *Asian Pac. J. Trop. Med.* 6: 689–669.
170 Kushwaha, S., Soni, V.K., Singh, P.K. et al. (2012). Withania somnifera chemotypes NMITLI 101R, NMITLI 118R, NMITLI 128R and withaferin A protect *Mastomys coucha* from *Brugia malayi* infection. *Parasite Immunol.* 34: 199–209.
171 van Wyk, B.E. (2008). A broad review of commercially important Southern African medicinal plants. *J. Ethnopharmacol.* 119: 342–355.
172 Wright, C.W. and Phillipson, J.D. (1990). Natural products and the development of selective antiprotozoal drugs. *Phytother. Res.* 4: 127–139.
173 Hoet, S., Opperdoes, F., Brun, R., and Quetin-Leclereq, J. (2004). Natural products active against African trypanosomes: a step towards new drugs. *Nat. Prod. Rep.* 21: 353–364.
174 Hoet, S., Stevigny, C., Block, S. et al. (2004). Alkaloids from Cassytha filiformis and related aporphines: antitrypanosomal activity, cytotoxicity, and interaction with DNA and topoisomerases. *Planta Med.* 70: 407–413.
175 Nok, A.J. (2002). Azaanthraquinone inhibits respiration and in vitro growth of long slender bloodstream forms of *T. congolense*. *Cell Biochem. Funct.* 20: 205–212.
176 Harvey, A.L., Edrada-Ebel, R., and Quinn, R.J. (2015). The re-emergence of natural products for drug discovery in the genomics era. *Nat. Rev. Drug Discov.* 14: 111–129.

177 Ibrahim, M.A., Mohammed, A., Isah, M.B., and Aliyu, A.B. (2014). Anti-trypanosomal activity of African medicinal plants: a review update. *J. Ethnopharmacol.* 154: 26–54.

178 Ntie-Kang, F., Zofou, D., Babiaka, S.B. et al. (2013). AfroDb: a select highly potent and diverse natural product library from African medicinal plants. *PLoS One* 8: e78085.

179 Lifongo, L.L., Simoben, C.V., Ntie-Kang, F. et al. (2014). A bioactivity versus ethnobotanical survey of medicinal plants from Nigeria, West Africa. *Nat. Prod. Bioprospect.* 4: 1–19.

180 Inada, A., Ogasawara, R., Koga, I. et al. (2008). Screening of Brazilian plants for antimicrobial and DNA damaging activities. *Chem. Pharm. Bull.* 56: 727–729.

181 Devkota, K.P., Wansi, J.D., Lenta, B.N. et al. (2010). *Planta Med.* 76: 1022–1025.

182 Cheesman, L., Nair, J.J., and van Staden, J. (2012). Antibacterial activity of crinane alkaloids from *Boophone disticha* (Amaryllidaceae). *J. Ethnopharmacol.* 140: 405–408.

183 Nnadi, C.O., Nwodo, N.J., Kaiser, M. et al. (2017). Steroid alkaloids from *Holarrhena africana* with strong activity against *Trypanosoma brucei* rhodesiense. *Molecules* 22: 1129.

184 Uzor, P.F., Ebrahim, W., Osadebe, P.O. et al. (2015). Metabolites from *Combretum dolichopetalum* and its associated endophytic fungus *Nigrospora oryzae* evidence for a metabolic partnership. *Fitoterapia* 105: 147–150. https://doi.org/10.1016/j.fitote.2015.06.018.

185 Bringmann, G., Dreyer, M., Faber, J.H. et al. (2004). Ancistrotanzanine C and related 5,1′- and 7,3′-coupled naphthylisoquinoline alkaloids from *Ancistrocladus tanzaniensis*. *J. Nat. Prod.* 67: 743–748.

186 Bringmann, G., Messer, K., Brun, R., and Mudogo, V. (2002). Ancistrocongolines A-D, new naphthylisoquinoline alkaloids from *Ancistrocladus congolensis*. *J. Nat. Prod.* 65: 1096–1101.

187 Cretton, S., Breant, L., Pourrez, L. et al. (2014). Antitrypanosomal quinoline alkaloids from the roots of Waltheria indica. *J. Nat. Prod.* 77: 2304–2311.

188 Ngantchou, I., Nyasse, B., Denier, C. et al. (2010). Antitrypanosomal alkaloids from *Polyalthia suaveolens* (Annonaceae): their effects on three selected glycolytic enzymes of *Trypanosoma brucei*. *Bioorg. Med. Chem. Lett.* 20: 3495–3498.

189 Montenegro, H., Gutierrez, M., Romero, L.I. et al. (2003). Aporphine alkaloids from *Guatteria* spp. with leishmanicidal activity. *Planta Med.* 69: 677–679.

190 Tekle, A.H., Zoure, H.G., Noma, M. et al. (2016). Progress towards onchocerciasis elimination in the participating countries of the African Programme for Onchocerciasis Control: epidemiological evaluation results. *Infect. Dis. Poverty* 5: 66.

191 Simoben, C.V., Ntie-Kang, F., Akone, S.H., and Sippl, W. (2018). Conrad compounds from African medicinal plants with activities against selected parasitic diseases: schistosomiasis, trypanosomiasis and leishmaniasis. *Nat. Prod. Bioprospect.* 8 (3): 151–169.

192 Leandro, L.M., Vargas, F.D., Barbosa, P.C.S. et al. (2012). Chemistry and biological activities of terpenoids from copaiba (*Copaifera* spp.) oleoresins. *Molecules* 17: 3866–3889.

193 Wube, A.A., Bucar, F., Gibbons, S. et al. (2010). Antiprotozoal activity of drimane and coloratane sesquiterpenes towards *Trypanosoma brucei* rhodesiense and *Plasmodium falciparum* in vitro. *Phytother. Res.* 24: 1468–1472.

194 Olila, D., Opuda-Asibo, J., and Olwa, O. (2001). Bioassay-guided studies on the cytotoxic and in vitro trypanocidal activities of sesquiterpenes (Muzigadial) derived from Uganda medicinal plant (*Warburgi ugandensis*). *Afr. Health Sci.* 1: 12–15.

195 Machumi, F., Yenesew, A., Midiwo, J.O. et al. (2012). Antiparasitic and anticancer carvotacetone derivatives of *Sphaeranthus bullatus. Nat. Prod. Commun.* 7: 1123–1126.

196 Ragasa, C.Y., Tsai, P.W., Galvez, C.T., and Shen, C.C. (2010). New carvotanacetone derivatives from *Sphaeranthus africanus. Planta Med.* 76: 146–151.

197 Jakupovic, J., Grenz, M., Bohlmann, F., and Mungai, G.M. (1990). Carvotacetone derivatives and eudesman-12,6β-olides from – *Sphaeranthus* species. *Phytochemistry* 29: 1213–1217.

198 Ebiloma, G.U., Igoli, J.O., Katsoulis, E. et al. (2017). Nigerian medicinal plants identifies new trypanocides with low toxicity and no cross-resistance to diamidines and arsenicals. *J. Ethnopharmacol.* 202: 256–264.

199 Nyasse, B., Ngantchou, I., Tchana, E.M. et al. (2004). Inhibition of both *Trypanosoma brucei* bloodstream form and related glycolytic enzymes by a new kolavic acid derivative isolated from *Entada abyssinica. Pharmazie* 59: 873–875.

200 Soh, D., Nkwengoua, E., Ngantchou, I. et al. (2013). Bioactive kaurane diterpenes from *Xylopia aethiopice* Daul (Annonaceae). *J. Appl. Pharm. Sci.* 3: 13–19.

201 Cheikh-Ali, Z., Okpekon, T., Roblot, F. et al. (2011). Labdane diterpenoids from *Aframomum sceptrum*: NMR study and antiparasitic activities. *Phytochem. Lett.* 4: 240–244.

202 Okpekon, T., Yolou, S., Gleye, C. et al. (2004). Antiparasitic activities of medicinal plants used in Ivory Coast. *J. Ethnopharmacol.* 90: 91–97.

203 Kamnaing, P., Tsopmo, A., Tanifum, E.A. et al. (2003). Trypanocidal diarylheptanoids from *Aframomum letestuianum. J. Nat. Prod.* 66: 364–367.

204 El-Seedi, H.R., El-Shabasy, R., Sakr, H. et al. (2011). Anti-schistosomiasis triterpene glycoside from the Egyptian medicinal plant *Asparagus stipularis. Rev. Bras. Farmacogn.* 22: 314–318.

205 Boulos, L. (1983). *Medicinal plants of North Africa*, 130. Algonac: Reference Publications, Inc.

206 Maregesi, S.M., Hermans, N., Dhooghe, L. et al. (2010). Phytochemical and biological investigations of *Elaeodendron schlechteranum*. *J. Ethnopharmacol.* 129: 319–326.

207 Bero, J., Ganfon, H., Jonville, M.C. et al. (2009). In vitro antiplasmodial activity of plants used in Benin in traditional medicine to treat malaria. *J. Ethnopharmacol.* 122: 439–444.

208 Bero, J., Hannaert, V., Chataigne, G. et al. (2011). In vitro antitrypanosomal and antileishmanial activity of plants used in Benin in traditional medicine and bio-guided fractionation of the most active extract. *J. Ethnopharmacol.* 137: 998–1002.

209 Bero, J., Beaufay, C., Hannaert, V. et al. (2013). Antitrypanosomal compounds from the essential oil and extracts of *Keetia leucantha* leaves with inhibitor activity on *Trypanosoma brucei* glyceraldehyde-3-phosphate dehydrogenase. *Phytomedicine* 20: 270–274.

210 Ngantchou, I., Nkwengoua, E., Nganso, Y. et al. (2009). Antitrypanosomal activity of polycarpol from *Piptostigma preussi* (Annonaceae). *Fitoterapia* 80: 188–191.

211 Tchinda, A.T., Tsopmo, A., Tane, P. et al. (2002). Vernoguinosterol and vernoguinoside, trypanocidal stigmastane derivatives from *Vernonia guineensis* (Asteraceae). *Phytochemistry* 59: 371–374.

212 Nwodo, N.J., Okoye, F.B.C., Lai, D.W. et al. (2014). Two trypanocidal dipeptides from the roots of *Zapoteca portoricensis* (Fabaceae). *Molecules* 19: 5470–5477.

213 Lenta, B.N., Vonthron-Senecheau, C., Weniger, B. et al. (2007). Leishmanicidal and cholinesterase inhibiting activities of phenolic compounds from *Allanblackia monticola* and *Symphonia globulifera*. *Molecules* 12: 1548–1557.

214 Ni, G., Yang, H.Z., Fu, N.J. et al. (2015). Cytotoxic taccalonolides and withanolides from *Tacca chantrieri*. *Planta Med.* 81: 247–256.

215 Dike, V.T., Vihiior, B., Bosha, J.A. et al. (2016). Antitrypanosomal activity of a novel taccalonolide from the tubers of. *Tacca leontopetaloides*. *Phytochem. Anal.* 27: 217–221.

216 Nwodo, N.J., Ibezim, A., Ntie-Kang, F. et al. (2015). Anti-trypanosomal activity of Nigerian plants and their constituents. *Molecules* 20: 7750–7771.

217 Kamboj, A. and Saluja, A.K. (2008). *Ageratum conyzoides* L.: a review on its phytochemical and pharmacological profile. *Int. J. Green Pharm.* 2: 59–68.

218 Beck, L., Favre, T.C., Pieri, O.S. et al. (2001). Replacing oxamniquine by praziquantel against *Schistosoma mansoni* infection in a rural community from the sugar-cane zone of northeast Brazil: an epidemiological follow-up. *Mem. Inst. Oswaldo Cruz* 96: 165–167.

219 El-Ansary, K.A., Ahmed, S.A., and Aly, S.A. (2007). Antischistosomal and liver protective effects of *Curcuma longaextract* in *Schistosoma mansoni* infect mice. *Indian J. Exp. Biol.* 45: 791–801.

220 Danso-Appiah, A. and de Vlas, S.J. (2002). Interpreting low praziquantel cure rates of *Schistosoma mansoni* infections in Senegal. *Trends Parasitol.* 18: 125–129.

221 Fallon, P.G. and Doenhoff, M.J. (1994). Drug-resistant schistosomiasis: resistance to praziquantel and oxamniquine induced in *Schistosoma mansoni* in mice is drug specific. *Am. J. Trop. Med. Hyg.* 51: 83–88.

222 Xiao, S.H., Tanner, M., N'Goran, E.K. et al. (2002). Recent investigations of artemether, a novel agent for the prevention of *Schistosomiasis japonica*, mansoni and haematobia. *Acta Trop.* 82: 175–181.

223 Bruneton, J. (1999). *Pharmacognosy Phytochemistry Medicinal Plants*, 2e, 463–730. Paris: Lavoisier Publishing.

224 Mirelman, D., Monheit, D., and Varon, S. (1987). Inhibition of growth of *Entamoeba histolytica* by allicin, the active principle of garlic extract (*Allium sativum*). *J Infect Dis* 156: 243–244.

225 Koch, H.P. and Lawson, L.D. (1996). *Garlic: The Science and Therapeutic Application of Allium sativum L. and Related Species*, 2e. Baltimore, MD: Lippincott Williams and Wilkins.

226 Kliks, M.M. (1985). Studies on the traditional herbal anthelmintic *Chenopodium ambrosioides* L.: ethnopharmacological evaluation and clinical field trials. *Soc. Sci. Med.* 21: 879–886.

227 Weiss, R.F. (1985). *Herbal Medicine*. Beaconsfield, UK: Beaconsfield Publishers Ltd.

228 Duke, J.A. (2002). *Handbook of Medicinal Herbs*, 2e. Boca Raton, FL, USA: CRC Press.

229 Ogboli, A.U., Nock, I.H., Obdurahman, E.M., and Ibrahim, N.D.G. (2000). Medicinal application of vernonia amygdalina del leaf extracts in the treatment of schistosomiasis in mice. *Nig. J. Nat. Prod. Med.* 4: 73–75.

230 Jellin, J.M., Batz, F., and Hitchens, K. (2000). *Natural Medicines Comprehensive Database*, 3e. Stockton, California: Therapeutic Research Faculty.

231 Santana, C.F., De-Almeida, E.R., Dos-Santos, E.R., and Souza, I.A. (1992). Action of *Mentha crispa* hydroethanolic extract in patients bearing intestinal protozoan. *Fitoterapia* 63: 409–410.

232 Sheir, Z., Nasr, A.A., Massoud, A. et al. (2001). A safe, effective, herbal anti-schistosomal therapy derived from myrrh. *Am. J. Trop. Med. Hyg.* 65: 700–704.

233 Hassan, M., El-Motaiem, M., Afify, H. et al. (2003). In vitro effect of Mirazid on *Schistosoma mansoni* in worms. *J. Egypt. Soc. Parasitol.* 33: 999–1008.

234 El-Baz, M.A., Morsy, T.A., El-Bandary, M.M., and Motawea, S.M. (2003). Clinical and parasitologicals studies on the efficacy of Mirazid treatment of *Schistosomiasis haematobium* in Tatoon, Etsa Center, El Fayoum Governorate. *J. Egypt. Soc. Parasitol.* 33: 761–776.

235 Massoud, A.M., El-Kholy, N.M., El-Shennawy, F.A., and Farag, R.E. (2004). Study of some immune aspects in patients with fascioliasis before and after *Chommiphora molmol* (Mirazid) treatment. *J. Egypt. Soc. Parasitol.* 34: 315–332.

236 Aly, H.F. and Aly, S.A. (2006). Essential role of *Citrus reticulata* and Mirazid in treatment of *Schistosoma mansoni* infected mice: biochemical and parasitological studies. *Pol. J. Food Nutr. Sci.* 15: 461–467.

237 Van Waarde, A., van den Thillart, G., Erkelens, C. et al. (1990). Functional coupling of glycolysis and phosphocreatine utilization in anoxic fish muscle. *J. Biol. Chem.* 265: 914–923.

238 Ali, S.A. and Hamed, M.A. (2006). Effect of *Ailanthus altissima* and *Zizyphus spina-christi* on bilharzial infestation in mice: histological and histopathological studies. *J. Appl. Sci.* 6: 1437–1446.

239 Aly, S.A., El-Rigal, N.S., and Rizk, M.Z. (2006). Nutritional supplementation with *Ailanthus altissima* and *Ziziphus spina christi* to compensate for some metabolic disorders in *Schistosoma mansoni* infected mice. *Pak. J. Biol. Sci.* 9: 1700–1706.

240 El-Rigal, N., Aly, S., Rizk, M.Z., and Said, A. (2006). Effect of *Ailanthus altissima* and *Ziziphus spina christi* extracts on some hepatic marker enzymes and antioxidants in *Schistosoma mansoni* infected mice. *Pol. J. Food Nutr. Sci.* 15: 199–202.

241 Egbuna, C. and Ifemeje, J.C. (2015). Biological functions and anti-nutritional effects of phytochemicals in living system. *IOSR-JPBS* 10 (2): 10–19. http://www.iosrjournals.org/iosr-jpbs/papers/Vol10-issue2/. . ./C010231019.pdf.

242 Egbuna, C., Mishra, A., and Goyal, M.R. (eds.) (2020). *Preparation of Phytopharmaceuticals for the Management of Disorders*, 1e. US: Academic Press, Elsevier.

243 Egbuna, C., Kumar, S., Ezzat, S.M. et al. (eds.) (2019). *Phytochemicals as Lead Compounds for New Drug Discovery*, 1e. Cambridge, USA: Elsevier.

244 Shashank, K. and Egbuna, C. (eds.) (2019). *Phytochemistry: In vitro and In Silico Updates*, 1e. Singapore: *Springer Nature.*

245 Egbuna, C., Ifemeje, J.C., Udedi, S.C., and Kumar, S. (eds.) (2018). *Phytochemistry, Volume 1: Fundamentals, Methods, and Applications*, 1e. New York: Apple Academic Press/CRC Taylor & Francis.

246 Egbuna, C., Kumar, S., Ifemeje, J.C., and Kurhekar, J.V. (eds.) (2018). *Phytochemistry, Volume 2: Pharmacognosy, Nanomedicine, and Contemporary Issues*, 1e. New York: Apple Academic Press/CRC Taylor & Francis.

247 Egbuna, C., Ifemeje, J.C., Kumar, S., and Sharif, N. (eds.) (2018). *Phytochemistry, Volume 3: Marine, Industrial, and Advances*, 1e. New York: Apple Academic Press/CRC Taylor & Francis.

Part II

Protozoan Infections

4

Herbal, Nutritional, and Traditional Remedies for Giardiasis: Phytochemicals as Drug Candidates

Samrat Bhattacharyya

Department of Botany, Sister Nibedita Government General Degree College for Girls, University of Calcutta, Kolkata, West Bengal, India

List of Abbreviations

DNA	Deoxyribonucleic acid
GGL	Galacto-glycerolipid
HIV	Human immunodeficiency virus
IC50	half-maximal inhibitory concentration
MTZ	Metronidazole
NAG	*N*-acetyl-D-glucosamine
WGA	wheat germ agglutinin
WHO	World Health Organization

4.1 Introduction

Giardiasis is an infectious intestinal disease caused by the protist *Giardia duodenalis* (synonyms: *Giardia lamblia*, *Giardia intestinalis*), which is a water-transmitted, anaerobic parasite. Global occurrence of giardiasis ranges within 20–60%, while only 2–7% incidence is observed in developed nations [1] Thus, there is a closer association of this disease with impoverished sectors that have limited access to medical care. For this reason, giardiasis has been incorporated in the Neglected Diseases Initiative of the World Health Organization (WHO) since 2006 [2, 3]. Giardia is a commonly identified etiological agent behind traveler's diarrhea with anthropo-zoonotic transmission potential and may also impair the

Neglected Tropical Diseases and Phytochemicals in Drug Discovery, First Edition.
Edited by Chukwuebuka Egbuna, Muhammad Akram, and Jonathan Chinenye Ifemeje.

health conditions of food-productive animals [4]. Certain parameters regarding the disease trajectories, biology of the causal organism, and course of remediation involving allopathic and herbal drugs need to be mentioned in order to have a clear understanding in order to proffer lasting solution.

4.1.1 Epidemiology and Symptoms of Giardiasis

Giardiasis is prevalent in the regions that have weak sanitary measures and low water quality management. The majority of infections take place during the summer and early autumn season. Epidemic instances often remain under-documented, because about 50–75% of infections do not show recognizable symptoms [5]. A two-peaked age distribution pattern can be seen in patients that are 1–9 years and 45–49 years old [6]. Although giardiasis spreads sporadically, infections from day-care crèches, contaminated water of swimming pools, inadequately filtered surface water can lead to frequent outbreaks [5, 7]. Edible items like salads from restaurants, raw vegetables, fresh fruits washed with untreated water often cause foodborne infections [8]. The most vulnerable ones are; allies of infected people, foreign travelers, individuals with compromised immune states (like immunoglobulin deficient and HIV infected people), patients with cystic fibrosis, low stomach acid level, and individuals in the vicinity of infected animals [5]. Symptom severity depends on factors like the age group, past instances of exposure, parasitic bulk, the virulence of the causal agent, and host response physiology. On a general note, symptoms get milder with recurrent infections [9]. The symptomatic range of giardiasis includes abdominal pain, diarrhea, anorexia, vomiting tendency, fatigue, malodorous flatulence, and erratic belching. Recognizable signs are loss of weight, distended abdomen, liquefied stools, and malabsorption [1, 10]. Some unusual cases lead to low fever, shivering, headache, urticarial symptoms, and polyarthritis. Mucous containing and bloody fecal discharge are rarely reported [1, 11]. For some patients, giardiasis can be very short-spanned and spontaneously cured, while in others prolonged infection requires proper medical assistance [10].

4.1.2 Life Cycle of *G. duodenalis*

Giardia is an ancient protozoan genus belonging to the order Diplomonadida and family Hexamitidae [12]. The life cycle (Figure 4.1) of this parasite has two distinct stages: (i) Motile, active trophozoite stage living in the host intestine with two anteriorly positioned nuclei, median body, four paired flagella, and a ventral spiral organelle or attachment disc. The trophozoite has a pyriform shape, 12–20 μm long and 5–10 μm wide [5]. (ii) Stationary, dormant cyst stage, containing four nuclei and a galactosamine rich cyst-wall with 0.3–0.5 μm thickness. Giardia cysts

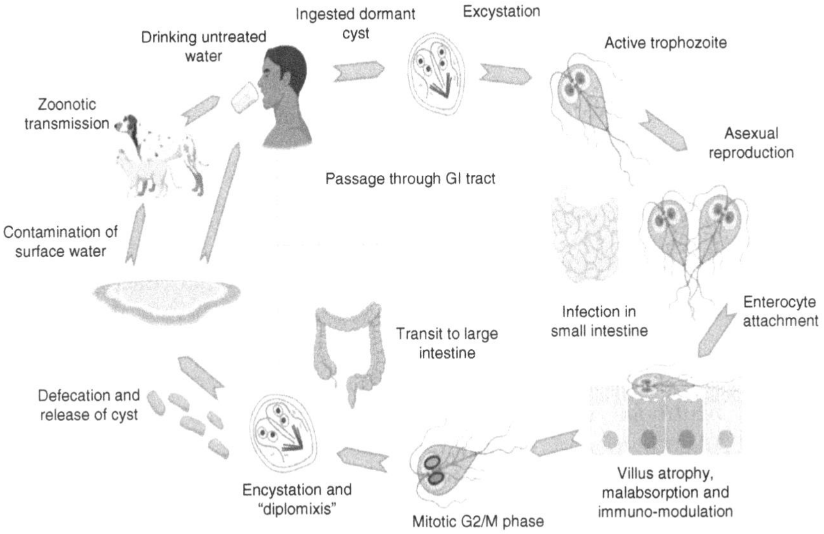

Figure 4.1 Life cycle and pathogenesis of *Giardia duodenalis*.

are ovoid, smooth-outlined, and can be 8–12 μm in length and 7–10 μm in width. Cysts are moderately resistant to disinfection by chlorine and are durable in the environment [5, 13]. Infectious cysts are ingested by human through contaminated water or food. Within the gastrointestinal tract, excystation process is initiated in contact with acidic gastric contents, which leads to the release of 1–2 active trophozoites in the duodenum [14]. The trophozoites can further reproduce asexually, through binary fission. The trophozoite mainly infects the upper portion of the small intestine, which has a conducive alkaline pH. As the trophozoites move across the small intestine and transit toward the colon, encystation process occurs and cysts are released during defecation [14]. It has been speculated, that the non-meiotic forms of genetic recombination and parasexuality may have a role in the life cycle of Giardia [15]. Nuclear fusion or genetic exchange is not visible in trophozoites, but mitotic recombination happens in cysts which can be called "diplomixis" [16]. Surprising rates of genetic exchange as high as about 20%, were recorded between the cyst nuclei and diplomixis involved homologous recombination-mediated by certain meiotic gene homologues [17].

4.1.3 Pathophysiology of Giardial Infection

Pathogenicity are caused by the active trophozoite stage of *Giardia*. After excystation, trophozoites swim to the surface of the duodenum and jejunum for attachment using their ventral disk [1]. Further, the superficial lectins of the parasite bind to the sugars of enterocyte surface, in turn causing villous atrophy, resulting in nutrient malabsorption. Rapid asexual propagation of trophozoites creates a physical barrier between the enterocytes and the intestinal lumen, alters intestinal permeability with hypersecretion of anion and fluids, ultimately causing apoptosis of enterocytes [1, 4]. Trophozoites scavenge nutrients in the intestinal lumen and glucose is their primary utilizable energy source. The amino acids like alanine and aspartate are also hijacked by the parasite, further magnifying the pathological syndrome [1, 4]. Damage of intestinal brush border and associated decimation of disaccharidase activity leads to increased disaccharide accumulation in the intestine, creating osmotic diarrhea [18]. Certain cathepsin-like cysteine proteases of the parasite, required for trophozoite encystation and excystation, also adds to the intestinal damage. These proteases further degrade interleukin-8, attenuate neutrophil chemotaxis, and disrupt the enterocytic cytoskeletal villin [19, 20]. Moreover, the arginine deaminase enzyme of the parasite depletes the luminal arginine pool and hampers interleukin-12 secretion from the dendritic cells [21]. Hence, the immunomodulatory factors secreted by trophozoites, inhibit pro-inflammatory host responses in infected intestinal tissues [4].

4.1.4 Resistance to Allopathic Drug and Relevance of Herbal Medication

Metronidazole (MTZ) is the most frequently prescribed allopathic drug, known to cure giardiasis all over the world. MTZ is orally administered up to a dose of 750 mg/day, roughly for 5–10 days [22]. MTZ becomes reduced at the nitro group by the ferredoxin electron-transport proteins of the trophozoites, unleashing the cytotoxic effect. The drug also creates oxidative stress by binding to thiol groups, generates cysteine adducts, and ultimately causes disruption of DNA. However, an emerging resistance to nitroheterocyclic drugs like nitrofuran furazolidone, 5-nitrothiazolide nitazoxanide has been recorded in *Giardia*, including 5-nitroimidazoles like MTZ itself [23]. The role of Nitroreductase 2 (NR-2) has also been characterized in *Giardia* and its overexpression causes almost threefold tolerance to MTZ, while in the case of *Escherichia coli* system, it might lead to an absolute insensitive state [24]. Adversely, metronidazole is associated with some prominent side-effects like nausea, pancreatitis, headache, neurotoxicity, neutropenia, and additionally, it is a potent carcinogen [14, 25].

On the other hand, naturally synthesized compounds derived from bacteria, animals, and plants can be promising anti-giardial drug candidates. Herbal products are especially being studied more and more over the years. Alcoholic or aqueous extracts of roots, bark, rhizomes, leaves, and inflorescences from traditionally described antidiarrheal plants have been tested worldwide on *Giardia* trophozoites [23]. Although many plant extracts with giardicidal activity have been validated, still the intrinsic mechanisms of actions are scarcely reported and isolation of active constituents have attained prime importance over time. Phytochemicals of great potentials include alkaloids, flavonoids, terpenoids, and certain others. For example, flavonoids like epicatechin, kaempferol, tiliroside isolated from certain Mexican medicinal plants exhibited anti-giardial activities comparable to standard drug MTZ [23]. Despite the constant trials for providing better healthcare access in developing countries, a substantial bulk of the population suffers from it. The disproportionate burden of disease is exacerbated by the constraints of financial accessibility, conditional acceptability, or poor quality of healthcare assistance [26]. As a measure of compensation, plant-based food supplements can also play a vital role to sustain the antiparasitic defense of infected patients and to curb down the giardiasis associated with nutritional inadequacies [1]. However, to counterbalance the risks of MTZ resistance, the discovery of new, side-effect free, and cost-effective plant-based giardicidal drugs, along with botanical food supplementation approach can prove to be fruitful. Rigorous toxicological assays, a better understanding of host-response physiology and molecular targets still need to be explored so that these herbal drugs or phytochemicals can be incorporated into modern mainstream clinical practice [1, 23].

4.2 Plant-Based Dietary Supplements for Giardiasis Management

Giardiasis can be successfully treated with appropriate nutritional interventions and phytotherapeutic measures, which could constitute the front-line approach to disease management. Nutritional management involves dietary supplements that not only inhibit giardial replication and adhesion to enterocytes but also promote host defense trajectories. In a holistic sense, the overall diet is channelized to reduce the acute symptomatology associated with giardiasis [1]. Figure 4.2 presents the dietary recommendations for giardiasis patients, focusing on the active principles of supplementation.

4.2.1 Oligosaccharide and Disaccharidase Supplementation

The pathogenicity of the protozoan parasite, as well as the inflammatory immuno-activation, can damage the enterocytes of the intestine. These pathophysiological actions can hamper digestion, leading to the prevalence of perceived food intolerance. Sensitivity to high and low FODMAP (fermentable oligosaccharides, disaccharides, monosaccharides, and polyols) foods were affirmed to be associated with giardial infection, along with a high frequency of IBS (irritable bowel

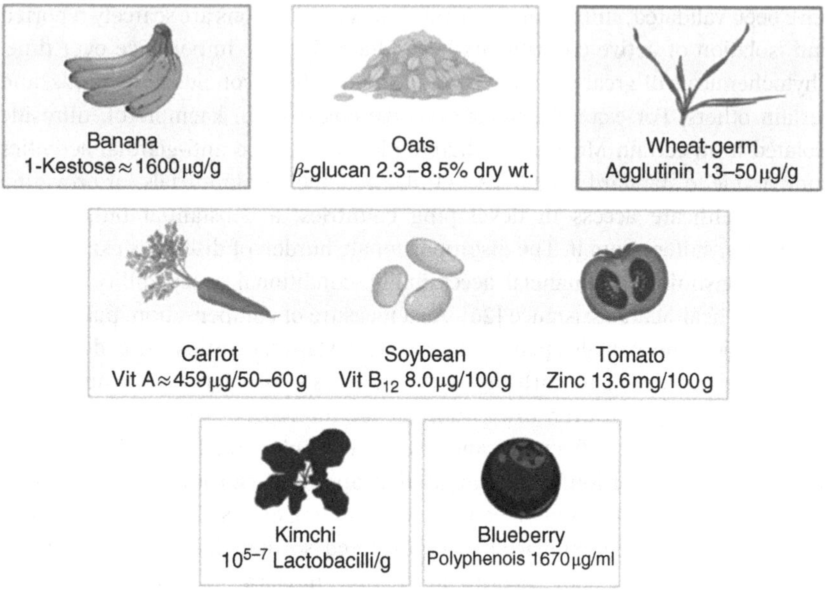

Figure 4.2 Proposed dietary supplementations for giardiasis management. *Source:* Based on [1].

syndrome). Generally, the treatment of abnormal sugar absorption associated with giardiasis, involves long-term avoidance of monosaccharides and disaccharides in the daily diet [27].

Prebiotic fructooligosaccharides might play a nominal role in giardiasis management, as they possess limited ability to influence the small intestinal environment [28]. Prebiotic fermentation elevates mucin production in the intestine which might fasten up giardial clearance [29]. Minimal application (about 2 g twice a day) of prebiotic sweeteners like "Neosugar" can be opted, in case of increased symptoms like abdominal bloating and flatulence [30]. Neosugar is a commercial mixture of fructo-oligosaccharides inclusive of 1-kestose, nystose, and 1F-1-fructofuranosyl nystose. Banana (*Musa* spp.) is a rich source of all these fruit sugars and its different cultivars contain up to 1600 μg of 1-kestose per gram weight [31, 32]. So, the consumption of bananas might be contextually helpful for giardiasis patients with sugar absorption imbalance related symptoms.

Moreover, the incurrence of disaccharidase malfunction has been previously noticed in the case of giardiasis it has not been correlated with reduced absorption of monosaccharides like fructose. The only link between these factors is that they occur in the same intestinal region [27]. Interestingly, a report has shown that the plant *Tinospora cordifolia* (Menispermaceae) is a reliable source of extracellular disaccharidase enzymes such as lactase, maltase, isomaltase, sucrase, and trehalase. The physicochemical nature and stabilities of these plant-derived enzymes are relatable with the mammalian ones and might be applied for disaccharide intolerance therapy. Dosage-wise, *T. cordifolia* enzyme was nontoxic up to 200 mg protein/kg, and in powdered form, 3–6 g amount is recommendable as herbal drug. Although there are no clinical trial reports on dietary supplementation of plant-disaccharidases for treating giardiasis, *Tinospora* should be considered for preclinical trials [33, 34].

4.2.2 Dietary Fiber-Enriched Food

Dietary fibers have an important role in bowel clearance during *Giardia* infection. The consumption of a high insoluble fiber-rich diet showed significant protective action. Rodent models fed with low-fiber (5%) food were more likely to contract giardiasis than in case of a high-fiber diet (20%). Biological examination revealed the clearance of trophozoites from the small bowel, caused by the decrease in the number of trophozoite attachments to the jejunal epithelium. Apparently, fiber content has a positive effect on mucus secretion and a negative impact on the mucosal colonization of trophozoites [35] Insoluble fiber intake notably increases the relative number of intestinal goblet cells and uplifts the luminal mucin levels [36]. Furthermore, lignins and soluble fibers can sequester bile salts away from the availability of the trophozoites, thereby delimiting their growth. Thus, insoluble and soluble dietary fibers play a significant role in giardial removal through multiple mechanisms [1, 37]. Whole grain oatmeal (*Avena sativa*) has received considerable attention for dietary fiber richness, and its β-glucan content can be up to 2.3–8.5% of dry weight. Studies reveal that oat possesses beneficial effects against certain gastrointestinal problems [38] which suggests, it might also be tried as a dietary aid for giardiasis infected people.

4.2.3 Agglutinin from Wheat

N-acetyl-D-glucosamine (NAG) moieties are the main building blocks of the cyst walls, as well as the trophozoites of *Giardia*. Wheat seedlings contains a lectin class of compound named wheat germ agglutinin (WGA) that precisely binds to NAG residues. Prior exposure of cysts to WGA inhibits excystation by 90%, through the proteolytic cleavage of cyst wall glycoproteins [39]. Additionally, WGA at a dose of 100 μg/day for two weeks, was reported to inhibit trophozoite growth *in vitro*. Parasites were arrested in the G2/M intermediary phase of the cell cycle and replication or further encystation was blocked, showing a 30% decrease in trophozoite numbers and 50% lesser cyst count in feces of mice [40]. In a clinical trial with infected human subjects, approximately 60% reduction of fecal cyst count was observed in asymptomatic patients treated with wheat germ supplement. Fecal antigen levels were also somewhat decreased after the wheat germ diet, with no signs of intolerance in the patients [41]. Considering the results of dose-dependent response, the inclusion of wheat germ into the diet (approximately one to two tablespoons, thrice daily) might show therapeutic benefit. Popular commercial preparations based on wheat germ contain WGA (about 13–53 μg/g) which can be of significant help in this aspect [1].

4.2.4 Vitamin and Mineral Supplements

There are quite a few instances of vitamin malabsorption and misbalance in mineral status associated with *Giardia* infection in humans, which deserve a discussion in the light of remedial supplementation.

4.2.4.1 Vitamin A Supplementation

Severe giardial infection-related malabsorption can highlight a selective deficiency of vitamin A, which imposes a traumatic effect on enterocytes [42]. Phrynoderma is a subtype of acquired follicular hyperkeratoses, usually linked to vitamin A imbalance, which exhibits symptoms like hyperpigmentation, follicular or keratotic papules arranged in symmetrical patches on the skin [43]. There is a report linking the instance of phrynoderma with chronic giardial infection, which has documented down-falling rates of retinol-binding protein (19 mg/l, normal: 37–56 mg/l) and vitamin A (0.76 mM/l; normal: 1.63–2.70 mM/l) [44]. In a study with Brazilian children infested with *Giardia*, retinol intervention significantly decreased intestinal infections, suggesting an immune regulatory modulation on the parasite [45]. These results have pointed to the necessity of vitamin A supplementation for patients showing clear symptoms of giardiasis associated avitaminosis. The plant sources of pro-vitamin A include carrots (retinol activity equivalent: 459 μg/carrot; daily value: 51%), broccoli, cantaloupe, and squash [46]. Common tropical fruits like mango and papaya, wild coastal fruits of *Pandanus* are extremely rich in carotenoids [47]. Red palm oil has a high retinol equivalency (5.7 : 1) and a daily intake of 5–10 ml fulfills the RDA of vitamin A for children and adults [48].

4.2.4.2 Compensation of Vitamin B_{12}

Vitamin B_{12} malabsorption has been observed in about 60% of giardiasis patients in several reports. Proposed mechanisms of vitamin B_{12} absorption anomalies can be linked to the competition between host and parasite and the incurrence of secondary bacterial overgrowth during giardial infection [49]. Severe vitamin B_{12} deficiency was documented in a congenitally agammaglobulinemic patient, suffering from prolonged giardial infection. Curing of the parasite load coupled with parenteral replacement of vitamin B_{12}, resolved the symptoms completely [50].

Vitamin B_{12}-enriched vegetarian diet might be particularly beneficial in this regard. A soybean-based Indonesian fermented food called "Tempeh" contains a significant amount of Vitamin B_{12}, up to 8.0 μg/100 g. An edible alga, Korean purple laver (*Porphyra* sp.), contains approximately 51.7 μg of B_{12} per 100 g dry weight [51].

4.2.4.3 Zinc Status Management

Zinc is a non-storable trace element and can decline during the development of infective diseases [52]. Giardiasis was reported as a risk factor for zinc malabsorption and metabolism in children which culminated into loss of Zn through urination, decimation of serum zinc carriers, lowering the serum-erythrocyte-zinc concentration [53]. The exact mechanism of zinc level compromise in giardiasis is not well known. There is a high turnover rate of mucosal epithelium and functional immaturity of enzymes associated with giardiasis and the Zn deficiency might be explained by the restoration of hampered mucosal lining as a result of the infection [54]. Another hypothesis suggests that zinc deficiency may result from redistribution of Zn from plasma to the liver, as an acute phase host-response and activation of metallothionein synthesis [55, 56]. Zinc supplementation has been found to reduce diarrhoeal symptoms caused by giardial infection. Experiments have shown, that infected mice fed on zinc supplemented diets included noticeable weight gain, quicker and higher *Giardia*-specific antibody production, and elevated serum zinc level during giardiasis [57]. If we think of a vegetarian diet for zinc supplementation in humans, there are plenty of promising options. WHO recommends 7 mg/day of Zn requirement for vegetarians. Zn enriched breakfast cereals have 1.9–7.8 mg/100 g and sun-dried tomatoes have an even higher Zn content, about 13.6 mg/100 g. Sunflower seeds, pumpkin seeds, and sesame have 5.8, 7.5, and 5.5 mg of Zn per 100 g weight [58].

4.2.5 Fermented Vegetables as Probiotic Source

Probiotics (live-microbes conferring health benefits) can reduce giardial infection through certain mechanisms like competition for limited adhesion sites on the enterocytes. Orally administered probiotics containing lactobacilli can influence the small intestinal micro-flora and may also directly inhibit giardial growth by inducing host immune response [1]. "Sauerkraut" can be considered as a vehicle for delivering probiotic bacteria, which is one of the most popular traditionally consumed fermented dish in Europe and Asia. It is prepared by the lactic acid fermentation of salted white cabbage (*Brassica oleracea* var. capitata). *Lactobacillus plantarum* and *Lactobacillus brevis* predominate dominate as the main microbial fermenter within 3.4–3.7 pH range [59]. "Kimchi" is another equivalent fermented Korean vegetable preparation. Radish is the mostly used raw material, but many other vegetables like cucumber and green onion can be used for the variants of kimchi. Kimchi fermentation is also dominated by *L. plantarum* (about 10^{5-7} bacilli/g), but *Leuconostoc mesenteroides* is also an active fermenter at low temperatures [60]. Mechanistically, strains of *L. plantarum* utilize a mannose-oriented adhesion pattern for intestinal adherence, which is also common for giardial attachment to epithelial cells [61]. Thus, *L. plantarum* from sauerkraut and kimchi can competitively inhibit giardial adhesion to enterocytes and can be consumed in small portions throughout the day [1].

4.2.6 Phenolic Compound Enriched Plant Food

Giardicidal activity of many flavonoid-rich herbs has been documented. *In vitro* activity of oregano (*Origanum vulgare*) and guava leaves (*Psidium guajava*) has been elucidated to be strongly anti-giardial, even more than tinidazole. These herbs have been used in traditional practices to treat diarrhea [1]. Intake of flavonoid supplements and botanical items with high flavonoid quotient like kale, French beans, parsley, and black currants might help in recovery from giardiasis [1, 62]. Intriguingly, polyphenol-rich pressed extracts of blueberry and a commercially packaged blueberry beverage have demonstrated considerable giardicidal activity. The polyphenols of blueberry extract and commercial juice, reduced trophozoite viability and altered their normal morphology at a dose of 167 μg/ml, as only $9.6 \pm 2.8\%$ trophozoites remained alive after 24 hours of incubation [63].

4.3 Traditional Treatment of Giardiasis with Herbal Ethnomedicine

In developing nations, 80% of rural residents rely on traditional medicines for primary health care [64]. The natural plant-derived preparations have proven to be sufficient for curing certain diseases and the active phytochemicals of these traditional crude medicines can be prospectively isolated as pharmaceutical leads [65] Ethnic plant-derived preparations with potential giardicidal properties are summarized in Table 4.1.

4.3.1 Hints of Antiprotozoal Medicines from Ancient Communities

Medicinal plants have been traditionally explored for numerous human diseases, from ancient times in many parts of the globe. In ancient Egypt, herbal sources played a major part in medicinal practices. The plants mentioned in the Ebers Papyrus include many herbs having antidiarrhoeal activity. Acacia plants (*Acacia nilotica*) were used as vermifuge and antidiarrhoeal, internal bleeding treating

Table 4.1 List of traditional medicines for giardiasis.

Categories of practice	Medicinal plants	Giardicidal dosage of crude extract	Reference
Ancient community medicines	*Acacia nilotica* (Mimosaceae)	500 μg/ml (100% inhibition)	[66, 67]
	Cuphea pinetorum (Lythraceae)	84.7 μg/ml (IC_{50})	[68]
	Casimiroa tetrameria (Rutaceae)	10 μg/ml (100% inhibition)	[69]
	Piper longum (Piperaceae) and *Butea monosperma* (Fabaceae)	900 mg/kg body weight (98% inhibition)	[70, 71]
Kitchen remedies	*Zingiber officinale* (Zingiberaceae) and *Cinnamomum zeylenicum* (Lauraceae)	20 mg/kg/day (90% and 100% inhibition)	[72]
	Anethum graveolens (Apiaceae)	1 ml thrice/day (100% inhibition)	[73]
	Allium sativum (Alliaceae)	400 μg/ml (up to 80% inhibition)	[74]
African ethnic medicine	*Trigonella foenum-graecum* (Fabaceae)	5 mg/ml (98–100% inhibition)	[75]
	Mentha piperata (Lamiaceae)	5 mg/ml (up to 92% inhibition)	[75]
	Solenostemma argel (Apocynaceae)	5 mg/ml (up to 59% inhibition)	[75]
South American ethno medicine	*Passiflora edulis* (Passifloraceae)	100 μg/ml (IC_{50})	[76]
	Chenopodium ambrosioides (Chenopodiacea)	250 μg/ml (IC_{50})	[76]
	Dorstenia contrajerva (Moraceae)	23.3 μg/ml (IC_{50})	[77]
	Senna villosa (Caesalpiniaceae)	32.1 μg/ml (IC_{50})	[77]

Source: Based on [65].

agent, which hints about its antiprotozoal activity. Bayberry (*Myrica cerifera*) was also prescribed to stop diarrhea, and for treating ulcers [66]. Contextually, during a research on Sudanese medicinal plants, *A. nilotica* was determined to be effective against giardial subcultures *in vitro*. Extracts caused 100% inhibition at a concentration of 500 μg/ml after 96 hours [67] which validated the medicinal use of acacia plants in ancient practice.

Cuphea pinetorum (Lythraceae), vernacularly called "Bakmomol" and "Vach'vet" by the Tzeltal and Tzotzils, is an important medicine considered by ancient Maya communities of Southern Mexico. In Mayan traditions, a decoction of the shoot is employed as an antidiarrhoeal and antidysenteric medicine. The active extract has a giardicidal IC_{50} (half maximal inhibitory concentration) value of 84.7 μg/ml. A bioassay-guided fractionation experiment showed the presence of eight active phytochemicals in the shoots, of which kaempferol might have the antidiarrhoeal activity in this Mayan traditional medicine [68]. The traditional practice of the Yucatec Mayan descendants of Mexico includes the plant *Casimiroa tetrameria* (Rutaceae), commonly called "Yuy," for the treatments of diarrhoea, dysentery, and severe gastrointestinal cramps. The alcoholic extract of the plant has shown antiprotozoal activity, at the concentration of 10 μg/ml [69].

"Pippali rasayana" (PR) is a traditional Ayurvedic (ancient Indian healing system) formulation consisting of extracts from *Piper longum* and *Butea monosperma*. The anti-giardial and immune-stimulatory effects of PR have been tried on mice and an oral dose of 900 mg/kg body weight, caused up to 98% parasite clearance ($p < 0.001$). PR significantly increased the macrophage phagocytic activity, under 225 mg/kg dosage. However, PR had no precise cytotoxic giardicidal effect *in vitro*, suggesting a different host response dependant clearance mechanism [70]. Further investigation involving a placebo-controlled trial with 50 symptomatic individuals, active treatment with 1 g of PR (thrice daily) caused complete cure in 92% of the subjects. Diarrhea and mucus level of stools dropped significantly and immune status was also improved [71].

4.3.2 Domestic Remedies and Alternative Medicine for Giardiasis

Some of the domestically available plant-parts, used traditionally as condiments and spices in the kitchens around the world, can also be helpful in minimization of giardiasis. *Zingiber officinale*, commonly named ginger, is a popular dietary spice with many medicinal traits. *Cinnamomum zeylanicum* or cinnamon is also a tropical spice used traditionally to cure various diseases. *In vivo* application of ginger and cinnamon's dichloromethane extracts with dosage of 20 mg/kg/day on *Giardia* infected rats, showed efficient cyst reduction up to 90.1 and 100%, whereas trophozoite reduction was 75.4 and 34.1%, respectively. The exposure to ginger and cinnamon extracts also altered the ultrastructure of trophozoites. Cinnamon extract was more effective, as it improved the histopathological consequences of the intestinal mucosal layer, possibly attributed by eugenol, as an immune-stimulant compound. On the other side, ginger was more efficient in decreasing trophozoites [72].

Anethum graveolens, commonly called dill, is a Mediterranean herb used as a spice, also traditionally consumed to heal gastrointestinal disturbances. Filtered and concentrated (from 20 g plant-parts) aqueous extract of dill was examined in pediatric subjects with the dosage of 1 ml thrice a day. The experimental dose led to a complete reduction of giardial cysts, as observed in stool samples after 14 days. However, the exact mechanism of giardicidal effect was not elaborated [73].

The bulbous spice garlic (*Allium sativum*) has been used as a folk medicine in different cultures. Freshly prepared aqueous extract with active garlic components in 400 μg/ml concentration caused up to 80% reduction in the number of giardial trophozoites with clear cytotoxic activity. Further fractionation confirmed the efficacy of thioallyl compound present in garlic, which was behind the giardicidal activity [74].

Ferula assafoetida is a gummy exudate-based strong condiment, extracts of which have been tested on *G. lamblia* cysts. The initial fecundity percentage of the alcoholic extract of asafoetida was 37%, with a concentration of 20 mg/ml, which reached up to 100% after several hours of application [78].

A natural product that has an insect–plant interaction-based derivation is "propolis" or "bee glue." Propolis, a resinous hive-stored mixture made by the efforts of worker bees, has been evaluated to have anti-giardial activity. Propolis from the Sonoran Desert of Mexico was shown to have active components like caffeic acid phenethyl ester, hesperetin, naringenin, and pinocembrin, effective against trophozoites between 222–680 μM concentrations [23]. In a Cuban clinical trial, giardiasis diagnosed patients were treated with a 10–30% solution of propolis. Efficacy evaluation by means of duodenal aspiration showed a 52% rate of cure amongst children treated with 20% solution. A 60% cure-rate was observed in adults who took 30% propolis, whereas tinidazole procured only 40% relief. No associated adverse effects were associated with propolis, which affirmed it as an alternative medicine [1, 79].

4.3.3 Applied Spectrum of Ethnopharmacology in Different Continents

Many plant-derived formulations with ethnomedicinal value are recorded in Asian, African, American, European, and other pharmacopeias, which are traditionally used to treat diarrhea. Some modern experiments revolving around medicinal crude extracts can be considered as an effort to validate the traditional use of medicinal plants with precise antiprotozoal activity. Traditional practices and their recent evaluations across different continents should be discussed schematically, as a basis of drug exploration [65].

4.3.3.1 Asian Medicinal Extracts

Extracts from indigenous medicinal plants of Indian subcontinent like *Aegle marmelos* (Rutaceae), *Acorus calamus* (Araceae), *Pongamia glabra* (Leguminosae), and *Strychnos nux-vomica* (Loganiaceae) have been proven to cure diarrhea in mice [65]. *Jussiaea suffruticosa* (Onagraceae) is another source of traditional medicine from India, where the shoot is pulped and mixed with fermented dairy-drink, as a relief for dysentery. An extract of this plant can cure castor oil-induced diarrhea and intestinal fluid accumulation [65, 80]. In traditional Chinese herbal medicine (CHM), *Codonopsis pilosulab* (Dang Shen), *Agastache rugosa* (Huo Xiang), *Glycyrrhiza uralensis* (Zhi Gan Cao), *Amomum villosum* (Sha Ren), *Perilla frutescens* (Zhi Su), *Magnolia officinalis* (Hou Po), *Angelica dahurica* (Bai Zhi) *Citrus unshiu* (Chen Pi), and *Curcuma wenyujin* (Yu Jin) are described as antidiarrhoeal plants. Orally administrable preparations like capsules, decoctions, and concentrated extract granules, were prepared from these plants in clinical trials, which cured abdominal pain, diarrhea, and visceral hypersensitivity with irritable bowel [81]. In Thailand, a self-medicating population of individuals was reported to utilize certain traditional plant species to treat giardiasis. As per an elaborate study including twelve such species, six exhibited noticeable anti-giardial activity, of which *Alpinia galanga* (Zingiberaceae) chloroform extract was the most efficient medication, with IC_{50} value 37.73 μg/ml [82].

4.3.3.2 African Traditional Medicines

A number of plants are used in Africa for various ailments [65]. One of the preliminary studies on giardicidal plants was carried out with 36 species, ethnically used for gastrointestinal disturbances by the Luo folks from Eastern Africa. The results revealed that methanolic extracts of 21 species caused mortality or growth reduction of trophozoites. Giardicidal capacities were not specified with IC_{50} values but were primitively categorized as active (A), strongly active (SA), weakly active (WA), and inactive (I) according to some unspecified criteria adopted in the pioneering work. Plants from Fabaceae, Asteraceae, Anacardiaceae, and Lamiaceae were mostly classified to be strongly active against *G. lamblia* [83]. *Trigonella foenum-graecum* (Fabaceae) and *Solenostemma argel* (Apocynaceae) are African plants respectively with anabolic and stomach colic properties. *Mentha piperata* (Lamiaceae) is traditionally used in the continent to treat diarrhea and indigestion. The giardicidal activity of the ethanolic extract of these three plants has been tested with 5 mg/ml concentration *in vitro*. Ethanolic extract of *T. foenum-graecum* was the most proficient against the trophozoites and showed 100% inhibition of the parasite after 96 hours of application, compared to 96% inhibition by control drug metronidazole at the same time-point. On the other face, the extract of *M. piperita* showed moderate activity against Giardia lamblia trophozoites with 92% parasite mortality after 96 hours. *S. argel* showed the lowest activity, having a trophozoite reduction rate of 59% after 96 hours [75].

4.3.3.3 South American Ethnomedicine

South American traditional practices encompass a huge variety of medicinal plants, which are particularly more important in the rural areas of Brazil and Mexico.

Following ethnopharmacological reports, there are assorted antidiarrhoeal, non-toxic, frequently available Mexican plants like *Anacardium occidentale* (Anacardiaceae), *Chenopodium ambrosioides* (Chenopodiaceae), *Passiflora edulis* (Passifloraceae), *Stachytarpheta cayennensis* (Verbenaceae), and *P. guajava* (Myrtaceae) have been clinically scrutinized (study group of Brazilian adults) to ascertain respective giardicidal values *in vitro*. Anti-giardial property was determined to be moderate in *A. occidentale* and *P. guajava* with $IC_{50} \leq 500\,\mu g/ml$. Giardicidal activity was higher for *C. ambrosioides* and *S. cayennensis* with $IC_{50} \leq 250\,\mu g/ml$ and even higher in *P. edulis* showing IC_{50} value $\leq 100\,\mu g/ml$ [76]. An antiprotozoal *in vitro* research, included a set of 26 Mexican traditional plants, ethnically used to cure gastrointestinal suffering. Methanolic extracts obtained from *Dorstenia contrajerva*, and *Senna villosa*, showed IC_{50} values 23.3 and 32.1 μg/ml respectively, in case of *G. lamblia*. Moreover, extracts from *Artemisia absinthium*, *Caesalpinia pulcherrima*, *Matricaria recutita*, *Chrysactinia mexicana*, *Lygodium venustum, Dichondra argentea*, *Lippia alba*, *Hippocratea excelsa,* and *Schinus molle* showed moderate activity toward either *Giardia* or *Entamoeba* with IC_{50} values ranging between 38.3 and 82.4 μg/ml. Decoctions of, *Artemisia ludoviciana*, *Bocconia frutescens*, *Chenopodium murale*, *Ocimum basilicum, Thymus vulgaris*, *Cocos nucifera,* and *A. sativum* exhibited moderate activity on *Giardia* with the IC_{50} range 44.1–99.8 μg/ml [77]. Relevance of all this traditional spheres can be correlated to the fundaments of chemically defined drug development to cure giardiasis.

4.4 Phytochemical Candidates for Advanced Giardicidal Drug Development

Plants generate a wide variety of secondary metabolite compounds, represented by the classes of phytochemicals such as terpenoids, alkaloids, flavonoids, steroidal saponins, hydroxycinnamate derivatives, and polyphenols. Few of these are responsible for the organoleptic properties of plants, such as color and fragrance, while others provide protection against pathogenic attack and herbivory [84]. Insightful studies on phytochemistry have guided the recognition and isolation of certain natural compounds with antiprotozoal activity. Strategic innovation of plant-derived drugs can result from chemical structure-based engineering of already known molecules or by the repurposing of medicines already used for other intestinal diseases. Combinatorial screening of chemical libraries and assessment phytochemical leads from traditional medicine can be helpful [85]. Phytochemical candidates of different compound classes, with prominent giardicidal activity are being discussed for advanced herbal drug development. Figure 4.3 shows some of the most promising phytochemical drug candidates.

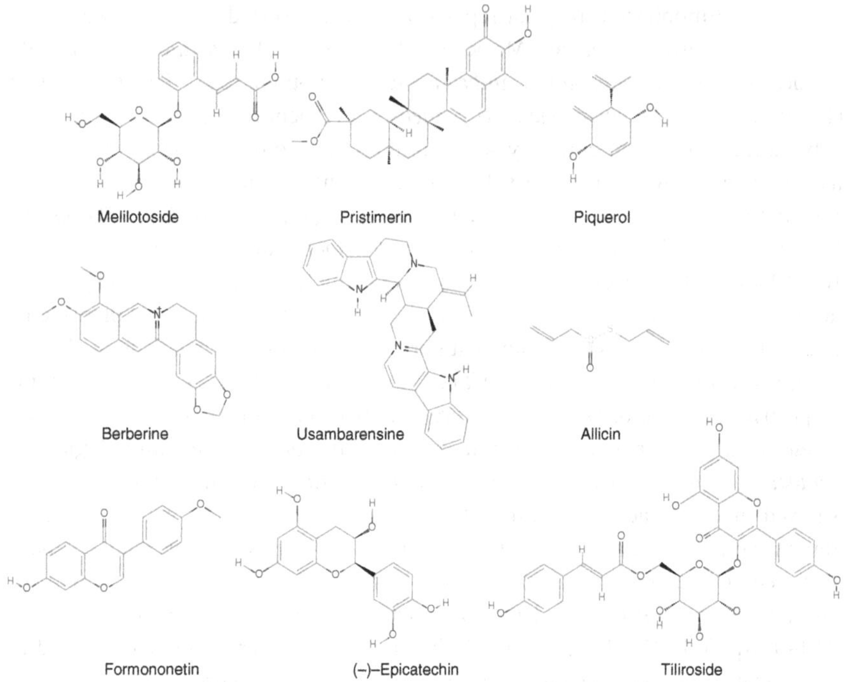

Figure 4.3 Candidate phytochemicals of different compound classes for advanced giardicidal drug development. *Source:* Based on [86–90].

4.4.1 Terpenoids

Melilotoside, a hydroxycinnamate glycoside terpenoid, was reported from the traditionally used plant *Teloxys graveolens* (Chenopodiaceae), which was also found to be the source of flavonoids like chrysin, pinocembrine, narcissin, pinostrobin, and rutin. Amongst these handful of isolated phytochemicals, Melilotoside demonstrated the highest antiparasitic activity against *G. lamblia*, showing 16.8 μg/ml IC50 values [91]. **Linearolactone** (a lactone diterpenoid) and **Polystachyne** (neoclerodane diterpene), were isolated from *Salvia polystachya* (Lamiaceae), which is a South American traditional plant (vernacular name "Chia") known to possess antidysenteric properties. Between the two mentioned diterpenoids, **Linearolactone** was examined to be a potent anti-amoebic and giardicidal phytochemical, with an IC_{50} value of 28.2 μM against *Giardia*. **Polystachyne A**, **B**, and **D** showed moderate giardicidal capacity with the IC_{50} values ranging between 107.5 and 134.7 μM [85, 92]. **Pristimerine** and **Tingenone** are two triterpenoids isolated from the roots of an antiprotozoal traditional plant, Hippocratea excelsa (Hippocrateaceae). The triterpenoids have exhibited IC_{50} values recorded to be 0.11 and 0.74 μM, respectively, against *Giardia* [93]. **7-hydroxy-3,4-dihydrocadalene** and **7-hydroxycalamenene** are two active antiparasitic sesquiterpenes found in *Heterotheca inuloides* (Asteraceae). Anti-giardial IC_{50} value for the cadalene was 64.42 ± 18.24 μg/ml and for the calamenene it was 13.5 ± 0.83 μg/ml. The electron micrographs revealed that the trophozoites of *G. intestinalis* exposed to the active constituents, underwent cellular ultrastructural disorganizations [94]. A monoterpene known as **Piquerol** and a diterpene **Trinervinol** are reported from another plant belonging Asteraceae named *Piqueria trinervia*. Giardicidal IC_{50} values for the compounds were 2.03 and 2.42 μg/ml, respectively. The pure phytochemical isolates exhibited very low toxicity on human fibroblasts, with serial CC_{50} (half-maximal cytotoxic concentration) values 590 and 510 μg/ml, which was comparable with the control drug metronidazole. A speculation suggested the scope of chemical modification of trinervinol, which might further enhance its giardicidal attribute [95]. Certain quassinoids (degraded triterpene lactones) like **Bruceine** (not to be confused with Brucine alkaloid) A, B, C, and D and brusatol were isolated from *Brucea javanica* fruits to examine anti-giardial properties. Amongst these, only **Bruceine A** showed feeble giardicidal activity with an IC_{50} value of 0.021 μM [96].

4.4.2 Aromatic Ether

Precocene I and **II** are chromene class of aromatic ethers obtained from a Thai weed *Ageratum conyzoides* (Asteraceae) extracts, which were perceived as active giardicidal chemicals. Leaf and flower extracts enriched with precocenes showed IC_{50} values 45.67 ± 0.51 and 96.00 ± 0.46 μg/ml, respectively. Electron microscopy showed the degeneration of flagella and anomalies of attachment discs of *G. duodenalis* trophozoites after the treatment of precocene rich extract [97].

4.4.3 Alkaloids

Emetine is an isoquinoline alkaloid incorporated in allopathic medicine, which was originally isolated from *Cephaelis ipecacuanha*. **Dehydroemetine** is an industrial derivative of the natural emetine and is a broad spectrum amoebicide. The conventionally prescribed dosage of dehydroemetine is 1–1.5 mg/kg body weight/day, but it has been reported to be extremely toxic and considered unsuitable due to some strong side effects like vomiting tendency [98]. There aren't many concrete reports of case-specific emetine administration for giardiasis patients. **Berberine** is one of the isoquinoline group alkaloids, found in quite a few medicinal plants. Berberine salts have exhibited *in vitro* antiparasitic activity against *Giardia*. Berberine sulfate was precisely found to alter the cytological features of the trophozoites, inciting vacuole formation, and anomalous glycogen deposition [1]. During an Indian clinical trial, patients were treated with berberine hydrochloride with an oral dose of 5 mg/kg weight/day, for about a week. There was a considerable disappearance of gastrointestinal symptoms, referred to be superior to metronidazole. Feces of the patients showed a 68% less giardial cyst load. According to the authors, the dose increment of berberine could have produced more proficient results against giardiasis [99]. An *in vitro* test, involving the alcoholic extract of Indian traditional medicine *P. longum* (Piperaceae) fruit powder, revealed 100% anti-giardial activity at the concentration of 125 mg/ml. The fractionated crystals of **Piperine** alkaloid showed giardicidal activity at dose of 1000 mg/ml. The number of Giardia trophozoites (initial count 2000) survived was 32.74 ± 0.71 after 24 hours of piperine treatment. The antigiardial activity of this alkaloid was extremely weaker than metronidazole [100]. Indole alkaloids reported from *Strychnos usambarensis* (Loganiaceae), including **Usambarensine** and **3',4'-dihydrousambarensine** are reportedly very strong antiprotozoal compounds. Usambarensine showed the giardicidal IC_{50} value 3.89 ± 1.41 μM and 3',4'-dihydrousambarensine had IC_{50} 5.14 ± 2.20 μM. The most conducive cytotoxic/giardicidal IC_{50} ratios were demonstrated by the same couple of alkaloids, for 3',4'-dihydrousambarensine it was 6.6 and for usambarensine it was 5.45. **Strychnopentamine** and **Isostrychnopentamine** were bis-indolic alkaloids from the same plant source, which exhibited IC_{50} values 7.30 ± 1.85 and 10.9 ± 4.5 μM, respectively for *Giardia* [101]. Phytochemical exploration has identified a pyrrolizidine alkaloid **Osyrisine**, from the Western Asian medicinal plant *Osyris alba* (Santalaceae). Alkaloid isolated from the plant extract showed remarkable *in vitro* giardicidal activity with IC_{50} value 7.57 ± 0.35 μg/ml and was also proved to have no non-cytotoxic effects on human cells, with IC_0 value of 596.66 ± 7.64 μg/ml on Vero cells *in vitro*. This antiparasitic activity and low human toxicity (although higher than metronidazole) indicated its therapeutic safety status for further clinical trials [102].

4.4.4 Organo-sulfur Compounds

Allicin is a thioallyl organo-sulfur compound, isolated from aqueous extract of garlic (*A. sativum*). It has demonstrated high giardicidal activity, potent thiol-modifying property and efficient cysteine proteases inhibitory effect against *Giardia* trophozoites. The notified IC_{50} value of allicin was 0.32 ± 0.02 mM. Moreover, a flow-cytometric investigation confirmed the pro-apoptotic effects of allicin on the trophozoites and exposed parasites also showed a G2 phase cell cycle arrest. *In vivo* treatment of allicin on gerbils with a dose of 36 mg/kg, exhibited a marked reduction of trophozoite count, declared to be better impactful than metronidazole [74]. **Methylgerambullin** is a sulfur-containing amide, reported from the shoot of *Glycosmis trichanthera* (Rutaceae). The high *in vitro* giardicidal efficiency of this compound was recorded with an EC_{50} (half-maximal effective concentration) value of 14.6 μM (6.14 μg/ml), after 24 hours of incubation. With positive efforts, methylgerambulin could be synthetically produced under laboratory conditions, signifying the scopes of further chemical derivation, with elevated biological activity. Methyldambullin and Sakambullin are two other sulfur-amides isolated from the same source which did not prove to have efficient giardicidal capacity. After 24 hours of incubation with 10 μg/ml concentration of the latter compounds, 54.8 ± 4.4 and $7.2 \pm 2.2\%$ parasite inhibition was noted, respectively [103].

4.4.5 Lipids and Steroids

A novel compound named **Galacto-glycerolipid** (GGL) was isolated from an Indian antidysenteric medicinal plant *Oxalis corniculata* (Oxalidaceae). This compound was found to have a firm giardicidal property, exhibiting an IC_{50} value of 3.7 ± 0.6 μg/ml, which was relatable with the control drug metronidazole. The other two saturated fatty acids obtained from the same plant source, had weak activity against the parasite as reflected in the IC_{50} values of 184 ± 0.3 and 206 ± 0.4, respectively. Light microscopic examination of GGL-exposed trophozoites showed hampered adhesion, blebbing, and cell-lysis. Toxicity experiment conducted on the mammalian HEK-293 cells did not show any inhibition by the anti-giardial concentration of GGL [104]. Extractions from the aerial parts of Rubus coriifolius (Rosaceae), a Mexican medicinal plant, revealed a **β-sitosterol 3-*O*-β-D-glucopyranoside** along with six other different active constituents. Giardicidal IC_{50} value recorded for the compound was 61.5 μg/ml [105]. The anti-giardial activity of *Yucca schidigera* (Asparagaceae) plant-powder was found to be quite strong, attributed by the **steroidal saponins**, previously reported to lyse protozoal plasma membrane. Butanolic solution of the streroid-rich yucca powder was subjected to an *in vivo* parasite attachment-inhibition assay on gerbils. Programming eight doses, summing up to 50 mg of powder administration in three days, with an alternative mode of daily doses delivering 18.3 mg powder divided in three days showed notable trophozoite reduction, ultimately leading parasite curing of the ileum.

However, the same dosage did not work out on experimental lambs, which suggested ultra-purification and concentration of the steroidal saponins of powder sample, to further magnify the giardicidal activity [106].

4.4.6 Phenolic Compounds

Asarinin is a phenolic lignan compound reported from an anti-amoebic South American plant *Zanthoxylum liebmannianum* (Rutaceae). Ethanolic crude extract from the shoot showed antiparasitic activity against *G. lamblia*, having the IC_{50} value of 58.00 μg/ml. Amongst the pure chemical isolates, asarinin was proved to be the most effective giardicidal compound, showing 35.45 μg/ml IC_{50} value [107]. Crude extracts of Ruta chalepensis (Rutaceae), with a strength of 150 μg/ml, had shown giardicidal activity with 60.07 μg/ml IC_{50} value. Chemical fractionation from the methanolic solution, led to the revelation of a furanocoumarin phenolic compound named **Chalepensin**, which showed the IC_{50} value 45.95 μg/ml against *Giardia*. Another pure fractionated phenolic compound from the same source was **Carvacrol**, which showed anti-giardial IC_{50} value to be 44.3 μg/ml [108]. **Kaempferol** and **Tiliroside** are two of the many phenolic compounds isolated from the Mexican plants, *Geranium mexicanum* (Geraniaceae) and *C. pinetorum* (Lythraceae). Both of these phytochemicals showed moderate to high anti-giardial activity (highest activity was shown by epicatechin), closer to metronidazole. The respective ED_{50} values obtained from *in vitro* assay were 2.057 and 1.429 μM/kg, respectively [109]. Also in a different study involving *Rubus coriifolius*, **(−)-Epicatechin** was fore-fronted as the most effective anti-giardial phytochemical with IC50 values of 1.9 μg/ml for *Entamoeba histolytica* and 1.6 μg/ml for *G. lamblia*. **Ellagic acid** was another potent giardicidal phenolic compound with 24.9 μg/ml IC_{50} value. These compounds have thus validated the traditional use of the source plant, in the treatment of dysentery and giardiasis [105]. In a chemical fractionation-based experiment involving the plant *Helianthemum glomeratum* (Cistaceae), five antiprotozoal flavonol glycosides were identified namely, **Tiliroside, Kaempferol-glucopyranoside**, **Astragalin**, **Quercitrin**, and **Isoquercitrin**. Tiliroside was revealed as the most proficient amoebicidal and giardicidal glycoside with IC_{50} values of 17.5 μg/ml against *Entamoeba* and 17.4 μg/ml for *Giardia*. Isoquercitrin was however only amoebicidal (IC50 14.7 μg/ml), with the weakest anti-giardial activity. On the other face, Quercitrin showed specific activity against *Giardia* with IC_{50} value 24.3 μg/ml. Astragalin exhibited IC_{50} 47.5 μg/ml and kaempferol-glucopyranoside showed a weaker IC_{50} value of 81.5 μg/ml. Any of these isolated compounds were not more or equally active as metronidazole [110]. Lignan compounds known as **Nordihydroguairetic acid** and **Demethoxyisoguaiacin,** were isolated from the extracts of *Larrea tridentata* (Zygophyllaceae), which is an antiprotozoal plant. Nordihydroguairetic acid was potently active against *G. lamblia* with EC_{50} value of 36 μM. Demethoxyisoguaiacin was also proven to be giardicidal having 49 μM EC_{50} value [111]. From the leguminous tree species *Dalbergia frutescens,* many antiprotozoal isoflavonoids could be fractionated. Amidst these isoflavonoids, **Formononetin** was the principal giardicidal compound, having a remarkable IC_{50} value of 0.03 μg/ml (about 0.1 μM), surprisingly higher than that of metronidazole (100 ng/ml, nearly 0.6 μM). Some more isoflavones were **Daidzein** with IC_{50} 3.75 ± 0.8 μg/ml, **Biochanin A** having that almost same of 3.50 ± 0.65 μg/ml and **Pseudobaptogenin** with IC_{50} of 0.56 ± 0.11. All the rest of the isoflavones are low in activity than metronidazole [112]. Table 4.2 has a synoptic arrangement of all the discussed information on active giardicidal plant-drug candidates.

4.5 Future Prospects

Different parameters for the whole-some management of giardiasis needs ample updates for a better future. These can be discussed as follows.

4.5.1 Extension of Medicinal Compound Screening

As per certain meticulous reviews, about 117 traditional medicinal extracts and about 101 chemically defined herbal drugs are known to have anti-giardial properties. Angiosperm families like Asteraceae, Fabaceae, Rutaceae are extensively studied and contain numerous species that can be a source of giardicidal drug candidates. Further extension of phytochemical explorations is needed with a strategic amalgamation of ethnobotanical, pharmacological, and biochemical studies [113]. Percentages of the different compound classes represented under this account are shown in Figure 4.4a. Moreover, modern high-throughput screening of biologically active anti-trophozoite compounds can unravel new possibilities regarding detailed systemic analysis, for a more accurate choice of drug candidates [90]. On this note of relevance, a high-throughput screening encompassing 4096 active bio-molecules and prescribed drugs, has shown 43 compounds with specific giardicidal properties. Among these drugs, 32 were previously reported and 11 were novel candidates. **Fumagillin**, a bioactive compound from fungal source, exhibited IC_{50} value of 10 nM during anti-giardial screening. Extending the quest for giardicidal phytochemicals toward cryptogams and fungi can thus turn out to be helpful. From a botanical point of view, the dicotyledonous angiosperm families like Asteraceae, followed by Fabaceae, Rutaceae, and Loganiaceae were seen to be enriched with the most effective giardicidal compounds, which is relatable with earlier assessments. Further explorations within these mentioned families (Figure 4.4b) or allied families of the respective orders, might reveal other effective anti-giardial phytochemicals.

4.5.2 Methodical Studies on Toxicology and Drug Adversity

The caveats in the understanding of plant-drug toxicity associated risk factors can be detrimental in the case of rushed therapeutic trails. Many traditional plants with apparent antiprotozoal activity, do not pass through cytotoxic

Table 4.2 Phytochemical candidates for advanced giardicidal drug development.

Class of compounds	Active phytochemicals	Plant sources	Giardicidal dosage	References
Terpenoids	Melilotoside	*Teloxys graveolens* (Chenopodiaceae)	IC_{50} 16.8 μg/ml	[91]
	Linearolactone	*Salvia polystachya* (Lamiaceae)	IC_{50} 28.2 μM	[85, 92]
	Polystachynes (A, B, and D)	"	IC_{50} 107.5 – 134.7 μM	[85, 92]
	Pristimerin	*Hippocratea excelsa* (Hippocrateaceae)	IC_{50} 0.11 μM	[93]
	Tingenone	"	IC_{50} 0.74 μM	[93]
	Trinervinol	*Piqueria trinervia* (Asteraceae)	IC_{50} 2.03 μg/ml	[95]
	Piquerol	"	IC_{50} 2.42 μg/ml	[95]
	Bruceine A	*Brucea javanica* (Simaroubaceae)	IC_{50} 0.021 μM	[96]
	7-Hydroxy-3,4-dihydrocadalene	*Heterotheca inuloides* (Asteraceae)	IC_{50} 64.42 ± 18.24 μg/ml	[94]
	7-Hydroxycalamenene	"	IC_{50} 13.5 ± 0.83 μg/ml	[94]
Aromatic ether	Precocenes I and II	*Ageratum conyzoides* (Asteraceae)	IC_{50} 45.67 ± 0.51 μg/ml	[97]
Alkaloids	Berberine	*Berberis vulgaris* (Berberidaceae)	5 μg/kg weight one day (Human)	[1, 99]
	Piperine	*Piper longum* (Piperaceae)	1000 mg/ml (pure) 250 mg/ml (crude)	[100]
	Usambarensine	*Strychnos usambarensis* (Loganiaceae)	IC_{50} 3.89 ± 1.41 μM	[101]
	3'-4' Dihydro-usambarensine	"	IC_{50} 5.14 ± 2.20 μM	[101]
	Strychnopentamine	"	IC_{50} 7.30 ± 1.85 μM	[101]
	Isostrychnopentamine	"	IC_{50} 10.9 ± 4.5 μM	[101]

Organo-sulphur compunds	Allicin	*Allium sativum* (Alliaceae)	IC_{50} 0.32 ± 0.02 μg/ml 36 mg/kg (Gerbils)	[74]
	Methylgerambulin	*Glycosmis trichanthera* (Rutaceae)	EC_{50} 14.6 μM (24 hours)	[103]
Lipids and steroids	Galacto-glycerolipid	*Oxalis carnniculata* (Oxalidaceae)	IC_{50} 3.7 ± 0.6 μg/ml	[98]
	β-Sitosterol 3-*O*-β-D-glucopyranoside	*Rubus coriifolius* (Rosaceae)	IC_{50} 61.5 μg/ml	[105]
	Steroidal saponins	*Yucca schidigera* (Asparagaceae)	18.3 mg/day (Gerbils)	[106]
Phenolics	Asarinin	*Zanthoxylum liebmannianum* (Rutaceae)	IC_{50} 35.45 μg/ml	[107]
	Chalepensin	*Ruta chalepensis* (Rutaceae)	IC_{50} 45.95 μg/ml	[108]
	Kaempferol and Tiliroside	*Geranium mexicanum* (Geraniaceae) *Cuphea pinetorum* (Lythraceae)	ED_{50} 2.057 μM/kg and 1.429 μM/kg	[109] [109]
	(−)-Epicatechin and Ellagic acid	*Rubus coriifolius* (Rosaceae)	IC_{50} 1.6 and 24.9 μg/ml	[105]
	Astragallin	*Helianthemum glomeratum* (Cistaceae)	IC_{50} 24.3 μg/ml	[110]
	Quercitrin	”	IC_{50} 47.5 μg/ml	[110]
	Formononetin	*Dalbergia frutescens* (Fabaceae)	IC_{50} 0.03 μg/ml	[112]
	Daidzein	”	IC_{50} 3.75 ± 0.8 μg/ml	[112]
	Biochanin A	”	IC_{50} 3.50 ± 0.65 μg/ml	[112]
	Pseudobaptogenin	”	IC_{50} 0.56 ± 0.11 μg/ml	[112]
	Norhydroguaiaretic acid and Demethoxyiso-guaiacin	*Larrea tridentata* (Zygophyllaceae)	EC_{50} 36 μM and 49 μM	[111]

Source: Based on [112].

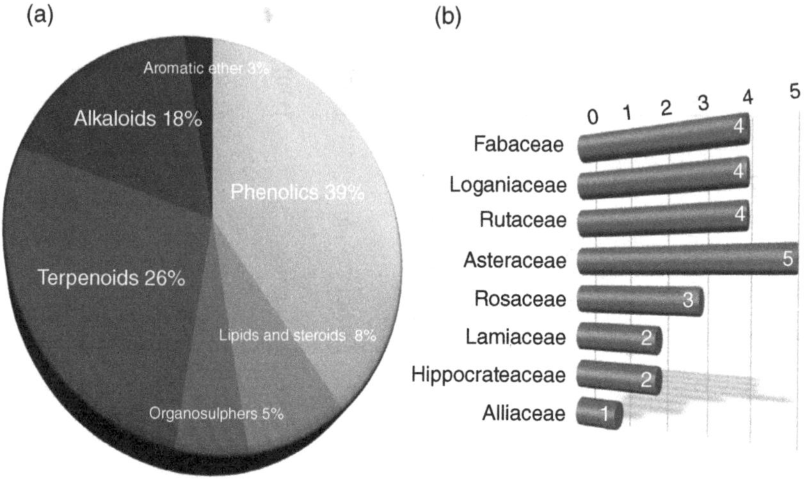

Figure 4.4 (a) Percentages of the discussed compound classes with giardicidal properties. (b) Angiosperm families with the numbers of phytochemical candidates for advanced anti-giardial herbal drug development (Note: Data are based on authors own projections).

examination. Over-dosage of unknown or known plants can lead to toxicity, due to misidentification, erroneous cultivation or harvesting, and preparation/preservation hazards. Toxicological assays should be managed by supervisory bodies, with stricter emphasis on pharmaco-vigilance in plant drug approval [76]. Certain kinds of drugs can be adverse and might propagate the symptoms even more. For example, a study has shown that injection of cortisone acetate in mice, can increase the trophozoite count of *Giardia muris,* till eight consecutive weeks [114]. Similarly, plant drugs with steroids might also cause such hazards, which need to be carefully studied before higher organism clinical trials.

4.5.3 Correlation of Herbal Drugs with Molecular Targets of *Giardia*

In the field of novel drug innovation, researchers have already started to hit on the molecular targets of *Giardia*, like its antioxidant enzymes, bumped kinases, and virulence factors like cysteine proteases, arginine deaminase, and carbamate kinase to name a few [23]. Disappointingly, there is not much effort of herbal drug application to point these molecular targets. Biotechnological and biochemical assays might find out such links between phytochemical candidates and the mentioned molecular targets of the parasite.

4.6 Summary, Prospects, and Conclusion

It could be fairly understood that dietary ingredients and domestic remedies also provide major support for the management of giardiasis and associated malabsorption syndrome. Blueberry polyphenols can be dually perceived as a supplement or alternative medicine. Allicin from garlic is perhaps the most easily accessible active component with strong giardicidal activity. The role of dietary fibers, agglutinins, prebiotics, and probiotics has been seen in the maintenance of small bowel, and vitamin and mineral-rich foods are also evaluated. If we think about the impact of traditional practices as well as ethnopharmacological investigation of giardicidal activity, tropical–subtropical countries like Sudan, China, India, Thailand, Cuba have made considerable efforts, but the largest contribution into this spectrum has come from Mexico and Brazil. A large number of antidiarrhoeal or antidysenteric plants have been followed up till the level of phytochemical characterization. All the leading giardicidal phytochemicals could be broadly grouped into compound classes but there are certain overlaps as well, as in the case of the monoterpene phenolic compound carvacrol. Phenolic compounds like (−)-Epicatechin, Formononetin are very promising candidates and more effective than metronidazole, whereas the alkaloid Piperine had almost no appreciable giardicidal activity. Human trials and cytotoxic assays were not reported uniformly, but the clinical applications of dill and the alkaloid Berberine was noticeable. Giardicidal IC_{50} dosages reviewed here were expressed in μM and μg/ml values, which can be easily interconverted (value of IC_{50} μg × 1000/Mol. Wt. of phytochemical = IC_{50} μM value) for comparative analyses. Even if the metronidazole resistance of the parasite becomes further strengthened, significant phytochemical exploration, backed up by toxicological and molecular studies can eradicate the adversities of giardiasis in the days to come.

References

1 Hawrelak, J. (2003). Giardiasis: pathophysiology and management. *Altern. Med. Rev.* 8 (2): 129–142.

2 Savioli, L., Smith, H., and Thompson, A. (2006). Giardia and cryptosporidium join the 'neglected diseases initiative'. *Trends Parasitol.* 22 (5): 203–208.

3 Ankarlev, J., Jerlstrom-Hultqvist, J., Ringqvist, E. et al. (2010). Behind the smile: cell biology and disease mechanisms of *Giardia* species. *Nat. Rev. Microbiol.* 8: 413–422.

4 Buret, A.G., Amat, C.B., Manko, A. et al. (2015). *Giardia duodenalis*: new research developments in pathophysiology, pathogenesis, and virulence factors. *Curr. Trop. Med. Rep.* 2 (3): 110–118. https://doi.org/10.1007/s40475-015-0049-8.

5 Leung, A.K.C., Leung, A.A.M., Wong, A.H.C. et al. (2019). Giardiasis: an overview. *Recent Patents Inflamm. Allergy Drug Discov.* 13 (2): 134–143.

6 Kimberlin, D.W., Brady, M.T., Jackson, M.A., and Long, S.S. (2015). *Giardia intestinalis (Formerly Giardia lamblia and Giardia duodenalis) Infections*. Illinois: American Academy of Pediatrics.

7 Lebwohl, B., Deckelbaum, R.J., and Green, P.H. (2003). Giardiasis. *Gastrointest. Endosc.* 57 (7): 906–913. https://doi.org/10.1016/S0016-5107(03)70028-5.

8 Adam, E.A., Yoder, J.S., Gould, L.H. et al. (2016). Giardiasis outbreaks in the United States, 1971–2011. *Epidemiol. Infect.* 144 (13): 2790–2801.

9 Halliez, M.C. and Buret, A.G. (2013). Extra-intestinal and long term consequences of *Giardia duodenalis* infections. *World J. Gastroenterol.* 19 (47): 8974–8985.

10 Langford, T.D., Housley, M.P., Boes, M. et al. (2002). Central importance of immunoglobulin A in host defense against *Giardia* spp. *Infect. Immun.* 70: 11–18.

11 Katz, D.E. and Taylor, D.N. (2001). Parasitic infections of the gastrointestinal tract. *Gastroenterol. Clin. N. Am.* 30: 797–815.

12 Plutzer, J., Ongerth, J., and Karanis, P. (2010). Giardia taxonomy, phylogeny and epidemiology: facts and open questions. *Int. J. Hyg. Environ. Health* 213 (5): 321–333.

13 Adam, R.D. (2001). Biology of *Giardia lamblia*. *Clin. Microbiol. Rev.* 14 (3): 447–475. https://doi.org/10.1128/CMR.14.3.447-475.2001.

14 Gardner, T.B. and Hill, D.R. (2001). Treatment of giardiasis. *Clin. Microbiol. Rev.* 14 (1): 114–128. https://doi.org/10.1128/CMR.14.1.114-128.2001.

15 Birky, C.W. Jr. (2010). Giardia sex? Yes, but how and how much? *Trends Parasitol.* 26 (2): 70–74. https://doi.org/10.1016/j.pt.2009.11.007.

16 Poxleitner, M.K. et al. (2008). Evidence for karyogamy and exchange of genetic material in the binucleate intestinal parasite *Giardia intestinalis*. *Science* 319: 1530–1533.

17 Carpenter, M.L., Assaf, Z.J., Gourguechon, S., and Cande, W.Z. (2012). Nuclear inheritance and genetic exchange without meiosis in the binucleate parasite *Giardia intestinalis*. *J. Cell Sci.* 125 (Pt 10): 2523–2532. https://doi.org/10.1242/jcs.103879.

18 Nain, C.K., Dutt, P., and Vinayak, V.K. (1991). Alterations in enzymatic activities of the intestinal mucosa during the course of *Giardia lamblia* infection in mice. *Ann. Trop. Med. Parasitol.* 85: 515–522.

19 Cotton, J., Bhargava, A., Ferraz, J. et al. (2014). *Giardia duodenalis* cathepsin B proteases degrade intestinal epithelial interleukin-8 and attenuate interleukin-8-induced neutrophil chemotaxis. *Infect. Immun.* 82 (7): 2772–2787.

20 DuBois, K.N., Abodeely, M., Sakanari, J. et al. (2008). Identification of the major cysteine protease of *Giardia* and its role in encystation. *J. Biol. Chem.* 283: 18024–18031.

21 Banik, S., Renner Viveros, P., Seeber, F. et al. (2013). *Giardia duodenalis* arginine deiminase modulates the phenotype and cytokine secretion of human dendritic cells by depletion of arginine and formation of ammonia. *Infect. Immun.* 81: 2309–2317.

22 Escobedo, A.A. and Cimerman, S. (2007). Giardiasis: a pharmacotherapy review. *Expert. Opin. Pharmacother.* 8: 1885–1902.

23 Argüello-García, R., Leitsch, D., Skinner-Adams, T., and Ortega-Pierres, M.G. (2020). Drug resistance in *Giardia*: mechanisms and alternative treatments for giardiasis. *Adv. Parasitol.* 107: 201–282.

24 Muller, J., Schildknecht, P., and Muller, N. (2013). Metabolism of nitro drugs metronidazole and nitazoxanide in *Giardia lamblia*: characterization of a novel nitroreductase (GlNR2). *J. Antimicrob. Chemother.* 68: 1781–1789.

25 IARC (1987). *Evaluation of Carcinogenic Risk to Humans*, 250–251. Lyon, France: International Agency for Research on Cancer Suppl. 7.

26 Peters, D.H., Garg, A., Bloom, G. et al. (2008). Poverty and access to health care in developing countries. *Ann. N. Y. Acad. Sci.* 1136: 161–171.

27 Litleskare, S., Wensaas, K.A., Eide, G.E. et al. (2015). Perceived food intolerance and irritable bowel syndrome in a population 3 years after a giardiasis-outbreak: a historical cohort study. *BMC Gastroenterol.* 15: 164. https://doi.org/10.1186/s12876-015-0393-0.

28 Cummings, J.H., Christie, S., and Cole, T.J. (2001). A study of fructo oligosaccharides in the prevention of traveller's diarrhoea. *Aliment. Pharmacol. Ther.* 15: 1139–1145.

29 Schneeman, B.O. (1999). Fiber, inulin and oligofructose: similarities and differences. *J. Nutr.* 129: 1424S–1427S.

30 Buddington, R.K., Williams, C.H., Chen, S.C., and Witherly, S.A. (1996). Dietary supplement of neosugar alters the fecal flora and decreases activities of some reductive enzymes in human subjects. *Am. J. Clin. Nutr.* 63: 709–716.

31 Tokunaga, T., Oku, T., and Hosoya, N. (1989). Utilization and excretion of a new sweetener, fructooligosaccharide (Neosugar), in rats. *J. Nutr.* 119 (4): 553–559. https://doi.org/10.1093/jn/119.4.553.

32 Agopian, R.G.D., Soares, C.A., Purgatto, E. et al. (2008). Identification of fructooligosaccharides in different banana cultivars. *J. Agric. Food Chem.* 56 (9): 3305–3310. https://doi.org/10.1021/jf073011l.

33 Sengupta, S., Mukherjee, A., Ray, L., and Sengupta, S. (2013). *Tinospora cordifolia*, a novel source of extracellular disaccharidases, useful for human disaccharidase deficiency therapy. *Phytother. Res. PTR* 27 (5): 725–730. https://doi.org/10.1002/ptr.4780.

34 Khan, M.M., Haque, M.S., and Chowdhury, M.S.I. (2017). Medicinal use of the unique plant Tinospora Cordifolia: evidence from the traditional medicine and recent research. *Asian J. Med. Biol. Res.* 2 (4): 508. https://doi.org/10.3329/ajmbr.v2i4.30989.

35 Leitch, G.J., Visvesvara, G.S., Wahlquist, S.P., and Harmon, C.T. (1989). Dietary fiber and giardiasis: dietary fiber reduces rate of intestinal infection by *Giardia lamblia* in the gerbil. *Am. J. Trop. Med. Hyg.* 41: 512–520.

36 Satchithanandam, S., Vargofcak-Apker, M., Calvert, R.J. et al. (1990). Alteration of gastrointestinal mucin by fiber feeding in rats. *J. Nutr.* 120: 1179–1184.

37 Roberfroid, M. (1993). Dietary fiber, inulin, and oligofructose: a review comparing their physiological effects. *Crit. Rev. Food Sci. Nutr.* 33: 103–148.

38 Rasane, P., Jha, A., Sabikhi, L. et al. (2015). Nutritional advantages of oats and opportunities for its processing as value added foods – a review. *J. Food Sci. Technol.* 52 (2): 662–675. https://doi.org/10.1007/s13197-013-1072-1.

39 Meng, T.C., Hetsko, M.L., and Gillin, F.D. (1996). Inhibition of *Giardia lamblia* excystation by antibodies against cyst walls and by wheat germ agglutinin. *Infect. Immun.* 64: 2151–2157.

40 Ortega-Barria, E., Ward, H.D., Keusch, G.T., and Pereira, M.E. (1994). Growth inhibition of the intestinal parasite *Giardia lamblia* by a dietary lectin is associated with arrest of the cell cycle. *J. Clin. Invest.* 94: 2283–2288.

41 Grant, J., Mahanty, S., Khadir, A. et al. (2001). Wheat germ supplement reduces cyst and trophozoite passage in people with giardiasis. *Am. J. Trop. Med. Hyg.* 65: 705–710.

42 Farthing, M.J. (1996). Giardiasis. *Gastroenterol. Clin. N. Am.* 25: 493–515.

43 Miller, S.J. (1989). Nutritional deficiency and the skin. *J. Am. Acad. Dermatol.* 21: 1–30.

44 Girard, C., Dereure, O., Blatiere, V. et al. (2006). Vitamin A deficiency phrynoderma associated with chronic giardiasis. *Pediatr. Dermatol.* 23 (4): 346–349.

45 Lima, A.A., Soares, A.M., Lima, N.L. et al. (2010). Effects of vitamin A supplementation on intestinal barrier function, growth, total parasitic, and specific *Giardia* spp infections in Brazilian children: a prospective randomized, double-blind, placebo-controlled trial. *J. Pediatr. Gastroenterol. Nutr.* 50 (3): 309–315. https://doi.org/10.1097/MPG.0b013e3181a96489.

46 Solomons, N.W. (2006). Vitamin A. In: *Present Knowledge in Nutrition*, 9e (eds. B. Bowman and R. Russell), 157–183. Washington, DC: International Life Sciences Institute.

47 Greiner, T. (2013). *Vitamin A: Moving the Food-Based Approach Forward.* FAO and WHO. http://www.fao.org/fileadmin/user_upload/agn/pdf/Greiner_VITAMIN_A_Final.pdf.

48 Loganathan, R., Subramaniam, K.M., Radhakrishnan, A.K. et al. (2017 Feb). Health-promoting effects of red palm oil: evidence from animal and human studies. *Nutr. Rev.* 75 (2): 98–113. https://doi.org/10.1093/nutrit/nuw054.

49 Ament, M.E. and Rubin, C.E. (1972). Relation of giardiasis to abnormal intestinal structure and function in gastrointestinal immunodeficiency syndromes. *Gastroenterology* 62: 216–226.

50 Springer, S.C. and Key, J.D. (1997). Vitamin B_{12} deficiency and subclinical infection with *Giardia lamblia* in an adolescent with agammaglobulinemia of bruton. *J. Adolesc. Health* 20 (1): 58–61. https://doi.org/10.1016/s1054-139x(96)00171-1.

51 Watanabe, F., Yabuta, Y., Bito, T., and Teng, F. (2014). Vitamin B_{12}-containing plant food sources for vegetarians. *Nutrients* 6 (5): 1861–1873. https://doi.org/10.3390/nu6051861.

52 Culha, G. and Sangun, M.K. (2007). Serum levels of zinc, copper, iron, cobalt, magnesium and selenium elements in children diagnosed with *Giardia* intestinalis and *Enterobiosis Vermicularis* in Hatay, Turkey. *Biol. Trace Elem. Res.* 118: 21–26.

53 Jendryczko, A., Sodowska, H., and Drozdz, M. (1993). Zinc deficiency in children infected with *Giardia lamblia*. *Wiad. Lek.* 46: 32–35.

54 Buret, A.G., Hardin, J., Olson, M., and Gall, D. (1992). Pathophysiology of small intestinal malabsorption in gerbils infected with *Giardia lamblia*. *Gastroenterology* 103: 506–513.

55 Cuevas, L.E. and Kovanagi, A. (2005). Zinc and infection: a review. *Ann. Trop. Paediatr.* 25: 149–160.

56 Schroeder, J.J. and Cousins, R.J. (1990). Interleukin 6 regulates metallothionein gene expression and zinc metabolism in hepatocyte monolayer cultures. *Proc. Natl. Acad. Sci. USA* 87: 3137–3141.

57 Astiazarán-García, H., Iñigo-Figueroa, G., Quihui-Cota, L., and Anduro-Corona, I. (2015). Crosstalk between zinc status and *Giardia* infection: a new approach. *Nutrients* 7 (6): 4438–4452.

58 Saunders, A.V., Craig, W.J., and Baines, S.K. (2012). Zinc and vegetarian diets. *MJA Open.* 1 (S2): 17–21. https://doi.org/10.5694/mja11.11493.

59 Penas, E., Martinez-Villaluenga, C., and Frias, J. (2017). *Chapter 24 – Sauerkraut: Production, Composition, and Health Benefits Fermented Foods in Health and Disease Prevention*, 557–576. Boston: Academic Press.

60 Cheigh, H., Park, K., and Lee, C.Y. (1994). Biochemical, microbiological, and nutritional aspects of kimchi (Korean fermented vegetable products). *Crit. Rev. Food Sci. Nutr.* 34 (2): 175–203. https://doi.org/10.1080/10408399409527656.

61 Sousa, M.C., Goncalves, C.A., Bairos, V.A., and Poiares-Da-Silva, J. (2001). Adherence of *Giardia lamblia* trophozoites to Int-407 human intestinal cells. *Clin. Diagn. Lab. Immunol.* 8: 258–265.

62 Wohlmuth, H. (1998). *Pharmacognosy and Medicinal Plant Pharmacology*, 77–89. Lismore, Australia: Southern Cross University Press.

63 Anthony, J.P., Fyfe, L., Stewart, D. et al. (2007). The effect of blueberry extracts on *Giardia duodenalis* viability and spontaneous excystation of *Cryptosporidium parvum* oocysts, *in vitro*. *Methods (San Diego, Calif.)* 42 (4): 339–348.

64 Kim, H.-S. (2005). Do not put too much value on conventional medicines. *J. Ethnopharmacol.* 100: 37–39.

65 Palombo, E.A. (2006). Phytochemicals from traditional medicinal plants used in the treatment of diarrhoea: modes of action and effects on intestinal function. *Phytother. Res. PTR* 20 (9): 717–724.

66 Abou El-Soud, N. (2010). Herbal medicine in ancient Egypt. *J. Med. Plants Res.* 4: 82–86.

67 Kabbashi, A.S., Osman, E.E., Garbi, M.I. et al. (2017). *In vitro* Antiprotozoal Activities and Cytotoxicity of Selected Sudanese Medicinal Plants. *International Journal of Biomedical Engineering and Clinical Science.* 3 (2): 6–13. https://doi.org/10.11648/j.ijbecs.20160302.11.

68 Calzada, F. (2005). Additional antiprotozoal constituents from *Cuphea pinetorum*, a plant used in Mayan traditional medicine to treat diarrhoea. *Phytother. Res.* 19 (8): 725–727. https://doi.org/10.1002/ptr.1717.

69 Heinrich, M., Heneka, B., Ankli, A. et al. (2005). Spasmolytic and antidiarrhoeal properties of the Yucatec Mayan medicinal plant *Casimiroa tetrameria*. *J. Pharm. Pharmacol.* 57 (9): 1081–1085.

70 Agarwal, A.K., Singh, M., Gupta, N. et al. (1994). Management of giardiasis by an immuno-modulatory herbal drug Pippali rasayana. *J. Ethnopharmacol.* 44: 143–146.

71 Agarwal, A.K., Tripathi, D.M., Sahai, R. et al. (1997). Management of giardiasis by a herbal drug 'Pippali rasayana': a clinical study. *J. Ethnopharmacol.* 56: 233–236.

72 Mahmoud, A., Attia, R., Said, S., and Ibraheim, Z. (2014). Ginger and cinnamon: can this household remedy treat giardiasis? Parasitological and histopathological studies. *Iran. J. Parasitol.* 9 (4): 530–540.

73 Sahib, A.S., Mohammed, I.H., and Sloo, S.A. (2014). Antigiardial effect of *Anethum graveolens* aqueous extract in children. *J. Intercult. Ethnopharmacol.* 3 (3): 109–112. https://doi.org/10.5455/jice.20140523104104.

74 Argüello-García, R., de la Vega-Arnaud, M., Loredo-Rodríguez, I.J. et al. (2018). Activity of thioallyl compounds from garlic against *Giardia duodenalis* trophozoites and in experimental giardiasis. *Front. Cell. Infect. Microbiol.* 8: 353. https://doi.org/10.3389/fcimb.2018.00353.

75 Najumudin, K., Ayubu, J., and Elnazeer, A.M. (2018). Antigiardial activity of some plant extracts used in traditional medicine in Sudan in comparison with metronidazole. *Microbiol. Curr. Res.* 2 (4): 75–82.

76 Neiva, V.D., Ribeiro, M.N., Nascimento, F.R. et al. (2014). Plant species used in giardiasis treatment: ethnopharmacology and *in vitro* evaluation of anti-Giardia activity. *Revista Brasileira De Farmacognosia-brazilian J. Pharmacognosy* 24: 215–224.

77 Calzada, F., Yépez-Mulia, L., and Aguilar, A. (2006). *In vitro* susceptibility of *Entamoeba histolytica* and *Giardia lamblia* to plants used in Mexican traditional medicine for the treatment of gastrointestinal disorders. *J. Ethnopharmacol.* 108 (3): 367–370. https://doi.org/10.1016/j.jep.2006.05.025.

78 Nazer, M.R., Abbaszadeh, S., Anbari, K., and Shams, M. (2019). A review of the most important medicinal herbs affecting giardiasis. *J. HerbMed. Pharmacol.* 8: 78–84.

79 Miyares, C., Hollands, I., Castaneda, C. et al. (1988). Clinical trial with a preparation based on propolis "propolisina" in human giardiasis. *Acta Gastroenterol. Latinoam.* 18: 195–201. [Article in Spanish].

80 Murugesan, T., Ghosh, L., Mukherjee, K. et al. (2000). Evaluation of antidiarrhoeal profile of *Jussiaea suffruticosa* Linn. extract in rats. *Phytother. Res.* 14: 381–383.

81 Zhu, J.J., Liu, S., Su, X.L. et al. (2016). Efficacy of Chinese herbal medicine for diarrhea-predominant irritable bowel syndrome: a meta-analysis of randomized, double-blind, placebo-controlled trials. Evidence-based complementary and alternative medicine. *eCAM* 2016: 4071260. https://doi.org/10.1155/2016/4071260.

82 Sawangjaroen, N., Subhadhirasakul, S., Phongpaichit, S. et al. (2005). The *in vitro* anti-giardial activity of extracts from plants that are used for self-medication by AIDS patients in southern Thailand. *Parasitol. Res. 95*: 17–21.

83 Johns, T., Faubert, G.M., Kokwaro, J.O. et al. (1995). Anti-giardial activity of gastrointestinal remedies of the Luo of East Africa. *J. Ethnopharmacol.* 46: 17–23.

84 Demain, A.L. and Fang, A. (2000). The natural functions of secondary metabolites. *Adv. Biochem. Eng. Biotechnol.* 69: 1–39.

85 Ramirez-Moreno, E., Soto-Sanchez, J., Rivera, G., and Marchat, L.A. (2017). Mexican medicinal plants as an alternative for the development of new compounds against protozoan parasites. *Nat. Remed. Fight Against Parasit.* https://doi.org/10.5772/67259.

86 National Center for Biotechnology Information (2020). PubChem Compound Summary for CID 5280759, Melilotoside. https://pubchem.ncbi.nlm.nih.gov/compound/Melilotoside (accessed 6 September 2020).

87 National Center for Biotechnology Information (2020). PubChem Compound Summary for CID 159516, Pristimerin. https://pubchem.ncbi.nlm.nih.gov/compound/Pristimerin (accessed 6 September 2020).

88 National Center for Biotechnology Information (2020). PubChem Compound Summary for CID 161734, Piquerol A. https://pubchem.ncbi.nlm.nih.gov/compound/Piquerol-A (accessed 6 September 2020).

89 National Center for Biotechnology Information (2020). PubChem Compound Summary for CID 2353, Berberine. https://pubchem.ncbi.nlm.nih.gov/compound/Berberine (accessed 6 September 2020).

90 National Center for Biotechnology Information (2020). PubChem Compound Summary for CID 72276, (−)-Epicatechin. https://pubchem.ncbi.nlm.nih.gov/compound/Epicatechin (accessed 6 September 2020).

91 Calzada, F., Velázquez, C., Cedillo-Rivera, R., and Esquivel, B. (2003). Antiprotozoal activity of the constituents of *Teloxys graveolens*. *Phytother. Res.* 17 (7): 731–732.

92 Calzada, F., Yepez-Mulia, L., Tapia-Contreras, A. et al. (2010). Evaluation of the antiprotozoal activity of neoclerodane type diterpenes from *Salvia polystachya* against *Entamoeba histolytica* and *Giardia lamblia*. *Phytother. Res.* 24 (5): 662–665. https://doi.org/10.1002/ptr.2938.

93 Mena-Rejón, G.J., Pérez-Espadas, A.R., Moo-Puc, R.E. et al. (2007). Antigiardial activity of triterpenoids from root bark of *Hippocratea excelsa*. *J. Nat. Prod.* 70 (5): 863–865.

94 Rodríguez-Chávez, J.L., Rufino-González, Y., Ponce-Macotela, M., and Delgado, G. (2015). *In vitro* activity of 'Mexican Arnica' *Heterotheca inuloides* Cass natural products and some derivatives against *Giardia intestinalis*. *Parasitology* 142 (4): 576–584.

95 Rufino-González, Y., Ponce-Macotela, M., Jiménez-Estrada, M. et al. (2017). *Piqueria trinervia* as a source of metabolites against *Giardia intestinalis*. *Pharm. Biol.* 55 (1): 1787–1791. https://doi.org/10.1080/13880209.2017.1325912.

96 Wright, C.W., Anderson, M.M., Allen, D. et al. (1993). Quassinoids exhibit greater selectivity against *Plasmodium falciparum* than against *Entamoeba histolytica*, *Giardia intestinalis* or *Toxoplasma gondii in vitro*. *J. Eukaryot. Microbiol.* 40 (3): 244–246. https://doi.org/10.1111/j.1550-7408.1993.tb04910.x.

97 Pintong, A.R., Ruangsittichai, J., Ampawong, S. et al. (2020). Efficacy of *Ageratum conyzoides* extracts against *Giardia duodenalis* trophozoites: an experimental study. *BMC Complement. Med. Ther.* 20 (1): 63.

98 Virk, A. (2017). Amebiasis, giardiasis, and other intestinal protozoan infections. In: *The Travel and Tropical Medicine Manual*, 5e (eds. C.A. Sanford, P.S. Pottinger and E.C. Jong), 429–443. https://doi.org/10.1016/b978-0-323-3750 6-1.00032-5.

99 Choudhry, V.P., Sabir, M., and Bhide, V.N. (1972). Berberine in giardiasis. *Indian Pediatr.* 9: 143–146.

100 Tripathi, D.M., Gupta, N., Lakshmi, V. et al. (1999). Antigiardial and immunostimulatory effect of *Piper longum* on giardiasis due to *Giardia lamblia*. *Phytother. Res. PTR* 13 (7): 561–565.

101 Wright, C.W., Allen, D., Cai, Y. et al. (1994). Selective antiprotozoal activity of some *Strychnos* alkaloids. *Phytother. Res.* 8: 149–152. https://doi.org/10.1002/ptr.2650080306.

102 Al-Jaber, H.I., Mosleh, I.M., Mallouh, A. et al. (2010). Chemical constituents of *Osyris alba* and their antiparasitic activities. *J. Asian Nat. Prod. Res.* 12: 814–820. https://doi.org/10.1080/10286020.2010.502892.

103 Drinić, M., Raninger, A., Zraunig, A. et al. (2019). Activity of methylgerambullin from *Glycosmis* species (Rutaceae) against *Entamoeba histolytica* and *Giardia duodenalis in vitro*. *Int. J. Parasitol. Drugs Drug Resist.* 10: 109–117.

104 Manna, D.K., Dutta, P.K., Achari, B., and Lohia, A. (2010). A novel galacto-glycerolipid from *Oxalis corniculata* kills *Entamoeba histolytica* and *Giardia lamblia*. *Antimicrob. Agents Chemother.* 54: 4825–4832.

105 Alanís, A.D., Calzada, F., Cedillo-Rivera, R., and Meckes, M. (2003). Antiprotozoal activity of the constituents of *Rubus coriifolius*. *Phytother. Res.* 17: 681–682. https://doi.org/10.1002/ptr.1150.

106 McAllister, T.A., Annett, C.B., Cockwill, C.L. et al. (2001). Studies on the use of *Yucca schidigera* to control giardiosis. *Vet. Parasitol.* 97 (2): 85–99. https://doi.org/10.1016/s0304-4017(01)00394-6.

107 Arrieta, J., Reyes, B., Calzada, F. et al. (2001). Amoebicidal and giardicidal compounds from the leaves of *Zanthoxylum liebmannianun*. *Fitoterapia* 72: 295–297.

108 Quintanilla-Licea, R., Mata-Cárdenas, B.D., Vargas-Villarreal, J. et al. (2014). Antiprotozoal activity against *Entamoeba histolytica* of plants used in northeast Mexican traditional medicine. Bioactive compounds from *Lippia graveolens* and *Ruta chalepensis*. *Molecules* 19 (12): 21044–21065. https://doi.org/10.3390/molecules191221044.

109 Barbosa, E., Calzada, F., and Campos, R. (2007). *In vivo* antigiardial activity of three flavonoids isolated of some medicinal plants used in Mexican traditional medicine for the treatment of diarrhea. *J. Ethnopharmacol.* 109 (3): 552–554.

110 Calzada, F. and Alanís, A.D. (2007). Additional antiprotozoal flavonol glycosides of the aerial parts of *Helianthemum glomeratum*. *Phytother. Res. PTR* 21 (1): 78–80. https://doi.org/10.1002/ptr.2031.

111 Bashyal, B., Li, L., Bains, T. et al. (2017). Larrea tridentata: a novel source for anti-parasitic agents active against *Entamoeba histolytica*, *Giardia lamblia* and *Naegleria fowleri*. *PLoS Negl. Trop. Dis.* 11 (8): e0005832. https://doi.org/10.1371/journal.pntd.0005832.

112 Khan, I.A., Avery, M.A., Burandt, C.L. et al. (2000). Antigiardial activity of isoflavones from *Dalbergia frutescens* bark. *J. Nat. Prod.* 63 (10): 1414–1416. https://doi.org/10.1021/np000010d.

113 National Center for Biotechnology Information (2020). PubChem Compound Summary for CID 5320686, Tiliroside. https://pubchem.ncbi.nlm.nih.gov/compound/Tiliroside (accessed 6 September 2020).

114 Nair, K.V., Gillon, J., and Ferguson, A. (1981). Corticosteroid treatment increases parasite numbers in murine giardiasis. *Gut* 22 (6): 475–480. https://doi.org/10.1136/gut.22.6.475.

5

Antiprotozoal Activity of Phytochemicals Against *Entamoeba histolytica*, the Causative Agent of Amoebiasis

Goutam Kumar Jena[1], Chinam Niranjan Patra[1], and Jnyanaranjan Panda[2]

[1] *Department of Pharmaceutics, Roland Institute of Pharmaceutical Sciences, Berhampur, Odisha, India*
[2] *Department of Chemistry, Roland Institute of Pharmaceutical Sciences, Berhampur, Odisha, India*

List of Abbreviations

ELISA	Enzyme-linked immunosorbent assay
LAMP	Loop-mediated isothermal amplification assay
POC	Point-of-Care
RT-PCR	Real-Time Polymerase Chain Reaction

5.1 Introduction

Amoebiasis is a common health problem globally but most predominant in the developing countries. It is caused by *Entamoeba histolytica* (*E. histolytica*) which is an enteric parasitic protozoan. Amoebiasis is estimated to claim 40 000–100 000 lives globally every year [1]. Mainly, the people of developing countries suffer from this infection due to limited sanitary conditions, where it is endemic. The historical background of amoebiasis can be summarized as follows:

Lambe in 1959 first reported the discharge of amoebae from a patient suffering from dysentery [1], but the similar parasite was also reported by Gros, a Russian author before 10 years of Lambe's report. Another classical report was presented by Losch (A Russian scientist) who treated a patient suffering from chronic dysentery. The feces of the patient was analyzed by Losch by using a microscope and observed a large number of amoebae but he thought these bacteria may be responsible for this disease. The patient who was treated by Losch was known to be the

Neglected Tropical Diseases and Phytochemicals in Drug Discovery, First Edition.
Edited by Chukwuebuka Egbuna, Muhammad Akram, and Jonathan Chinenye Ifemeje.

first case of death from amoebiasis. The organism was named *Entamoeba coli* by Losch and on further experiments on dogs, he confirmed that it was responsible for colonic ulceration as well as dysentery [1]. In 1883, Koch illustrated presence of amoebae in the ulcerated section of bowel of human and correlated the parasite and the intestinal lesion [2]. As this parasite causes tissue lysis, Schaudinn named this organism as *E. histolytica* in 1903 [1, 3]. In 1913, Walker and Sellard conducted a historical experiment using 40 number of human volunteers [1]. He divided all the volunteers into two groups containing 20 volunteers in each group. The first group was fed with cysts of pathogenic *E. histolytica*. On the other hand, the second group was fed with nonpathogenic *Entaamoeba coli*. Four volunteers out of 20, in group 1 suffered from acute dysentery and recovered from this but remained as carriers, other fourteen volunteers remained as carriers without suffering from dysentery whereas rest two volunteers neither suffered from dysentery nor became carriers and so taken as negative control. In group 2, no volunteers developed disease. The results of the experiment conducted by Walker and Sellard, concluded that *E. histolytica* is the causative agent of amoebic dysentery and there is existence of two distinct species of amoeba i.e. pathogenic and nonpathogenic and all people may not develop acute dysentery rather carrying the pathogenic species due to their immune condition as well as other factors.

The three nonpathogenic species of amoeba i.e. *Entaamoeba disper*, *Entaamoeba Bangladeshi* and *Entaamoeba moshkovskii* were recently reported which are closely related to *E. histolytica* and cause asymptomatic amoebiasis. On observing by light microscope, their cysts were found to be very identical as that of *E. histolytica*. Among these three nonpathogenic species of amoeba, *E. disper* is widely studied.

Normally, the parasite is maintained under control inside the intestinal lumen of the host but intestinal amoebiasis occurs with varying degree of severity by *E. histolytica*, due to some unknown signals. Amoebiasis caused by *E. histolytica* depends on many factors as well as the interaction among host, amoeba, and pathogenic microorganisms.

Nitroimidazoles are the first-line treatment option for amoebiasis. Development of resistance to Nitroimidazoles and their derivatives followed by side effects are the key issues nowadays [4]. Hence as an alternative treatment option, some phytochemicals like flavonoids, alkaloids, terpenoids, iridoids, quassinoids, and phenolic compounds obtained from various plant sources can be used for amoebiasis to address the key issues arisen by Nitroimidazoles and their derivatives [5]. Although the use of phytochemicals in the treatment of amoebiasis is a traditional approach, it has fewer side effects and less chance of developing resistance. Hence the natural products have also made an important impact in the treatment of amoebiasis. According to World Health Organization (WHO), approximately 80% of the people in less-developed countries still rely on traditional medicine for their healthcare needs.

5.2 Life Cycle of *Entamoeba histolytica*

The *E. histolytica* exhibits a very simple life cycle (Figure 5.1) with two distinct phases' i.e. infective cyst and vegetative trophozoites [6]. It seldom needs a vector for its transmission. Generally, the infection starts due to ingestion of food or

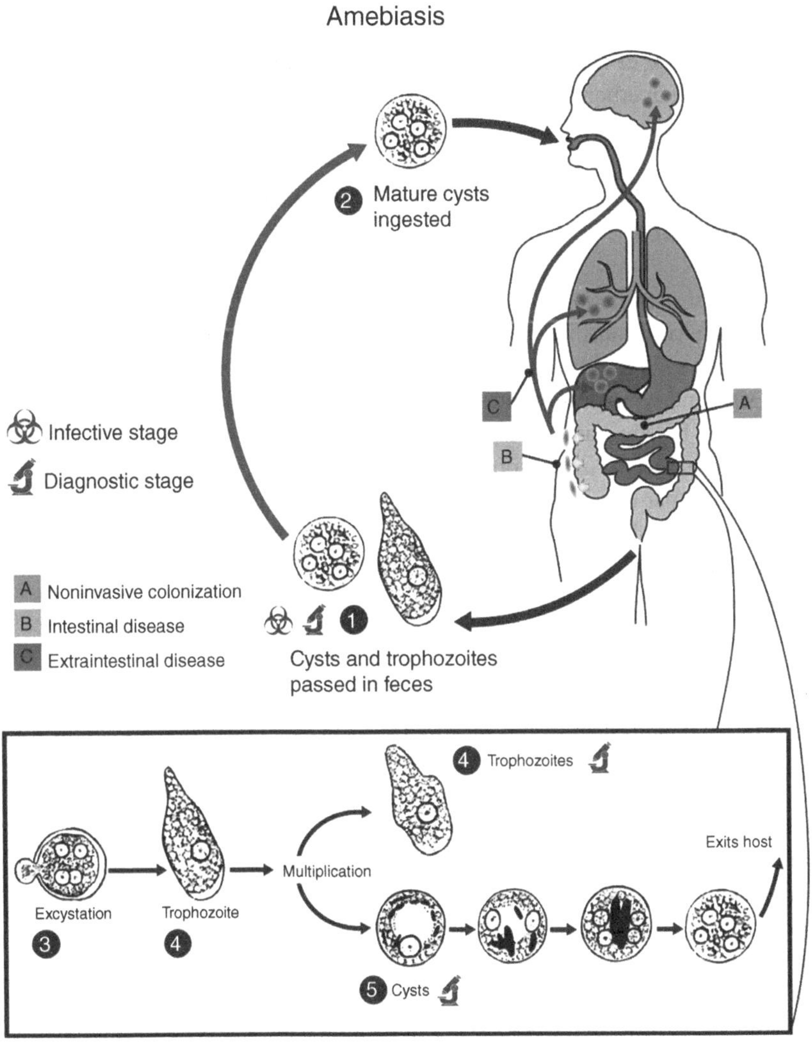

Figure 5.1 Key phytochemicals effective against *Entamoeba histolytica*, the causative agent of amoebiasis. *Source:* DPDx – Amebiasis, Centers for Disease Control and Prevention, 16 September 2020.

water contaminated with cyst-resistant form of *E. histolytica*. The excystation of cysts takes place in the terminal ileum of the host under favorable conditions and trophozoites (amoebae) are produced, which are motile and potentially invasive in nature. The trophozoites generally undergo division by binary fission and they are encysted under adverse conditions and finally passed out in feces. The amoebae proliferate and accumulate as polyploidy cells in the terminal part of ileum of intestine. These polyploid cells adhere to the mucosal surface and attack the large intestine, leading to diarrhea as well as colitis in the host [7]. The amoebae are transformed back to resistant form of cysts which are excreted with feces and repeat the cycle in other host under suitable conditions.

5.3 Diagnostic Techniques for Identification of *E. histolytica*

Parasitological, immunological, and molecular techniques are the basis of identification of *E. histolytica* causing amoebiasis. The examination of parasites in stool, body fluid or tissue sample is recommended as "gold standard" in diagnosis [4, 8]. The clinical sign and symptoms of intestinal amoebiasis are gastrointestinal discomfort and watery or bloody diarrhea. The intestinal amoebiasis is diagnosed by using different methods such as microscopy, culture and isoenzyme analysis, antigen detection test, molecular-based test and point-of-care test. The intestinal amoebiasis differs from extra-intestinal amoebiasis by two ways. Firstly, there is no concurrent amoebic colitis in case of most of the patients with extra-intestinal amoebiasis. Secondly, there is no development of IgG, which is most common in case of intestinal amoebiasis.

5.3.1 Intestinal Amoebiasis

The different diagnostic tests for intestinal amoebiasis are summarized in the following section

5.3.1.1 Microscopy

Microscopic examination is considered as "gold standard" in diagnosis of intestinal amoebiasis. This technique is commonly used to visualize the cysts and/or trophozoites of *E. histolytica* present in the stool or colonic mucosa of patients. To observe the motile trophozoites, the stool sample should be examined within one hour of sample collection. The maximum possibility of getting trophozoites is in loose stool rather than in formed stools but the cysts can be observed both in loose as well as in formed stool. To observe the morphology, size and number of nuclei clearly, permanent stool smear should be investigated. The recommended stains for preparing stool smear for routine use are haematoxylin and Wheatley's trichrome. Apart from the recommended stains, methylene blue, iodine-trichrome, Giemsa, etc. can also be used. Although the microscopic examination helps in visualizing the parasite but it fails to distinguish different species of amoeba.

5.3.1.2 Culture and Isoenzyme Analysis

Earlier, culture of stool followed by analysis of isoenzyme was utilized as a benchmark to distinguish *E. histolytica* and *E. disper*. Rectal biopsy, liver abscess aspirate and stool sample are used to culture *E. histolytica*. Zymodeme enzymes are used as marker to analyze isoenzyme from the cultured amoeba [9]. A zymodeme is a group of strains having the same electrophoretic pattern for a few enzymes. Out of total 24 established zymodemes, 21 are from human isolates and rest three is obtained from experimentally cultured amoeba strains. *E. histolytica* and *E. disper* are discriminated reliably due to presence of genetically different hexokinase enzymes. These two amoeba species can also be distinguished by presence of different zymodeme bands in them (*E. histolytica* has three bands i.e. II, XIV, and XIX whereas *E. disper* has only one band i.e. I). However, cultured amoeba trophozoites are required for analysis of isoenzyme. The isoenzyme analysis of *E. histolytica* culture from clinical samples may often show pseudo negative result.

5.3.1.3 Antigen Detection Test

Due to the shortcomings of traditional diagnostic tests for amoebiasis, modern technique i.e. enzyme-linked immunosorbent assay (ELISA) is used. ELISA technique is suitable for clinical as well as epidemiological studies when molecular assays are not available. This technique is well known for its simplicity and rapidity of diagnosis. The *E. histolytica* TechLab kit is the most commonly used kit for detection of *E. histolytica*. It is used specifically to detect *E. histolytica* Gal/GalNac lectin in stool samples [9]. The limitation of first generation leads to second generation kit known as TechLab *E. histolytica* II which is superior to first generation kit in terms of sensitivity as well as specificity.

5.3.1.4 Molecular-Based Test

Molecular diagnostic tests have achieved significant success in diagnosing almost all infectious diseases including amoebiasis. In order to detect and distinguish different species of amoeba in stool, tissues, and liver lesion aspirates, different DNA amplification techniques are available which include conventional PCR, real-time PCR, nested PCR, and loop-mediated isothermal amplification assay (LAMP).

5.3.1.5 Point-of-Care (POC) Test

Currently, point-of-care test is highly recommended as a rapid diagnostic test for detection of amoebiasis. It can be used for mass screening of endemic population and monitoring of control programmes for quick diagnosis and reducing transmission of disease. The most significant role of this test is to reduce gap between diagnosis and treatment by providing quick result to physicians. It is highly economical as compared to PCR and ELISA due to no requirement of sophisticated and costly devices. Hence it is the preferred diagnostic tool in developing countries. Monoclonal antibodies are used in this test specific to a particular antigen.

5.4 Treatment of Amoebiasis

Amoebiasis can be prevented by maintaining hygienic conditions as well as by using tap water and toilet. Generally, the people either living in endemic areas or frequently travelling to these areas are prone to amoebiasis. Once a person is diagnosed positive for *E. histolytica*, he or she has to follow a well-established treatment plan. So many synthetic compounds are found to be effective for the treatment of intestinal amoebiasis but 5-Nitroimidazoles like metronidazole, tinidazole, omidazole, and secnidazole are found to be most effective for both intestinal as well as tissue levels. However, these drugs produce severe side effects like diarrhea, nausea, vomiting, gastritis, leucopenia, headache, and allergy. The treatment of amoebiasis solely depends on metronidazole due to its easy availability in commercial scale with low price [9]. But the treatment with metronidazole is often discarded or insufficient to eradicate the infection completely due to the severe adverse effect it produced, resulting resistant parasites [4, 10]. Apart from this, the genotoxicity, mutagenicity, and carcinogenicity are reported in human, bacteria, and rodents, respectively [9]. Encephalopathy as well as neuropathy is also reported. Tinidazole is reported to be more effective with fewer side effects as compared to metronidazole.

5.5 Alternative Treatment for Amoebiasis

It is reported that 80% of the people in the developing countries use traditional medicine to treat amoebiasis yet [5]. Generally, medicinal plants are traditionally used in countries where cholera, diarrhea, and dysentery are major cause of morbidity and mortality [4]. Phytochemicals obtained from plant extracts having antispasmodic effect can reduce gut motility and increase water absorption. Hence they can be used in the treatment of amoebiasis. All phytochemicals are not safe and may be toxic so proper evaluation test is to be conducted for the safety of traditional plant extracts [11]. Although some phytochemicals are the potential candidates against *E. histolytica* but rare information is available for their chemistry, mechanisms, toxicity, and clinical trials. The major class of active phytochemicals against *E. histolytica* is presented in Figure 5.2. Several medicinal plants effective against *E. histolytica* are reported in this context, few of them are portrayed in Table 5.1.

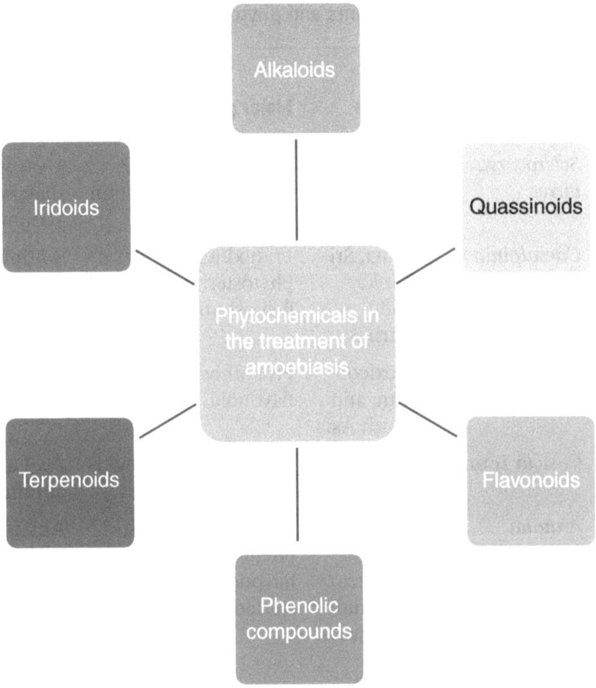

Figure 5.2 Key phytochemicals effective against *Entamoeba histolytica*, the causative agent of amoebiasis.

5.6 Chemical Structure of Phytochemicals Effective Against *E. histolytica*

Some of the important chemical structures of the phytochemicals active against *E. histolytica* such as flavonoids (general structure), flavones, Proanthocyanidin, flavan-3-ols, and flavanones were portrayed as Figure 5.3–5.7, respectively.

5.7 Possible Mechanisms of Phytochemicals as Antiamoebic

The exact mechanism of phytochemicals as antiamoebic for the treatment of amoebiasis is yet to be made out but different possible mechanisms are inhibition of metabolism, modulation of amoebic migration, inhibition of phagocytosis, alteration of DNA replication and induction of apoptosis, alteration of cytoskeleton, decrease of cytotoxicity, inhibition of protein kinase, inhibition of cytoadherence, inhibition of proliferation, modulation of inflammatory response, and inhibition of detoxifying enzymes.

Table 5.1 Medicinal plants and phytochemicals with antiamoebic properties.

Name of medicinal plants	Source	Major phytochemicals	Uses	References
Sclerocarya birrea	South Africa and Zimbabwe	Flavonol glycoside, epicatechin, quercetin	Dysentery, diarrhea, rheumatism	[12]
Caesalpinia bonducella	India, Sri Lanka, and Burma.	Franoditerpenes, phytosterinin, β-sitosterol, flavonoids	Antimicrobial, antidiarrhoeal, cytotoxicity	[13]
Punica granatum	Mexico, Iran, and South Asia	Proanthocyanidin, flavonoids	Antiamoebic and antioxidant	[14]
Castela Texana	Mexico,	Chaparrin,quassinoids and sesquiterpenes	Antiamoebic and antimutagenic	[15]
Annona purpurea	Mexico, Central America, and South America	Tannins, alkaloids, phenols, glycosides, flavonoids, and steroids	Aniamoebic, anthelmintic, analgesic, anti-inflammatory, antipyretic, wound healing	[16]
Voacanga globosa	Phillipines	Alkaloids and saponins	Anticancer and antiamoebic	[17]
Cnidoscolus chayamansa	Mexico	Kaempferol-3,7-dimethyl ether and 5-hydroxy-7-3′,4′-trimethoxy-flavanone	Anti-inflammatory and antiamoebic	[18]
Rauwolfia obscura	India	Indole alkaloids	Antihypertensive and antiamoebic	[19]
Carica Popaya	India	Flavonoids and papain, kaempferol	Antiamoebic, antioxidant, and anti-hypertensive	[19, 20]

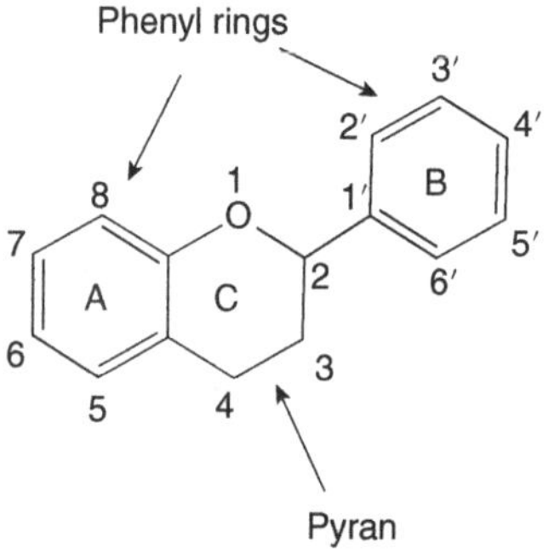

Figure 5.3 General structures of flavonoids.

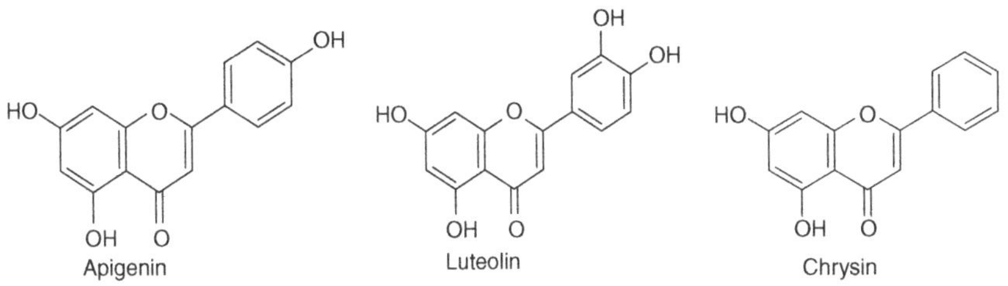

Figure 5.4 Structures of flavones (apigenin, luteolin, and chrysin).

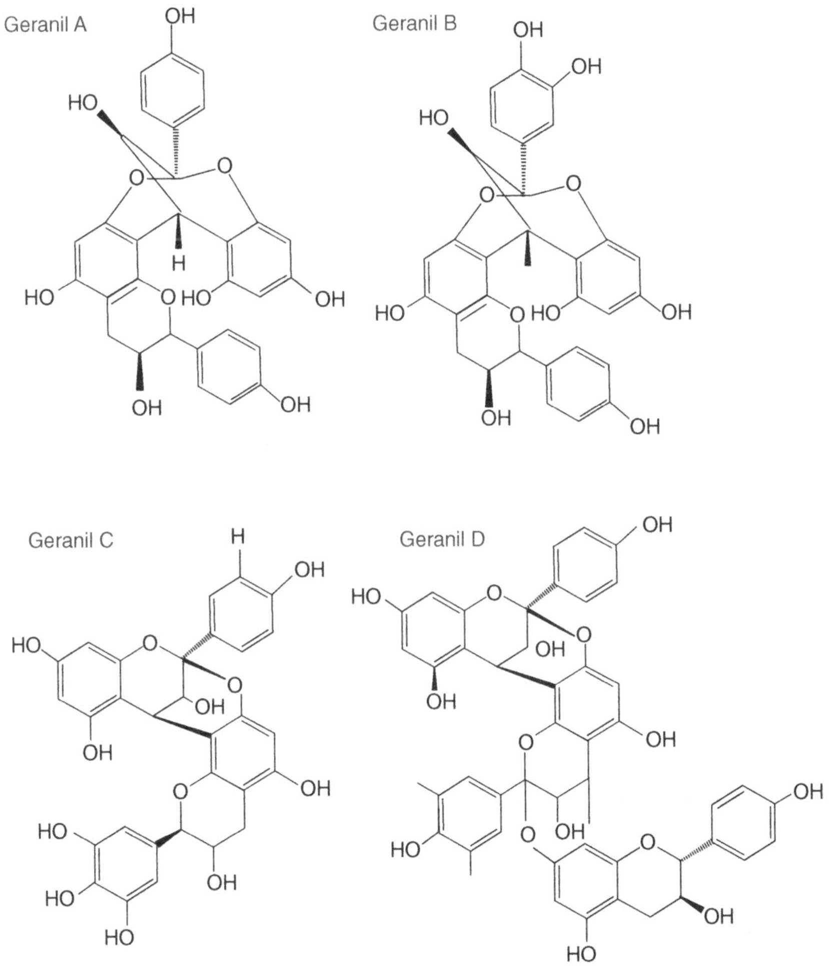

Figure 5.5 Structures of Proanthocyanidin (Geranil A, Geranil B, Gerenil C, and Gerenil D).

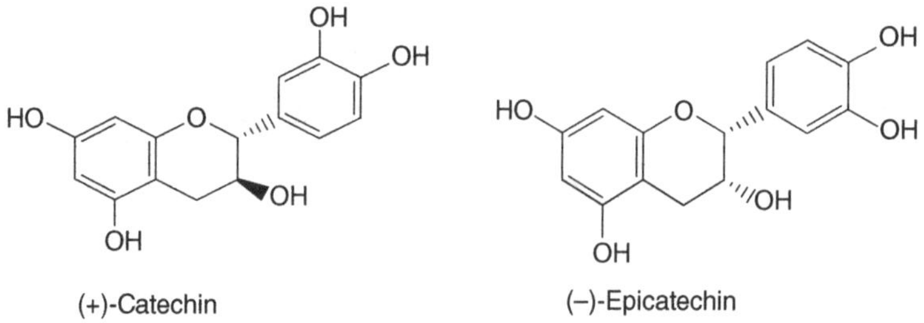

Figure 5.6 Structures of Flavan-3-ols [(+)-Catechin and (−)-Epicatechin].

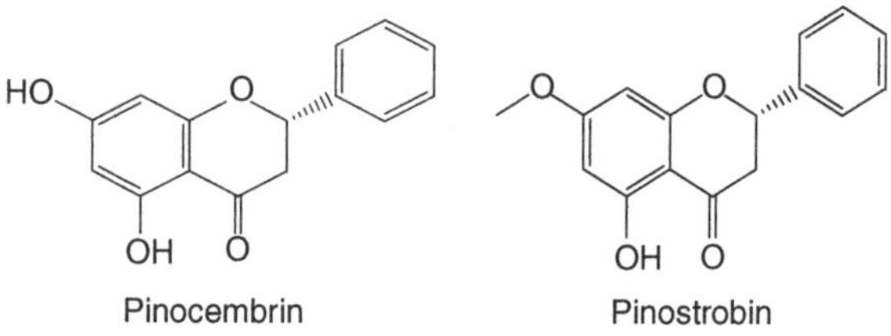

Figure 5.7 Structures of flavanones (Pinocembrin and Pinostrobin).

5.8 Future Projections

Even in the twenty-first century, amoebiasis remains a common health problem worldwide especially in the developing countries due to lack of standard sanitary conditions for large section of population. Different approaches have been made to eradicate amoebiasis. The researchers have been trying their level best to develop diagnostic methods for quick and accurate diagnosis and effective vaccine as well as appropriate treatment against amoebiasis without any side effect in near future. The isolation and screening of suitable phytochemicals against *E. histolytica* from different plant sources have been tried to find out alternative and safe treatment for amoebiasis.

5.9 Conclusion

Entamoeba histolytica, the causative organism of amoebiasis, remains as either a commensal or an invasive form of disease in the human intestine. In most of the individuals, the parasite remains as commensal as asymptomatic and continues to multiply and spread. In few cases, it shows the symptoms either as invasion in intestinal tissues or in some extra-intestinal sites. Hence the key issue in amoebiasis is to understand the basic principle behind the mechanism of invasion of

E. histolytica, produced selectively in few individuals but not in all cases. Another problem in this field is to detect the parasite accurately in case of infected individuals. Once the parasite is identified, the proper treatment protocol has to be followed. The best treatment option for amoebiasis is the use of traditional medicines i.e. major phytochemicals extracted from different plant sources. After the extraction, isolation, and purification of the phytochemicals which are found to be effective against entamoeba histolytica as discussed can be used in the treatment of amoebiasis as an alternative and safe option as compared to Nitroimidazoles and their derivatives.

References

1 Carrero, J.C., Reyes-López, M., Serrano-Luna, J. et al. (2020). Intestinal amoebiasis: 160 years of its first detection and still remains as a health problem in developing countries. *Int. J. Med. Microbiol.* 310 (1): 151358.

2 Pritt, B.S. and Graham Clark, C. (2008). Amebiasis. *Mayo Clin. Proc.* 83 (10): 1154–1160.

3 Rossignol, J.F., Kabil, S.M., El-Gohary, Y., and Younis, A.M. (2007). Nitazoxanide in the treatment of amoebiasis. *Trans. R. Soc. Trop. Med. Hyg.* 101 (10): 1025–1031.

4 Madura, H.T. (1948). Treatment of amoebiasis. *J. Christ. Med. Assoc. India* 23 (4): 179.

5 Martínez-Castillo, M., Pacheco-Yepez, J., Flores-Huerta, N. et al. (2018). Flavonoids as a natural treatment against *Entamoeba histolytica*. *Front. Cell Infect. Microbiol.* 8: 1–14.

6 Shirley, D., Hung, C., and Moonah, S. (2015). Intestinal and genital infections. In: *Tenth Edit. Entamoeba histolytica (Amebiasis)*, 699–706. Elsevier Inc.

7 Samie, A., Mahlaule, L., Mbati, P. et al. (2020). Prevalence and distribution of *Entamoeba* species in a rural community in northern South Africa. *Food Waterborne Parasitol.* 18: e00076.

8 Visser, L.G., Verweij, J.J., Van Esbroeck, M. et al. (2006). Diagnostic methods for differentiation of *Entamoeba histolytica* and *Entamoeba dispar* in carriers: performance and clinical implications in a non-endemic setting. *Int. J. Med. Microbiol.* 296 (6): 397–403.

9 Proctor, E.M. (1991). Laboratory diagnosis of amebiasis. *Clin. Lab. Med.* 11 (4): 829–859.

10 Pehrson, P. and Bengtsson, E. (1983). Treatment tinidazole of non-invasive A comparison alone and in combination with diloxanide between furoate. *Trans. R. Soc. Trop. Med. Hyg.* 77 (6): 845–846.

11 Cao, H., Chai, T.T., Wang, X. et al. (2017). Phytochemicals from fern species: potential for medicine applications. *Phytochem. Rev.* 16: 379–440.

12 Mariod, A.A. and Abdelwahab, S.I. (2012). *Sclerocarya birrea* (Marula), an African tree of nutritional and medicinal uses: a review. *Food Rev. Int.* 28 (4): 375–388.

13 Billah, M.M., Islam, R., Khatun, H. et al. (2013). Antibacterial, antidiarrhoeal, and cytotoxic activities of methanol extract and its fractions of *Caesalpinia bonducella* (L.) Roxb leaves. *BMC Complement. Altern. Med.* 13: 1–7.

14 Yan, L., Zhou, X., Shi, L. et al. (2017). Phenolic profiles and antioxidant activities of six Chinese pomegranate (*Punica granatum* L.) cultivars. *Int. J. Food Prop.* 20: S94–S107.

15 Reyes-López, M., Villa-Treviño, S., Arriaga-Alba, M. et al. (2005). The amoebicidal aqueous extract from *Castela texana* possesses antigenotoxic and antimutagenic properties. *Toxicol. Vitr.* 19 (1): 91–97.

16 Díaz-Godínez, C., Ontiveros-Rodríguez, J.C., Ríos-Valencia, D.G. et al. (2020). Anti-amoebic activity of leaf extracts and aporphine alkaloids obtained from *Annona purpurea*. *Planta Med.* 86 (6): 425–433.

17 Vital, P.G. and Rivera, W.L. (2011). Antimicrobial activity, cytotoxicity, and phytochemical screening of *Voacanga globosa* (Blanco) Merr. leaf extract (Apocynaceae). *Asian Pac. J. Trop. Med.* 4 (10): 824–828.

18 Pérez-González, M.Z., Gutiérrez-Rebolledo, G.A., Yépez-Mulia, L. et al. (2017). Antiprotozoal, antimycobacterial, and anti-inflammatory evaluation of *Cnidoscolus chayamansa* (Mc Vaugh) extract and the isolated compounds. *Biomed. Pharmacother.* 89: 89–97.

19 Tona, L., Kambu, K., Ngimbi, N. et al. (1998). Antiamoebic and phytochemical screening of some Congolese medicinal plants. *J. Ethnopharmacol.* 61 (1): 57–65.

20 Vij, T. and Prashar, Y. (2015). A review on medicinal properties of *Carica papaya* Linn. *Asian Pacific J. Trop. Dis.* 5 (1): 1–6.

6

Antileishmanial Potentials of Phytochemicals

Shahira M. Ezzat[1,2], *Mohamed A. Salem*[3], *and Ahmed Zayed*[4]

[1] *Department of Pharmacognosy, Faculty of Pharmacy, Cairo University, Cairo, Egypt*
[2] *Department of Pharmacognosy, Faculty of Pharmacy, October University for Modern Sciences and Arts (MSA), Cairo, Egypt*
[3] *Department of Pharmacognosy, Faculty of Pharmacy, Menoufia University, Menoufia, Egypt*
[4] *Department of Pharmacognosy, Faculty of Pharmacy, Tanta University, Tanta, Egypt*

6.1 Introduction

Leishmaniases are a group of neglected tropical diseases (NTD) which affects about 12–15 million people in 88 countries around the world [1]. The protozoan species belonging to Leishmania is the main cause of the disease which is transmitted by sandflies of the genus *Phlebotomus* and *Lutzomyia* [2].

The infection begins when an infected sandfly bites a person, here the promastigote form of the *leishmania* enters the human body. The promastigotes are then up-taken by macrophages where they are transformed into the amastigote which is the intracellular form of *leishmania* [3]. There are three main forms of Leishmaniases, according to its clinical manifestations: Cutaneous leishmaniasis (CL), which affects only localized parts of the skin and is the most common form of the disease; mucocutaneous leishmaniasis (MCL), which has the ability to destroy mucous tissue and is exclusively present in America; and visceral leishmaniasis (VL) which is the less common type of leishmaniasis but causes liver and spleen distention and can be fatal if it does not receive prompt treatment [4].

Pentavalent antimony-based drugs are the main chemotherapeutic treatment against leishmaniases [5]. These drugs not only have serious side effects such as cardiotoxicity, hepatotoxicity, and nephrotoxicity, but also the parasite has acquired resistance to such drugs in many countries, in which leishmaniases is a major public health problem [2]. The WHO has recommended the use of

Neglected Tropical Diseases and Phytochemicals in Drug Discovery, First Edition.
Edited by Chukwuebuka Egbuna, Muhammad Akram, and Jonathan Chinenye Ifemeje.

medicinal plants and traditional treatments for chronic diseases such as leishmaniasis, where patients are often impaired to seek other medical treatment [6, 7]. In Mexico, approximately 3000 plant species are employed in the practice of traditional medicine and many for the treatment of protozoal infections [8, 9].

6.2 Antileishmanial Activity of Propolis

Propolis is a chemical compound produced by worker bees through combining exudates or parts of the plants, with beeswax and secretions from the salivary glands of the worker bee rich in enzymes. It varies widely in color due to the variation in chemical composition [10]. Due to its antimicrobial power, Propolis is considered the "chemical weapon" that the bees use to maintain an aseptic internal environment in beehives [11, 12]. In addition to its use in dietary supplements and cosmetic formulations, Propolis has many reported biological activities such as anti-inflammatory, antiulcer, antitumor, immunostimulant, hepatoprotective, antibacterial, antifungal, and potential against protozoa [12–14].

The *in vitro* antileishmanial activity of propolis from different regions such as Turkey [15], Brazil [16, 17], Portugal [18], and Cuba [19] have been reported.

6.2.1 Propolis from Turkey

The ethanolic extract of Adana propolis was tested against *Leishmania at* five concentrations (25, 50, 100, 50, 500, and 750 μg/ml). The number of promastigotes in each concentration was calculated using a hemocytometer slide at 24, 48, and 72 hours after being harvested. The Adana propolis samples produced inhibitory effect only at concentrations of 250, 500, and 750 μg/ml through reducing the proliferation of *Leishmania tropica* parasites [20].

Another study by the same authors evaluated the antileishmanial activities of "Bursa" and "Hatay" propolis samples against *Leishmania infantum* and *L. tropica* strains at 50, 100, 250, 500, 750, and 1000 μg/ml concentration of each sample. The growth of leishmania parasites was significantly reduced in the presence of 500, 750, and 1000 μg/ml of Hatay propolis. Bursa propolis was more effective as it showed the inhibitory effects at lower concentrations (250, 500, 750, and 1000 μg/ml). Thus, the *in vitro* results showed that the Hatay and Bursa propolis samples decreased significantly the proliferation of *L. infantum* and *L. tropica* parasites [15].

6.2.2 Propolis from Brazil

The antileishmanial activity and cytotoxicity of brown propolis originating from the semiarid region of Piauí, Brazil were evaluated [16]. The propolis significantly inhibited the growth of *Leishmania amazonensis promastigotes* and reduced the infection of murine macrophages and the number of internalised amastigotes in these cells. The dichloromethane fraction was the most active and showed the best selectivity index. The study showed that fractionation of the propolis had an improvement in the activity without increasing the cytotoxicity against mammalian cells. Thus the Brazilian brown propolis is a potential source for development of apitherapeutic products for the treatment of leishmaniasis.

6.2.3 Propolis from Portugal

The *in vitro* activity of two different Portuguese propolis extracts; one from the North-eastern Portugal (Bragança county), while the other was collected from the centre of Portugal (Leiria county) together with its major plants *Populus* × *Canadensis* (male and female plants) and *Cistus ladanifer* were evaluated against *Plasmodium falciparum*, *L. infantum*, *Trypanosoma brucei*, and *Trypanosoma cruzi* using the WHO Special Program for Research and Training in Tropical Diseases (TDR). The tested samples showed moderate activity against *T. brucei*. For *L. infantum*, propolis samples and male and female *Populus* showed a similar IC_{50} value of 8.11 μg/ml. *C. ladanifer* showed the lowest activity with an IC_{50} value of 32.46 μg/ml [18].

6.2.4 Propolis from Cuba

In Cuba, 18 *Cuban propolis* extracts of different types such as brown, red, and yellow types collected from various locations were tested for antileishmanial activity using *L. amazonensis* as a model of intracellular protozoa and *Trichomonas vaginalis* as a model of extracellular protozoa. The results showed no correlation between the types of propolis and the obtained antiprotozoal; samples were active against intracellular amastigotes of *L. amazonensis*, but caused toxicity on peritoneal macrophages from BALB/c mice. However, only five samples have inhibitory effect on *T. vaginalis trophozoites* at concentration less than 10 μg/ml [19].

6.2.5 Propolis from Ecuador

In Ecuador, the alcoholic extract of propolis is a common homemade remedy and also 10–20% ethanolic tinctures or powder mixed with honey are sold in the market as antioxidant, anti-inflammatory, and antimicrobial remedies [21]. One study was done by Cuesta-Rubio et al. [21], in which the methanol extract of three propolis samples were collected from three localities of Ecuador (Quito, Guayaquil, and Cotacachi). Based on their chemical composition they were divided into two main samples of propolis: Cotacachi propoli sample (CPS), rich in flavonoids and Quito and Guayaquil samples (QPS and GPS) rich in triterpenic alcohols and acetyl triterpenes.

The three methanol extracts of propolis showed inhibitory activity against *L. amazonensis* growth, where CPS was the most active ($IC_{50} = 17.1 \pm 1.7\,\mu g/ml$) may be due to its high flavonoid content. The samples were also tested *in vitro* against promastigotes of *L. amazonensis* and macrophage from BALB/c mice where all the tested samples were able to inhibit parasite growth, out of which Cotacachi sample was the most active. This propolis sample was also the most promising, displaying a CC_{50} sevenfold higher than the IC_{50}value on parasite. CPS contains the flavonoid sakuranetin, which is reported to be active against *Leishmania* spp. [22]. So it is thought that this flavonoid and its derivatives may be responsible for the observed antiparasitic activity.

6.3 Antileishmanial Activity of Wild Mushrooms

Six mushrooms from Gangetic plane of West Bengal, India which are used as food and for medicinal purposes in traditional folklore by local Santal tribes in India were tested for the anti-proliferating effect against *Leishmania donovani* (MHOM/IN/83/AG83) [23]. These species are *Astraeus hygrometricus* (Pers.) Morgan (Local name – Putko/Kurkure Chatu); *Russula laurocerasi* Melzer (Local name – Jhaal Patra); *Russula albonigra* (Krombh.) Fr. (Local name – Kalo/Kend Patra); *Termitomyces eurhizus* (Berk.) Heim (Local name – Parab Chatu); and *Russula delica* Fr. (Local name – Sada Patra) with them being native to Lateritic zones of five districts of Birbhum, Murshidabad, Burdwan, Bankura, and West Midnapore and *Tricholoma giganteum* Massee (Local name – Dhoodh Chatu). Three extracts were prepared from the milled and freeze dried mature fruitbodies (Basidiocarps) of each species. Eighty percent (80%) ethanol extract, water-soluble polysaccharide fraction and polyphenolic fraction were prepared and tested for their antileishmanial ability against the promastigotes and amastigotes. The 80% Ethanol extract of *A. hygrometricus* and *T. giganteum* significantly inhibited the growth of *L. donovani* promastigotes and interfered in lipid biosynthesis at concentration of 250 μg/ml. These two extracts also induced apoptosis in promastigotes. Water-soluble polysaccharide fraction of *A. hygrometricus, R. laurocerasi, R. albonigra, T. eurhizus, R. delica,* and polyphenolic fraction of *R. laurocerasi* showed dose-dependent inhibition of the intracellular amastigotes replication in macrophages. Significantly, 50% inhibitory concentration of the active extracts against intracellular amastigotes induced release of nitric oxide and interleukin-12 (IL-12) in murine macrophages and dendritic cells assay and also found considerably nontoxic on murine splenocytes.

6.4 Antileishmanial Activity of Medicinal Plants from Various Flora

Herbal remedies for leishmaniasis has attracted great interest especially in different countries of Latin America, where it is a public health concern, as in Peru [24], Bolivia [25], Colombia [26], and Brazil [27, 28]. These medicinal plants are native to the tropical Andes located in South America, tracing the Andes Path and run along Venezuela, Colombia, Ecuador, Peru, and Bolivia. They are characterized by complex geography including different altitudes, and wide range of climatic features in the Andes and Amazonian. Therefore, they support an enormous heterogeneity of ecological conditions and rich diversity of flora and fauna, as well [29]. Hence, indigenous peoples of these areas have a long history of treatment of different diseases by medicinal plants and transmitted these practical traditions to following generations [30]. Interestingly, these plants may be native and common in many cases in different flora of these countries.

6.4.1 Peruvian Flora

The flora of Peru is among the principal and most diverse of the world flora. Approximately 25 000 plant species are native and endemic in Peru representing 10% of all plant species in the world. This biodiversity owes to the different geographical conditions found in coastal desert, mountains, jungle, and rainforest providing 28 of the 32 possible climates and 84 of the 103 ecological zones existing in the world [31]. Examples include the best known in South America *Uncaria tomentosa* (cat's claw), which has been traditionally used for 2000 years in in treating inflammation, arthritis, bone pain, asthma, deep wounds, and cancer. In addition, *Phyllantus niruri* (chanca piedra) or *Lepidium meyenii* (maca) is used for fertility improvement and protection of cells against oxidative stress [32].

Specifically, several Peruvian medicinal plants have been reported to have potential *in vitro* leishmanicidal activity mainly against *L. amazonensis* promastigote and amastigote stages. Most of these medicinal plants are traditionally used by the Chayahuita, an ethnic group from the Peruvian Amazon, that treats symptoms of leishmaniasis [33, 34]. However, compounds responsible for this activity have not been identified and isolated yet completely. Examples include;

- ***Piper hispidum* var. hispidum Sw. and *Piper strigosum* (Fam. Piperaceae)**
Leaves are crushed and applied on skin wounds and swellings as a dressing for treating symptoms of CL including inflammation. Chalcones were shown to be the bioactive compounds, but in other related *Piper* species (e.g., *Piper elongatum*) against extracellular promastigotes of *Leishmania braziliensis in vitro* [35]. These

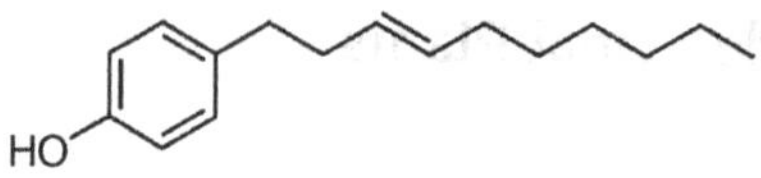

Figure 6.1 The alkenylphenol chemical structure (Gibbilimbol B) isolated from *Piper malacophyllum*. *Source:* de Oliveira et al. [38].

observations were confirmed by synthesis of chalcone analogues, which demonstrated improved and selective activity against the parasites comparing with the natural chalcones [36, 37]. However, in others (e.g., *Piper malacophyllum*), an alkenylphenol (Gibbilimbol B, Figure 6.1) was found to have a plasma membrane disruption activity [38]. These findings have supported the traditional use of *P. hispidum* and *P. strigosum* in treatment of leishmaniasis.

- ***Caladium steudneriifolium* Engl. (Fam. Araceae)**

Caladium steudneriifolium is commonly known as the elephant ear. Its root could be finely grated and applied locally on cutaneous lesions, until it is healed.

- ***Capirona decorticans* Spruce (Fam. Rubiaceae)**

For small lesions and at the very beginning of the disease, the bark aqueous extract could be taken twice daily until cicatrisation.

- ***Siparuna* sp. (Fam. Monimiaceae)**

The woody stem is grated and boiled. The preparation is drunk three times daily for eight days.

- ***Smilax papyracea* Duhamel (Fam. Smilacaceae)**

Roots extract is drunk many times daily, until relieved of cutaneous symptoms.

- ***Tabernaemontana sananho* Ruiz & Pav. (Fam. Apocynaceae)**

The root is rasped and applied locally on the affected area. Several monomeric anddimeric indole alkaloids have been isolated from *Tabernaemontana* species. The dimeric alkaloids were shown in the Bolivian species *Tabernaemontana van heurckii* Mull. Arg. to be responsible for activity against *L. amazonensis in vitro* and *In vivo* investigations [39].

- ***Euterpe oleracea* Maritus (açai palm) (Fam. Arecaceae)**

Euterpe oleracea Maritus is a palm tree and historically known as açai. It is native to the Varzea area of the Amazon, especially in Brazil, Peru, Suriname, and Trinidad and Tobago. Its global importance came from the fruit juice, which is marketed as an energy drink. It is rich in polyphenolic compounds with high antioxidant and anti-inflammatory properties. These compounds include orientin, isoorientin, and vanillic acid, as well as anthocyanins cyanidin-3-glucoside and cyanidin-3-rutinoside [40]. Recently, Da Silva, et al., evaluated the *in vitro* selective effects of clarified açai palm juice on the promastigotes and amastigotes of *L. amazonensis* (MHOM/BR/26361) and *L. infantum* (=*L. chagasi*) (MHOM/BR/27840).

The results showed that the açai juice induced cellular morphological changes and reduced the number of *L. amazonensis* promastigotes (39.68–96.05%) and *L. infantum* promastigotes (59.12–93.88%) in 72 and 48 hours, respectively and without causing harmful effects to the host macrophages cells [41].

6.4.2 Ecuadorean Flora

Ecuador's natural wealth provides the richest source of plants in South America, in proportion to its area (283 561 km^2). It contains approx. 270 families of vascular plants and 20 000 species [42].

Based on interviews with healers using traditional medicine, recognized there as Agents of Traditional Medicine (ATMs) and that found in the literature, Gachet et al. [43] assessed the *in vitro* antiprotozoal potential of plants traditionally usedin Ecuador in the treatment of leishmaniasis. The authors of this study found that, out of 431 use-records, 145 plants are used for the treatment of leishmaniasis and the 10 most frequently reported taxa accounted for 37.7% of all records. From a representative selection of 39 species, a total of 140 extracts were screened *in vitro* against *L. donovani*. The leishmanicidal activity was observed in 14 plant species, among them were *Elephantopus mollis*, *Minquartia guianensis*, *Bocconia integrifolia*, *Gouania lupuloides*, *Scoparia dulcis*, an as-yet-unidentified species of *Piper* and *Brugmansia* [43].

- ***Elephantopus mollis* (Fam. Astraceae)**

Elephantopus mollis is reported to be used in folk medicine for the treatment of various diseases, including nephritis, edema, dampness, chest pain, fever, scabies, joint pain due to wound and cough [44]. In addition, an *in vitro* leishmanicidal activity has proven against *Leishmania major*. For this activity, seven sesquiterpenoid lactones were isolated, identified and investigated from the whole plant of the related Peruvian and Brazilian collections. They were named as 2,5-epoxy-2β-hydroxy-8α-(2-methylpropenoyloxy)-4(15),10(14),11(13)-germacratrien-12,6α-olide, (4βH)-8α-(2-methylpropenoyloxy)-2-oxo-1(5),10(14),11(13)-guaiatrien-12,6α-olide and (4βH)-5α-hydroxy-8α-(2-methylpropenoyloxy)-1(10),11(13)-guaiadiene-12,6α-olide, together with molephantin, elephantopin, isoelephantopin, and 2-deethoxy-2β-methoxyphantomolin. It was found that the α-methylene-γ-butyrolactone moiety was essential for the leishmanicidal activity [45].

- ***Minquartia guianensis* (Fam. Olacaceae)**

Ecuador natives use the bark infusion for intestinal infections treatment caused by parasites, against muscular pain and cutaneous inflammation including leishmaniasis [46, 47]. Triterpenes (e.g., lupeol, taraxerol, lupenona, and squalene), xanthone and minquartynoic acid were isolated in previous phytochemical studies on *M. guianensis*. Specifically, the minquartynoic acid isolated from bark showed activity against *P. falciparum* and *L. major* [48].

- ***Bocconia integrifolia* (Fam. Paperveraceae)**

Rocha et al. [49] have reviewed medicinal plants and bioactive natural products with antileishmanial activity [49]. It was shown that the alkaloidal fraction of the leaves and stem bark of *B. integrifolia* were responsible for the activity against *L. amazonensis* and *L. braziliensis* [25].

6.4.3 Mexican Flora

Mexico is a unique biogeographic area reflecting a diverse endemic flora. It comprises 24 360 vascular plant species, among them 10 235 (42%) are endemic. The biodiversity in Mexico and Central America is odd, owing to its transition zone between the Nearctic and Neotropical biotas [50].

The aerial parts of *Cleoserrata serrata* (Jacq.) Iltis are commonly applied in South-Central Mexico to treat wounds and bacterial skin infections and in Panama by Kuna, Ngöbe-Buglé, and Teribe Indians for tropical warm baths and by Kunas in the form of "Ina kuamakalet" for snakebites. Recently, *C. serrata* CH_2Cl_2/MeOH extract was incubated at different concentration *with L. Mexicana* amastigotes and promastigotes *in vitro*, as well as with bacteria that usually coinfect skin ulcers. The results showed a significant inhibition of the parasite survival (which was more effective in infective amastigotes) and additionally inhibited growth of the coinfective bacteria *Staphylococcus aureus* and *Pseudomonas aeruginosa* [51].

Delgado-Altamirano et al. [52] has investigated recently a few Mexican plants extracts against promastigotes and intracellular amastigotes of *L. amazonensis*. This study found that CH_2Cl_2 and CH_2Cl_2/MeOH *Schinus molle*, CH_2Cl_2 of *Lantana camara* and aqueous *Prosopis laevigata* extracts were active and selective against *L. amazonensis* promastigotes. Moreover, the lower IC_{50} and higher selectivity index of these extracts indicated presence of valuable candidates with leishmanicidal activity [52].

Furthermore, in other related studies, 20 Mexican medicinal plants have been evaluated against *Leishmania Mexicana*. Among them, *Tridax procumbens*, *Lonchocarpus xuul*, and *Pentalinon andrieuxii* showed potential activity and from these plants, active compounds have been isolated, such as the 3(*S*)-16,17-didehydrofalcarinol or Oxylipin, cholestra-4,20,24-trien-3-one or pentalinosterol, 24-methylcholest-4-24(28)-dien-3-one, cholest-4-en-3-one, 6,7-dihydroneridie-none, neridienone, cholest-5,20,24-trien-3β-ol, and isocordoin. One of these compounds; pentalinonsterol, has been synthesized and assayed experimentally in a VL caused by *L. donovani* using BALB/c mice as a model. In addition, the I.V administration of liposomal formulation of pentalinonsterol demonstrated a significant reduction of parasite load in mouse liver and spleen [53].

6.4.4 Bolivian Flora

Like other ethnic groups around the world, Tacana people living in the northern area of the Department of La Paz have their traditional medicinal plants. They used these plants to treat skin problems, including leishmania ulcers, skin infections, inflammation, and wound healing. Evaluation of 38 Tacana medicinal plants extracts was performed *in vitro* against promastigotes of *L. amazonensis* and *L. braziliensis*. The results revealed that about 60% demonstrated leishmanicidal activity, showing relatively higher species activity against the *L. braziliensis* strain than *L. amazonensis*. Among these plants, a total of five crude extracts can be considered as promising leishmanicidal species: *Jacaranda glabra* (Fam. Bignoniaceae), *S. dulcis* (Fam. Plantaginaceae), *Tessaria integrifolia* (Fam. Astraceae), *Thalia geniculata* (Fam. Marantaceae), and Ejije Bid'u leaves [54].

6.4.5 Iranian Flora

Since leishmaniasis is endemic in Iran and the CL is the most common type, a total of 68 articles including 188 experiments (140 *in vitro* and 48 *in vivo*) in nine databases between 1999 and 2015 have discussed herbal treatment of leishmaniasis and around 98 types of plants were examined against three genera of *Leishmania* sp. Systematic and meta-analysis review of this study showed that the most Iranian plants with antileishmanial activity were *Artemisia* sp., *Allium sativum*, *Achilleamille folium*, *Peganum harmala*, and *Thymus vulgaris* [55].

Experimentally, 17 ethanolic extract of different Iranian plants were investigated *in vitro* against *L. major* promastigotes and the mouse macrophage cell line J774 by Manjili, H. K., et al. in 2012 based on the Iranian ethnobotanical history. Only four plants, *Caesalpinia gilliesii* (Fam. Fabaceae), *Satureia hortensis* (Fam. Labiatae), *Carum copticum heirm* (Fam. Umbellifererae), and *Thymus migricus* (Fam. Lamiaceae), exhibited higher selectivity against the leishmania promastigotes relative to the macrophage cell line, based on the measured IC_{50} values. These data indicated that these plants contain not yet known active ingredient which require further investigations for isolation and identification [56].

Moreover, methanolic extracts of *Calendula officinalis* (Fam. Astraceae) flowers, *Datura stramonium* (Fam. Solanaceae) seeds, and *Salvia officinalis* leaves were evaluated *in vitro* against *L. major* promastigotes (strain MROH/IR/75/IR). The antipromastigote and anti-amastigote activities of *D. stramonium* seeds and *S. officinalis* leaves were less potent than that of the *C. officinalis* flower extract. However, all extracts reduced the infectivity of *L. major* amastigotes in macrophages [57].

6.5 Antileishmanial Activity of Chemical Constituents

The current available synthetic drugs for treatment of leishmaniasis, including miltefosine, pentavalent antimonials, and amphotericin B, are highly toxic. Additionally, these drugs are highly expensive and parasite resistance is also a major concern. Natural products constitute a sustainable, cost-effective, and effective source for promising candidates with leishmanicidal potential. In this context, several classes of secondary metabolites have been reported as antileishmanial chemical constituents. The following section introduces different classes and examples of phytochemicals that have antileishmanial activities (Figure 6.2).

6.5.1 Alkaloids

The crude extract as well as subfractions of *Guatteria latifolia* (family Annonaceae) were reported to have an *in vitro* antileishmania amazonensis activity [58]. This activity was connected to reduction of nitric oxide (NO) production in gamma interferon (IFN-γ) and lipopolysaccharide (LPS)-stimulated macrophages [58]. The activity was attributed to the presence of puterine, oxoputerine, and lysicamine in from *G. latifolia* [58] (Figure 6.2).

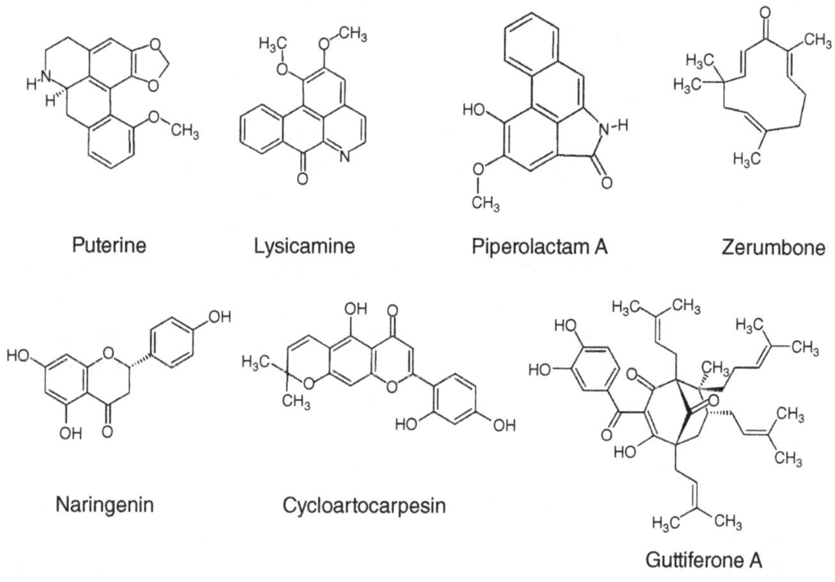

Figure 6.2 Examples of phytoconstituents with antileishmania activity. *Source:* Based on Ferreira et al. [58].

Piperolactam A (Figure 6.2), an aristolactam alkaloid isolated from *Piper betle* root extract, showed promising antileishmanial activity against *L. donovani* wild-type, and drug-resistant strains (sodium stibogluconate and paromomycin) and field isolated (GE1) resistant strains [59]. Inclusion of Piperolactam A in hydroxypropyl-β-cyclodextrin (HPBCD) nanoparticles resulted in significant increase in selectivity leishmanicidal potential [59].

6.5.2 Flavonoids

The extract of *Codonopsis clematidea* and its bioactive flavonoid naringenin have been evaluated for their antileishmanial potential [60]. This study showed that naringenin can be considered a potential candidate against VL, owing to its ability to activate $CD4^+$ and $CD8^+$ T cells as well as trigger NFκB and iNOS production [60].

Among the prenylated flavonoids, cycloartocarpesin has been isolated from the rhizomes of *Dorstenia contrajerva*. This compound showed a promising *in vitro* antileishmanial activity [61]. Cycloartocarpesin has been also found in many fruits.

6.5.3 Terpenes

Macrocyclic lathyrane diterpenoids, isolated from *Jatropha multifidi*, owed a promising *in vitro* antileishmanial activity [62]. Among diterpenoids, 14-deoxy-1β-hydroxy-4(4*E*)-jatrogrossidentadione and 15-deoxy-1β-hydroxy-4(4*E*)-jatrogrossidentadione showed promising activity [62].

Zerumbone, a monocyclic sesquiterpene isolated from *Zingiber zerumbet*, showed a promising *in vitro* antileishmanial activity [63]. This activity was attributed to its ability to induce ROS mediated apoptosis in *L. donovani* promastigotes [63].

6.5.4 Glycoglycerolipids

Glycoglycerolipids are known as ubiquitous structural and signaling molecules in living organisms [64]. Among glycoglycerolipids, 1-*O*-linolenoyl-2-*O*-stearoyl-3-*O*-β-D-galactopyranosyl glycerol has been isolated from the rhizomes of *D. contrajerva*. This compound showed a promising *in vitro* antileishmanial activity [61].

6.5.5 Guttiferone A as a Lead Compound for Semisynthesis of Antileishmanial Molecules

Guttiferone A, a catechol pharmacomodulation, has been isolated from Clusiaceae family including *Symphonia globulifera*, *Garcinia livingstonei*, and *Garcinia macrophylla* [65]. *In vitro* evaluation of Guttiferone A toward *L. donovani* on the promastigote form showed a promising antileishmanial activity for Guttiferone A (IC_{50} = 5.83 μM) [65].

Guttiferone A belongs to the PolyPrenylated AcylPhloroglucinols (PPAPs) family. The synthesis of ether and ester derivatives of the catechol moiety revealed that the monoacetyl derivative produced similar antileishmanial activity [65].

6.6 Conclusion

As a conclusion, nearly all the experiments were performed *in vitro*, based on traditional folklore medicine and did not reveal the exact mechanism by which these extracts and/or compounds act. Additionally, in comparison with the chemical standard treatment (e.g., Miltefosine), the natural or herbal treatment showed less activity and potency. In other cases, the leishmanicidal activity of several herbal treatments are not completely clarified, due to unknown or unidentified composition. Therefore, further *in vivo* investigations, isolation and identification of responsible compounds involved in the leishmancidal activity should be carried out and focused on. These investigations may lead to hypotheses upon which theories and structure activity relationships could be constructed and consequently much potent plants-derived antileishmanial medicaments.

References

1 WHO (2017). *Leishmaniasis*. WHO Regional Office for Africa. http://wwwafrowhoint/health-topics/Leishmaniasis.

2 Tiuman, T.S., Santos, A.O., Ueda-Nakamura, T. et al. (2011). Recent advances in leishmaniasis treatment. *Int. J. Infect. Dis.* 15 (8): e525–e532.

3 Murray, W.H., Berman, D.J., Davies, R.C., and Saravia, G.N. (2005). Advances in leishmaniasis. *Lancet* 366: 1561–1577.

4 Akhoundi, M., Kuhls, K., Cannet, A. et al. (2016). A historical overview of the classification, evolution, and dispersion of leishmania parasites and sandflies. *PLoS Negl. Trop. Dis.* 10 (3): e0004349.

5 Bifeld, E. and Clos, J. (2015). The genetics of *Leishmania virulence. Med. Microbiol. Immunol.* 204 (6): 619–634.

6 WHO (2010). Control of the leishmaniases. *World Health Organ. Tech. Rep. Ser.*: 22–26.

7 WHO (2013). Traditional medicine strategy 2014–2023. *Altern. Integr. Med.*: 1–76.

8 Alonso-Castro, A., Dominguez, F., José Maldonado-Miranda, J. et al. (2016). Use of medicinal plants by health professionals in Mexico. *J. Ethnopharmacol.* 198: 81–86.

9 Atlas INI (1994). *Atlas INI de las Plantas de la Medicina Tradicional Mexicana.* Insitiuto Nacional Indigenista. Tomo I,II, III.

10 Cuesta-Rubio, O., Piccinelli, A.L., Campo Fernandez, M. et al. (2007). Chemical characterization of *Cuban propolis* by HPLC – PDA, HPLC – MS, and NMR: the brown, red, and yellow *Cuban* varieties of propolis. *J. Agricult. Food Chem.* 55 (18): 7502–7509.

11 Sforcin, J.M. and Bankova, V. (2011). Propolis: is there a potential for the development of new drugs? *J. Ethnopharmacol.* 133 (2): 253–260.

12 Bankova, V., Popova, M., and Trusheva, B. (2014). Propolis volatile compounds: chemical diversity and biological activity: a review. *Chem. Central J.* 8 (1): 28.

13 Farooqui, T. and Farooqui, A.A. (2012). Beneficial effects of propolis on human health and neurological diseases. *Front. Biosci. (Elite Ed.)* 4: 779–793.

14 Catchpole, O., Mitchell, K., Bloor, S. et al. (2015). Antiproliferative activity of New Zealand propolis and phenolic compounds vs human colorectal adenocarcinoma cells. *Fitoterapia* 106: 167–174.

15 Duran, N., Muz, M., Culha, G. et al. (2011). GC-MS analysis and antileishmanial activities of two Turkish propolis types. *Parasitol. Res.* 108 (1): 95–105.

16 Santana, L.C., Carneiro, S.M.P., Caland-Neto, L.B. et al. (2014). Brazilian brown propolis elicits antileishmanial effect against promastigote and amastigote forms of *Leishmania amazonensis. Nat. Prod. Res.* 28 (5): 340–343.

17 Ayres, D.C., Marcucci, M.C., and Giorgio, S. (2007). Effects of Brazilian propolis on *Leishmania amazonensis. Mem. Inst. Oswaldo Cruz* 102: 215–220.

18 Falcao, S.I., Vale, N., Cos, P. et al. (2014). *In vitro* evaluation of Portuguese propolis and floral sources for antiprotozoal, antibacterial and antifungal research. *Phytother. Res. PTR* 28 (3): 437–443.

19 Monzote Fidalgo, L., Sariego Ramos, I., Garcia Parra, M. et al. (2011). Activity of *Cuban propolis* extracts on *Leishmania amazonensis* and *Trichomonas vaginalis. Nat. Prod. Commun.* 6 (7): 973–976.

20 Duran, G., Duran, N., Culha, G. et al. (2008). *In vitro* antileishmanial activity of *Adana propolis* samples on *Leishmania tropica*: a preliminary study. *Parasitol. Res.* 102 (6): 1217–1225.

21 Cuesta-Rubio, O., Fernández, M.C., Hernández, I.M. et al. (2017). Chemical profile and anti-leishmanial activity of three *Ecuadorian propolis* samples from Quito, Guayaquil and Cotacachi regions. *Fitoterapia* 120: 177–183.

22 Grecco, S.S., Reimão, J.Q., Tempone, A.G. et al. (2012). *In vitro* antileishmanial and antitrypanosomal activities of flavanones from *Baccharis retusa* DC. (Asteraceae). *Exp. Parasitol.* 130 (2): 141–145.

23 Mallick, S., Dutta, A., Dey, S. et al. (2014). Selective inhibition of *Leishmania donovani* by active extracts of wild mushrooms used by the tribal population of India: an *in vitro* exploration for new leads against parasitic protozoans. *Exp. Parasitol.* 138: 9–17.

24 Kvist, L.P., Christensen, S.B., Rasmussen, H.B. et al. (2006). Identification and evaluation of Peruvian plants used to treat malaria and leishmaniasis. *J. Ethnopharmacol.* 106 (3): 390–402.
25 Fournet, A., Barrios, A.A., and Muñoz, V. (1994). Leishmanicidal and trypanocidal activities of Bolivian medicinal plants. *J. Ethnopharmacol.* 41 (1): 19–37.
26 Weniger, B., Robledo, S., Arango, G.J. et al. (2001). Antiprotozoal activities of Colombian plants. *J. Ethnopharmacol.* 78 (2): 193–200.
27 Muzitano, M.F., Cruz, E.A., de Almeida, A.P. et al. (2006). Quercitrin: an antileishmanial flavonoid glycoside from *Kalanchoe pinnata*. *Planta Med.* 72 (1): 81–83.
28 Muzitano, M.F., Tinoco, L.W., Guette, C. et al. (2006). The antileishmanial activity assessment of unusual flavonoids from *Kalanchoe pinnata*. *Phytochemistry* 67 (18): 2071–2077.
29 Maldonado Fonkén MS. An introduction to the bofedales of the Peruvian High Andes. *Mires Peat.* 2014;15: Art. 5, 1–13.
30 Gonzales, G. and Valerio, L. (2006). Medicinal plants from Peru: a review of plants as potential agents against cancer. *Anti Cancer Agents Med. Chem.* 6 (5): 429–444.
31 Cermal-Editions (2014). Typical Peruvian plants. https://www.peruinformation.org/typical-plants (accessed 18 March 2019).
32 Berłowski, A., Zawada, K., Wawer, I., and Paradowska, K. (2013). Antioxidant properties of medicinal plants from Peru. *Food Nutr. Sci.* 04 (08): 71–77.
33 Odonne, G., Valadeau, C., Alban-Castillo, J. et al. (2013). Medical ethnobotany of the Chayahuita of the Paranapura basin (Peruvian Amazon). *J. Ethnopharmacol.* 146 (1): 127–153.
34 Estevez, Y., Castillo, D., Pisango, M.T. et al. (2007). Evaluation of the leishmanicidal activity of plants used by Peruvian Chayahuita ethnic group. *J. Ethnopharmacol.* 114 (2): 254–259.
35 Hermoso, A., Jiménez, I.A., Mamani, Z.A. et al. (2003). Antileishmanial activities of dihydrochalcones from *Piper elongatum* and synthetic related compounds. Structural requirements for activity. *Bioorg. Med. Chem.* 11 (18): 3975–3980.
36 Boeck, P., Bandeira Falcao, C.A., Leal, P.C. et al. (2006). Synthesis of chalcone analogues with increased antileishmanial activity. *Bioorg. Med. Chem.* 14 (5): 1538–1545.
37 de Mello, T.F., Bitencourt, H.R., Pedroso, R.B. et al. (2014). Leishmanicidal activity of synthetic chalcones in *Leishmania* (Viannia) *braziliensis*. *Exp. Parasitol.* 136: 27–34.
38 de Oliveira, A., Mesquita, J.T., Tempone, A.G. et al. (2012). Leishmanicidal activity of an alkenylphenol from *Piper malacophyllum* is related to plasma membrane disruption. *Exp. Parasitol.* 132 (3): 383–387.

39 Munoz, V., Moretti, C., Sauvain, M. et al. (1994). Isolation of bis-indole alkaloids with antileishmanial and antibacterial activities from *Peschiera van heurkii* (syn. *Tabernaemontana van heurkii*). *Planta Med.* 60: 455–459.

40 Yamaguchi, K.K.L., Pereira, L.F.R., Lamarão, C.V. et al. (2015). Amazon acai: chemistry and biological activities: a review. *Food Chem.* 179: 137–151.

41 Da Silva, B.J.M., Souza-Monteiro, J.R., Rogez, H. et al. (2018). Selective effects of *Euterpe oleracea* (acai) on *Leishmania* (*Leishmania*) *amazonensis* and *Leishmania infantum*. *Biomed. Pharmacother.* 97: 1613–1621.

42 Flora-of-Ecuador (2018). Dept of Biological and Environmental Sciences, University of Gothenburg, in co-operation with Pontificial Catholic University of Ecuador, Q. Flora of Ecuador. https://bioenv.gu.se/forskning/huvudforskningsomraden/systematik (accessed 20 March 2019).

43 Gachet, M.S., Lecaro, J.S., Kaiser, M. et al. (2010). Assessment of anti-protozoal activity of plants traditionally used in Ecuador in the treatment of leishmaniasis. *J. Ethnopharmacol.* 128 (1): 184–197.

44 Kabiru, A. (2013). Elephantopus species: traditional uses, pharmacological actions and chemical composition. *Adv. Life Sci. Technol.* 15: 6–14.

45 Fuchino, H., Koide, T., Takahashi, M. et al. (2001). New sesquiterpene lactones from *Elephantopus mollis* and their leishmanicidal activities. *Planta Med.* 67 (7): 647–653.

46 Cursino, L.M.C., Nunez, C.V., RCd, P. et al. (2012). Triterpenes from *Minquartia guianensis* (Olacaceae) and *in vitro* antimalarial activity. *Química Nova* 35: 2165–2168.

47 Atta-ur-Rahman (2012). *Studies in Natural Products Chemistry (Bioactive Natural Products (Part P))*, 1e. Elsevier Academic Press.

48 Cristina, R., Paula, D., Fernanda, M. et al. (2012). Triterpenes from *Minquartia guianensis* (Olacaceae) and *in vitro* antimalarial activity. *Química Nova* 35 (11): 2165–2168.

49 Rocha, L.G., Almeida, J.R.G.S., Macêdo, R.O., and Barbosa-Filho, J.M. (2005). A review of natural products with antileishmanial activity. *Phytomedicine* 12 (6): 514–535.

50 Sosa, V., De-Nova, J.A., and Vásquez-Cruz, M. (2018). Evolutionary history of the flora of Mexico: dry forests cradles and museums of endemism. *J. Syst. Evol.* 56 (5): 523–536.

51 Alamilla-Fonseca, L.N., Delgado-Domínguez, J., Zamora-Chimal, J. et al. (2018). *Leishmania mexicana* cell death achieved by *Cleoserrata serrata* (Jacq.) Iltis: learning from Maya healers. *J. Ethnopharmacol.* 211: 180–187.

52 Delgado-Altamirano, R., Monzote, L., Pinon-Tapanes, A. et al. (2017). *In vitro* antileishmanial activity of Mexican medicinal plants. *Heliyon* 3 (9): e00394.

53 Gutierrez-Rebolledo, G.A., Drier-Jonas, S., and Jimenez-Arellanes, M.A. (2017). Natural compounds and extracts from Mexican medicinal plants with anti-leishmaniasis activity: an update. *Asian Pac. J. Trop. Med.* 10 (12): 1105–1110.

54 Arevalo-Lopez, D., Nina, N., Ticona, J.C. et al. (2018). Leishmanicidal and cytotoxic activity from plants used in Tacana traditional medicine (Bolivia). *J. Ethnopharmacol.* 216: 120–133.

55 Soosaraei, M., Fakhar, M., Hosseini Teshnizi, S. et al. (2017). Medicinal plants with promising antileishmanial activity in Iran: a systematic review and meta-analysis. *Ann. Med. Surg. (Lond.)* 21: 63–80.

56 Kheiri Manjili, H., Jafari, H., Ramazani, A., and Davoudi, N. (2012). Anti-leishmanial and toxicity activities of some selected Iranian medicinal plants. *Parasitol. Res.* 111 (5): 2115–2121.

57 Nikmehr, B., Ghaznavi, H., Rahbar, A. et al. (2014). *In vitro* anti-leishmanial activity of methanolic extracts of *Calendula officinalis* flowers, *Datura stramonium* seeds, and *Salvia officinalis* leaves. *Chin. J. Nat. Med.* 12 (6): 423–427.

58 Ferreira, C., Passos, C.L., Soares, D.C. et al. (2017). Leishmanicidal activity of the alkaloid-rich fraction from *Guatteria latifolia*. *Exp. Parasitol.* 172: 51–60.

59 Bhattacharya, P., Mondal, S., Basak, S. et al. (2016). *In vitro* susceptibilities of wild and drug resistant *Leishmania donovani* amastigotes to piperolactam A loaded hydroxypropyl-beta-cyclodextrin nanoparticles. *Acta Trop.* 158: 97–106.

60 Kaur, G., Chauhan, K., and Kaur, S. (2018). Immunotherapeutic potential of *Codonopsis clematidea* and naringenin against visceral leishmaniasis. *Biomed. Pharmacother.* 108: 1048–1061.

61 Peniche-Pavía, H.A., Medrano-Nahuat, D., Torres-Tapia, L.W. et al. (2016). Metabolites isolated from the rhizomes of *Dorstenia contrajerva* with anti-leishmanial activity. *Phytochem. Lett.* 18: 140–143.

62 Falodun, A., Imieje, V., Erharuyi, O. et al. (2014). Isolation of antileishmanial, antimalarial and antimicrobial metabolites from *Jatropha multifida*. *Asian Pac. J. Trop. Biomed.* 4 (5): 374–378.

63 Mukherjee, D., Singh, C.B., Dey, S. et al. (2016). Induction of apoptosis by zerumbone isolated from *Zingiber zerumbet* (L.) Smith in protozoan parasite *Leishmania donovani* due to oxidative stress. *Braz. J. Infect. Dis.* 20 (1): 48–55.

64 Manzo, E., Ciavatta, M.L., Pagano, D., and Fontana, A. (2012). An efficient and versatile chemical synthesis of bioactive glyco-glycerolipids. *Tetrahedron Lett.* 53 (7): 879–881.

65 Fromentin, Y., Gaboriaud-Kolar, N., Lenta, B.N. et al. (2013). Synthesis of novel guttiferone A derivatives: in-vitro evaluation toward *Plasmodium falciparum*, *Trypanosoma brucei* and *Leishmania donovani*. *Eur. J. Med. Chem.* 65: 284–294.

Part III

Helminth Infections

7

Prospects of Phytochemicals for the Treatment of Helminthiasis

Muhammad Akram[1], Ejaz Mohiuddin[2], Charles Oluwaseun Adetunji[3], Tolulope Olawumi Oladosun[4], Phebean Ozolua[3], Frances N. Olisaka[5], Chukwuebuka Egbuna[6,7], Olugbenga Samuel Micheal[8], Juliana Bunmi Adetunji[9], Leena Hameed[2], Chinaza Godswill Awuchi[10], Kingsley C. Patrick-Iwuanyanwu[7,11], and Olugbemi Tope Olaniyan[12]

[1] *Department of Eastern Medicine and Surgery, Government College University Faisalabad, Faisalabad, Pakistan*
[2] *Faculty of Eastern Medicine, Hamdard University Karachi, Karachi, Pakisthan*
[3] *Applied Microbiology, Biotechnology and Nanotechnology Laboratory, Department of Microbiology, Edo State University Uzairue, Iyamho, Nigeria*
[4] *Department of Microbiology, Federal University of Agriculture, Makurdi, Nigeria*
[5] *Environmental and Public Health Microbiology Laboratory, Department of Biological Sciences, Faculty of Science, Benson Idahosa University, Benin City, Nigeria*
[6] *Department of Biochemistry, Faculty of Natural Sciences, Chukwuemeka Odumegwu Ojukwu University, Uli, Nigeria*
[7] *Nutritional Biochemistry and Toxicology Unit, World Bank Africa Centre of Excellence, Centre for Public Health and Toxicological Research (PUTOR), University of Port Harcourt, Port Harcourt, Nigeria*
[8] *Cardiometabolic Research Unit, Department of Physiology, College of Health Sciences, Bowen University, Iwo, Nigeria*
[9] *Nutritional and toxicological Research Laboratory, Department of Biochemistry Sciences, Osun State University, Osogbo, Nigeria*
[10] *Department of Physical Sciences, Kampala International University, Kampala, Uganda*
[11] *Department of Biochemistry, Faculty of Science, University of Port Harcourt, Port Harcourt, Nigeria*
[12] *Laboratory for Reproductive Biology and Developmental Programming, Department of Physiology, Edo State University Uzairue, Iyamho, Nigeria*

List of Abbreviations

BLAST	Basic Local Alignment Search Tool
DNA	Deoxyribonucleic acid
IPIs	Intestinal parasitic infections
NCBI	National Centre for Biotechnology Information
QPCR	Quantitative polymerase chain reaction
STH	Soil-transmitted helminthiasis

Neglected Tropical Diseases and Phytochemicals in Drug Discovery, First Edition.
Edited by Chukwuebuka Egbuna, Muhammad Akram, and Jonathan Chinenye Ifemeje.

7.1 Introduction

Helminthiasis is a common intestinal worm infection found amongst urban and rural populations [1]. From the beginning of human existence, medicinal plants have been used to treat diseases [2–4]. Recently, there has been a growing revival to treat helminthiasis with medicinal plants [5–8]. Various medicinal plants have an ancient history in terms of helminthiasis treatment in traditional systems of medicine. Intestinal worm infections are increasing worldwide. The most serious helminthiasis infections are prevalent in the developing countries, but can also be found across the developed countries. The most common soil-transmitted helminths are parasitic worms from the phyla Nematoda (roundworms) and Platyhelminthes (flatworms); which includes Hookworms (*Ancylostoma duodenale* and *Necator americanus*), roundworm (*Ascaris lumbricoides*), and whipworm (*Trichuris trichiura*). Approximately 4.5 billion people are at risk of being infected by helminths [9, 10]. All ages are affected by intestinal parasitic infections (IPIs) but children, preschool, and school-going children are the high-risk group and this disease is responsible for poor health and malnutrition. Recurrent infections lead to excess morbidity that can continue from generation to generation in the already poor group of people who are at risk of this disease because of lack of sanitation, lack of access to safe water, and improper hygiene. So, it is most common wherever there are poverty and economic instability [11]. *A. lumbricoides* is the most common helminth and currently infects about 1 billion people worldwide [12]. Figure 7.1 depicts an *Ascaris* worm being removed from a patient's bile duct in South Africa.

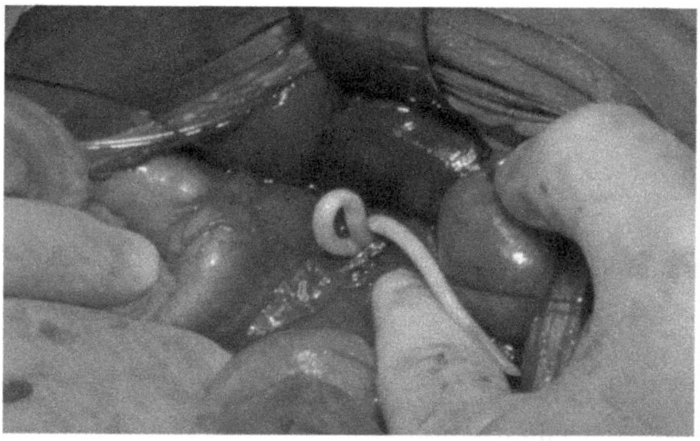

Figure 7.1 Adult *Ascaris* worms being removed from the bile duct of a patient in South Africa. *Source:* Larry Hadley. https://en.wikipedia.org/wiki/Soil-transmitted_helminthiasis#/media/File:Adult_ascaris_worms_being_removed_from_the_bile_duct_of_a_patient_in_South_Africa.png. License under CC AS-BY 2.0.

7.1.1 History

The origin of the word Helminths is Greek which means worms [13]. It has a very old history and has been found in the feces of mummified humans. It has also been described by Hippocrates, Egyptian medical papyri, and in the Bible as infections that have plagued humans since before the era of our earliest recorded history. The eggs of intestinal helminths can be found in the mummified feces of humans dating back thousands of years [1] and can be recognized by the characteristic clinical features of helminth infections from the ancient writings of Hippocrates [1]. The history of treatment with herbal medicine dates back to 3500 years BC and is still widely practiced in India and China which are rich in medicinal plants resources [14].

7.1.2 Prevalence

It is estimated that over 1 billion people in developing regions such as sub-Saharan Africa (SSA), Asia, and the Americas are infected with one or *more* species of helminths [1]. The prevalence noted in Karachi by a study was 52.8%. The most common intestinal parasites were *Giardia lamblia* (28.9%) followed by *A. lumbricoides* (16.5%), *Blastocystis hominis* (10.1%), *Hymenolepis nana* (0.9%), *Endolimax nana* (1.8%), *Entamoeba coli* (2.3%), and *Iodoamoeba butschlii* (3.2%). Coinfection was noted in 43% of samples which comprised of a single parasite while 10% were of multiple parasites [15]. The age group most commonly affected was one to five years of age [12]. Children of preschool age are more affected. There is a high degree of host-parasite tolerance. Roundworms contribute to malnutrition in children and may lead to growth retardation. The prevalence of pinworm infestation is high in children and preschool children of both sexes. Overcrowding, contaminated clothing, and shared bedding favor reinfection and spread. *A. duodenale* is prevalent in Mediterranean countries, Europe, Egypt, and India while *N. Americanus* is prevalent in America and throughout tropical East Africa. Its susceptibility is general, but it is more frequent in Whites than in Negroes. Though some immunity develops with infections, the infected persons remain potential spreaders of infection, so long as they are infected. The common group affected is between 5 and 40 years.

7.2 Molecular Characteristics of Soil-transmitted Helminthiasis

In times past, soil-transmitted helminthiasis (STH) was diagnosed by microscopic examination of stool samples for the presence of the parasites. However, this technique is often regarded to be insensitive. Molecular characterization of STH on the other hand, begins with the extraction of the DNA from the parasite's eggs or larvae in the sample using a DNA extraction kit for stool or soil samples. In situations where the stool samples cannot be processed at the point (or immediately) of collection, they should be preserved in a solution of potassium dichromate preceding onward transfer to the laboratory for analysis [16]. Quantitative PCR (QPCR) is a viable molecular technique used in the molecular diagnosis of STH. The specific nucleotide sequences (regions) often targeted with respect to diagnosis for STH parasites include internal transcribed spacers 1 and 2 (ITS-1, ITS-2) for primer design usually applicable to all STH parasites (*Strongyloides stercoralis, Necator americanus, Ascaris lumbricoides, Trichuris trichiura,* and *A. duodenale*), while Cytochrome oxidase 1 and/or 5.8S regions are targets in *S. stercoralis, A. duodenale, N. americanus,* and *A. lumbricoides* [16–20]. Having extracted the target regions using the designed primers, QPCR is used for the amplification of these regions with the designated amplification protocol. Amplicons obtained should be sent for sequencing and alignment with existing sequences on the National Centre for Biotechnology Information (NCBI) platform through the Basic Local Alignment Search Tool (BLAST) would be useful for identifying the particular parasite present in the sample.

7.3 Clinical Features and Pathogenesis

The clinical features may vary from being asymptomatic to very vague. Heavy worm infection can cause digestive problems, abdominal pain, diarrhea, vomiting, disturbed sleep, restlessness, and malnutrition [21]. Intestinal obstruction, obstruction of pancreatic and biliary duct, and liver abscess are also known to occur [22–24]. Many times, a live worm in the stool is the first detected sign of roundworm infection. Climate, food preferences, hygiene, and contact with vectors all influence the host's exposure to infection. Host defense mechanisms can eliminate potential infections, but most times once an infection becomes established, it may remain for many years. Sometimes, even infections are asymptomatic while leading to severe pathological changes over time. Worms are large in size and can migrate inside the body causing tissue damage of patients. The host defense mechanism may also damage the body indirectly [25]. Intestinal worms enter the human body via different pathways.

Some intestinal worms such as *Ascaris, Echinococcus, Enterobius,* and *Trichuris* enter by accidental ingestion of contaminated food and water. Other worms may enter the body via the skin. The infected human with mature worms serves as a reservoir for infection of other hosts. The immediate source of infection is the soil containing embryonated eggs. Many helminths affect children than adults, and children living under unhygienic community conditions with a weak immune system are the most vulnerable. Some children are genetically stronger to resist infection. Lifestyle and dietary conditions may also affect the immune system of the body [26]. Large *Ascaris* can cause blockage in the intestine and bile duct. *Wuchereria* blocks the lymph flow and this condition is called elephantiasis. *Echinococcus granulosus* causes pressure atrophy, a huge fluid-filled cyst. The multilocular hydatid cysts caused by *Echinococcus multilocularis* have a different growth form, migrate to different organs of the body causing widespread damage.

Intestinal worms can produce various abnormalities in the intestinal mucosa, some exhibit physical and chemical damage to the organs, and immunopathologic responses also cause damage to organs. Hookworms known as *Ancylostoma* and *Necator* suck blood from the capillaries of the mucosa. Worms secrete anticoagulants that cause bleeding and blood loss, which leads to anemia and loss of protein. Inflammatory changes occur due to worm infestations which cause protein-losing enteropathies. Vitamin B_{12} deficiency has been observed in patients with roundworm infestations [27].

The skin, lungs, liver, and intestines are most affected sites of the human body. During the migratory phase of worms in the body, some signs and symptoms are produced, including granulomatous lesions, organomegaly, pruritis, pneumonia, petechial hemorrhages, eosinophilia, and urticaria. Worms obtain nutrition from the host, which is the most common etiology of pathogenesis, as evidenced by metaplastic and hyperplastic variations in the epithelium of the host. Hyperplasia in the bile duct epithelium has been observed in association with liver fluke infections. Neoplastic changes may occur due to chronic inflammation caused by some parasitic infection, but the exact etiology is not known. Worms release excretory–secretary materials and cause direct damage to host tissues that are responsible for the pathology [28].

It is alarming that intestinal infections are associated with a degree of immune suppression. Immune suppression has been shown by various factors such as antigen overload, antigenic competition, and induction of suppressor cells, and the production of lymphocyte-specific suppressor factors. Roundworm, fluke or tapeworm may manifest as abdominal pain. Children are particularly susceptible to nutritional deficiencies because of infestations by intestinal roundworms. The poor sanitary facilities, open field defecation, and promiscuous defecation lead to the transmission of roundworm infection. Growth retardation and malnutrition have been observed in ancylostomiasis, strongyloidiasis, trichuriasis, and ascariasis. Severe cases of ancylostomiasis and strongyloidiasis may also cause weight loss. Anemia and fatigue often occur in cases of ancylostomiasis and strongyloidiasis. Cyanocobalamine deficiency has been observed in cases of fish tapeworm (*Diphyllobothrium latum*), which results in hemoglobin deficiency. Fatigue may be a symptom of *fish tapeworm*. Pulmonary eosinophilia occurs due to immature worms present in the lungs. Dry coughing, shortness of breath, fever, chest pain,

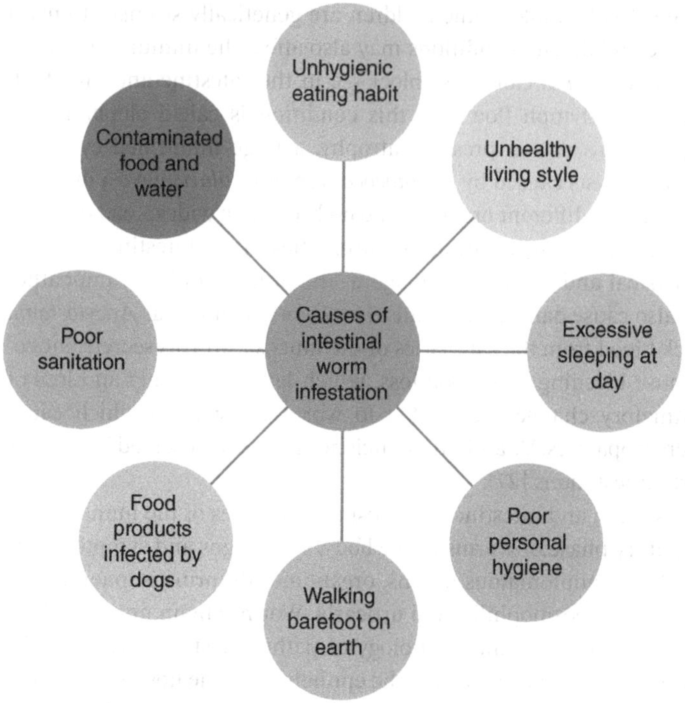

Figure 7.2 Causes of intestinal worm infestation.

and wheezing are often observed in roundworm infestations. Ascariasis is a problem throughout the tropical world and is the major cause of ill health in developing countries. Helminthiasis is prevalent in rural areas of developing countries. Overcrowding and poor housing especially in parasite-favorable climatic conditions keep the transmission rate at high levels (Figure 7.2).

7.4 Prevention

Soil-transmitted helminths are a big source of infection especially in endemic areas and it is difficult to achieve a significant elimination of infection with anthelminthic drugs only so good initiatives should be taken to control the transmission of the disease also along with the morbidity control [29]. Provision should be made for dug well and borehole or any other type of latrines for villagers and persons living in urban areas where there is no facility of private latrines. Vegetables and fruits must be washed, if possible in running water before using them. Washing hands thoroughly before taking food and after defecation, especially children must follow this rule strictly. Adequate facilities should be made to avoid overflow of ascarid eggs from the drains and latrines, etc. Habits of nail-biting and scratching bare anal area should be discouraged. In selected endemic areas and not in all parts of the world, it is a reportable disease. Vegetables and fruits must be washed properly before eating. Special care is required when these are taken raw. People should have a habit of wearing chappals, shoes and should not go out barefoot. In the endemic region, people must cultivate the habit of washing their feet regularly to prevent infection. Also, programs should be put in place by the healthcare system on educating the masses about the disease and its preventive measures [30].

7.5 Treatment

Anthelmintics are drugs that are used to treat intestinal worms. There are various treatment options that are available in the allopathic system of medicine, but allopathic drugs exert side effects (Table 7.1). A wide number of herbal products are now employed in the treatment of intestinal worms for their purported better efficacy and safety compared to synthetic medicine [37, 38]. Herbal medicine is an integral part of the development of modern civilization. A literature review indicates activities of de-worming drugs that have the potential to treat helminthiasis. The overall findings of the current review show that allopathic drugs and some documented medicinal plants have demonstrable *in vivo/in vitro* anthelmintic activity.

Table 7.1 Side effects of anthelmintic drugs.

Drugs	Side effects	References
Praziquantel	Nausea, vomiting, abdominal pain, giddiness, drowsiness	[31]
Mebendazole	Nausea, vomiting, dizziness, hypotension	[32]
Thiabendazole	Vomiting, diarrhea, anorexia, dizziness, bradycardia, pruritis, rashes, headache, drowsiness, hypotension	[33]
Albendazole	GI disturbance	[34]
Levamisole	Nausea, vomiting, abdominal pain, dizziness, diarrhea, skin rashes, transient neutropenia	[35]
Pyrantel pamoate	Minimal side effects. Neurotoxicity with very large dose, proteinuria	[36]

7.6 Plants and Phytochemicals with Anthelmintic Activities

Some well-known plants validated for their anthelmintic activity are: *Acacia albida, Anogeissus leiocarpus, Areca catechu, Azadirachta indica, Bixa orellana, Butea frondosa, Butea monosperma, Caesalpina crista, Carica papaya, A. lumbricoides, Carum copticum, Cassia alata, Cucurbita maxima, Cucurbita moschata, Diospyros scabra, Embelica ribes, Hyoscyamus niger, Lagenaria siceraria, Lantana camara, Lansium domensticum, Leucaena lucocephala, Mangifera indica, Melia azedarach, Moringa olefera, Nigella sativa, Peganum harmala, Pisitacia integrrima, Quisqualis indica, Randia dumetorum, Vernonia anthelmintica,* and *Xylopia aethiopica*. The plants commonly used by herbalist as anthelmintic agents are *Bidens pilosa, Tamarindus indica, Combretum collinum, Solanecio mannii, Leonotis nepetifolia, Sclerocarya birrea, Albizia coriaria, Euclea divinorum, Aloe secundiflora, Plectranthus barbatus, Rotheca myricoides, Ximenia Americana, Vernonia amygdalina, Hypitis suaveolens, Erythrina abyssinica, Eclipta prostrate, Cucumis aculeatus, Harrisonia abyssinica, C. papaya, Searsia natalensis,* and *Kigelia africana*. *Ferula assafetida* L. (Apiaceae), *Foeniculum vulgare* (Apiaceae), *Trachyspermum ammi* (Apiaceae), *Calotropis procera* (Asclepiadaceae), *Artemisia brevifolia* (Astraceae), *Brassica campestris* (Bracicaceae), *Matricaria chamomilla* (Astraceae), *Bombax ceiba* (Bumbacaceae), *Eruca sativa* (Bracicaceae), *Aloe vera* (Lillaceae), *Piper nigrum* (Pipreceae), *Withania somnifera* (Solanaceae), *Cuscuta reflexa* (Cuscutaceae), and *Mallotus philippinensis* (Euphorbiaceae) [39].

Phytochemicals are fast gaining attention as alternative sources of therapy in the treatment of helmintic infections. Phytochemicals from medicinal plants (Table 7.2) with proven efficacy in the treatment of STH are usually beneficial to the human body beyond their anthelminthic properties, with negligible side effects when administered within the safe limits. According to [66], *Punica granatum, Clausena anisata,* and *Zanthoxylum zanthoxyloides* are examples of plants containing phytochemicals with anthelminthic activities for the treatment of STH. The various phytochemicals obtained from these plant extracts include alkaloids, terpenes, flavonoids, and tannins. Biocidal activity of Elligatannins from extracts of *P. granatum* against *Cryptosporidium parvum* – a parasitic protozoan has also been reported [67], which is also believed to contribute to the anthelminthic properties of *P. granatum* in the study. Anthelminthic activities of immature fruits of *Magnifera indica* (mango) against *S. stercoralis* were evaluated by [68] as being linked to the presence of tannins in immature mangoes. Other metabolites detected in the immature mango fruit reported are saponins, hydrolyzable tannins, proanthocyanidins, and triterpenes.

Table 7.2 Medicinal plants having anthelmintic activity.

Plant name	Family	Parts used	Chemical constituents	Medicinal uses	Pharmacological activity	References
Carica papaya	*Caricacea*	Seeds, leaves, milky juice, ripe, and unripe fruit	Papain, pectins, carotenoid pigment, carpasemine, carpine, carposide, chemopapain, vitamin C, caricin, protein, carbohydrate, and fatty acids	It is used in stomach ache, dysentery and rheumatism	Digestive, diuretic, anthelmintic, antibacterial, resolvent, antiphlegmatic, antiasthmatic, anti-inflammatory, and antihypertensive	[40, 41]
Butea monosperma Lam	*Fabaceae*	Gums, seeds, flowers, leaves, and bark	Polypeptides, plant proteinase, lipolytic enzymes, proteolytic enzymes, pyrocatecin, mucilaginous material, tannins, gum, phenylalanine, alanine, aspartic acid, histidine, fructose, glucose, lignoceric acid, arachidic acid, palmitic acid, stearic acid, isobutyine, aurones, chalcones, isomonospermoside, monospermide, sulphurine, butin 7 glucoside, isocoreopsin, coreopsin, isobutrin, butin, several flavonoids	It is used in intestinal worms, obesity, fungal infection, cancer, and hyperglycemia	Anthelmintic, antiobesity, antifungal, chemopreventive, and nephroprotective	[42–44]
Baliospermum montanum Muell	*Euphorbiaceae*	Seeds	Montanin, terpenoids, saponins, sterols, glycosides, and flavonoids	It is used in indigestion, urinary tract infection, helminthiasis, inflammation, asthma, constipation, and cancer	Diuretic, purgative, tonic, antiasthmatic, anti-inflammatory, antitumor, antibacterial, anthelmintic, and hepatoprotective	[45]

(*Continued*)

Table 7.2 (Continued)

Plant name	Family	Parts used	Chemical constituents	Medicinal uses	Pharmacological activity	References
Adhatoda vasica	*Acanthaceae*	Flowers, bark, roots, and flowers	Adhatodic acid, vasicinol, vasicine, volatile odorous principle, essential oil, and alkaloids	It is used in asthma, tuberculosis, cough, bronchitis, bronchiectatsis, and lung cancer	Bronchodilator, anthelmintic, and antitussive	[46–48]
Cansjera rheedii	*Opiliaceae*	Aerial parts	Flavonoids, saponins, fats, fixed oils, amino acids, proteins, tannins, steroids, glycosides, carbohydrates, and alkaloids	It is used in helminthiasis	Anti-inflammatory, anthelmintic, antioxidant, and antidiabetic	[49]
Ficus benghalensis	*Moraceaea*	Bark, leaves and seeds	Meso-inositol, tiglic acid ester, furocoumarin, triterpenoids, triterpenes, sterols, flavonol, ketones, glucose, beta sitosterol, 6-heptatriacontene, and tetratriacontene-2-one	It is used in inflammatory bowel disease and helminthiasis	Hypoglycemic, anti-inflammatory, hypolipidemic, antibacterial, immunomodulant, anti-allergic, antipyretic, analgesic, anti-stress, antioxidant, antiatherogenic, growth promoter, ameliorative, wound healer, and antidiarrheal	[50–52]
Rumex abyssinicus	*Polygonaceae*	Aerial parts	Betulone and oleic acid	It is used in inflammation and microbial infections	Antimicrobial, anti-inflammatory, and analgesic, diuretic	[53–55]
Calotropis procera	*Asclepiadaceae*	Roots and leaves	Calotropin, uscharin, calotropagenin, calotoxin, benzoyllineolone, benzoyl isolineolone, syriogenin, beta amyrin, evanidine-3-rhamnoglucoside, *O*-pyrocatechuic acid, urcharidin, uzarigenin, and proceroside	It is used in heminthiasis	Antiasthmatic, antirheumatic, antiseptic, anti-inflammatory, expectorant, antispasmodic, nervine tonic, stomachic, antiphlegmatic, analgesic, diaphoretic, febrifuge, diuretic, anticancer, anthelmintic, and antinociceptive	[56–59]

Vitex negundo	*Lamiaceae*	Whole plant	Ethyl-hexadecenoate, carryophyllene epoxide, δ-guaiene, (*E*)-nerolidol, germacren-4-ol, α-selinene, β-selinene, hexadecanoic acid, germacrene D, guaia-3, 7-diene, and valencene	It is used in filarial and inflammation	Antimicrobial, anti-inflammatory, antinociceptive, anticonvulsant, and bronchodilator	[39]
Mimosa pudica	*Fabaceae*	Leaves	C-glycosyl flavones, aromatic amino acids, phenolic ketones, flavonoids, alkaloids, and essential oil	It is used in wounds, dysentery and piles	Anthelmintic, hypolipidemic, and antidepressant	[60, 61]
Verbascum thapsus	*Scrophulariaceae*	Aerial parts	Flavonoids, carbohydrates, saponins, tannins, terpenoids, proteins, glycosides, fats, and fixed oils	It is used in cough, asthma, migraine headache, inflammatory diseases, pulmonary problems, and earache	Antiasthmatic, anthelmintic, and antiviral	[62, 63]
Cynodon dactylon	*Poaceae*	Aerial parts	Carotenoids and flavonoids	It is used in diabetes mellitus, cardiovascular disorders, helminthiasis, and cancer	Anti-inflammatory, chemopreventive, antidiabetic, and antiarthritic	[64, 65]

7.6.1 Modes of Action of Phytochemical Against the Soil-transmitted Helminthiasis

For a phytochemical to qualify as an anthelminthic agent, it should successfully target at least one of the stages involved in the life cycle of the helminth. Primarily, this is essential in eradicating the parasite's existence and thereby halting the spread of parasitic infections. The modes of action exhibited by phytochemicals against soil-transmitted helminths as observed are cuticular damage in adult worms, larval development inhibition (LDI), and egg-hatching inhibition (EHI).

Luteolin was reported to be responsible for cuticular damage in adult worms. Tannins were reported to induce EHI and motility in certain stages of larval development of *Haemonchus contortus* [69]. Tannins interact with proteins present in the cuticle of nematode, thereby altering the physical and chemical properties of this region [70]. In a research finding reported by [71], there was 95.7% efficacy of the ethanolic fraction of hexane extract of mango seeds in EHI of *H. contortus*.

7.7 Scientific Reports of Medicinal Plants with Anthelmintic Properties

Some scientific studies have found evidence which is in support of the traditional use of plants against helminths. This section discusses some of these plants.

7.7.1 *Adhatoda vasica*

Adhatoda vasica belongs to the family Acanthaceae. It contains vitexin, quercetin, kaemferol, apigenin, triterpenes, steroids, betaine, vasicolinone, vasicol, and vasicine [72, 73]. Yadav and Tangpu [74] reported the anticestodal activity of *A. vasica* extract against *Hymenolepis diminuta* infections in rats. *A. vasica* is commonly prescribed to treat intestinal worm infestation. Eggs per gram of feces and percentage recovery rates were counted to estimate the efficacy of extract. The extract was administered to rats infected with immature and mature *H. diminuta*. The assay showed that 800 mg/kg double dose of the water extract was effective against mature worms. Eggs per gram count decreased to 79.57%. This efficacy was better than 5 mg/kg of praziquantel, the standard drug used to treat worm infections. There was significant activity against immature worms also. The study indicated that *A. vasica* has anticestodal activity and is useful in intestinal worm infections [74].

7.7.2 *Allium sativum*

Garlic is very rich in selenium. The sulfur compound allicin, produced by crushing or chewing fresh garlic, in turn, produces other sulfur compounds: ajoene, allyl sulfides, and vinyldithiins [75]. Oil of *Alium sativum* has also been reported to possess anthelmintic activity and expels out the parasites from the intestine [2, 76].

7.7.3 *Baliospermum montanum* Muell

Anthelmintic activity of *Baliospermum montanum* Muell was tested *in vitro* against *Pheretima posthuma* and *Ascardia galli*. The concentration of the drug was 10–100 mg/ml and the time of death and time of paralysis was determined. Significant inhibitory activity was observed at 100 mg/ml of *B. montanum* Muell [77].

7.7.4 *Butea monosperma*

Butea monosperma has antihyperglycemic, anthelmintic, antidiarrheal, wound healer, anticonvulsant, antifungal, anti-inflammatory, hepatoprotective, antistress, and antimicrobial activities [78]. The seeds of *B. monosperma* possess oil, proteolytic and lypolytic enzymes, plant proteinase and polypeptidase, a nitrogenous acidic compound, along with palasonin. The *in vitro* anthelmintic activity of methanolic extract of the seeds of *B. monosperma* has also been reported [79, 80].

7.7.5 *Calotropis procera*

Shivkar and Kumar [81] reported the anthelmintic activity of the latex of *Calotropis procera*. Adult earthworms were used for the assay. Extracts exhibited dose-dependent response. The extract was able to inhibit spontaneous motility (paralysis) and evoked responses to pin-prick [81]. This effect of the extract was comparable to the commonly used drug piperazine, indicating significant anthelmintic activity.

7.7.6 *Carica papaya*

Benzylisothiocyanate present in the seeds of *Carica papaya* is the chief or sole antihelmintic bioactive [82]. Okeniyi et al. [41] reported the effectiveness of dried *C. papaya* seeds against human intestinal parasitosis. For this study, 60 asymptomatic children with intestinal parasites in Nigeria received an elixir (20 ml) of *C. papaya* seeds with honey or honey alone (placebo group) in randomized clinical trials. After seven days of therapeutic intervention, microscopic examination of stool was done to check for the presence of intestinal worms. There was a significantly different degree of reduction of intestinal parasites in stools of children having received *C. papaya* elixir than those receiving honey alone. No side effects were reported in both treatment groups.

7.7.7 *Ficus benghalensis*

Aswar et al. [83] investigated the anthelmintic activity of petroleum ether, chloroform, water, and methanol extracts of *Ficus benghalensis*. All the extract paralyzed and killed the earthworm.

7.7.8 *Mimosa pudica*

Mimosa pudica belongs to the family Fabaceae. *Mimosa pudica* is also usually recognized as Lajjalu. It is cultivated as a garden plant or grows as a weed in fields. The whole plant is used for medicinal purposes. It is found in India, South and Central America. The water, ethanol, and petroleum ether (100, 200, and 500 mg/kg) of *M. pudica* seeds were investigated for anthelmintic activity against *P. posthuma*. The time of paralysis and death of worms was noted. Normal saline was used as control and albendazole as a standard reference. The aqueous and alcoholic extract caused paralysis and death of worms. The effect of aqueous and alcoholic extract was comparable to the standard and effect of petroleum ether was weaker than standard [84].

7.7.9 *Punica granatum*

Bendgude et al. [85] reported the antihelmintic activity of the pomegranate peel extract of 25, 50, and 75% against *Ascaris suum* females *in vitro*. The 25 and 50% peel extracts were not superior to mebendazole while the 100% was equivalent to it. The chemical constituents responsible for the anthelmintic activity are alkaloids and tannins.

7.7.10 *Verbascum thapsus*

Verbascum thapsus L. belongs to the family Scrophulariaceae. It is an erect biennial plant distributed in all provinces of Pakistan, Kashmir, and Gilgit Baltistan [86]. It contains flavonoids, phenylethanoid glycosides, isocatalpol, aucubin, iridoid, thapsuine, and saponins [87–90]. Ali et al. [91] reported the anthelmintic and relaxant activities of *V. thapsus* Mullein. Anthelmintic activities of this plant were investigated. Crude aqueous methanolic extracts of the plant were used against *Ascaridia galli* and *Raillietina spiralis*. There were two treatment groups. The albendazole treatment group was negative control. Parameters for measurement were time taken for paralysis and death to occur. This extract demonstrated appreciable anthelmintic activity.

7.7.11 *Zingiber officinale*

Zingiber officinale contains shogaols, gingerols, bisabolene, and zingiberene [92]. Up to 1–4% volatile oils are present in the dried rhizome of *Z. officinale*, which are responsible for the characteristic odor and taste and they are also the active constituents of *Z. officinale*. Zingiberene and bisabolene are the aromatic ingredients, while gingerols and shogaols are the pungent compounds [93]. *Z. officinale* possess anthelmintic activity against a wide range of helminthes. The alcoholic extracts of rhizome have good activity against *Ascaris lumbricoldes* [94]. In an *in vitro* study against the adult worm *H. contortus*, the undiluted extracts of *Z. officinale* caused the death of all worms within two hours of the experiment. The total duration of the experiment was 6 hours and observations were noted on 0, 2, 4.6 hours postexposure while the maximum number of worms remained alive in normal saline after 4 hours postexposure [76].

7.8 Future Prospects

A regular and periodic routine of treatment of children below and within the elementary school age and individuals within the at-risk group should be adopted based on the baseline prevalence recorded for these parasites in the various localities. The synthetic drugs of choice include Albendazole which was described by [68], a vermicide that works by keeping the larva from absorbing sugar (glucose), so that the worm loses energy and dies. Mebendazole, levamisole, oxantel pamoate composite, and pyrantel pamoate composite are also drugs of choice. A general mechanism through which these drugs work is by the reduction in egg production of the parasites. However, the shortcoming of synthetic drugs for STH is the increasing rate of drug resistance in parasites. At this juncture, it is important to embrace what nature has in stock through the screening of medicinal plants and phytochemicals which has proven over time that they are the chief source of efficacious chemicals for man's use as food, medicine and other purposes [95–103]. The combined efforts of preventive programs and aggressive search for new lead compounds from medicinal plants will not only help eliminate the tragedies caused by STH, but make life especially those in endemic regions a little better and meaningful. Also, since STH is associated with malnutrition in children, there is need for adequate nutritional intake through proper dieting. Nutritional information on what to recommend is available in the edited book (*Functional Foods and Nutraceuticals – Bioactive Components, Formulations and Innovations*) by Egbuna and Tupas [100].

7.9 Conclusion

The prevalence of intestinal worms in children is rising day-by-day due to poor education and health awareness about the prevention of intestinal worm infestation. Existing approaches to intervene have met with limited success. It is recommended to start a campaign to control intestinal worms by organizing public health seminars, social media campaigns to show the health hazards of unhygienic practices and a focus on improved sanitation, to educate school children and their parents. This should be done on a priority basis to take preventive measures, especially in underdeveloped cities. Personal hygiene and washing of hands after defecation and before eating is necessary. Fingernails of children should be kept clean. Scratching of the perianal area should be avoided. A literature review showed that medicinal plants are effective in the treatment of intestinal worms in children. However, further isolation of active constituents should be done, and the mechanism of action should be studied.

References

1 Hotez, P.J., Brindley, P.J., Bethony, J.M. et al. (2008). Helminth infections: the great neglected tropical diseases. *The Journal of clinical investigation* 118 (4): 1311–1321.

2 Egbuna, C. and Ifemeje, J.C. (2015). Comparative studies on the phytochemical properties of five Nigerian medicinal plants. *Journal of Advances in Medical and Pharmaceutical Sciences* 6 (2): 1–12.

3 Egbuna, C., Kumar, S., Ezzat, S.M. et al. (eds.) (2019). *Phytochemicals as Lead Compounds for New Drug Discovery*, 1e. Cambridge, USA: Elsevier. 378 pages.

4 Mtewa, A.G., Egbuna, C., and Rao, G.M.N. (2020). *Poisonous Plants and Phytochemicals in Drug Discovery*, 1e. USA: Wiley. ISBN: 978-1119650232. 416 Pages.

5 Heinrich, M., Rimpler, H., and Barrera, N.A. (1992). Indigenous phytotherapy of gastrointestinal disorders in a lowland Mixe community (Oaxaca, Mexico): ethnopharmacologic evaluation. *Journal of Ethnopharmacology* 36 (1): 63–80.

6 McGaw, L., Jäger, A., and Van Staden, J. (2000). Antibacterial, anthelmintic and anti-amoebic activity in South African medicinal plants. *Journal of Ethnopharmacology* 72 (1–2): 247–263.

7 Togola, A., Diallo, D., Dembélé, S. et al. (2005). Ethnopharmacological survey of different uses of seven medicinal plants from Mali,(West Africa) in the regions Doila, Kolokani and Siby. *Journal of Ethnobiology and Ethnomedicine* 1 (1): 7.

8 Guarrera, P.M. (1999). Traditional antihelmintic, antiparasitic and repellent uses of plants in Central Italy. *Journal of Ethnopharmacology* 68 (1–3): 183–192.

9 Horton, J. (2003). Human gastrointestinal helminth infections: are they now neglected diseases? *Trends in Parasitology* 19 (11): 527–531.

10 Utzinger, J. and Keiser, J. (2004). Schistosomiasis and soil-transmitted helminthiasis: common drugs for treatment and control. *Expert Opinion on Pharmacotherapy* 5 (2): 263–285.

11 Steketee, R.W. (2003). Pregnancy, nutrition and parasitic diseases. *The Journal of Nutrition* 133 (5): 1661S–1667S.

12 Bethony, J., Brooker, S., Albonico, M. et al. (2006). Soil-transmitted helminth infections: ascariasis, trichuriasis, and hookworm. *The Lancet* 367 (9521): 1521–1532.

13 Faust, E.C., Russell, P.F., and Jung, R.C. (1970). Craig and Faust's clinical parasitology. In: *Craig and Faust's Clinical Parasitology*, 8e. Philadelphia/London: Lea and Febiger/Henry Kimpton.

14 Tandon, V., Yadav, A., Roy, B., and Das, B. (2011). Phytochemicals as cure of worm infections in traditional medicine systems. In: *Emerging Trends in Zoology*, 351–378. New Delhi: Narendra Publishing House.

15 Mehraj, V., Hatcher, J., Akhtar, S. et al. (2008). Prevalence and factors associated with intestinal parasitic infection among children in an urban slum of Karachi. *PloS one* 3 (11): e3680.

16 O'Connell, E.M. and Nutman, T.B. (2016). Molecular diagnostics for soil-transmitted helminths. *American Journal of Tropical Medicine and Hygiene* 95 (3): 508–513. https://doi.org/10.4269/ajtmh.16-0266.

17 Nilforoushan, M., Mirhendi, H., Rezaie, S. et al. (2007). A DNA-based identification of *Strongyloides stercoralis* isolates from Iran. *Iranian Journal of Public Health* 36: 16–20.

18 Verweij, J.J., Canales, M., Polman, K. et al. (2009). Molecular diagnosis of *Strongyloides stercoralis* in faecal samples using real-time PCR. *Transactions of the Royal Society of Tropical Medicine and Hygiene* 103: 342–346.

19 Moghaddassani, H., Mirhendi, H., Hosseini, M. et al. (2011). Molecular diagnosis of *Strongyloides stercoralis* infection by PCR detection of specific DNA in human stool samples. *Iranian Journal of Parasitology* 6: 23–30.

20 Sharifdini, M., Mirhendi, H., Ashrafi, K. et al. (2015). Comparison of nested polymerase chain reaction and real-time polymerase chain reaction with parasitological methods for detection of *Strongyloides stercoralis* in human fecal samples. *American Journal of Tropical Medicine and Hygiene* 93: 1285–1291.

21 Haque, R. (2007). Human intestinal parasites. *Journal of Health, Population, and Nutrition* 25 (4): 387.

22 Villamizar, E., Mendez, M., Bonilla, E. et al. (1996). Ascaris lumbricoides infestation as a cause of intestinal obstruction in children: experience with 87 cases. *Journal of Pediatric Surgery* 31 (1): 201–205.

23 Burkhart, C.N. and Burkhart, C.G. (2005). Assessment of frequency, transmission, and genitourinary complications of enterobiasis (pinworms). *International Journal of Dermatology* 44 (10): 837–840.

24 Wasadikar, P. and Kulkarni, A. (1997). Intestinal obstruction due to ascariasis. *British Journal of Surgery* 84 (3): 410–412.

25 De Silva, N.R., Brooker, S., Hotez, P.J. et al. (2003). Soil-transmitted helminth infections: updating the global picture. *Trends in Parasitology* 19 (12): 547–551.

26 Degarege, A., Animut, A., Legesse, M., and Erko, B. (2010). Malaria and helminth co-infections in outpatients of Alaba Kulito Health Center, southern Ethiopia: a cross sectional study. *BMC Research Notes* 3 (1): 143.

27 Chen, M., Chang, K., Lin, K. et al. (2012). Retinopathy in a patient with Fanconi anemia and vitamin B_{12} deficiency. *Eye* 26 (2): 331.

28 Wakid, M. (2006). Distribution of intestinal parasites among food handlers in Jeddah, Saudi Arabia. *Journal of Parasitic Diseases* 30 (2): 146–152.

29 Knopp, S., Mohammed, K.A., Stothard, J.R. et al. (2010). Patterns and risk factors of helminthiasis and anemia in a rural and a peri-urban community in Zanzibar, in the context of helminth control programs. *PLoS Neglected Tropical Diseases* 4 (5): e681.

30 Phuc, T.Q., Mihrshahi, S., Casey, G.J. et al. (2009). Lessons learned from implementation of a demonstration program to reduce the burden of anemia and hookworm in women in Yen Bai Province, Viet Nam. *BMC Public Health* 9 (1): 266.

31 Erko, B., Degarege, A., Tadesse, K. et al. (2012). Efficacy and side effects of praziquantel in the treatment of Schistosomiasis mansoni in schoolchildren in Shesha Kekele Elementary School, Wondo Genet, Southern Ethiopia. *Asian Pacific Journal of Tropical Biomedicine* 2 (3): 235.

32 Joseph, S.A., Montresor, A., Casapía, M. et al. (2016). Adverse events from a randomized, multi-arm, placebo-controlled trial of mebendazole in children 12–24 months of age. *The American Journal of Tropical Medicine and Hygiene* 95 (1): 83–87.

33 Igual-Adell, R., Oltra-Alcaraz, C., Soler-Company, E. et al. (2004). Efficacy and safety of ivermectin and thiabendazole in the treatment of strongyloidiasis. *Expert Opinion on Pharmacotherapy* 5 (12): 2615–2619.

34 Jevtić, M., Mikić, D., Arsić-Komljenović, G. et al. (2008). Adverse effects of long term, continual administration of high doses of albendazole in the treatment of echinococcal disease. *Vojnosanitetski Pregled* 65 (7): 539–544.

35 Sümegi, V., Haszon, I., Iványi, B. et al. (2004). Long-term effects of levamisole treatment in childhood nephrotic syndrome. *Pediatric Nephrology* 19 (12): 1354–1360.

36 Ferrara, P., Bersani, I., Bottaro, G. et al. (2011). Massive proteinuria: a possible side effect of pyrantel pamoate? *Renal Failure* 33 (5): 534–536.

37 Ali, S.S., Kasoju, N., Luthra, A. et al. (2008). Indian medicinal herbs as sources of antioxidants. *Food Research International* 41 (1): 1–15.

38 Ahmad, K., Usmanghani, K., Akhtar, N., and Nazar, H. (2015). Clinical assessment of coded Unani formulation D-worm and mebandazole for the treatment of hook worm, roundworm and whip worm. *Pakistan Journal of Pharmaceutical Sciences* 28 (6): 2115–2118.

39 Jabbar, A., Raza, M.A., Iqbal, Z., and Khan, M.N. (2006). An inventory of the ethnobotanicals used as anthelmintics in the southern Punjab (Pakistan). *Journal of Ethnopharmacology* 108 (1): 152–154.

40 Singh, O. and Ali, M. (2011). Phytochemical and antifungal profiles of the seeds of Carica papaya L. *Indian Journal of Pharmaceutical Sciences* 73 (4): 447.

41 Okeniyi, J.A., Ogunlesi, T.A., Oyelami, O.A., and Adeyemi, L.A. (2007). Effectiveness of dried Carica papaya seeds against human intestinal parasitosis: a pilot study. *Journal of Medicinal Food* 10 (1): 194–196.

42 Chokchaisiri, R., Suaisom, C., Sriphota, S. et al. (2009). Bioactive flavonoids of the flowers of *Butea monosperma*. *Chemical and Pharmaceutical Bulletin* 57 (4): 428–432.

43 Ahmed, F., Siddaraju, N., Harish, M., and Urooj, A. (2012). Effect of *Butea monosperma* Lam. leaves and bark extracts on blood glucose in streptozotocin-induced severely diabetic rats. *Pharmacognosy Research* 4 (1): 33.

44 Sonkar, N., Ganeshpurkar, A., Yadav, P. et al. (2014). An experimetal evaluation of nephroprotective potential of Butea monosperma extract in albino rats. *Indian Journal of Pharmacology* 46 (1): 109.

45 Cherian, A.M., Snima, K., Kamath, C.R. et al. (2015). Effect of *Baliospermum montanum* nanomedicine apoptosis induction and anti-migration of prostate cancer cells. *Biomedicine & Pharmacotherapy* 71: 201–209.

46 Soni, S., Anandjiwala, S., Patel, G., and Rajani, M. (2008). Validation of different methods of preparation of Adhatoda vasica leaf juice by quantification of total alkaloids and vasicine. *Indian Journal of Pharmaceutical Sciences* 70 (1): 36.

47 Claeson, U.P., Malmfors, T., Wikman, G., and Bruhn, J.G. (2000). Adhatoda vasica: a critical review of ethnopharmacological and toxicological data. *Journal of Ethnopharmacology* 72 (1–2): 1–20.

48 Dhuley, J.N. (1999). Antitussive effect of *Adhatoda vasica* extract on mechanical or chemical stimulation-induced coughing in animals. *Journal of Ethnopharmacology* 67 (3): 361–365.

49 Ramjith, U., Roopitha, B., and Jacob, C. (2013). Isolation anti-diabetic and antioxidant evaluation of aqueous extract of Cansjera rheedii leaves. *Asian Journal of Pharmacutical and Clinical Research* 6 (3): 228–234.

50 Ahmad, S., Rao, H., Akhtar, M. et al. (2011). Phytochemical composition and pharmacological prospectus of *Ficus bengalensis* Linn. (Moraceae)—a review. *Journal of Medicinal Plants Research* 5 (28): 6393–6400.

51 Garg, V.K. and Paliwal, S.K. (2011). Wound-healing activity of ethanolic and aqueous extracts of *Ficus benghalensis*. *Journal of Advanced Pharmaceutical Technology & Research* 2 (2): 110.

52 Patel, M.A., Patel, P.K., and Patel, M.B. (2010). Effects of ethanol extract of *Ficus bengalensis* (bark) on inflammatory bowel disease. *Indian Journal of Pharmacology* 42 (4): 214.

53 Eguale, T., Tadesse, D., and Giday, M. (2011). *In vitro* anthelmintic activity of crude extracts of five medicinal plants against egg-hatching and larval development of *Haemonchus contortus*. *Journal of Ethnopharmacology* 137 (1): 108–113.

54 Getie, M., Gebre-Mariam, T., Rietz, R. et al. (2003). Evaluation of the anti-microbial and anti-inflammatory activities of the medicinal plants *Dodonaea viscosa, Rumex nervosus* and *Rumex abyssinicus*. *Fitoterapia* 74 (1–2): 139–143.

55 Mekonnen, T., Urga, K., and Engidawork, E. (2010). Evaluation of the diuretic and analgesic activities of the rhizomes of *Rumex abyssinicus* Jacq in mice. *Journal of Ethnopharmacology* 127 (2): 433–439.

56 Murti, Y., Yogi, B., and Pathak, D. (2010). Pharmacognostic standardization of leaves of *Calotropis procera* (Ait.) R. Br.(Asclepiadaceae). *International Journal of Ayurveda Research* 1 (1): 14.

57 Moustafa, A., Ahmed, S., Nabil, Z. et al. (2010). Extraction and phytochemical investigation of *Calotropis procera*: effect of plant extracts on the activity of diverse muscles. *Pharmaceutical Biology* 48 (10): 1080–1190.

58 Alencar, N., Figueiredo, I., Vale, M. et al. (2004). Anti-inflammatory effect of the latex from *Calotropis procera* in three different experimental models: peritonitis, paw edema and hemorrhagic cystitis. *Planta Medica* 70 (12): 1144–1149.

59 Soares, P.M., Lima, S.R., Matos, S.G. et al. (2005). Antinociceptive activity of *Calotropis procera* latex in mice. *Journal of Ethnopharmacology* 99 (1): 125–129.

60 Ahmad, H., Sehgal, S., Mishra, A., and Gupta, R. (2012). *Mimosa pudica* L.(Laajvanti): an overview. *Pharmacognosy Reviews* 6 (12): 115.

61 Molina, M., Contreras, C., and Tellez-Alcantara, P. (1999). *Mimosa pudica* may possess antidepressant actions in the rat. *Phytomedicine* 6 (5): 319–323.

62 Turker, A.U. and Gurel, E. (2005). Common mullein (*Verbascum thapsus* L.): recent advances in research. *Phytotherapy Research: An International Journal Devoted to Pharmacological and Toxicological Evaluation of Natural Product Derivatives* 19 (9): 733–739.

63 Escobar, F.M., Sabini, M.C., Zanon, S.M. et al. (2012). Antiviral effect and mode of action of methanolic extract of *Verbascum thapsus* L. on pseudorabies virus (strain RC/79). *Natural Product Research* 26 (17): 1621–1625.

64 Muthukrishnan, S.D., Kaliyaperumal, A., and Subramaniyan, A. (2015). Identification and determination of flavonoids, carotenoids and chlorophyll concentration in *Cynodon dactylon* (L.) by HPLC analysis. *Natural Product Research* 29 (8): 785–790.

65 Albert-Baskar, A. and Ignacimuthu, S. (2010). Chemopreventive effect of *Cynodon dactylon* (L.) Pers. extract against DMH-induced colon carcinogenesis in experimental animals. *Experimental and Toxicologic Pathology* 62 (4): 423–431.
66 Williams, A.R., Soelberg, J., and Jager, A.K. (2016). Anthelmintic properties of traditional African and Caribbean medicinal plants: identification of extracts with potent activity against *Ascaris suum in vitro*. *Parasite* 23: 24. https://doi.org/10.1051/parasite/2016024.
67 Al-Mathal, E.M. and Alsalem, A.A. (2013). Pomegranate (*Punica granatum*) peel is effective in a murine model of experimental *Cryptosporidium parvum* ultrastructural studies of the ileum. *Experimental Parasitology* 134 (4): 482–494.
68 El-Sherbini, G.T. and Osman, S.M. (2013). Anthelmintic activity of unripe *Mangifera indica* L. (Mango) against *Strongyloides stercoralis*. *International Journal of Current Microbiology and Applied Sciences* 2 (5): 401–409.
69 Engstrom, M.T., Karonen, M., Ahern, J.R. et al. (2016). Chemical structures if plant hydrolysable tannins reveal their *in vitro* activity against egg hatching and motility of *Haemonchus contortus* nematodes. *Journal of Agricultural and Food Chemistry* 64: 840–851.
70 Athanasiadou, S., Kyriazakis, I., Jackson, F., and Coop, R.L. (2001). Direct anthelmintic effects of condensed tannins towards different gastrointestinal nematodes of sheep: *In vitro* and *in vivo* studies. *Veterinary Parasitology* 99: 205–219.
71 Costa, C.T.C., Morais, S.M., Bevilaqua, C.M.L. et al. (2002). Ovicidal effect of *Mangifera indica* L. seeds extracts on *Haemonchus contortus*. *Brazilian Journal of Veterinary Parasitology* 11: 57–60.
72 Kumar, A., Ram, J., Samarth, R., and Kumar, M. (2005). Modulatory influence of *Adhatoda vasica* Nees leaf extract against gamma irradiation in Swiss albino mice. *Phytomedicine* 12 (4): 285–293.
73 Joshi, B.S., Bai, Y., Puar, M.S. et al. (1994). 1H-and 13C-NMR assignments for Some Pyrrolo {2, 1b} quinazoline alkaloids of *Adhatoda vasica*. *Journal of Natural Products* 57 (7): 953–962.
74 Yadav, A.K. and Tangpu, V. (2008). Anticestodal activity of *Adhatoda vasica* extract against *Hymenolepis diminuta* infections in rats. *Journal of Ethnopharmacology* 119 (2): 322–324.
75 Otunola, G.A., Oloyede, O.B., Oladiji, A.T., and Afolayan, A.J. (2010). Comparative analysis of the chemical composition of three spices – *Allium sativum* L. *Zingiber officinale* Rosc. and *Capsicum frutescens* L. commonly consumed in Nigeria. *African Journal of Biotechnology* 9 (41): 6927–6931.
76 Iqbal, Z., Nadeem, Q.K., Khan, M. et al. (2001). *In vitro* anthelmintic activity of *Allium sativum*, *Zingiber officinale*, *Curcurbita mexicana* and *Ficus religiosa*. *International Journal of Agriculture and Biology* 3 (4): 454–457.
77 Mali, R. and Wadekar, R. (2008). *In vitro* anthelmintic activity of *Baliospermum montanum* muell arg roots. *Indian Journal of Pharmaceutical Sciences* 70 (1): 131.

78 Gupta, P., Chauhan, N., Pande, M., and Pathak, A. (2012). Phytochemical and pharmacological review on *Butea monosperma* (Palash). *International Journal of Agronomy and Plant Production* 3 (7): 255–258.

79 Iqbal, Z., Lateef, M., Jabbar, A. et al. (2006). *In vivo* anthelmintic activity of *Butea monosperma* against *Trichostrongylid nematodes* in sheep. *Fitoterapia* 77 (2): 137–140.

80 Prashanth, D., Asha, M., Amit, A., and Padmaja, R. (2001). Anthelmintic activity of *Butea monosperma*. *Fitoterapia* 72 (4): 421–422.

81 Shivkar, Y.M. and Kumar, V.L. (2003). Anthelmintic activity of latex of *Calotropis procera*. *Pharmaceutical Biology* 41 (4): 263–265. https://doi.org/10.1076/phbi.41.4.263.15666.

82 Farnsworth, N.R., Bingel, A.S., Cordell, G.A. et al. (1975). Potential value of plants as sources of new antifertility agents I. *Journal of Pharmaceutical Sciences* 64 (4): 535–598.

83 Aswar, M., Aswar, U., Watkar, B. et al. (2008). Anthelmintic activity of *Ficus benghalensis*. *International Journal of Green Pharmacy* 2 (3): 170–172.

84 Bendgude, R., Maniyar, M., Kondawar, M. et al. (2012). Anthelmintic activity of leaves of *Mimosa pudica*. *International Journal of Institutional Pharmacy and Life Sciences* 2: 120–125.

85 Amelia, M., Jasaputra, D.K., and Tjokropranoto, R. (2017). Effects of Pomegranate Peel (*Punica granatum* L.) extract as an anthelmintic. *Journal of Medicine & Health* 1 (5): 409–416.

86 Shinwari, Z.K. and Gilani, S.S. (2003). Sustainable harvest of medicinal plants at Bulashbar Nullah, Astore (northern Pakistan). *Journal of Ethnopharmacology* 84 (2–3): 289–298.

87 Bileflimi, V.T.K. (2004). Chemical constituents of *Verbascum* L. species. *FABAD Journal of Pharmaceutical Sciences* 29: 93–107.

88 Seifert, K., Schmidt, J., Lien, N., and Johne, S. (1985). Iridoide aus Verbascum-Arten. *Planta Medica* 51 (05): 409–411.

89 Warashina, T., Miyase, T., and Ueno, A. (1991). Iridoid glycosides from *Verbascum thapsus* L. *Chemical and Pharmaceutical Bulletin* 39 (12): 3261–3264.

90 Souleles, C. and Geronikaki, A. (1989). Flavonoids from *Verbascum thapsus*. *Scientica Pharmaceutica* 57 (1): 59–61.

91 Ali, N., Shah, S.W.A., Shah, I. et al. (2012). Anthelmintic and relaxant activities of *Verbascum thapsus* Mullein. *BMC Complementary and Alternative Medicine* 12 (1): 29.

92 Ali, B.H., Blunden, G., Tanira, M.O., and Nemmar, A. (2008). Some phytochemical, pharmacological and toxicological properties of ginger (*Zingiber officinale* Roscoe): a review of recent research. *Food and Chemical Toxicology* 46 (2): 409–420.

93 Tyler, V.E. (1994). *Herbs of Choice: The Therapeutic Use of Phytomedicinals*. Pharmaceutical Products Press. (imprint of Haworth Press, Inc.).

94 Raj, R.K. (1975). Screening of indigenous plants for anthelmintic action against human *Ascaris lumbricoides*: part-II. *Indian Journal of Physiology and Pharmacology* 19 (1): 47–49.

95 Egbuna, C., Ifemeje, J.C., Udedi, S.C., and Kumar, S. (eds.) (2018). *Phytochemistry*. In: *Fundamentals, Methods, and Applications*, 1e, vol. *1*. New York: Apple Academic Press/CRC Taylor & Francis. ISBN: 978-1771887595. 684 pages, part of a 3 volume set.

96 Egbuna, C., Kumar, S., Ifemeje, J.C., and Kurhekar, J.V. (eds.) (2018). *Phytochemistry*. In: *Pharmacognosy, Nanomedicine, and Contemporary Issues*, 1e, vol. *2*. New York: Apple Academic Press/CRC Taylor & Francis. ISBN: 978-1771887601. 620 pages, part of a 3 volume set.

97 Egbuna, C., Ifemeje, J.C., Kumar, S., and Sharif, N. (eds.) (2018). *Phytochemistry*. In: *Marine, Industrial, and Advances*, 1e, vol. *3*. New York: Apple Academic Press/CRC Taylor & Francis. ISBN: 978-1771887618. 502 pages, part of a 3 volume set.

98 Saravanan, K., Egbuna, C., Averal, H.I. et al. (eds.) (2020). *Drug Development for Cancer and Diabetes*, 1e. Apple Academic Press/Taylor & Francis. ISBN: 9781771888608. 392 pages.

99 Egbuna, C. and Sawicka, B. (2019). *Natural Remedies for Pest, Disease and Weed Control*, 1e. USA: Academic Press (Elsevier). ISBN: 9780128193051. 268 pages, 978-0-12-819304-4.

100 Egbuna, C. and Tupas, G.D. (eds.) (2020). *Functional Foods and Nutraceuticals - Bioactive Components, Formulations and Innovations*, 1e. Cham, Switzerland: Springer Nature. ISBN: 978-3-030-42318-6. 646 pages.

101 Egbuna, C. and Mtewa, A.G. (eds.) (2021). *Phytochemistry, the Military and Health: Phytotoxins and Natural Defenses*, 1e. Cambridge, USA: Elsevier. ISBN: 9780128232309. 615 pages.

102 Shashank, K. and Egbuna, C. (eds.) (2019). *Phytochemistry: In vitro and in silico updates*, 1e. Singapore: Springer Nature. ISBN: 978-981-13-6919-3. 589 pages.

103 Moser, W., Schindler, C., and Keiser, J. (2017). Efficacy of recommended drugs against soil transmitted helminthes: systematic review and network meta-analysis. *British Medical Journal* 358: j4307. https://doi.org/10.1136/bmj.j4307.

8

Efficacy of Phytochemicals of Medicinal Plants for the Treatment of Human Echinococcosis

Echinococcal Disease, Hydatidosis, or Hydatid Disease Drug Discovery

Charles Oluwaseun Adetunji[1], *Chukwuebuka Egbuna*[2,3], *Tolulope Olawumi Oladosun*[4], *Muhammad Akram*[5], *Olugbenga Samuel Micheal*[6], *Frances N. Olisaka*[7], *Phebean Ozolua*[1], *Juliana Bunmi Adetunji*[8], *Goddidit Esiro Enoyoze*[9], *and Olugbemi Tope Olaniyan*[10]

[1] *Applied Microbiology, Biotechnology and Nanotechnology Laboratory, Department of Microbiology, Edo State University Uzairue, Iyamho, Nigeria*
[2] *Department of Biochemistry, Faculty of Natural Sciences, Chukwuemeka Odumegwu Ojukwu University, Uli, Nigeria*
[3] *Nutritional Biochemistry and Toxicology Unit, World Bank Africa Centre of Excellence, Centre for Public Health and Toxicological Research (PUTOR), University of Port-Harcourt, Port Harcourt, Nigeria*
[4] *Department of Microbiology, Federal University of Agriculture, Makurdi, Nigeria*
[5] *Department of Eastern Medicine and Surgery, Government College University Faisalabad, Faisalabad, Pakistan*
[6] *Cardiometabolic Research Unit, Department of Physiology, College of Health Sciences, Bowen University, Iwo, Nigeria*
[7] *Environmental and Public Health Microbiology Laboratory, Department of Biological Sciences, Faculty of Science, Benson Idahosa University, Benin City, Nigeria*
[8] *Nutritional and toxicological Research Laboratory, Department of Biochemistry Sciences, Osun State University, Osogbo, Nigeria*
[9] *Department of Biological Sciences, Plant Biology and Biotechnology Unit, Edo University Iyamho, Iyamho, Nigeria*
[10] *Laboratory for Reproductive Biology and Developmental Programming, Department of Physiology, Edo State University Uzairue, Iyamho, Nigeria*

List of Abbreviations

ADA	Adenosine deaminase
ADME	Absorption, distribution, metabolism, and elimination
ALP	Alkaline phosphatase
BLAST	Basic Local Alignment Search Tool
CD68	Cluster of differentiation 68

Neglected Tropical Diseases and Phytochemicals in Drug Discovery, First Edition.
Edited by Chukwuebuka Egbuna, Muhammad Akram, and Jonathan Chinenye Ifemeje.

DDBJ	Data Bank of Japan
DNA	Deoxyribonucleic acid
EBI	European Bioinformatics Institute
HCFA	Hydatid cyst fluid antigen
HFFs	Human foreskin fibroblasts
iNOS	Inducible nitric oxide synthase
NADH	Nicotinamide adenine dinucleotide
NCBI	National Centre for Biotechnology Information
NF-Kb	Nuclear factor kappa-light-chain-enhancer of activated B cells
PBS	Phosphate-buffered saline
PCR	Polymerase chain reaction
RH	Rat hepatoma
TNF-α	Tumor necrosis factor alpha

8.1 Introduction

It has been observed that several types of parasites have evolved during the evolution of mankind and they normally use human beings as their host organism. Moreover, most of these parasites constitute nuisance and cause high level of unpleasant or discomfort situations such as flea and lice infestation of the body which may seriously affect our health, especially the internal parasites. Most of the ectoparasites could be handled mechanically while removal of the internal parasites is complicated. Several medicinal plants have been utilized by mankind for the treatment of several illnesses and health disorders for several years. Most especially, the application of plant parts, secondary metabolites, and plant products as antiparasitic drugs. The infections by endoparasites can scarcely be inhibited by vaccination [1]. Moreover, several synthetic drugs have been produced for the management of endoparasites but there are still several challenges encountered in their utilization. Most of these synthetic drugs are not effective due to resistance coupled with the fact that the development of new and novel antiparasitic drugs has not been the top priority of most pharmaceutical industry because numerous diseases occur in most developing countries where a larger percentage of the population could not afford the exorbitant price of the antiparasitic drugs.

Therefore, in view of the aforementioned facts, an alternative to these is adequate research into the application of medicinal plants, their phytochemical components, or natural product derivatives [2–7]. Moreover, most of the new drugs that are derived from medicinal plants possess pharmacophores which implies they have functional groups with enhanced pharmacological activity obtained from natural products [8]. Also, there is a need to assess several traditional medicinal plants for the development of new drugs that could be utilized for the treatment of these parasitic infections. Moreover, several interesting results have been documented from several results that have established the biological activities of these medicinal plants when tested in an *in vitro* assay against their several vectors or their actual causative parasites.

Cystic echinococcosis (CE) has been recognized as a zoonotic parasitic disease that could be caused by the larval stage of some small tapeworm of some animals. The incidence of CE could lead to death and high morbidity in livestock and human being [9]. Moreover, CE has been documented as one of the neglected tropical diseases by the World Health Organization [10]. The incidence of CE, hydatid cysts, matures in the internal organs, mainly in the lungs or liver of their intermediate hosts and also in humans such as unilocuslar fluid-filled bladder [11]. It has been observed that this infection tends to appear as asymptomatic at the outset for numerous years or permanently [12]. There may be a malfunction of several organs which may lead to the death or continued development of cysts without effectual treatment [13]. It has been highlighted that chemotherapy does have influence but may lead to complications. Hence, the application of an efficient protoscolicidal agent is important to reduce the challenges of intraoperative spillage of the cysts filling in the course of surgery and uninterruptedly reoccurrence of CE in almost 10% of the postoperative circumstances [14]. Also, different types of protoscolicidal agents have been documented to incapacitate the substances of the cysts, nevertheless generally are not nontoxic for the reason that might be linked to their high side effects comprising liver necrosis, methemoglobinemia, and sclerosing cholangitis [15].

In view of the aforementioned facts, this chapter intends to document the application of phytochemicals and medicinal plants that could be used for the management and treatment of *Echinococcus* infection.

The nucleotide sequencing of the mitochondrion in various species of *Echinococcus* is of immense diagnostic benefits as it is a veritable source of molecular markers, which in turn provides valuable information on its molecular epidemiology [16]. The disease phenotypes, host specificities, distribution, genetic variation, and evolution of these parasites are linkable to the differences occurring in specific regions of their corresponding mitochondrial genomes. Next-generation sequencing (NGS) technology has been of great advantage in solving a myriad of problems around genome sequencing. These analyses are relatively affordable, with high-throughput in mitochondrial genomic studies. There are basically six species of *Echinococcus*. Pathogenic types with global public health burden include *Echinococcus granulosus*, *Echinococcus multilocularis, Echinococcus oligarthrus,* and *Echinococcus vogeli*. Reports on newly identified species from rodents and lions from Africa namely *Echinococcus shiquicus* and *Echinococcus Felidis,* respectively, have been documented. By genotypic characterization of the mitochondrion, there exist not less than ten (10) strains of *E. granulosus* [17].

8.2 Molecular Analysis

Upon obtaining endocyst samples under aseptic conditions, they should be fixed in 95% ethanol, and the cyst fluids centrifuged at 500g for 30 minutes. The centrifuged sample should be stored at 20 °C in preparation for molecular analysis. A small portion of the endocyst (of about 11–16 mg) should be sampled. The ethanol used to fix it must then be removed, and this sample should be washed twice in phosphate-buffered saline (PBS) for 15 minutes. The sample could then be centrifuged at 3500g for five minutes and PBS removed in the process. The mitochondrial DNA could then be extracted from the sample using a designated DNA extraction kit following the manufacturer's instructions [18]. This could also be done traditionally using the glass bead method [19]. Mitochondrial genes – cytochrome c oxidase subunit 1 (*cox I*) and NADH dehydrogenase subunit 1 (*nd I*) – are the targets in this assay, they are to be amplified by polymerase chain reaction (PCR), with the specific primers (designed according to sequences obtained from the GenBank). DNA amplification should be performed in a 20 μl consisting of DNA constituting 1–10 ng, the premixed solution (0.2 mM dNTPs), 1 μM each of the primers, 1×TBE buffer, and 0.5 U Taq polymerase enzyme. The PCR protocol ideal for the amplification of both genes (*cox I* and *nd I*) is two minutes (two minutes) at 95 °C, 45 seconds of 40 cycles at 95 °C, 45 seconds at 57 °C, 90 seconds at 72 °C, and finally 10 minutes at 72 °C [18]. The PCR products are then subjected to gel electrophoresis, purification, and sequencing. The nucleotides sequences obtained would then be analyzed with the Basic Local Alignment Search Tool (BLAST) and the databases from the National Centre for Biotechnology Information (NCBI), European Molecular Biology Laboratory – EBI (European Bioinformatics Institute), or DNA Data Bank of Japan (DDBJ).

8.3 Life Cycle of Echinococcosis

The two main species of *Echinococcus* responsible for echinococcosis with major public health concerns in human populations are *E. granulosus* and *E. multilocularis*. *E. granulosus* which is the causative agent of CE has canines such as dog as its definitive host, which houses the adult worms (Figure 8.1). In these hosts, they are not seen to cause infections in the organs. The adult worm lays eggs, shed in the feces of the definitive hosts, and are ingested by herbivores such as sheep while grazing, which serve as intermediate hosts. These eggs that are being ingested, hatch in the small intestine of the intermediate host into oncospheres (which are the embryonic forms of the worm usually enclosed in a spherical membrane having hooks). This oncosphere navigates through the intestinal wall, via the circulatory system to the liver predominantly, where it

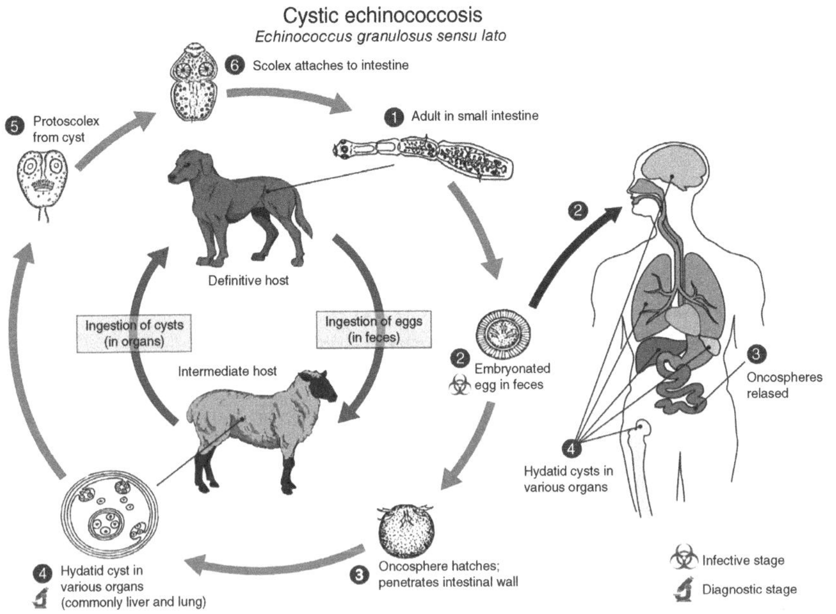

Figure 8.1 Life cycle of *Echinococcus granulosus*. *Source:* DPDx – Echinococcosis, Centers for Disease Control and Prevention, 16 September 2020.

develops into cyst. The cyst begins to enlarge and produces protoscoleces and daughter cysts within itself [17].

The cysts are finally ingested by the primary (definitive) host through the consumption of infected organs from these herbivores [20]. The protoscoleces from the cysts escape, evaginate, and attach to the intestinal mucosa of the definitive host, where they develop into adults within 80 days, and hence the cycle is complete. Humans are usually considered to be a dead-end host and also an accidental intermediate host in the life cycle of *E. granulosus* [11]. The infection could be by direct contact with egg(s) from feces of the definitive host or contaminated water, food, and/or soil. These eggs mature into oncospheres which then pass through the small intestine to infect vital internal organs (such as the liver, bones, lungs, and rarely, kidneys, heart, and the brain) via the circulatory system. The cyst is usually round and its diameter in CE ranges between 1 and 20 cm, consisting of two layers of membranes – the inner membrane is nucleated, which is referred to as the germinal membrane, and the germinal layer contains clear yellowish fluid. The appearance of the scoleces in the hydatid fluid shares resemblance with sand grains and hence they are called hydatid sand, while the outer membrane is acellular, and appears laminated. The formation of a calcified fibrous capsule around the cyst is the immune response staged by the host to CE; this is the layer often captured mostly in imaging studies [21].

Unlike the CE, alveolar echinococcosis which is caused by *E. multilocularis* has a life cycle similar to that of *E. granulosus,* except that the intermediate hosts involved therein are rodents whereas the definitive hosts are foxes predominantly. The infection in humans results in the formation of hydatid cysts also, having close similarities with symptoms usually presented in human *E. granulosus* infections. Chamber formation in alveolar echinococcosis is multiple in nature as against the unilocular pattern of cyst formation in CE.

8.4 Previous Studies on the Positive Effects of Medicinal Plants and Phytochemicals

A lot of complications are still been recorded due to improper diagnosis and treatment of human cystic *Echinococcus*, regardless of the developments in therapeutic strategies and modern imaging procedures. It is important that this disease is detected early enough for proper management to prevent great numbers of motility and morbidity [22].

Gangwar et al. [23] demonstrated the *in vitro* efficiency of the fruit glandular hair extract of the plant *Mallotus philippinensis* for its protoscolicidal action against hydatid cysts of *E. granulosus* which were collected using aseptic techniques from the infected liver and lungs of cattle. Their studies showed a mortality rate of up to 99% after treatment for 60 minutes with a concentration of 20 mg/ml of the extract, while treatments with the same concentration for more than 120 minutes showed a 100% mortality rate, as well as the disintegration of the protoscoleces. The results obtained were compared with the recognized standard drug Praziquantel, an anticestodal drug. It was also observed that Praziquantel had a comparable scolicidal effect. Meanwhile, the fruit glandular hair extract of the plant *M. philippinensis* did so without the distressing side effects observed in patients treated with Praziquantel. In another study, Albani et al. [24] established the *in vitro* effect of thymol (extracted from the plants *Origanum vulgare* and *Thymus vulgare*) and terpene-containing essential oils extracted from *Mentha piperita*, *Mentha pulegium*, and *Rosmarinus officinalis* on the proliferation of *E. granulosus* larval cells. It was observed that the action of thymol or the other essential oils led to the seizure of cell division and also triggered a considerable decrease in the number of cells after the second day of incubation, with the highest inhibitory effect observed in the essential oil of *M. pulegium* (82%) after seven days. As a result of their findings, it was suggested that these plant extracts could be used as an antihelminthic in the treatment of Echinococcosis.

Almalki et al. [25] evaluated the *in vitro* scolicidal effects of various concentrations of the ethanolic extracts of *Curcuma longa* and *Zingiber officinale* on *Echinococcus protoscoleces*, isolated aseptically from sheep livers. They observed that the *Z. officinale* extract showed the strongest scolicidal effect at a concentration of 30 mg/ml after 20 minutes (100%) while the maximum scolicidal effect of *C. longa* was 93.2% at a concentration of 50 mg/ml after 30 minutes. Their study concluded that the extracts from *Z. officinale* and *C. longa* could be used efficiently as a protoscolicidal agent against *E. granulosus* that causes hydatidosis, by inactivating the pathogens to prevent multiple secondary echinococcosis as a result of recurrence after surgical treatment of the initial infection. In an older study, Al-Mayah et al. [26] carried out a study to assess the *in vivo* effects of *Nigella sativa* aqueous seeds extract against metacestodes of *E. granulosus*. They studied the morphological and histopathological variations in cysts and the organs infected in white albino mice which were injected with protoscoleces of the parasite. They also observed the enzymic activities of adenosine deaminase (ADA) and alkaline phosphatase (ALP) in each mouse. Their studies showed that mice treated with the *N. sativa* seeds extracts had a lower number and diameter of cysts and a higher cyst decrease percentage than the control group which was infected. The authors reported deterioration and necrosis of cysts in histological sections of the liver from mice treated with various concentrations of the extract. An increase in ADA and a decrease in ALP activities were also observed in the treated mice. Their study revealed that a 25 mg concentration of the aqueous extract of *N. sativa* seed was significant in inhibiting the metacestodes of *E. granulosus*.

Verma et al. [27] investigated the anticestodal action of an endophytic fungi *Pestalotiopsis* sp. from Neem plant on the protoscoleces of hydatid cysts of *E. granulosus*. Various mortality rates were noticed at diverse concentrations with respect to the periods of exposure. *Pestalotiopsis* sp. (581 ± 15.0) proved a scolicidal activity of up to 97% mortality rate within 30 minutes of incubation. As compared with standard medications used, endophytic *Pestalotiopsis* sp. of Neem plant displayed meaningful anticestodal actions. Also, Labsi et al. [28] investigated the action of pomegranate (PGE) peel aqueous extract on the progress of secondary experimental echinococcosis on the viability of protoscoleces of *E. granulosus*, and the immunomodulatory characteristics of PGE. In their experiment, Swiss mice were injected intraperitoneally with viable protoscoleces of the helminth. Then, PGE was administered intragastrically daily during CE growth. The development of cysts and damage to the liver were examined macroscopically and histologically. Nitric oxide production (observed in patients infected with the parasite), TNF-α in plasma, and the hepatic expression of iNOS, NF-kB, and the marker CD68 were evaluated. After the treatment of the infected mice with PGE, a significant decrease in nitric oxide, TNF-α levels, iNOS, CD68, and NF-kB expression was recorded. This decrease was in a strong relationship with the inhibition of cyst development (63.08%). Hence, it was concluded that the use of PGE against the development of experimental echinococcosis had both anthelminthic and immunomodulatory results.

Moazeni and Mohseni [29] investigated the *in vitro* scolicidal effect of methanolic extract of Sumac (*Rhus coriaria*). They aseptically obtained protoscoleces from sheep livers infected with hydatid cysts of the parasite. The protoscoleces were treated with several concentrations of the extract, leading to varying rates of inhibition. One hundred percent (100%) mortality rate was observed at a concentration of 50 mg/ml after 10 minutes of exposure. This *in vitro* study revealed that methanolic extracts of *R. coriaria* may be considered as an efficient natural scolicidal agent. While in a more recent study, Vakili et al. [30] analyzed the effects of the plant *Artemisia sieberi* on *E. granulosus* protoscoleces, collected from livestock, and the impact of three concentrations of aqueous extract of *A. sieberi* was assessed over three different periods of exposure. Their results revealed various mortality rates of the protoscoleces as a function of the periods of exposure and concentrations of the extract used. The highest mortality rate ($92.6 \pm 1.27\%$) was observed after treatment at a concentration of 75 mg/ml for 10 minutes. It was concluded that the aqueous extract *A. sieberi* had significant protoscolicidal activity and could be used as a natural agent against hydatid cyst protoscoleces which causes Echinococcosis.

Niazi et al. [31] carried out a study to assess the *in vitro* and *ex vivo* scolicidal effects of *Olea europaea* L. leaf extract on hydatid cyst protoscoleces, which were collected from the livers of infected sheep, after which they were treated with several concentrations of *O. europaea* L. extract *in vitro* and *ex vivo*. The mortality rate of protoscoleces was evaluated by the eosin exclusion test (0.1% eosin staining). The *in vitro* studies showed a mean mortality rate of 100% after 10 minutes of incubation with a concentration of 300 mg/ml of the *O. europaea* L. extract and a mean mortality rate of 100% after 20 minutes of incubation with a concentration of 150 mg/ml. After the inoculation of *O. europaea* L. extract directly into the hydatid cyst (*ex vivo*), the mean of the mortality of protoscoleces was 100% after 12 minutes of incubation with a concentration of 300 mg/ml or 25 minutes of incubation with a concentration 150 mg/ml of the extract. Based on their findings, it was concluded that the extract of the *O. europaea* L. had a substantial scolicidal effect on hydatid cyst protoscoleces.

Lv et al. [32] investigated the *in vitro* and *in vivo* efficiencies of chemotherapy against *E. granulosus* by albendazole liposome (L-ABZ), Huaier aqueous extract, and a Huaier aqueous extract/L-ABZ combination. Phosphate buffer was used for washing the protoscoleces which were collected. They observed that more than 95% of the *E. granulosus* protoscoleces parasites were still alive while more than 65% exhibited discrete movement with the L-ABZ, Huaier aqueous extract treatment. They observed a damaging effect of the cyst protoscoleces at 2 mg/ml Huaier aqueous extract and addition of 10 mg/ml L-ABZ. They concluded that Huaier aqueous extract had a significant reduction in viability of protoscoleces by 50% after incubation for 24 hours and there was no significance in the viable protoscoleces in the mammalian cell line used. They established that Huaier could improve the role of L-ABZ in the treatment and management of CE.

Haghani et al. [33] studied the scolicidal action of methanolic extract of the plants *Ocimum bacilicum* and *Allium cepa*. Protoscoleces of *E. granulosus* were aseptically isolated from sheep livers containing hydatid cysts. The protoscoleces were exposed to the leaf extracts of the 2 plants at 2.5, 5, and 10% concentrations for a maximum of 60 minutes. The scolicidal action of the leaf extracts of the plants was not adequate. The highest concentration used (10% concentration) of *O. bacilicum* and *A. cepa* extracts after 60 minutes of exposure led to 24.1 and 16.8% of mortality of the proscoleces, respectively. It was observed that although these plants have high antibacterial activity, they did not have an equally high scolicidal effect and hence are not suitable to be used in surgical operations carried out to treat hydatid cysts. Yones et al. [34] investigated the *in vitro* action of alcoholic extracts of *Salvia officinalis*, *Thymus vulgaris*, and pure compounds thymol and menthol on *E. granulosus* protoscoleces. Various concentrations were used to test for the viability of protoscoleces within the first seven days post-treatment (PT). Total loss of viability of protoscoleces was observed with 500 μg/ml concentrations of *S. officinalis* and *T. vulgaris* extracts six and seven days PT, respectively. The pure compounds thymol and menthol exhibited strong effects with a concentration of 50 μg/ml two and five days PT, respectively. These results were similar to those shown by albendazole sulfoxide (800 μg/ml) and it was concluded that these plant extracts exhibited an even more effective scolicidal action than the synthetic albendazole sulfoxide.

Zibaei et al. [35] investigated the *in vitro* scolicidal action of hydroalcoholic extracts of *khuzestanica* leaves and aqueous extracts of *O. europaea* leaves on *E. granulosus* protoscoleces obtained from the liver of sheep diseased with the parasite. Different concentrations of the extracts were used over various exposure periods to check for the mortality rate of the proscoleces. A mortality rate of 96.7% was observed after 120 minutes of exposure to 0.01% of *O. europaea* extracts while 0.1% *Satureja khuzestanica* had very potent scolicidal effects, as a 100% scolicidal effect was observed. It was inferred that *S. khuzestanica* exhibited a stronger scolicidal effect than *O. europaea* against cystic *Echinococosis protoscoleces*. Rahimi-Esboei et al. [36] performed a study to assess the scolicidal activity of ultrasonic methanol extract of *Allium sativum* flower. Protoscoleces of the parasite were collected using aseptic techniques from sheep livers infected with hydatid cyst and were treated by various concentrations of the extract over different exposures periods. The highest mortality rate (98%) was observed when the protoscoleces were exposed to 100 mg/ml of the extract for a period of 180 minutes as confirmed by 0.1% eosin staining. This experiment revealed that ultrasonic *A. sativum* flower extract has a strong scolicidal effect and could be used in presurgical therapy to prevent a secondary recurrence of hydatid cyst after initial treatment.

Moazeni et al. [37] carried out an *in vivo* study on the preventive and therapeutic action of methanolic extracts of the plant *Zataria multiflora* on protoscoleces of *E. granulosus* which were isolated using aseptic techniques from the liver of the infected sheep. Mice were then infected intraperitoneally and after a period of eight months, they were treated with the extracts of *Z. multiflora* (8 g/l) for a period of 30 days. After the treatment period, the hydatid cysts were retrieved from the mice, weighed, and their sizes were measured. The results obtained were compared with those of Albendazole (ABZ), which was used on another class of mice among the originally infected lot. A significant decrease was recorded, and scanning electron microscopy was used to observe for damage to the germinal layer of the cysts at the ultrastructural level. A similar result was observed in a class of mice which was under prophylactic therapy of *Z. multiflora* extracts. These results led them to the conclusion that methanolic extracts of *Z. multiflora* could be used both in treatment and prevention of hydatidosis, though they pointed out the need for an evaluation of the viability of developing this natural agent into an efficient medication for the treatment of human hydatidosis.

8.4.1 *In vitro* and *in vivo* Effect of Phytochemicals Against *Echinococcus* Infection

The *in vitro* and *in vivo* activities of carvacrol against the metacestodes of *E. granulosus* were investigated [38]. For the *in vitro* assay, Carvacrol was dissolved in dimethyl sulphoxide (DMSO) at concentrations of 10, 5, and 1 μg/ml, while for the *in vivo*, 2.5 mg/ml ABZ was dissolved in deionized water and mixed vigorously for 12 hours, while carvacrol at 4 mg/ml drug concentration was dissolved in olive oil. Then the mixture, after shaking, was given intragastrically to the mice. The authors observed that 10 μg/ml of carvacrol had the greatest protoscolicidal effect. After six days post-incubation (p.i), viability was lowered to $17.62 \pm 13.93\%$ and after 60 days, it was further lowered to 0% while the 5 and 1 μg/ml concentrations of carvacrol, later triggered a protoscolicidal effect, lowering viability of protoscoleces after 60 days p.i to 16.17 ± 1.26 and $31.24 \pm 10.52\%$, respectively, showing a reduction in cyst weight. Also, upon treatment of the germinal layer of cyst with carvacrol, there was a loss of the multicellular structure features, suggesting damage of the protoscoleces. They therefore established that carvacrol could be used as a substitute for the treatment of Human Echinococcosis.

Elissondo et al. [39] reported the *in vitro* efficiency of thymol against the protoscoleces of *E. granulosus*. Protoscoleces obtained aseptically from cyst collected from naturally infected cattle were cultured in medium 199 adding 100 IU penicillin, 100 μg/ml streptomycin, and 4 mg/ml glucose. Thymol at a drug concentration of 10 mg/ml was dissolved in DMSO and added to the medium bringing it to a final concentration of 10, 5, and 1 μg/ml with a control having protoscoleces incubated with 10 μl DMSO in culture medium. The authors noted that the highest protoscolicidal effect was observed with thymol at 10 μg/ml, with viability lowered after 12 days of incubation to $53.5 \pm 11.9\%$ and further reduced to 0% in 80 days. Meanwhile, at 5 and 1 μg/ml concentrations of thymol, there was an after-protoscolicidal effect, thereby lowering of protoscoleces viability after 42 days of incubation to near 50%. They observed that the thymol-treated protoscoleces showed ultrastructural damage at the tegument of the parasite. The authors concluded that thymol is effective in the treatment against *E. granulosus* protoscoleces.

Echinococcus multilocularis, E. granulosus, and *Taenia solium* are known to be responsible for the most serious and life-threatening helminth infections. Hemphill et al. [40] carried a review on the use of the laboratory models of *E. multilocularis* and *E. granulosus* in the quest for new drugs that could be adopted for chemotherapy of echinococcosis. Their study revealed that the *E. multilocularis* model, unlike other cestode models, fulfilled the conditions that would permit for intensified drug screening processes as a result of its rapid growth and proliferation *in vitro*, easy access to comprehensive genomic information and EST databases, among other advantages.

Küster et al. [41] evaluated the *in vitro* action of amino ozonides against the metacestode of *E. multilocularis*. Metacestodes were dissected from mice which have been infected experimentally and were incubated in 20 mg/l each of levofloxacin and ciprofloxacin and PBS overnight. The metacestodes were collected after culturing for about two months. All neutral and acidic ozonides were screened before their use against *E. multilocularis,* while nitazoxanide and artesunate were used as positive control. They observed that all neutral and acidic ozonides tested had no effect on *E. multilocularis* metacestodes and was toxic at high concentrations in human foreskin fibroblasts (HFFs) and rat hepatoma (RH) cells. They also revealed that the primary amines with an aminopropylether substructure were most effective (with variance of c. 75–100%) in metacestode damage as shown by high levels of phosphoglucose isomerase (PGI) which is a main constituent of the metacestode vesicle fluid. They concluded that ozonides induced significant metabolic damage in metacestodes after exposure but was less effective than Nitazoxanide.

Gonźalez-Sapienza et al. [42] carried out a study to improve the already existing immunodiagnostic techniques used to determine the presence of hydatid infection in a patient. *E. granulosus* Antigen B which possesses the highest diagnostic value among other peptide antigens (such as Ag5) was studied further and a 38-mer peptide (p176) was delineated from the parent protein. The peptide, p176, was observed to have a higher diagnostic sensitivity (80%) and specificity (94%) when tested against the same panel of sera. It was suggested that p176 could be used more efficiently in the laboratory diagnosis of hydatidosis.

Barbieri et al. [43] compared a synthetic peptide (GU4) collected from Antigen-B with crude hydatid cyst fluid antigen (HCFA), immunopurified AgB, antigen 5 (Ag5), peptides 65 (AgB mimotope), and 89–122 (Ag5 mimotope) for efficiency in diagnosis of cystic hydatidosis. From their study, it was observed that AgB showed a higher diagnostic efficiency (82%) than Ag5 (74%) and HCFA (71%). It was also observed that peptides 89–122 and AgB when combined, exhibited a higher sensitivity (82%) as opposed to the sensitivity of AgB alone with specificity equal to that of AgB (86%). It was suggested that hydatid serology might be more efficient by combining several defined antigens during the diagnosis of cystic hydatidosis.

Horton [44] investigated the use of ABZ in the chemotherapy of echinococcosis infections. 253 patients with *E. granulosus* cysts were treated with ABZ, of which 79.5% were either cured or had improved. His findings provided adequate proof that ABZ was effective and could be used as maintenance therapy for inoperable cases.

Moreno et al. [45] experimented on the *in vivo* combined use of Praziquantel and ABZ in the treatment of hydatid disease. It was observed that the combination was highly effective (100% efficiency) as a type of prophylactic therapy in the prevention of the development of hydatid cysts as a result of the propagation of protoscoleces during surgical procedures. This combination, however, was unsuccessful for the treatment of experimental hydatidosis, as there was no substantial difference between the treated and control mouse groups. The combination of Praziquantel and ABZ was therefore suggested for implementation in chemoprophylaxis against human hydatidosis.

Wang et al. [46] studied the use of recombinant *E. granulosus* ferritin (rEgferritin, cloned from *E. granulosus* and expressed in *Escherichia coli*) as a vaccine to confer immunity on mice, which were exposed intraperitoneally to brood capsules collected aseptically from fertile *E. granulosus* cysts from the livers of diseased humans. The vaccinated and exposed mice displayed significant protective efficiency of up to 85.6%. Thus, they inferred that rEgferritin could be a potential candidate as an efficient vaccine to confer immunity to echinococcosis.

Küster et al. [47] evaluated the *in vitro* and *in vivo* efficiency of mefloquine, a synthetic quinine analog, against *E. multilocularis* metacestodes responsible for alveolar echinococcosis. The parasites were cultured *In vitro* in the presence of 24 μM mefloquine for 10 days after which a parasiticidal dose-dependent effect of the drug was observed and confirmed by murine bioassays. *in vivo,* mice were infected with *E. multilocularis* and 25 mg/kg of mefloquine was administered twice a week for eight weeks intraperitoneally. The results observed were similar to the decline seen with ABZ (200 mg/kg/day) applied orally. On the other hand, when mefloquine was administered orally, it was unsuccessful in attaining any decrease in parasite weight. In summary, they observed that mefloquine was an effective parasiticide that could be used in the treatment of alveolar echinococcosis.

Spicher et al. [48] demonstrated the *in vitro* and *in vivo* treatment of *E. multilocularis* and *E. granulosus* larval stages with the antimalarials and artesunate. While 6 weeks intragastric treatment of mice infected with metacestodes of the parasites had no influence, *in vitro* treatment of the larval stages of the parasites with the antimalarial drugs dihydroartemisinin (DHA) and artesunate showed promising results. Scanning and transmission electron microscopy revealed that treatment with artesunate and DHA led to intense tissue modifications and loss of the multicellular structure of the germinal layer. They also observed that the artesunate-albendazole and dihydroartemisinin-albendazole combination treatments led to lower average parasite weight values compared with ABZ treatments alone, though the differences were just short of statistical significance.

Mohammadnejad et al. [49] investigated the *in vitro* efficiency of the antihelminthic drugs artemether, artemisinine, ABZ, and their combinations against protoscoleces of *E. granulosus* which were aseptically collected from the livers of sheep infected with hydatid cysts of the parasite and then treated with various concentrations of the drugs. The protoscolicidal action of artemether and its combinations was observed to be significantly stronger than the other combinations. The maximum protoscolicidal action was observed with 200 μg/ml of artemether four days PT, while ABZ killed protoscoleces seven days PT. Artemisinin exhibited favorable results against evaginated protoscoleces nine days PT. It was observed that the drug combination with the greatest efficiency was artemether and artemisinin. The desirable action of artemether and the surprising result displayed by artemisinin portrays them as potential scolicidal agents that could be used in the treatment of human hydatidosis.

Pensel et al. [50], in an *in vitro* study, evaluated the efficiency of 5-fluorouracil (5-FU) and paclitaxel (PTX) against *E. granulus* larvae and cells, collected from the liver and lungs of diseased cattle. The results showed that 5-FU inhibited the protoscoleces in a time and dose-dependent manner and was stronger in effect than PTX. Studies found that 5-FU led to total loss of viability after 24 days. It was concluded that both 5-FU and PTX would produce desirable results at realizable concentrations to hinder the persistence of protoscoleces and metacestodes of the parasite.

Spicher et al. [48] performed an *in vitro* and *in vivo* experiment with Artemisinin and Artemisinin derivatives for the treatment of the protoscoleces and metacestodes of *Echinococcus*. They observed that exposure of *E. multilocularis* metacestodes *in vitro* to artesunate and DHA caused an obvious tissue variation and damage of the typical multicellular structure of the germinal layer. Also, artesunate-albendazole and DHA-albendazole treatments showed low mean parasite weights than were found with only the ABZ treatment. They concluded that combination treatment with ABZ led to a lower parasite weight than when ABZ is taken alone for the treatment of protoscoleces and metacestodes of *Echinococcus*.

8.5 Synthetic Drugs Previously Used for Management of Echinococcosis Disease

Antiparasitic drug therapy employed in the treatment of echinococcosis includes the use of benzimidazoles, ABZ, and mebendazole (MBZ) exclusively for use in humans which are licensed having gone through empirical tests in detecting their efficacy and safety for use. The pharmacokinetics and pharmacodynamics of the drugs are also very important points to note in the therapeutic administration of these drugs. Pharmacokinetics which is concerned with the dose–concentration of the therapeutic agent and pharmacodynamics which relates to the concentration-effect will come into play by evaluating the vital processes of Absorption, Distribution, Metabolism, and Elimination (ADME) – Pharmacokinetics and the pharmacodynamics concepts of maximum response and sensitivity are involved in the determination of the magnitude of the effect the drug exerts at specific concentrations. Search into novel sources of therapy such as synthetic molecules with proven effects in the treatment of other kinds of infections such as parasitic, mycotic, and bacterial infections are also being tested for their possible adoption for treatment of echinococcosis [51]. Below is an appraisal of each of the existing antiparasitic drugs listed above being used for the treatment of echinococcosis:

In cases where a surgical operation is not possible, treatment is usually achieved through administering benzimidazole for a prolonged duration. It was first discovered as an antiparasitic drug against parasitic worms in the gastrointestinal tract of domestic animals and farm animals. Today, they are widely in use in both human and veterinary medicine. Their activities extend beyond anthelminthic capabilities to anti-tumorigenic, antibacterial, anti-inflammatory, antimycotic, antioxidant, and antidiabetic potentialities [52]. Benzimidazoles interfere with the assembly of the worm's cytoskeletal protein tubulin into polymers of microtubule and hence subsequent inhibition of microtubule-mediated metabolic cellular processes, such as cell division. It disrupts the mechanisms involved in the uptake of glucose by the developing larva of the parasite which is necessary for survival and metamorphosis, thereby causing a drastic reduction in the level of glycogen available in the larva and the collapse of organelles such as mitochondria and the endoplasmic reticulum [53]. BMZ has been reported to exert inhibitory effects on *Echinococcus* sp. rather than the destruction of parasites [54].

Studies revealed that the age of patients to which BMZ is administered also determines the therapeutic effect. It was discovered that it was more effective in children and adolescents rather than adult patients. The target organ also is perceived to contribute to the outcome of treatment. MBZ and ABZ are derivatives of BMZ. MBZ was reported with high efficacy, with proven broad-spectrum activity. Its biocidal effect on *E. granulosus* has been documented. However, a very small fraction is always absorbed at the intestinal level due to its insolubility and thereby passed to the liver for metabolism.

8.6 Conclusion and Future Prospects

Echinococcosis is a parasitic infection which manifests in two distinct forms in humans. The first being CE which is also known as hydatidosis while the second is alveolar echinococcosis. Both are caused by the tapeworms *E. granulosus* and *E. multilocularis*, respectively. The devastating effects of these parasites to humans in endemic areas have caused severe losses and poor livelihood. Efforts to treat the infection are complemented by traditional knowledge of herbs. However, due to the shift in attention to other diseases affecting people in the developed world, research for drugs in the area of echinococcosis is limited, hence, leaving the affected populace to rely on hearsay herbs or old drugs which often come with severe side effects. To this end, it is recommended that research be redirected to investigating medicinal plants, their phytochemical components or derivatives as potential source of new drugs for the treatment of echinococcosis.

References

1 Van Wyk, B.E. and Wink, M. (2004). *Medicinal Plants of the World: An Illustrated Scientific Guide to Important Medicinal Plants and Their Uses*. Portland, OR, USA: Timber Press.

2 Adetunji, C.O., Olaleye, O.O., Adetunji, J.B. et al. (2011a). Studies on the antimicrobial properties and phytochemical screening of methanolic extracts of *Bambusa vulgaris* leaf. *International Journal of Biochemistry* 3 (1): 1–8.

3 Adetunji, C.O., Arowora, K.A., Afolayan, S.S. et al. (2011b). Evaluation of antibacterial activity of leaf extract of Chromolaena odorata. *Science Focus* 16 (1): 1–6.

4 Adetunji, C.O., Olaleye, O.O., Umanah, J.T. et al. (2011c). *In vitro* antibacterial properties and preliminary phytochemical of *Kigelia africana*. *Journal of Research in Physical Sciences* 7 (1): 8–11.

5 Adetunji, C.O., Kolawole, O.M., Afolayan, S.S. et al. (2011d). Preliminary phytochemical and antibacterial properties of *Pseudocedrela kotschyi*: a potential medicinal plant. *Journal of Research in Bioscience. African Journal of Bioscience* 4 (1): 47–50.

6 Adetunji, C.O. and Olaleye, O.O. (2011). Phytochemical screening and antimicrobial activity of the plant extracts of *Vitellaria paradoxa* against selected microbes. *Journal of Research in Biosciences* 7 (1): 64–69.

7 Adetunji, C.O., Olatunji, O.M., Ogunkunle, A.T.J. et al. (2014). Antimicrobial activity of ethanolic extract of *Helianthus annuus* stem. *Sikkim Manipal University Medical Journal* 1 (1): 79–88.

8 Newman, D.J. and Cragg, G.M. (2007). Natural products as sources of new drugs over the last 25 years. *Journal of Natural Products* 70: 461–477.

9 Neumayr, A., Tamarozzi, F., Goblirsch, S. et al. (2013). Spinal cystic echinococcosis – a systematic analysis and review of the literature: part 2. Treatment, follow-up and outcome. *PLoS Neglected Tropical Diseases* 7: e2458. https://doi.org/10.1371/journal.pntd.0002458.

10 WHO (2012). *Accelerating work to overcome the global impact of neglected tropical diseases: a roadmap for implementation: executive summary*. World Health Organisation pp 22. WHO/HTM/NTD/2012.1. http://www.who.int/neglected_diseases/NTD_RoadMap_2012_Fullversion.pdf.

11 Thompson, R.C. and McManus, D.P. (2002). Towards a taxonomic revision of the genus *Echinococcus*. *Trends in Parasitology* 18: 452–457.

12 Liu, C., Zhang, H., Lei, W. et al. (2014). An alternative mebendazole formulation for cystic echinococcosis: the treatment efficacy, pharmacokinetics and safety in mice. *Parasites & Vectors* 7: 589–598.

13 Adas, G., Arikan, S., Kemik, O. et al. (2009). Use of albendazole sulfoxide, albendazole sulfone and combined solutions as scolicidal agents on hydatid cysts (*in vitro* study). *World Journal of Gastroenterology* 15: 112–116.

14 Junghanss, T., da Silva, A.M., Horton, J. et al. (2008). Clinical management of cystic echinococcosis: state of the art, problems, and perspectives. *American Journal of Tropical Medicine and Hygiene* 79: 301–311.

15 Gholami, S.H., Rahimi-Esboei, B., Ebrahimzadeh, M.A., and Pourhajibagher, M. (2013). *In vitro* effect of *Sambucus ebuluson* scolices of hydatid cysts. *European Review for Medical and Pharmacological Sciences* 17: 1760–1765.

16 Jex, A.R., Littlewood, D.T., and Gasser, R.B. (2015). Sequencing and annotation of mitochondrial genomes from individual parasitic helminths. In: *Parasite Genomics Protocols, Methods in Molecular Biology*, vol. 1201 (ed. C. Peacock), 51. New York: © Springer Science+Business Media https://doi.org/10.1007/978-1-4939-1438-8_2.

17 Bhutani, N. and Kajal, P. (2018). Hepatic echinococcosis: a review. *Annals of Medicine and Surgery* 36: 99–105.

18 Piccoli, L., Bazzocchi, C., Brunetti, E. et al. (2012). Molecular characterization of Echinococcus granulosus in south-eastern Romania: evidence of G1–G3 and G6–G10 complexes in humans. *Clinical Microbiology and Infection* 19: 578–582.

19 Ibrahim, E.A.A., Yara, M.A., and Moustafa, A.A. (2016). Prevalence and molecular characterization of hydatid cyst isolates from cattle in Egypt. *Asian Journal of Animal and Veterinary Advances* 11: 794–804.

20 Lissandrin, R., Tamarozzi, F., Piccoli, L. et al. (2016). Factors influencing the serological response in hepatic *Echinococcus granulosus* infection. *American Journal of Tropical Medicine and Hygiene* 94: 166–171. https://doi.org/10.4269/ajtmh.15-0219.

21 Gottstein, B. (2000). Hydatid disease, major tropical syndromes by body system. In: *Systemic Infections*. Chap 2000: 169 section 6.

22 Siracusano, A., Teggi, A., and Ortona, E. (2009). Human cystic echinococcosis: old problems and new perspectives. *Interdisciplinary Perspectives on Infectious Diseases* 2009: 474368. https://doi.org/10.1155/2009/474368 7 pages.

23 Gangwar, M., Verma, V.C., Singh, T.D. et al. (2013). In-vitro scolicidal activity of *Mallotus philippinensis* (Lam.) Muell Arg. fruit glandular hair extract against hydatid cyst *Echinococcus granulosus*. *Asian Pacific Journal of Tropical Medicine* 6 (8): 595–601.

24 Albani, C.M., Denegri, G.M., and Elissondo, M.C. (2014). Effect of different terpene-containing essential oils on the proliferation of *Echinococcus granulosus* larval cells. *Interdisciplinary Perspectives on Infectious Diseases* 2014: 746931. https://doi.org/10.1155/2014/746931.

25 Almalki, E., Al-Shaebi, E.M., Al-Quarishy, S. et al. (2017). *In vitro* effectiveness of *Curcuma longa* and *Zingiber officinale* extracts on *Echinococcus protoscoleces*. *Saudi Journal of Biological Sciences* 24: 90–94.

26 Al-Mayah, K.S., Al-Bashir, N.M., and Al-Azzawi, B.M. (2012). *In vivo* efficacy of *Nigella Sativa* aqueous seed extract against metacestode of *Echinococcus granulosus*. *Medical Journal of Babylon* 9 (1): 140–151.

27 Verma, V.C., Gangwar, M., Yashpal, M., and Nath, G. (2013). Anticestodal activity of endophytic *Pestalotiopsis* sp. on protoscoleces of hydatid cyst *Echinococcus granulosus*. *BioMed Research International* 2013: 308515. https://doi.org/10.1155/2013/308515. 11 pages.

28 Labsi, M., Khelifi, L., Mezioug, D. et al. (2016). Antihydatic and immunomodulatory effects of *Punica granatum* peel aqueous extract in a murine model of echinococcosis. *Asian Pacific Journal of Tropical Medicine* 9 (3): 211–220.

29 Moazeni, M. and Mohseni, M. (2012). Sumac (*Rhus coriaria* L.): scolicidal activity on hydatid cyst protoscolices. *Surgical Science* 3: 452–456.

30 Vakili, Z., Radfar, M.H., Bakhshaei, F., and Sakhaee, E. (2019). *In vitro* effects of *Artemisia sieberi* on *Echinococcus granulosus* protoscolices. *Experimental Parasitology* 197: 65–67. https://doi.org/10.1016/j.exppara.2018.10.011.

31 Niazia, M., Sakib, M., Sepahvanda, M. et al. (2019). *In vitro* and *ex vivo* scolicidal effects of *Olea europaea* L. to inactivate the protoscolecs during hydatid cyst surgery. *Annals of Medicine and Surgery* 42: 7–10.

32 Lv, H., Jiang, Y., Liao, M. et al. (2013). *In vitro* and *in vivo* treatments of *Echinococcus granulosus* with Huaier aqueous extract and albendazole liposome. *Parasitology Research* 112: 193–198.

33 Haghani, A., Roozitalab, A., and Safi, S.N. (2014). Low scolicidal effect of *Ocimum bacilicum* and *Allium cepa* on protocoleces of hydatid cyst: an *in vitro* study. *Comparative Clinical Pathology* 23: 847–853. https://doi.org/10.1007/s00580-013-1701-0.

34 Yones, D.A., Taher, G.A., and Ibraheim, Z.Z. (2011). *In vitro* effects of some herbs used in egyptian traditional medicine on viability of protoscolices of hydatid cysts. *Korean Journal of Parasitology* 49 (3): 255–263. https://doi.org/10.3347/kjp.

35 Zibaei, M., Sarlak, A., Delfan, B. et al. (2012). Scolicidal effects of *Olea europaea* and *Satureja khuzestanica* extracts on protoscolices of hydatid cysts. *Korean Journal of Parasitology* 50 (1): 53–56. https://doi.org/10.3347/kjp.2012.50.1.53.

36 Rahimi-Esboei, B., Ebrahimzadeh, M.A., Fathi, H., and Anzahaei, F.R. (2016). Scolicidal effect of *Allium sativum* flowers on hydatid cyst protoscolices. *European Review for Medical and Pharmacological Sciences* 20: 129–132.

37 Moazeni, M., Larki, S., Oryan, A., and Saharkhiz, M.J. (2014). Preventive and therapeutic effects of *Zataria multiflora* methanolic extract on hydatid cyst: an *in vivo* study. *Veterinary Parasitology* 205 (1–2): 107–112. https://doi.org/10.1016/j.vetpar.2014.07.006.

38 Fabbri, J., Maggiore, M.A., Pensel, P.E. et al. (2016). *in vitro* and *In vivo* efficacy of carvacrol against *Echinococcus granulosus*. *Acta Tropica* 164: 272–279. https://doi.org/10.1016/j.actatropica.2016.09.001.

39 Elissondo, M.C., Albani, C.M., Gende, L. et al. (2008). Efficacy of thymol against *Echinococcus granulosus* protoscoleces. *Parasitology International* 57: 185–190.

40 (a) Hemphill, A., Stadelmann, B., Scholl, S. et al. (2010). *Echinococcus metacestode* as laboratory models for the screening of drugs against cestodes and trematodes. *Parasitology* 137: 569–587. (b) McManus, D.P., Gray, D.J., Zhang, W., and Yang, Y. (2012). Diagnosis, treatment and management of echinococcosis. *British Medical Journal* 344: e3866. https://doi.org/10.1136/bmj.e3866.

41 Küster, T., Kriegel, N., Stadelmann, B. et al. (2014). Amino ozonides exhibit *in vitro* activity against *Echinococcus multilocularis* metacestodes. *International Journal of Antimicrobial Agents* 43: 40–46.

42 Gonzalez-Sapienza, G., Lorenzo, C., and Nieto, A. (2000). Improved immunodiagnosis of cystic hydatid disease by using a synthetic peptide with higher diagnostic value than that of its parent protein, *Echinococcus granulosus* antigen B. *Journal of Clinical Microbiology* 38 (11): 3979–3983.

43 Barbieri, M., Fernandez, V., Gonzalez, G. et al. (1988). Diagnostic evaluation of a synthetic peptide derived from a novel antigen B subunit as related to other available peptides and native antigens used for serology of cystic hydatidosis. *Parasite Immunology* 20: 51–61.

44 Horton, R.J. (1989). Chemotherapy of echinococcus infection in man with albendazole. *Transactions of the Royal Society of Tropical Medicine and Hygiene* 83: 97–102.

45 Moreno, M.J., Urrea-Paris, M.A., Casado, N., and Rodriguez-Caabiero, F. (2001). Praziquantel and albendazole in the combined treatment of experimental hydatid disease. *Parasitology Research* 87: 235–238.

46 Wang, Y., Li, Z., Li, Z. et al. (2009). Recombinant ferritin protects mice against challenge with *Echinococcus granulosus*. *Acta Parasitologica* 54 (4): 335–340.

47 Küster, T., Stadelmann, B., Hermann, C. et al. (2011). *In vitro* and *in vivo* efficacies of mefloquine-based treatment against alveolar echinococcosis. *Antimicrobial Agents and Chemotherapy* 55 (2): 713–721.

48 Spicher, M., Roethlisberger, C., Lany, C. et al. (2008). *In vitro* and *in vivo* treatments of echinococcus protoscoleces and metacestodes with artemisinin and artemisinin derivatives. *Antimicrobial Agents and Chemotherapy* 52 (9): 3447–3450. https://doi.org/10.1128/AAC.00553-08.

49 Mohammadnejad, F., Ghaffarifar, F., Dalimi, A., and Hassan, Z.M. (2016). *In vitro* effects of artemether, artemisinine, albendazole, and their combinations on *Echinococcus granolosus* protoscoleces. *Jundishapur Journal of Natural Pharmaceutical Products* 11 (1): e30565. https://doi.org/10.17795/jjnpp-30565.

50 Pensela, P.E., Albania, C., Gambo, G.U. et al. (2014). *In vitro* effect of 5-fluorouracil and paclitaxel on *Echinococcus granulosus* larvae and cells. *Acta Tropica* 140: 1–9.

51 Siles-Lucas, M., Casulli, A., Cirilli, R., and Carmena, D. (2018). Progress in the pharmacological treatment of human cystic and alveolar echinococcosis: compounds and therapeutic targets. *PLoS Neglected Tropical Diseases* 12 (4): e0006422. https://doi.org/10.1371/journal.pntd.0006422.

52 Keri, R.S., Hiremathad, A., Budagumpi, S., and Nagaraja, B.M. (2015). Comprehensive review in current developments of benzimidazole – based medicinal chemistry. *Chemical Biology & Drug Design* 86: 19–65. https://doi.org/10.1111/cbdd.12462.

53 Wu, X.W., Chen, X.L., Zhang, S.J. et al. (2011). Pericyst may be a new pharmacological and therapeutic target for hydatid disease. *China Medical Journal* 124: 2857–2862.

54 Hemphill, A., Stadelmann, B., Rufener, R. et al. (2014). Treatment of echinococcosis: albendazole and mebendazole – what else? *Parasite* 21: 70. https://doi.org/10.1051/parasite/2014073.

9

Filaricidal Activity of Phytochemicals Against Lymphatic Filariasis

Jayashree V. Hanchinalmath[1], Darcia D'mello[2], Kirankumar Shivasharanappa[2], T. Pramod[3], and Sharangouda J. Patil[4]

[1] *Department of Life Sciences, School of Sciences, Jain University, Bangalore, Karnataka, India*
[2] *Department of Life Sciences, Garden City University, Bangalore, Karnataka, India*
[3] *Department of Microbiology, The Oxford College of Science, Bangalore, Karnataka, India*
[4] *Department of Zoology, NMKRV College for Women (Autonomous), Bangalore, Karnataka, India*

List of Abbreviations

ADL	Adeno lymphangitis
ATP	Adenosine tri phosphate
DEC	Diethyl carbamazine
DPPH	2,2-Diphenyl-1 picryhydrazyl
g/kg	Gram/kilogram
GPELF	Global Programme to Eliminate Lymphatic Filariasis
hrs	Hours
IC_{50}	Inhibitory concentrations
L1, L2, and L3	Larva 1, larva 2, and larva 3
LC_{50}	Lethal concentration
MDA	Mass drug administration
mg/kg	Milligram/kilogram
mg/mL	Miligram/millilitres
MIC	Minimum inhibitory concentration
min	Minutes
ml/kg	Millilitres/kilogram
MTT	3-(4,5-Dimethylthiazol-2-yl)-2,5-diphenyltetrazolium bromide
ng/ml	Nanograms/litre
NTDs	Neglected tropical diseases

Neglected Tropical Diseases and Phytochemicals in Drug Discovery, First Edition.
Edited by Chukwuebuka Egbuna, Muhammad Akram, and Jonathan Chinenye Ifemeje.

RM	Relative mobility
ROS	Reactive oxygen species
WHO	World Health Organization
μM	Micro molar
μg/ml	Microgram/millilitres

9.1 Introduction

One of the major neglected tropical diseases (NTDs) that has a significant impact on the global public health is lymphatic filariasis, also known as Elephantiasis. Lymphatic filariasis is a well-known vector-borne, chronic disease that is prevalent in Asia, America, Western Pacific, and Africa, particularly in tropical and subtropical regions [1]. The parasites that cause lymphatic filariasis are *Brugia malayi, Wuchereria bancrofti,* and *Brugia timor*. The parasite transmits to humans by means of mosquito bite which include *Aedes, Anopheles, Culex,* and *Mansonia* genera [2]. However, in 91% of cases, filariasis is caused by *W. bancrofti* while the remaining 9% is caused by *B. malayi* and *B. timori* among South and South East Asia [3].

Normally, the adult worms disrupt the normal functions of the lymphatic vessels and thereby thrive in it [4]. Further, these worms scan thrive for approximately six to eight years and produce loads of microfilariae that eventually start circulating in the blood stream. However, the death of the adult parasite within the host may lead to inflammation, thereby inducing the first step of pathogenicity. This inflammation can lead to abnormal enlargement of body parts, thereby inducing Lymphedema, Hydrocoele, and Elephantiasis. Only a few percentages of these patients will show lymphedema [5]. The abnormal functioning of lymphatic system can induce the swelling and edema which is often seen in the legs, but may also occur in the arms, breast, and gentiles while the infected men may develop hydrocele or swelling of the scrotum [2]. In lymphatic filariasis, the lymphatic systems are unable to protect against the infections due to decline in activities of lymphatic components. Thus, the affected patients are more susceptible to bacterial infections in the skin and lymphatic system. This will lead to hardening of the skin, which is known as Elephantiasis [6]. Apart from this, filariasis may be acquired during childhood and may still persist over a period of time and may eventually cause a lifetime of an impaired lymphatic system. Thereby, suppressing the host's first line of defense immunity as well as increases the risk of debilitating episodes of filarial fever known as adenolymphangitis (ADL) [7].

Lymphatic filariasis is the second largest disease in the globe that results in permanent or long-term disability [8]. However, it still persists as an epidemic in Asia, Africa, Pacific Island nations, and South America. Over 856 million individuals need instant chemotherapy to prevent the spreading of the disease in about 52 countries altogether [9]. As per the global baseline estimate, over 15 million people are associated with Lymphedema and Elephantiasis, while on the other hand, lymphatic filariasis is commonly seen in both male and female, with over 10% of women suffering from swollen limbs, and 50% of men with mutilating genital symptoms [8].

9.2 Life Cycle of the Parasite

The development and maturation of filaricidal parasite occurs in two distinct phases that are essential for the life cycle of the parasite (Figure 9.1). The first phase involves the mosquito that serves as a biological vector and an intermediate host that is required for the developmental cycle and transmission of parasitic agent. Further, the mosquito feeds on human blood and ingests microfilariae that circulates in the blood stream [10]. Over a span of 7–21 days, within the mosquito, the microfilariae shed sheaths, penetrate the midgut, and then migrates to the

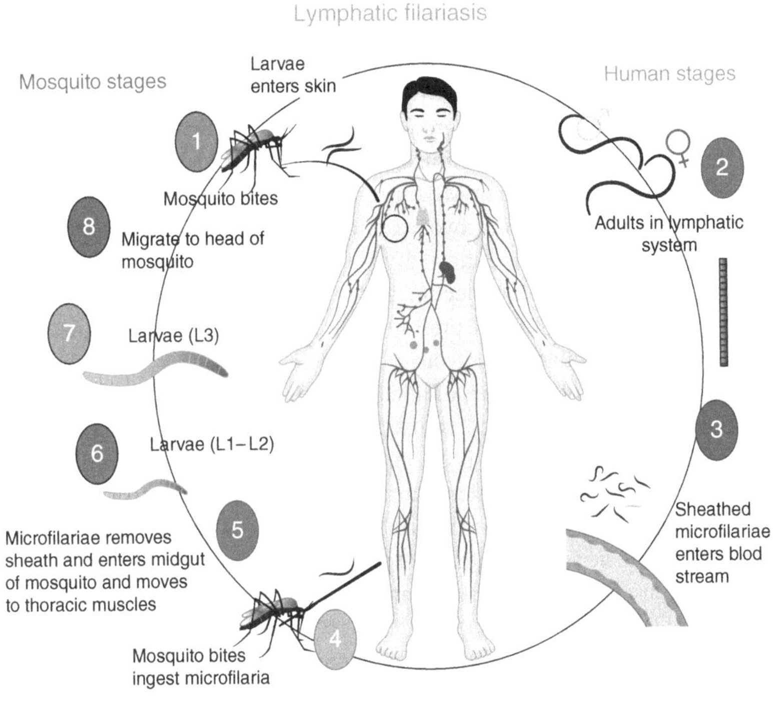

Figure 9.1 Life cycle of the parasite of lymphatic filariasis.

thoracic muscles where the microfilariae increase in its size, molt, and eventually develops into an infectious larva (L1 and L3). However, no multiplication or sexual reproduction of microfilariae occurs inside the mosquito. The infective larva (L3) migrates to the salivary glands, enters the proboscis and escapes onto human skin when the mosquito takes another blood meal [11].

Later, in the second phase, the development of the infective larvae (L3) takes place within the human body including the sexual reproduction and egg production. The infective larva (L3) actively penetrates through the skin via the bite hole and develops into an adult worm in the lymphatic system over a span of six months. These adult worms can survive in the lymphatic system for up to 5–15 years [12]. After the mating of the male and female adult worms, the females produce an average of 10 000 sheathed eggs (microfilaria) daily. These microfilariae penetrate into the blood stream and exhibit the classic nocturnal periodicity and super periodicity. In order to continue the cycle, another mosquito has to take up a blood meal and ingests the microfilariae [13].

9.3 Synthetic Drugs Used for Treatment of Lymphatic Filariasis

Ever since the World Health Organization (WHO) has recognized lymphatic filariasis as one of the major neglected tropical disease, various research schemes were launched such as Global Programme to Eliminate Lymphatic Filariasis (GPELF) through mass drug administration (MDA) [1, 6]. This treatment procedure consisted of one dose of MDA per year, total of four to six doses to all with exception for children up to two years of age and also pregnant women. Currently, the chemotherapeutic drugs to treat filariasis includes Diethylcarbamazine (DEC), Ivermectin, and Albendazole that are known to efficiently bring down microfilariae present in the blood stream [14]. But, most of these drugs have undesirable side effects.

A synthetic derivative of piperazine called Diethylcarbamazine (*N*,*N*-diethyl-4-methyl-1-piperazine carboxamide) is used as an anthelmintic drug for the treatment of filariasis in humans, dogs, and cats. DEC serves as an inhibitor of arachidonic acid metabolism in microfilaria. Thus, resulting in vasoconstriction leading to immobilization of microfilarial parasites as well as enhances the adherence and cytotoxic activity of the host platelets and granulocytes. As a result, the microfilariae are more susceptible to innate immune attack [15]. However, the adverse effects of DEC can lead to Mazzotti reaction, a complex, acute inflammatory response characterized by fever, tachycardia, hypotension, adenitis, and an ocular inflammatory response, usually resulting from the death of microfilarial load. Also, a decrease in the circulating eosinophils was profoundly seen in all cases just before or during the clinical and physiological changes [16]. However, it is sometimes difficult to determine whether these reactions are due to the death of microfilariae or by DEC itself.

The next preferred broad-spectrum anthelmintic agent is Albendazole (methyl 5-(propylthio)-2-benzimidazole carbamate) that serves as a useful drug for giardiasis, trichuriasis, filariasis, neurocysticercosis, hydatid disease, pinworm disease, and ascariasis. Albendazole mainly binds to the colchicine-sensitive site of tubulin, thereby causing the degenerative alterations of the intestinal cells of the worms, thus inhibiting its polymerization into the microtubules. The loss of the cytoplasmic microtubules leads to impaired uptake of glucose by the larval and adult stages of the susceptible parasites, while the degenerative changes in the endoplasmic reticulum decreases the ATP production [17]. Due to the diminished energy production, the parasite is immobilized and eventually dies off. However, the side effects associated with Albendazole may cause abdominal pain, dizziness, headache, fever, nausea, vomiting, or temporary hair loss in sensitive patients. In rare cases, it can even cause persistent sore throat, severe headache, seizures, vision problems, yellowing of eyes or skin, dark urine, stomach pain, easy bruising, mental/mood changes, very stiff neck, or changes in the amount of urine. Also, it has rarely been reported whether Albendazole can cause marrow suppression, agranulocytosis, or aplastic anemia. However, the risk of developing these side effects seems to be increased in patients with liver disease, including echinococcal cysts [18].

The third currently used synthetic drug is Ivermectin, an antiparasitic agent. Ivermectin acts on the parasites and kills them by interfering with the nervous system and muscle functioning by enhancing the inhibitory neurotransmission activity. This drug binds to glutamate-gated chloride channels in the membranes of invertebrate nerve and muscle cells, thereby causing an increased permeability to chloride ions, resulting in cellular hyperpolarization, followed by paralysis and even death [19]. The adverse effect associated with Ivermectin includes fever, myalgia, headache, sore throat, and cough that are usually more prevalent in individuals with higher microfilaremic levels. Unusual side effects were also noted in a few patients with high-density microfilaremia such as itchy rashes, lymphatic nodules, and raised alkaline phosphatase level. These exaggerated unusual reactions were possibly due to allergic response of the susceptible host to rapid killing of a large number of microfilariae [20].

A polytherapy treatment that includes Ivermectin with DEC or Albendazole is more effective than a single drug alone. Apart from this, physical barriers such as mosquito net, chemical-like insect repellent, or mass chemotherapy are some techniques which are applied to prevent the transmission of the disease [21]. Thus, it reduces the spreading through mosquitoes during their subsequent bites. Due to various side effects related to these synthetic drugs, there is an urgent need to find and alternative nontoxic drug to treat filariasis.

9.4 Phytochemicals Used for the Treatment of Lymphatic Filariasis

The current synthetic drugs that were used for MDA implementation by national programmes are DEC, Ivermectin, and Albendazole that are effective in reducing microfilariae counts but not effective enough to kill the adult worms [6]. Therefore, such drugs are of only limited advantage to the patients and are usually related with side effects. Due to lack of vaccines and adverse side effects of the synthetic drugs, there is an urgent need for the development of a unique, nontoxic, and low-cost antifilarial drug [22]. Phytochemical compounds used as traditional drugs are obtained from various plants [23]. Many of these medicinal plants contain pentacyclic triterpene and oleanolic acid that have antifilarial activity. A list of plants and phytochemicals with antifilarial activities are presented in Table 9.1 and Figure 9.2.

9.4.1 *Acacia auriculiformis*

Acacia auriculiformis (Family: Fabaceae), commonly known as Ear leaf Acacia, is an evergreen tree that grows in tropical and subtropical areas. It is reported that the acaciaside – A and B which are triterpenoid saponins isolated from *A. auriculiformis* stem are efficient at a concentration of 4 mg/ml and kills 97% of *Setaria cervi* microfilaria and in about 100 minutes kills all adult worms [24]. However, when the ethanolic extract of stem was orally administered to *Dirofilaria immitis*-infected dogs, it has shown to be effective against both microfilaria and adult worms [25]. This produces physiological stress in worms which leads to the death resulting in leakage and increasing of lipid peroxidation in the cell membrane [26]. Formation of free radicals occurs through the unsaturated system present in saponins which stimulates damage in cell membrane of helminths through peroxidation [27].

9.4.2 *Aegle marmelos*

Aegle marmelos (Family: Rutaceae), also called as Bael, is a plant that native to the Indian subcontinents and Southeast Asia. According to various studies that have been reported, it was noticed that at 100 mg/ml concentration of leaf methanolic extract, *B. malayi* microfilariae lost its motility completely after 48 hours, resulting in halt of the larval physiological process. IC_{50} of 70 mg/ml was found in leaf methanolic extract [28]. This is because the leaves of *A. marmelos* were reported to consist of polyphenolic and coumarins compounds which resulted in a dose-dependent increase in the level of lipid peroxidation and protein carbonylation. Thus, it results in the loss of microfilarial motility [29].

Table 9.1 List of plant-derived phytochemicals: plant name, common name, family, part used, and references for the application of lymphatic filariasis.

Sr. no	Plant name	Common name	Family	Part used	Phytochemicals	References
1	*Acacia auriculiformis*	Earleaf acacia	Fabaceae	Stem	Acaciaside-A and acaciaside-B	[24]
2	*Aegle marmelos*	Bael	Rutaceae	Leaves	Polyphenolic and coumarins	[28]
3	*Alnus nepalensis*	Nepalese alder	Betulaceae	Leaves	Platyphyllenone, alusenone, hirustenone, and hirsutanonol	[30, 31]
4	*Andrographis paniculata*	Green chiretta	Acanthaceae	Leaves	Andrographolide, andrographiside, and β-sitosterolglucoside	[32]
5	*Azadirachta indica*	Neem tree	Meliaceae	Leaves	Azadirachtin	[35]
6	*Bauhinia racemosa*	Mountain ebony	Fabaceae	Leaves	(2*S*)-1,2-di-*O*-linolenoyl-3-*O*-a-galactopyranosyl(1/6)-*O*-b-galactopyranosyl glycerol	[37]
7	*Butea monosperma*	Flame-of-the-forest	Fabaceae	Leaves	Euphane trieterpenoid and pterocrpan	[39]
8	*Caesalpinia bonducella*	Nicker bean	Caesalpiniaceae	Seeds	Caesalpin	[43]
9	*Cardiospermum halicacabum*	Balloon plant	Sapindaceae	Stem	Saponin, L-amino acid, quebrachitol, capric acid, and arachidic fatty acid	[45]
10	*Centratherum anthelminticum*	Kalijiri	Asteraceae	Seeds	3-*O*-α-L-Arabinopyranosyl-hederagenin and 3-*O*-α-L-rhamnopyranosyl-(1 → 2)-α-L-arabinopyranosyl]-hederagenin	[21]

(*Continued*)

Table 9.1 (Continued)

Sr. no	Plant name	Common name	Family	Part used	Phytochemicals	References
11	*Excoecaria agallocha*	Milky mangroove	Euphorbiaceae	Stem	2,2-Diphenyl-1 picryhydrazyl	[48]
12	*Ficus racemosa*	Indian fig tree	Moraceae	Fruits	Gluanol and β sitosterol	[49]
13	*Glycyrrhiza glabra*	Licorice	Fabaceae	Roots	Glycyrrhetinic acid	[50]
14	*Hibiscus sabdariffa*	Roselle	Malvaceae	Leaves	Hydroxy citric acid	[51]
15	*Lantana camara*	Lantana	Verbenaceae	Stem	Oleanonic acid and oleanolic acid	[52]
16	*Leucas cephalotes*	Dronpushpi	Labiatae	Flowers	Labellenic acid (Octadeca – 5,6-dienoic acid)	[53]
17	*Mallotus philippensis*	Kamala tree	Euphorbiaceae	Leaves	Phorbic acid and bergeni	[54]
18	*Morinda citrifolia*	Noni	Rubiaceae	Fruits	β-D-glucopyranose pentaacetate, ascorbic acid, asperuloside tetraacetate, ethyl caprylate, hexanoic acid, octanoic acid, and caprylic acid	[55]
19	*Moringa oleifera*	Drumstick tree	Moringaceae	Gum	Galacturonic acid, galactose arabinose	[56]
20	*Plumbago indica*	Indian leadwort	Plumbaginaceae	Roots	Plumbagin	[57]
21	*Pongamia pinnata*	Pongam tree	Fabaceae	Fruits	Glucosides, pongamosides A–C and pongamoside D	[58]
22	*Psoralea corylifolia*	Babchi	Fabaceae	Leaves	Psoralen and isopsoralen	[21]
23	*Ricinus communis*	Castor oil plant	Euphorbiaceae	Seeds	Stearic acids, ricinoleic acid, and isoricinoleic acid	[46]

24	*Saxifraga stracheyion*	Himalayan Bergenia	Saxifragaceae	Roots	b-Sitosterol, (+) catechin-3-gallate, and bergenin	[59]
25	*Sphaeranthus indicus*	East Indian globe	Asteraceae	Leaves	5,4′-Dimethoxy-3′-prenylbiochanin 7-*O*-β-D-galactoside	[60]
26	*Streblus asper*	Sand paper tree	Moraceae	Stem	Asperoside and strebloside	[61, 62]
27	*Trachyspermum ammi*	Ajwain	Apiaceae	Fruits	2-Isopropyl-5-methyl phenol	[63]
28	*Vitex negundo*	Chinese chaste tree	Verbenaceae	Roots	Negundoside, agnuside, and vitegnoside	[64, 65]
29	*Xylocarpus granatum*	Puzzlenut tree	Meliaceae	Fruits	Gedunin and photogedunin	[66]
30	*Zingiber officinale*	Ginger	Zingeberaceae	Rhizome	Mono- and sesquiterpenes, borneol, geraniol, limonene, β-elemene, zingiberol, and linalool	[67]

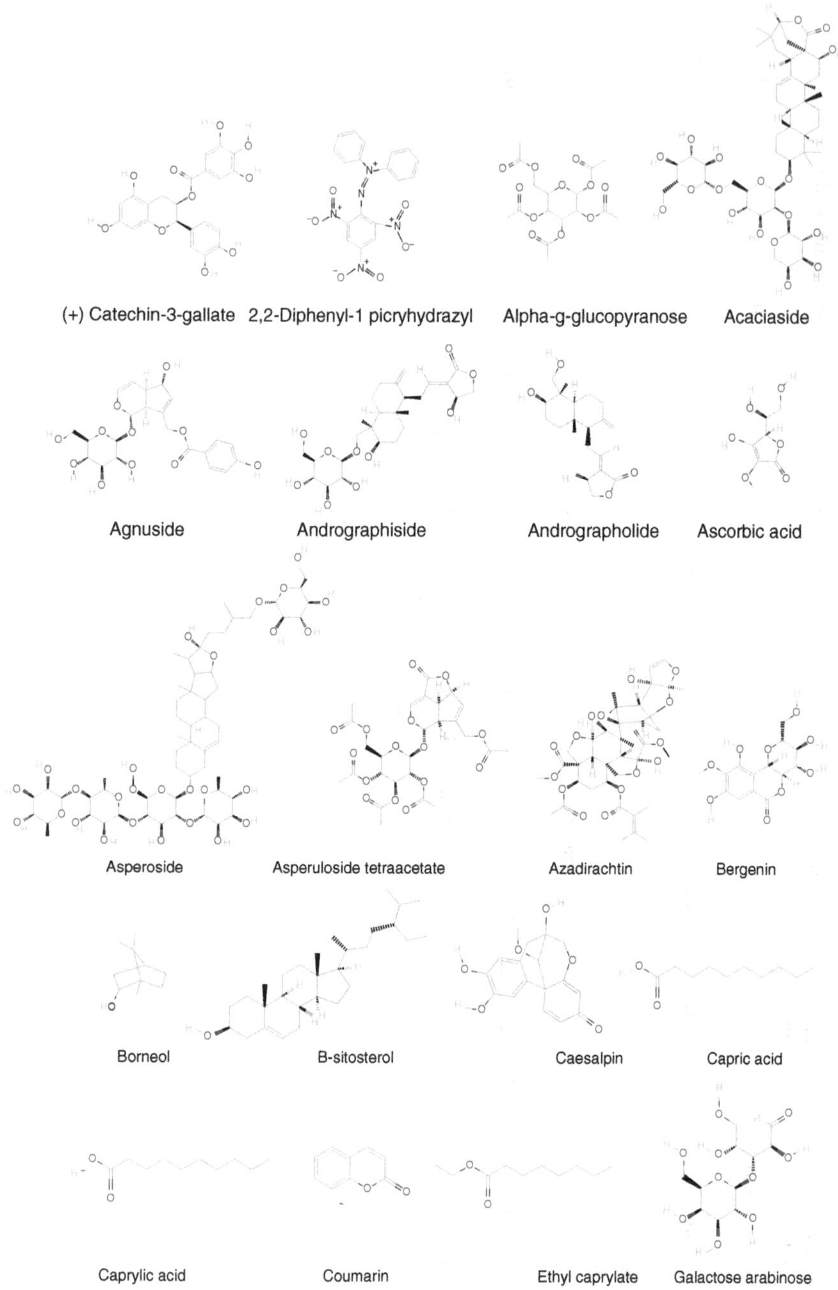

Figure 9.2 Structures of bioactive principles reported against lymphatic filariasis.

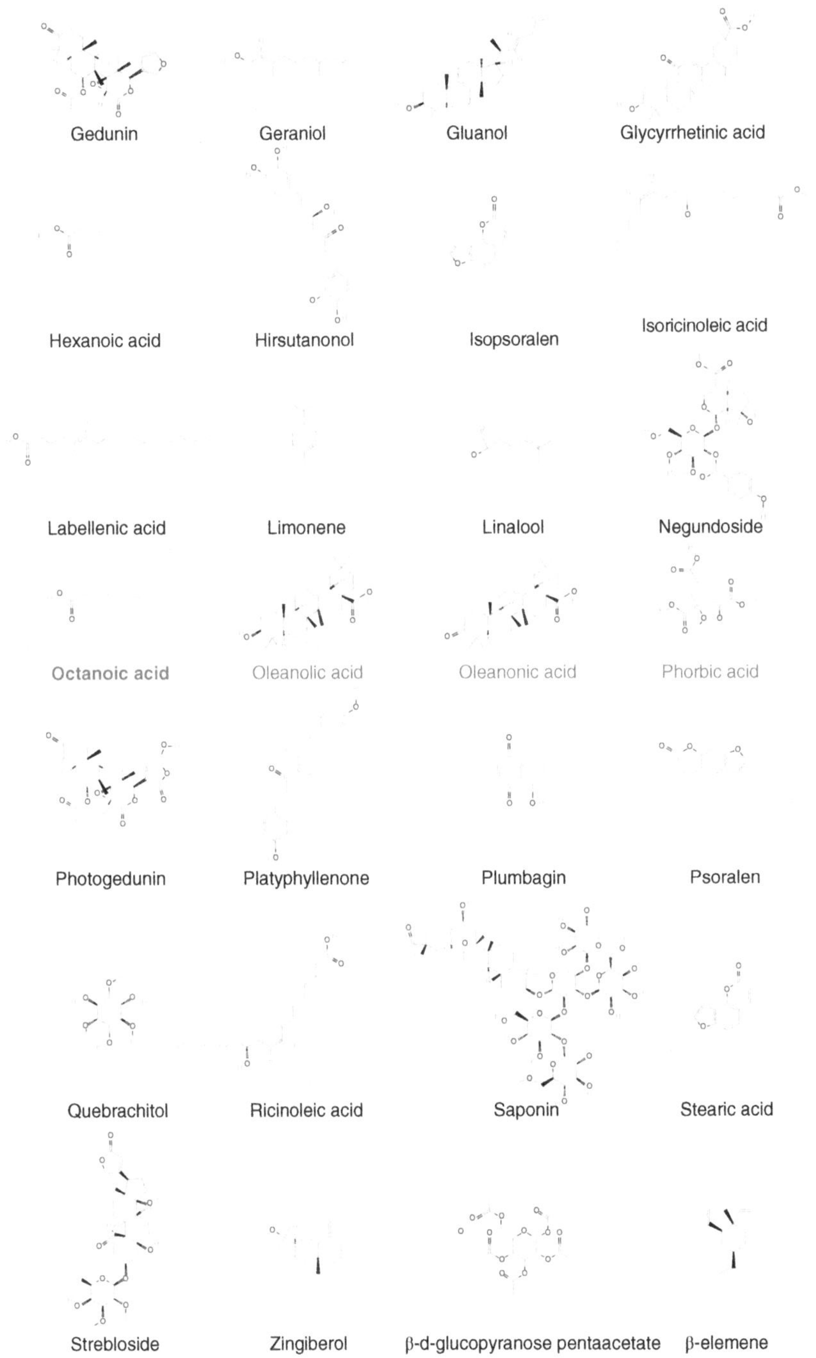

Figure 9.2 (Continued)

9.4.3 *Alnus nepalensis*

Alnus nepalensis (Family: Betulaceae), also called as Nepalese Alder, is found in the Himalayan subtropical high land. Phytochemicals from leaves of *A. nepalensis* were isolated and tested against human lymphatic filariasis both *in vitro* and *in vivo* [30]. The researchers found the isolated compounds were platyphyllenone, alusenone, hirustenone, and hirsutanonol. Hirustenone isolated from *A. nepalensis* has shown effective macro and microfilarial activity against *B. malayi* worms, which was obtained by mean inhibition percentage of control drug DEC and MTT [31].

9.4.4 *Andrographis paniculata*

Andrographis paniculata (Family: Acanthaceae) is an annual herbaceous plant that is native to Southeast Asia, especially China, India, and Sri Lanka. According to a study, the leaf aqueous extract containing andrographolide, andrographiside, and β-sitosterolglucoside in *A. paniculata* revealed *in vitro* and *in vivo* microfilaricidal activity (*Dipetalonema recondium*) in 40 minutes [32]. However, the administration of the extract (0.06 ml/kg body weight) reduces the number of the microfilariae in infected dogs by more than 85% [33]. Earlier, Zaridah et al. analyzed the extract for its antifilarial activity in worms by relative movability (RM) value for 24 hours [23]. At 5 and 10 mg/ml concentration of extract, RM of 0% was seen indicating parasite death.

9.4.5 *Azadirachta indica*

Commonly known as "Neem Tree", *A. indica* (Family: Meliaceae) is found in hot tropical regions such as Africa and India [34]. Due to the presence of a triterpenoid biochemical component called azadirachtin, the leaves and the bark of the tree often taste bitter. According to a study focused on *Teladorsagia circumcincta* helminths carried out by Al-Rofaai et al., it was found that the first-stage larvae-L1 were sensitive with low value of LC_{50} 7.15 mg/ml of the extracted organic solvent from the leaves of *A. indica* compared with 24.91 mg/ml on infective-stage larvae-L3 [35]. However, other studies showed that the LC_{50} value of flower extracts of alcohol and water was found to be 15 and 18 ng/ml with respect to the microfilariae *S. cervi* motility. Apart from this, the ethanolic extract of leaves also has a considerable inhibition against microfilariae reduction and worm viability evidenced by reactive oxygen species (ROS) upregulation [36].

9.4.6 *Bauhinia racemosa*

Bauhinia racemosa (Family: Fabaceae), also known as Mountain Ebony, is native to the tropical regions of Southeast Asia. A study carried out by Sashidhara et al. [37]

focused on the isolation of galactolipids compounds of *B. racemosa* leaves from ethanolic extract and its filaricidal activity. The results indicated that the *n*-butanol fraction containing a galactolipid called as (2*S*)-1,2-di-*O*-linolenoyl-3-*O*-a-galactopyranosyl(1/6)-*O*-b-galactopyranosyl glycerol yielded IC_{50} of 5.46 mg/ml for adulticidal activity and IC_{50} of 4.89 mg/ml microfilaricidal activity with MIC value of 15.6 mg/ml.

9.4.7 *Butea monosperma*

Butea monosperma (Family: Fabaceae) commonly known as the Flame-of-the-Forest is native to tropical and subtropical parts of the Indian Subcontinent and Southeast Asia is traditionally used as a medicinal plant in many Ayurvedic formulations in India [38]. However, the leaves of *B. monosperma* against adult *S. cervi* revealed potential antifilarial activity with an IC_{50} value of 1.25 for methanol and 3.6 mg/ml for hexane–ethanol extract [39]. In one more experiment, leaf and root aqueous extract containing euphane triterpenoid and pterocrpan showed significant activity. In comparison to leaf extract, it was noticed that the roots extracts exhibit a dose-dependent inhibition of motility of microfilariae [40]. The leaf extract containing polyphenolic compounds results in mortality through rise in lipid peroxidation level and also leads to carbonylation of proteins [41].

9.4.8 *Caesalpinia bonducella*

Caesalpinia bonducella (Family: Caesalpiniaceae) commonly known as Nicker Bean is a climbing perennial shrub distributed worldwide in tropical and subtropical regions of Asia, Africa, and Indo-Pakistan subcontinent. However, the leaves are known to have antifilarial effects [42]. *C. bonducella* seed kernel extract containing caesalpin 2 g/kg tested in animal model infected with *B. malayi* and *L. sigmodontis* shows potential macrofilaricidal, microfilaricidal, and also female-sterilizing activity [43].

9.4.9 *Cardiospermum halicacabum*

Cardiospermum halicacabum (Family: Sapindaceae) found in tropical and subtropical parts is a climbing perennial shrub [44]. Khunkitti et al. [45] previously reported the *in vitro* filaricidal activity of ethanolic and aqueous extracts mainly comprising chemical constituents such as saponin, L-amino acid, β sitosterol, quebrachitol, capric acid, and arachidic fatty acid of *C. halicacabum* against *Brugia pahangi*. The researchers found that using 500 μg/ml of the methanolic extract results in motility reduction of adult significantly after exposure for 24 hours and 2 days for ethanolic extract. Furthermore, the ethanol extract at 2 mg/ml concentration inhibited motility of adult worms and also released the microfilariae from females [45].

9.4.10 *Centratherum anthelminticum*

Centratherum anthelminticum (Family: Asteraceae), also known as Kalijiri, is an ethnomedicinal plant in India and a common ingredient in Ayurvedic medicine. The fruit contains the active compound which has anthelmintic activity. The seed methanolic and water extracts resulted in instant motility inhibition of *S. cervi* that was characterized by reduction in tone, amplitude, and rate of contractions. The isolated glycosides, 3-*O*-α-L-arabinopyranosyl-hederagenin and 3-*O*-α-L-rhamnopyranosyl-(1→2)-α-L-arabinopyranosyl]-hederagenin from the methanol extract, showed 65.64% reduction in formazan formation at a concentration of 2 mg/ml at an incubation period of four hours against *Setaria digitata* [46].

9.4.11 *Excoecaria agallocha*

Excoecaria agallocha (Family: Euphorbiaceae), also called as Milky Mangrove, is an habitat of Southeast Asian mangrove swamps. The plant consists of triterpenoids, diterpenoids, and flavonoids which are proven to have a chemical defense mechanism [47]. Patra et al. [48] reported that *E. agallocha* leaf methanolic extract has antifilarial activities at 10 μg/ml concentration and also proposed that it resulted in *S. digitata* parasite death in a dose-dependent manner. The aqueous extract exhibited reducing power, DPPH radical scavenging, and also hydrogen scavenging activity that eventually increased with the increase in extract concentration. Damage in cell membrane structure in eggs and microfilariae occurs and decrease of microfilarial motility further results in the death of the parasite. Therefore, it shows potentiality against filarial parasite which can block the embryogenesis developmental process, lowering the oxidative stress that is produced in the course of long-term symptoms of lymphatic filariasis in human beings [48].

9.4.12 *Ficus racemosa*

Ficus racemosa (Family: Moraceae) commonly known as Indian fig tree is a large deciduous tree found all over India. In an *in vitro* study carried out, it was noticed that the β-sitosterol and gluanol isolated from the fruit ethanolic and water extract showed immediate inhibition of adult worm motility. Both of these fruit extracts tested *in vitro* resulted in microfilarial death with 21 mg/ml at LC_{50} and 35 mg/ml at LC_{90} for ethanol extract; 27 mg/ml at LC_{50} and 42 mg/ml at LC_{90} for water extract [49].

9.4.13 *Glycyrrhiza glabra*

Glycyrrhiza glabra (Family: Fabaceae), commonly known as Licorice, is a perennial herb that has been used for medicinal remedies for thousands of years. The liquorice roots is known to contain a unique triterpenic acid known as glycyrrhetinic acid that is found to be potential compound *in vitro* against *B. malayi* microfilariae with inhibition values LC_{100} at 12.5 μM and IC_{50} at 1.20 μM; however, it was not potential against the adult worms. The glycyrrhetinic acid, which is a synthetic amide analog, is found to have significant potency against filarial worms both *in vivo* and *in vitro* with the removal of a symbiont bacterium called *Wolbachia* which helps parasites' survival [50].

9.4.14 *Hibiscus sabdariffa*

Hibiscus sabdariffa (Family: Malvaceae), native plant distributed in the tropical regions throughout, is used basically as infusion preparation for beverages. Recently, the hydroxy citric acid isolated from *H. sabdariffa* leaf crude extract was assessed *in vitro* against *B. malayi* which is a human filarial parasite. Significant *in vivo* microfilaricidal potential against *B. malayi* was yielded by butanol fraction with 100% activity at 250 μg/ml concentration whereas at 500 mg/kg concentration, only 30% activity [51].

9.4.15 *Lantana camara*

Lantana camara (Family: Verbenaceae), commonly known as Lantana, is a native to the American tropics. It is known that the crude extract prepared from the stem of *L. camara* exhibits a considerable amount of antifilarial activity against *B. malayi* and in sterilization of female worms *in vivo*. Apart from this, the five-day administration of stem extract of *L. camara* resulted in 43% death of adult worms and also 76% of living female worms were sterilized due to the two isolated phytochemical compounds such as oleanonic acid and oleanolic acid. At *in vitro* conditions, most of the adult worms of *B. malayi* were killed at 0.031 mg/ml of extract concentration [52].

9.4.16 *Leucas cephalotes*

Leucas cephalotes (Family: Labiatae), also known as Dronapushpis, is an erect, scaberulous or pubescent, stout, annual weed plant seen in Africa, Asia, and India. The flower and stem extracts from water and alcohol yielded significant inhibition of *S. cervi* filarial parasite mobility. Further, the stem and flower extracts from alcohol yielded labellenic acid (Octadeca – 5,6-dienoic acid) which has shown some changes in motility at earlier stages and complete of adult worm at the later stage. The stem extract has shown irreversible activity whereas the flower extract gave reversible activity [53].

9.4.17 *Mallotus philippensis*

Mallotus philippensis (Family: Euphorbiaceae), known as the Kamala tree, is a widely distributed perennial shrub or small tree in tropical and subtropical regions. The leaves of *M. philippensis* contain compounds such as phorbic acid and bergeni. The aqueous and alcoholic extracts of the leaves have shown to have potential antifilarial effect against *S. cervi* (LC_{50} and LC_{90} were 18 and 20 ng/ml for aqueous extract; 12 and 15 ng/ml for alcoholic extract) because of changes in permeability of cell membrane [54].

9.4.18 *Morinda citrifolia*

Morinda citrifolia (Family: Rubiaceae) is a plant that is widely distributed in and around volcanic terrains, lava-strewn coasts, and in clearings or lime stone outcrops. The medicinal applications of this plant extracts span over all kinds of ailments involving both modern and complementary medicines. The fruit extract of *M. citrifolia* constitutes of β-D-glucopyranose pentaacetate, ascorbic acid, asperuloside tetraacetate, ethyl caprylate, hexanoic acid, octanoic acid, and caprylic acid. As per the *in vitro* study on the filaricidal activity of *M. citrifolia* against *W. bancrofti*, it was depicted that on the supplement of the culture media consisting of microfilariae *W. bancrofti* with the fruit extract results in the death of parasites in about 20 hours in comparison to the group without extract (control group), which lived till 60 hours [55].

9.4.19 *Moringa oleifera*

Moringa oleifera (Family: Moringaceae), also known as Drumstick tree, is indigenous to South Asia. The gum extract (125 mg/ml) containing galacturonic acid, galactose, and arabinose is known to cause permanent damage in microfilarial motility and also 56% of MTT inhibition in *B. malayi* female adult worms. *In vivo* primary screening of extract resulted in about 69% mortality of adult worms and sterilization of 83% female worms. In experiment using *Mastomys coucha,* about 44% of *B. malayi* worms were destroyed [56].

9.4.20 *Plumbago indica*

Plumbago indica (Family: Plumbaginaceae), also known as the Indian leadwort, is a plant that is native to Southeast Asia, Indonesia, the Philippines, and Southern China. The plant *P. indica* was analyzed for antifilarial activity against *S. digitata* using methanol extract of root. Also, immobilization of worms was achieved at 0.0006 mg/ml by the compound 5-hydroxy-2-methyl-1,4-naphthalenedione also known as 5-hydroxy-2-methyl-1,4-naphthoquinone isolated from *P. indica* [57].

9.4.21 *Pongamia pinnata*

Pongamia pinnata (Family: Fabaceae) is native to southern Japan and China, Southeast Asia, Northern Australia and grows on a variety of land types from dry, rocky sea shores to island waterway shores in tropical and subtropical regions. The fruit of *P. pinnata* contains three important compounds such as glucosides, pongamosides A–C, and pongamoside D. The inhibitory effect was studied in *S. cervi* worm motility using fruit aqueous and alcohol extracts at a concentration of 250 and 120 μg/ml, respectively. Further, both the extracts were analyzed for their consequent outcome in the nerve–muscle interaction [58].

9.4.22 *Psoralea corylifolia*

Psoralea corylifolia (Family: Fabaceae), also known as Babchi, is an annual plant found throughout India. Qamaruddin et al. [21] reported that the aqueous and alcohol extracts of the leaves and seeds containing psoralen and isopsoralen of *P. corylifolia* possessed significant antifilarial activity against *S. cervi*. The extracts yielded the spontaneous motility inhibition of worm and also *S. cervi* nerve muscle preparation. For both the alcohol extracts of leaves and seeds, 160 and 30 μg/ml are the concentrations which resulted in the motility of worms [21].

9.4.23 *Ricinus communis*

Ricinus communis (Family: Euphorbiaceae) is normally seen in the temperate and tropical parts. However, it was observed that the efficacy of the filaricidal effect of organic solvent extract containing stearic acids, ricinoleic acid, and isoricinoleic acid of *R. communis* seeds against filarial parasite *B. malayi*. Also, Nisha et al. [46] research on the extract of seed from *R. communis* using 1 mg/ml ethanolic fraction resulted in the control of growth of *S. digitata* microfilaria in 1 hour. Based on the motility of worms and MTT assay, it was noticed that in four hours of exposure, the seed extract resulted in 72.39% of inhibition in the growth of worms [46].

9.4.24 *Saxifraga stracheyion*

Saxifraga stracheyion (Family: Saxifragaceae) is a plant species found commonly around the Western Himalayas, Afghanistan, and Tajikistan. It was found that *S. stracheyion* root extract has some important components (+) catechin-3-gallate, b-sitosterol, and bergenin. Through various studies, it was reported that the root extracts using aqueous and alcohol inhibited *S. cervi* worm motility significantly which was analyzed by upsurge of amplitude and also decrease in rate of contraction. At 30 μg/ml of water extract and 20 μg/ml of alcoholic extract concentration, there was a significant effect in nerve–muscle preparation which suggests the permeability barrier in the cuticle [59].

9.4.25 *Sphaeranthus indicus*

Sphaeranthus indicus (Family: Asteraceae), also known as East Indian Globe, is a herb with a wide range of benefits that is used for the treatment of pruritus, edema, arthritis, filariasis, gout, and cervical adenopathy [60]. Apart from this, a novel isoflavone glycoside, 5,4′-dimethoxy-3′-prenylbiochanin 7-*O*-β-D-galactoside, has been isolated from leaves of *S. indicus* which has hindered about 61.20% in formazan formation against *S. digitata* after four hours of exposure at 1 mg/ml concentration [60].

9.4.26 *Streblus asper*

Streblus asper (Family: Moraceae), commonly known as Sand paper tree, is indigenous to many tropical countries. It has been reported that the movement of the adult worm was hindered significantly by *S. asper* water and alcohol extracts along with decrease in the nerve–muscle preparation with respect to amplitude and also rate of contraction *in vitro*. Further results in killing of the microfilariae where LC_{50} was 90 ng/ml and LC_{90} was 33.5 ng/ml [61]. However, the *in vitro* and *in vivo* activity of the aqueous extract of the stem consisting of asperoside and strebloside demonstrated considerable antifilarial activity against *B. malayi*, thereby killing the worms by inhibiting the motility and glucose uptake of the parasites [62].

9.4.27 *Trachyspermum ammi*

Trachyspermum ammi (Family: Apiaceae) is found in Egypt as a native species and mostly as cultivated plant in India, Iraq, Afghanistan, Iran, and Pakistan. The research focusing on MTT assay with respect to worm movement and reduction of formazan development was done on *S. digitata*. The results obtained showed that the phenolic monoterpene isolated from fraction hexane: ethyl acetate had adulticidal potential. The isolated component was further analyzed in *B. malayi* using *M. coucha* rodent *in vivo*. The results showed good activity with respect to female sterility and also macrofilaricidal [63].

9.4.28 *Vitex negundo*

Vitex negundo (Family: Verbenaceae), commonly known as the Chinese chaste tree, is native to tropical Eastern and Southern Africa and Asia. In an *in vitro* study, the root extract of *V. negundo* having phyto-constituents such as negundoside, agnuside, and vitegnoside exhibited reduction in microfilarial movement in *B. malayi* at a dose of 100 ng/ml after incubation for about 48 hours [64]. However, the extract had a significant effect in a dose-dependent manner; the treated worms were completely immobilized. In the MTT assay, when the extract was given to the adult worms, there was considerable influence in viability due to reduction of MTT into formazan. The changes occur usually by the reduction in enzymatic levels of mitochondria at the cellular point [65].

9.4.29 *Xylocarpus granatum*

Xylocarpus granatum (Family: Meliaceae), commonly known as the Puzzlenut tree, is native to the tropical and subtropical western Indo-Pacific region. Gedunin and photogedunin are the two major phytochemicals isolated from the fruits of *X. granatum* that exhibited both *in vitro* and *in vivo* anti-filaricidal activity with an IC_{50} value of 0.239 and 0.213 μg/ml, respectively, resulting in 70–80% inhibition of the adult *B. malayi* [66].

9.4.30 *Zingiber officinale*

Zingiber officinale (Family: Zingiberaceae), commonly known as Ginger, have originated as part of the ground flora of tropical lowland forests of Southeast Asia or India. The rhizome consists of essential oils as well as limonene, borneol, β-elemene, geraniol, linalool, and zingiberol. However, the alcoholic extract of *Z. officinale* rhizomes at 100 mg/kg reduces the microfilarial concentration by a maximum of 98% in dogs that were naturally infected with *D. immitis* [67].

9.5 Future Perspective

Filariasis still persists as one of the chief infections among the public causing socioeconomic problems across the globe. Even though a lot of work is being undertaken to downgrade the infection load and also to eradicate the disease by means of programmes which administer the drug in mass regionally. The major drawback is drug resistance and side effects. However, massive challenge is ought to be tackled in incapacitating these problems. Lot of research we mentioned above is already done to prevent the disease; more research on high-throughput technologies need to be employed in validating the medicinal plants to be used as a drug to cure the filariasis infection.

9.6 Conclusion

This chapter emphasizes on the role of natural or plant-derived products, particularly phyto-constituents, used in curing the lymphatic filariasis. Some of the chemically synthesized drugs that are used as therapeutic agents to manage the disease were found to contain few carcinogenic impurities as well as show some adverse effects which are the major drawbacks preventing their usage. However, the natural plant-derived phyto-constituents are found to be valuable agents for parasites causing filariasis. This kind of natural medication provides an avenue for therapeutic treatment with plant materials that are easily available and assessable. Thus, the conventional drugs with respect to safety, efficacy, and also cost have significant potency to be used as an alternative in the treatment of lymphatic filariasis.

References

1 World Health Organization (2010). Progress report 2000–2009 and strategic plan 2010–2020 of the global programme to eliminate lymphatic filariasis: halfway towards eliminating lymphatic filariasis. No. WHO/HTM/NTD/PCT/2010.6.
2 Taylor, M.J., Hoerauf, A., and Bockarie, M. (2010). Lymphatic filariasis and onchocerciasis. *Lancet* 376 (9747): 1175–1185. https://doi.org/10.1016/S0140-6736(10)60586-7. Epub 2010 Aug 23.
3 Ndjonka, D., Rapado, L.N., Silber, A.M. et al. (2013). Natural products as a source for treating neglected parasitic diseases. *Int. J. Mol. Sci.* 14 (2): 3395–3439. https://doi.org/10.3390/ijms14023395. Published online 2013 Feb 6.
4 Bockarie, M.J., Pedersen, E.M., White, G.B., and Michael, E. (2009). Role of vector control in the global program to eliminate lymphatic filariasis. *Annu. Rev. Entomol.* 54: 469–487. https://doi.org/10.1146/annurev.ento.54.110807.090626.
5 Dreyer, G. and Piessens, W.F. (2000). Worms and microorganisms can cause lymphatic disease in residents of filariasis-endemic areas. In: *Tropical Medicine: Science and Practice Lymphatic Filariasis* (ed. B.T. Nutman), 245–264. World Scientific Publishing https://doi.org/10.1142/9781848160866_0010.
6 Ottesen, E.A. (2000). The global programme to eliminate lymphatic filariasis. *Trop. Med. Int. Health* 5 (9): 591–594. https://doi.org/10.1046/j.1365-3156.2000.00620.x.
7 Gyapong, J.O., Gyapong, M., and Adjei, S. (1996). The epidemiology of acute adenolymphangitis due to lymphatic filariasis in northern Ghana. *Am. J. Trop. Med. Hyg.* 54 (6): 591–595. https://doi.org/10.4269/ajtmh.1996.54.591.
8 World Health Organization (2009). Global programme to eliminate lymphatic filariasis. *Wkly Epidemiol. Rec.* 84: 437–444.

9 World Health Organization (2012). Provisional strategy for interrupting lymphatic filariasis transmission in loiasis-endemic countries: report of the meeting on lymphatic filariasis, malaria and integrated vector management, Accra, Ghana, 5–9 March, No. WHO/HTM/NTD/PCT/2012.6.

10 Singh, G. and Urhekar, A.D. (2013). Advanced techniques for detection of filariasis—a review. *Int. J. Res. Stud. Biosci.* 2013, 1 (1): 17–22. www.arcjournals.org.

11 Weil, G.J., Lammie, P.J., and Weiss, N. (1997). The ICT filariasis test: a rapid-format antigen test for diagnosis of bancroftian filariasis. *Parasitol. Today* 13 (10): 401–404. https://doi.org/10.1016/s0169-4758(97)01130-7.

12 Morehead, C. (1860). *Clinical Researches on Disease in India*, vol. 2, 1–824. Longman, Green, Longman, and Roberts.

13 Nutman, T.B. (2001). Lymphatic filariasis: new insights and prospects for control. *Curr. Opin. Infect. Dis.* 14 (5): 539–546. https://doi.org/10.1097/00001432-200110000-00006.

14 Molyneux, D. (2003). Lymphatic filariasis (elephantiasis) elimination: a public health success and development opportunity. *Filaria J.* 2 (1): 13. https://doi.org/10.1186/1475-2883-2-13.

15 Maizels, R.M. and Denham, D.A. (1992). Diethylcarbamazine (DEC): immunopharmacological interactions of an anti-filarial drug. *Parasitology* 105 (Suppl): S49–S60. https://doi.org/10.1017/s0031182000075351.

16 Francis, H., Awadzi, K., and Ottesen, E.A. (1985). The Mazzotti reaction following treatment of onchocerciasis with diethylcarbamazine: clinical severity as a function of infection intensity. *Am. J. Trop. Med. Hyg.* 34 (3): 529–536. https://doi.org/10.4269/ajtmh.1985.34.529.

17 Serbus, L.R., Landmann, F., Bray, W.M. et al. (2012). A cell-based screen reveals that the albendazole metabolite, albendazole sulfone targets, Wolbachia. *PLoS Pathog.* 8 (9): e1002922. https://doi.org/10.1371/journal.ppat.1002922. Published online 2012 Sep 20.

18 Babu, B.V., Rath, K., Kerketta, A.S. et al. (2006). Adverse reactions following mass drug administration during the programme to eliminate lymphatic filariasis in Orissa state, India. *Trans. R. Soc. Trop. Med. Hyg.* 100 (5): 464–469. https://doi.org/10.1016/j.trstmh.2005.07.016. Published online 2006 May 01.

19 Crump, A. and Omura, S. (2011). Ivermectin,'wonder drug'from Japan: the human use perspective. *Proc. Jpn. Acad. Ser. B Phys. Biol. Sci.* 87 (2): 13–28. https://doi.org/10.2183/pjab.87.13. Published online 2011 Feb 10.

20 Yates, D.M. and Wolstenholme, A.J. (2016). Dirofilaria immitis: identification of a novel ligand-gated ion channel-related polypeptide. *Int. J. Parasitol. Drugs Drug Resist.* 6 (3): 343–355. https://doi.org/10.1016/j.ijpddr.2016.06.003. Published online 2016 Jul 1.

21 Qamaruddin, A., Parveen, N., Khan, N.U., and Singhal, K.C. (2002). Potential antifilarial activity of the leaves and seeds extracts of *Psoralea corylifolia* on cattle filarial parasite *Setaria cervi*. *J. Ethnopharmacol.* 82 (1): 23–28. https://doi.org/10.1016/s0378-8741(02)00141-1.

22 Dhananjeyan, M.R., Milev, Y.P., Kron, M.A., and Nair, M.G. (2005). Synthesis and activity of substituted anthraquinones against a human filarial parasite, *Brugia malayi*. *J. Med. Chem.* 48 (8): 2822–2830. https://doi.org/10.1021/jm0492655.

23 Zaridah, M.Z., Idid, S.Z., Omar, A.W., and Khozirah, S. (2001). *In vitro* antifilarial effects of three plant species against adult worms of subperiodic *Brugia malayi*. *J. Ethnopharmacol.* 78 (1): 79–84. https://doi.org/10.1016/s0378-8741(01)00286-0.

24 Ghosh, N.K., Babu, S.S., Sukul, N.C., and Ito, A. (1996). Cestocidal activity of *Acacia auriculiformis*. *J. Helminthol.* 70 (2): 171–172. https://doi.org/10.1017/s0022149x00015340.

25 Chakraborty, T., Sinhababu, S.P., and Sukul, N.C. (1995). Antifilarial effect of a plant *Acacia auriculiformis* on canine dirofilariasis. *Trop. Med.* 37 (1): 35–37. https://jglobal.jst.go.jp/en/detail?JGLOBAL_ID=200902128365887185.

26 Babu, S.P.S., Sarkar, D., Ghosh, N.K. et al. (1997). Enhancement of membrane damage by saponins isolated from *Acacia auriculiformis*. *Jpn. J. Pharmacol.* 75 (4): 451–454. https://doi.org/10.1254/jjp.75.451.

27 Nandi, B., Roy, S., Bhattacharya, S., and Sinha Babu, S.P. (2004). Free radicals mediated membrane damage by the saponins acaciaside A and acaciaside B. 18 (3): 191–194. https://doi.org/10.1002/ptr.1387. Published online 2004 April 15.

28 Sahare, K.N., Anandhraman, V., Meshram, V.G. et al. (2008). Anti-microfilarial activity of methanolic extract of *Vitex negundo* and *Aegle marmelos* and their phytochemical analysis. *Indian J. Exp. Biol.* 46 (2): 128–131. http://nopr.niscair.res.in/handle/123456789/4443.

29 Sharma, R.D., Veerpathran, A.R., Dakshinamoorthy, G. et al. (2010). Possible implication of oxidative stress in anti filarial effect of certain traditionally used medicinal plants *in vitro* against *Brugia malayi* microfilariae. *Pharm. Res.* 2 (6): 350–354. https://doi.org/10.4103/0974-8490.75453.

30 Yadav, D., Kushwaha, V., Saxena, K. et al. (2013). Diarylheptanoid compounds from *Alnus nepalensis* express *in vitro* and *in vivo* antifilarial activity. *Acta Trop.* 128 (3): 509–517. https://doi.org/10.1016/j.actatropica.2013.07.015.

31 Yadav, D., Singh, S.C., Verma, R.K. et al. (2012). Antifilarial diarylheptanoids from *Alnus nepalensis* leaves growing in high altitude areas of Uttarakhand, India. *Phytomed. Int. J. Phyto Phytopharmacol.* 20 (2): 124–132. https://doi.org/10.1016/j.phymed.2012.10.017.

32 Akbar, S. (2011). *Andrographis paniculata*: a review of pharmacological activities and clinical effects. *Altern. Med. Rev.* 16 (1): 66–77.

33 Dutta, A. and Sukul, N.C. (1982). Filaricidal properties of a wild herb, *Andrographis paniculata* [on *Dipetalonema reconditum* from a dog]. *J. Helminthol.* 56 (20): 81–84.

34 Koh, H.L., Kian, C.T., and Tan, C.H. (2009). *Guide to Medicinal Plants—A: An Illustrated Scientific and Medicinal Approach*, 1–297. Hackensack, NJ, USA: World Scientific Publishing Co, Inc https://doi.org/10.1080/10496470903379027. Published online 14 Dec 2009.

35 Al-Rofaai, A., Rahman, W.A., Sulaiman, S.F., and Yahaya, Z.S. (2012). *In vitro* activity of neem (*Azadirachta indica*) and cassava (*Manihot esculenta*) on three pre-parasitic stages of susceptible and resistant strains of *Teladorsagia* (Ostertagia) *circumcincta*. *Vet. Parasitol.* 188 (1–2): 85–92. https://doi.org/10.1016/j.vetpar.2012.03.002.

36 Mishra, V., Parveen, N., Singhal, K.C., and Khan, N.U. (2005). Antifilarial activity of Azadirachta indica on cattle filarial parasite *Setaria cervi*. *Fitoterapia* 76 (1): 54–61. https://doi.org/10.1016/j.fitote.2004.10.010.

37 Sashidhara, K.V., Singh, S.P., Misra, S. et al. (2012). Galactolipids from *Bauhinia racemosa* as a new class of antifilarial agents against human lymphatic filarial parasite, *Brugia malayi*. *Eur. J. Med. Chem.* 50: 230–235. https://doi.org/10.1016/j.ejmech.2012.01.057. Published online 2012 Feb 3.

38 Mukerji, A.K. and Bhaduri, N.V. (1947). The treatment of intestinal worms with the indigenous drugs Butea, Embelia and Kamala. *Ind. Med. Gaz.* 82 (2): 66–69.

39 Deshmukh, M., Sahare, K.N., Patidar, R.K. et al. Antifilarial activity of *Butea monosperma* L. leaves extracts against *Setaria cervi*. *Trends Vector Res. Parasitol.* 1: 1–5. https://doi.org/10.7243/2054-9881-1-1.

40 Sahare, K.N., Anandhraman, V., Meshram, V.G. et al. (2012). Antifilarial potential of *Butea monosperma* L. against microfilaria *in vitro*. *Int. J. Pharm. Tech. Res.* 4 (3): 1181–1184.

41 Sahare, K.N., Anandharaman, V., Meshram, V.G. et al. *In vitro* effect of four herbal plants on the motility of *Brugia malayi* microfilariae. *Ind. J. Med. Res.* 127 (5): 467–472. Published online 2011 Feb 25.

42 Datte, J.Y., Traore, A., Offoumou, A.M., and Ziegler, A. (1998). Effects of leaf extract of *Caesalpinia bonduc* (Caesalpiniaceae) on the contractile activity of uterine smooth muscle of pregnant rats. *J. Ethnopharmacol.* 60 (2): 149–155. https://doi.org/10.1016/s0378-8741(97)00144-x.

43 Gaur, R.L., Sahoo, M.K., Dixit, S. et al. (2008). Antifilarial activity of *Caesalpinia bonducella* against experimental filarial infections. *Ind. J. Med. Res.* 128 (1): 65–70. Published online 2011 Feb 25.

44 Deepan, T., Alekhya, V., Saravanakumar, P., and Dhanaraju, M.D. (2012). Phytochemical and anti-microbial studies on the leaves extracts of *Cardiospermum halicacabum* Linn. *Adv. Biol. Res.* 6 (1): 14–18. https://doi.org/10.5829/idosi.abr.2012.6.1.56393.

45 Khunkitti, W., Fujimaki, Y., and Aoki, Y. (2000). *In vitro* antifilarial activity of extracts of the medicinal plant *Cardiospermum halicacabum* against *Brugia pahangi*. *J. Helminthol.* 74 (3): 241–246. https://doi.org/10.1017/s0022149x00000342.

46 Nisha, M., Kalyanasundaram, M., Paily, K.P. et al. (2007). *In vitro* screening of medicinal plant extracts for macrofilaricidal activity. *Parasitol. Res.* 100: 575–579.

47 Zou, J.H., Dai, J., Chen, X., and Yuan, J.Q. (2006 Jun). Pentacyclic triterpenoids from leaves of *Excoecaria agallocha*. *Chem. Pharm. Bull.* 54 (6): 920–921. https://doi.org/10.1248/cpb.54.920.

48 Patra, J.K., Mohapatra, A.D., Rath, S.K. et al. (2009). Screening of antioxidant and antifilarial activity of leaf extracts of *Excoecaria agallocha* L. *Int. J. Integrat. Biol.* 7 (1): 9–15.

49 Mishra, V., Khan, N.U., and Singhal, K.C. (2005). Potential antifilarial activity of fruit extracts of *Ficus racemosa Linn* against *Setaria cervi in vitro*. *Ind. J. Exp. Biol.* 43 (4): 346–350.

50 Kalani, K., Kushwaha, V., Verma, R. et al. (2013). Glycyrrhetinic acid and its analogs: a new class of antifilarial agents. *Bioorg. Med. Chem. Lett.* 23 (9): 2566–2570. https://doi.org/10.1016/j.bmcl.2013.02.115.

51 Saxena, K., Dube, V., Kushwaha, V. et al. Antifilarial efficacy of *Hibiscus sabdariffa* on lymphatic filarial parasite *Brugia malayi*. *Med. Chem. Res.* 20 (9): 1594–1602. https://doi.org/10.1007/s00044-010-9419-x.

52 Misra, N., Sharma, M., Raj, K. et al. (2007). Chemical constituents and antifilarial activity of *Lantana camara* against human lymphatic filariid *Brugia malayi* and rodent filariid *Acanthocheilonema viteae* maintained in rodent hosts. *Parasitol. Res.* 100 (103): 439–448. https://doi.org/10.1007/s00436-006-0312-y.

53 Parveen, N., Khan, N.U., and Singhal, K.C. (2002). *In vitro* antifilarial potential of the flower and stem extracts of *Leucas cephalotes* on cattle filarial parasite *Setaria cervi*. *J. Nat. Rem.* 2 (2): 155–163.

54 Singh, R., Singhal, K.C., and Khan, N.U. (1997). Antifilarial activity of *Mallotus philippensis* Lam. on *Setaria cervi* (Nematoda: *Filarioidea*) *in vitro*. *Ind. J. Physiol. Pharmacol.* 41 (4): 397–403.

55 Al-Abd, N.M., Nor, Z.M., Al-Adhroey, A.H. et al. (2013). Recent advances on the use of biochemical extracts as filaricidal agents. *Evid.-based Comp. Alternat. Med.* 986573 https://doi.org/10.1155/2013/986573.

56 Kushwaha, V., Saxena, K., Verma, S.K. et al. (2011). Antifilarial activity of gum from *Moringa oleifera* Lam. on human lymphatic filaria *Brugia malayi*. *Chron. Young Scient.* 2 (4): 201. https://doi.org/10.4103/2229-5186.93025.

57 Mathew, N., Paily, K.P., Vanamail, P. et al. (2002). Macrofilaricidal activity of the plant *Plumbago indica/rosea in vitro*. *Drug Dev. Res.* 56 (1): 33–39. https://doi.org/10.1002/ddr.10056.

58 Uddin, Q., Parveen, N., Khan, N.U., and Singhal, K.C. (2003). Antifilarial potential of the fruits and leaves extracts of *Pongamia pinnata* on cattle filarial parasite *Setaria cervi*. *Phytother. Res.* 17 (9): 1104–1107. https://doi.org/10.1002/ptr.1269.

59 Singh, R., Singhal, K.C., and Khan, N.U. (2000). Exploration of antifilarial potential and possible mechanism of action of the root extracts of *Saxifraga stracheyion* on cattle filarial parasite *Setaria cervi*. *Phytother. Res.* 14 (1): 63–66. https://doi.org/10.1002/(sici)1099-1573(200002)14:1<63::aid-ptr567>3.0.co;2-5.

60 Galani, V.J., Patel, B.G., and Rana, D.G. (2010). *Sphaeranthus indicus* Linn.: a phytopharmacological review. *Int. J. Ayurveda Res.* 1 (4): 247–253. https://doi.org/10.4103/0974-7788.76790.

61 Parveen, N., Singhal, K.C., Khan, N.U., and Singhal, P. (1989). Potential antifilarial activity of *Streblus asper* against *Setaria cervi* (Nematoda: Filarioidea). *Ind. J. Pharamcol.* 21 (2): 16–21.

62 Chatterjee, R.K., Fatma, N., Murthy, P.K. et al. Macrofilaricidal activity of the stembark of *Streblus asper* and its major active constituents. *Drug Develop Res.* 26 (1): 67–78. https://doi.org/10.1002/ddr.430260106.

63 Mathew, N., Misra-Bhattacharya, S., Perumal, V., and Muthuswamy, K. (2008). Antifilarial lead molecules isolated from *Trachyspermum ammi*. *Molecules* 13 (9): 2156–2168. https://doi.org/10.3390/molecules13092156.

64 Parveen, N. (1991). Antifilarial activity of *Vitex negundo* against *Setaria cervi*. *Fitoterapia* 62 (2): 163–165.

65 Sahare, K.N. and Singh, V. (2013). Antifilarial activity of ethyl acetate extract of *Vitex negundo* leaves *in vitro*. *Asian Pac. J. Trop. Med.* 6 (9): 689–692. https://doi.org/10.1016/S1995-7645(13)60119-4.

66 Misra, S., Verma, M., Mishra, S.K. et al. (2011). Gedunin and photogedunin of *Xylocarpus granatum* possess antifilarial activity against human lymphatic filarial parasite *Brugia malayi* in experimental rodent host. *Parasitol. Res.* 109 (5): 1351–1360. https://doi.org/10.1007/s00436-011-2380-x.

67 Datta, A. and Sukul, N.C. (1987). Antifilarial effect of *Zingiber officinale* on *Dirofilaria immitis*. *J. Helminthol.* 61 (3): 268–270. https://doi.org/10.1017/s0022149x00010142.

10

Dracunculiasis (Guinea Worm Disease) and Phytochemicals in Drug Discovery

Muhammad Akram[1], *Chukwuebuka Egbuna*[2,3], *Mehwish Iqbal*[4], *Zarrin Basharat*[5], *Mithun Rudrapal*[6], *Kingsley C. Patrick-Iwuanyanwu*[3,7], *Johra Khan*[8,9], *and Chukwuemelie Zedech Uche*[10]

[1] *Department of Eastern Medicine and Surgery, Government College University Faisalabad, Faisalabad, Pakistan*
[2] *Department of Biochemistry, Faculty of Natural Sciences, Chukwuemeka Odumegwu Ojukwu University, Uli, Nigeria*
[3] *Nutritional Biochemistry and Toxicology Unit, World Bank Africa Centre of Excellence, Centre for Public Health and Toxicological Research (PUTOR), University of Port Harcourt, Port Harcourt, Nigeria*
[4] *Institute of Public Health, Dow University of Health Sciences, Karachi, Pakistan*
[5] *Jamil-ur-Rahman Center for Genome Research, Dr. Panjwani Center for Molecular Medicine and Drug Research, International Center for Chemical and Biological Sciences, University of Karachi, Karachi, Pakistan*
[6] *Department of Pharmaceutical Chemistry, Rasiklal M. Dhariwal Institute of Pharmaceutical Education and Research, Chinchwad, Pune, Maharashtra, India*
[7] *Department of Biochemistry, Faculty of Science, University of Port Harcourt, Port Harcourt, Nigeria*
[8] *Department of Medical Laboratory Sciences, College of Applied Medical Sciences, Majmaah University, Saudi Arabia*
[9] *Health and Basic Sciences Research Center, Majmaah University, Majmaah, Saudi Arabia*
[10] *Department of Medical Biochemistry and Molecular Biology, Faculty of Basic Medical Sciences, University of Nigeria, Enugu Campus, Nigeria*

List of Abbreviations

CDC	Centers for Disease Control and Prevention
GWD	Guinea worm disease
WHO	World Health Organization

10.1 Introduction

Guinea worm disease (GWD), also known as dracunculiasis, is an agonizing and incapacitating disease caused by the nematode parasite, *Dracunculiasis medinensis* L. GWD is a disease mostly found among socioeconomically

Neglected Tropical Diseases and Phytochemicals in Drug Discovery, First Edition.
Edited by Chukwuebuka Egbuna, Muhammad Akram, and Jonathan Chinenye Ifemeje.

deprived people and countryside villagers, particularly in Saharan African nations that do not have proper access to potable drinking water. The disease has a little fatality rate, but causes a lot of discomforts typically in patients with secondary contamination and then leaving the spot of the worm susceptible to tetanus or sepsis (Figure 10.1), with its accompanying socioeconomic burden [1–6].

James Africanus Horton, a West African physician, described the infection [7] but erroneously assumed that it affects through the feet. The harmful water was identified earlier as the source of infection. However, in the year 1870, a Russian natural scientist, Alexei Fedchenko [8] established that larvae driven out from appearing female parasites in the extremities of patients growing in freshwater copepods or existing in pools are subsequently swallowed through drinking water. An adult female *D. medinensis* is among the extensive nematodes estimated to be 100 cm long and 1–2 mm broad. The external part of the genitalia of the female parasite is central throughout the body. In females of parasitic nematodes, the microfilariae newly hatch immature larvae appearing from this cavity through the existence of the parasite. In Guinea worm, though it is sealed with a plug, moreover, the entire cavity of the body is loaded by the uterus, which expands both anteriorly as well as dorsally and encloses from one to three million. The gastrointestinal tract is too comprehensively horizontal and nonpurposeful. Intense red parasites have been reported in Pakistan [9]; however, their consequence is unknown.

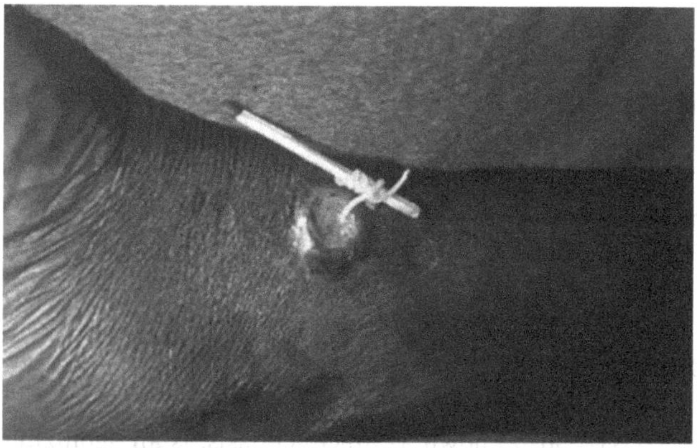

Figure 10.1 Dracunculosis/GWD. A method used to extract a Guinea worm from the leg vein of a human patient. *Source:* Adapted from Centers for Disease Control and Prevention - DPDx; Global Health, Division of Parasitic Diseases. Freely taken from https://www.cdc.gov/parasites/Guineaworm/gwep.html.

10.2 Historical Insights of Dracunculiasis

Dracunculiasis was reported primarily in the fifteenth century BC in Egypt. An explanation of the infection is given in the "Turin Papyrus," which expresses the very old fallacy of the Egyptian sun supernatural being. These records were recently endorsed through the pathological assessment of an Egyptian mummy in which a calcified parasite was recognized as *D. medinensis* [10]. Similar records have been made in or to have come into being in eleventh century anno Domini. It was also established in Persia and Israel. In 1980, one of the objectives put forward for the global supply of water and hygiene up to 10 years (1981–1990), by the WHO in association with the UN, was the elimination of GWD [11]. According to the CDC, the 44th global Assembly of Health conducted in May 1991 supported the objective of eliminating GWD by the end of 1995 [12]. This objective has been attained in many regions of the world, including Pakistan (last patient recorded in 1993), India (last patient confirmed in 1997), and Iran (last patient confirmed in 1970) [13, 14].

10.3 Prevalence, Burden, and Distribution of Disease

In the year 1947, over 48 000 000 inhabitants were approximated to be infected by GWD in India, Middle East, and Africa [15]. Throughout the 1970s, the ailment was reported in India, the Islamic states of Pakistan, Iran, Yemen, Saudi Arabia, East, and West Africa. In 1976, the WHO approximated a frequency of 10 000 000 dracunculiasis patients worldwide [16–18]. The infection was prevalent at the commencement of the twentieth century. In the 1980s, the transmission of the disease was limited to only 20 countries. By the end of 2009, only 4 countries namely Mali, Ethiopia, Ghana, and Sudan had aboriginal cases, and the yearly frequency was 3190 cases contrasted to 3.5 million in 1986. Sudan was alone accounted for 86% of total clinical cases reported in 2009. In recent years, the largest burden of Guinea worm infection has moved in Ghana, Sudan, and Nigeria. These three countries collectively account for 93% of total infective cases globally, with Sudan accounting for 73% of the total cases.

10.4 Life Cycle, Pathogenesis, and Clinical Manifestations of Dracunculiasis

The parasite *D. medinensis* comes under the category of Spirurida. It produces eggs holding larvae or discharge-free larvae and needs arthropods as transitional hosts (Figure 10.2). The grown-up female Guinea worm is one of the extensive

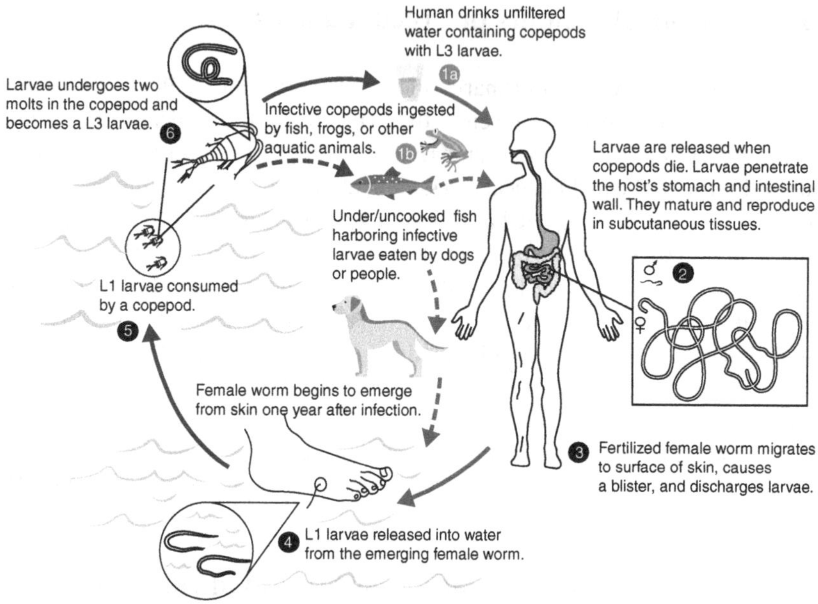

Figure 10.2 The life cycle of *Dracunculiasis medinensis*. *Source:* Department of Health and Human Services [20].

nematodes, counting up to 1 m in length and is only 1–2 mm thick. There is no identified reservoir of infection for animals, which makes elimination much more possible if harmless drinking water can be assured. People turn out to be infected when they drink water containing copepods or pathogenic larvae. Human beings are the main authoritative host, whereas contaminations have also been confirmed in a range of animals ranging from herbivores, monkeys, household as well as undomesticated Felidae and Canidae [19]. The swallowed copepods are destroyed by the digestive fluids of the stomach. The liberated larvae then pass into the small intestine, where they pierce the wall of the intestine and move to the connective tissues of the thoracic cavity and the abdominal wall. The larvae of both females and males grow fully and mate about 60–90 days after contamination. The male parasite passes away soon after breeding, while the female growth takes up to 10–14 months. In this process, the female parasite initiates an excruciating eruption on the skin, typically from the lower limbs; the eruption ruptures, larvae are discharged on exposure with water by the mature female parasite, slowly migrate to the surface of the body, and emerge through the skin (Figure 10.2).

In the first stage (Figure 10.1), the locomotive larvae are swallowed by the copepods in which they go through development to the third larval infective phase in two weeks, consequently being responsible for the transmission cycle [20]. When the concerned parts of the body are inundated in water, the female parasite liberates larvae, which are swallowed by copepods and consequently undergo the life cycle [2]. The parasite measures 23 m with a sharpened tail and a completely produced gut, even though they do not feed. They change positions dynamically in the water, similar to a free-existing parasite (nematode), and can survive for a few days in water. Pre-developing female parasites can move effortlessly throughout the connective tissues; however, as soon as they come out to the surface, a small number of larvae escape under the surface of the skin through a sudden break at the frontal end. The response of the host results in the development of an intense, excruciating blister, which ruptures in a few days to present as a superficial sore followed by a distinct inflammatory reaction in opposition to the epidermis of the whole parasite, averting its elimination. The microbiologically disinfected fluid of eruption encloses larvae which encompasses primarily polymorphonuclear neutrophils amid eosinophils, lymphocytes, and macrophages [21]. In one research in Ghana, 28% of cases had enduring pain 12–18 months after the appearance of parasites and 0.5% had an unending physical disability in the manifestation of "fastened" joints [22]. In a further research in Benin, there was 0.3% fatality caused by tetanus and sepsis in blood [23].

10.5 Prevention and Eradication of Disease

Though no drugs or vaccines exist for the prophylaxis or treatment of dracunculiasis, the cycle of transmission can be targeted at various points. The most preventable method of infection is by staying away from infected sources of drinking water, straining infected water with fabric, and delicate-mesh filters before consumption. The disease is on the edge of being eliminated from the globe, with only 126 patients being accounted for in 2014 [13]. The accomplishment of the worldwide campaigns to wipe out the ailment did not only minimize the number of patients reported yearly, but also diminished the burden of disease. In 2014, a typical of 1.3 parasites per patient were accounted (maximum of four parasites in one patient) in contrast to a typical of 1.4 parasites per patient (as high as nine parasites in one patient) in the year 2013 and in 2012, 1.7 parasites per patient [1, 2, 13, 20, 24, 25]. However, the spread of infection had decreased considerably to 152 814 patients in 1995. Pakistan terminated the spread of the disease in 1993 and Kenya had its final aboriginal patient in 1994. India had broken off the spread of the disease in 1996, followed by Senegal, Cameroon, and Yemen in 1997. In 2004, the 12 enduring prevalent countries promised to eradicate the infection by the year 2009 [10]. In the last period of 2009, 8 of these 12 left behind countries, declared to be free from dracunculiasis. The number of patients turned decreased to 3190 throughout the year with 80% of the patients recorded in the Southern part of Sudan [11].

10.6 Medicinal Plants and Phytochemicals Active Against Dracunculiasis

Medicinal plants are no doubt repositories of both known and unknown chemical compounds referred to as phytochemicals or secondary metabolites [26–35]. Plant extracts or their products have been useful in treating dracunculiasis (Table 10.1), either applied topically or taken by the oral route. The use of *Aloë barteri, Calotropis procera,* and *Cassia occidentalis* against GWD dates back to 1937 [36]. Leaves of *Guiera senegalensis* and several parts of *Annona senegalensis* have been used as vermifuge. The fruit of *A. senegalensis* has been found to be beneficial against Guinea worm [36–38]. Some African traditional medicine plants such as *Combretum mucronatum* and *Jatropha curcas* have been used for treating dracunculiasis [39]. Root decoction of *Combretum micranathum* not only works against Guinea worms in infected patients, but also decreases the inflammation around

Table 10.1 Some medicinal plants and phytochemicals active against dracunculiasis.

Plant	Family	Constituents/bioactive components	References
Combretum mucronatum (Schummach.)	Combretaceae	Procyanidins, cinnamtannins, isovixetin, vixetin, isoorientin, isoquercitrin	[49]
Elaeophorbia drupifera (Thonn.)	Euphorbiaceae	Terpenoids (eupha-8,24-diene-3-ol)	[50]
Jatropha curcas (Linn.)	Euphorbiaceae	Phorbolesters	[51]
Lablab purpureus (Linn.)	Fabaceae	Rotenoids (deguelin, dehydrodeguelin, rotenol, rotenone, tephrosin, and sumatrol)	[46]
Moringa Olifera (Lam.)	Moringaceae	Tannins, saponins moringyne, niazirin, quercetin, kaempferol, β-sitosterol, niazimisin, benzyl isothiocyanate	[52]
Sesbania pachycarpa (DC.)	Fabaceae	Alkaloids, coumarins, flavonoids, tannins, triterpenes, steroids, saponosides	[53]

sores [40, 41]. *Richiea caparoides* and *Stachys parviflora* have also been reported to have nematicidal effect [42, 43]. Leaves of *Elaeophorbia drupifera* are effective against dracunculiasis [44]. The leaves of *E. drupifera* and *Hilleria latifolia* in the form of soup in amalgamation with a palm preparation are also wormicidal [36]. Some other plants such as *Moringa oleifera* Lam. [45], *Sesbania pachycarpa* [46], and *Lablab purpureus* [47, 48] have also found usefulness in dracunculiasis.

10.7 Conclusion

Dracunculiasis is a neglected tropical disease, yet a scary communicable disease. Despite significant development in drug discovery, unsafe drinking water and proper hygiene are some of the major health problems, which cause the disease, particularly in the developing countries, to persist. Though phytotherapy remains to be an effective alternative treatment option, it has been practiced on a limited scale. The best preventive measure to eradicate this communicable disease is the provision of safe clean water and the maintenance of proper hygiene in areas where it is endemic. Since phytotherapy could be a reliable therapeutic option, several plants have been demonstrated to possess therapeutic efficacy against dracunculiasis. Bioactive phytochemicals are, of course, responsible for the therapeutic potential of medicinal plants. However, to increase bioactive chemical space, more medicinal plants need to be screened against dracunculiasis and modern approaches such as plant-derived nanoparticulate formulations can also be tested. Data and information regarding disease burden, eradication strategies, and useful plants should be disseminated systematically and curated in annotated databases.

Acknowledgments

Authors would like to thank the Deanship of Scientific Research at Majmaah University for supporting this work. Authors are also thankful to World Bank, University of Port Harcourt, Majmaah University, National Universities Commission, the Association of African Universities and other research institutes for their support.

References

1 Greenaway, C. (2004). Dracunculiasis (Guinea worm disease). *Can. Med. Assoc.* 170 (4): 495–500.
2 Cairncross, S., Muller, R., and Zagaria, N. (2002). Dracunculiasis (Guinea worm disease) and the eradication initiative. *Clin. Microbiol. Rev.* 15 (2): 223–246.

3 Chippaux, J.P., Banzou, A., and Agbede, K. (1992). Impact social et économique de la dracunculose: une étude longitudinale effectuée dans deux villages du Bénin. *Bull. World Health Organ.* 70 (1): 73–78.

4 Ilegbodu, V.A., Kale, O.O., Wise, R.A. et al. (1986). Impact of Guinea worm disease on children in Nigeria. *Am. J. Trop. Med. Hyg.* 35 (5): 962–964.

5 Nwosu, A.B.C., Ifezulike, E.O., and Anaya, A.O. (1980). Endemic dracontiasis in Anambra State of Nigeria: geographical distribution, clinical features, epidemiology and socio- economic impact of the disease. *Ann. Trop. Med. Parasitol.* 76: 187–200.

6 Smith, G.S., Blum, D., Huttley, S.R.A. et al. (1989). Disability from dracunculiasis: effect on mobility. *Ann. Trop. Med. Parasitol.* 83 (2): 151–158.

7 Watts, S.J., Brieger, W.R., and Yacoob, M. (1989). Guinea worm: an in-depth study of what happens to mothers, families and communities. *Soc. Sci. Med.* 29 (9): 1043–1049.

8 Fedchenko, A.P. (1971). Concerning the structure and reproduction of the Guinea-worm (*Filaria medinensis* L.). *Am. J. Trop. Med. Hyg.* 20: 511–523.

9 Eberhard, M.L., Rab, M.A., and Dilshad, M.N. (1989). Red *Dracunculus medinensis*. *Am. J. Trop. Med. Hyg.* 41: 479–481.

10 World Health Assembly. Eradication of dracunculiasis. In: *Fifty-seventh World Health Assembly*. Geneva: World Health Organization (WHO).

11 World Health Organization (WHO) (2015a). Dracunculiasis eradication: global surveillance summary. *Wkly Epidemiol. Rec.* 82: 133–140.

12 World Health Organization (WHO) (2015b). Dracunculiasis eradication: global surveillance summary. *Wkly Epidemiol. Rec.* 90 (19): 201–215.

13 World Health Organization (WHO) (2015). Dracunculiasis. http://www.who.int/dracunculiasis/disease/disease_more/en/ (accessed 20 July 2015).

14 Markell, E. (1999). *Medical Parasitology*, vol. 354. Philadelphia: W. B. Saunders Company.

15 Stoll, N. (1947). This wormy world. *J. Parasitol.* 33: 1–18.

16 Belcher, D. (1985). Opportunities for control of dracunculiasis: transmission of epidemiology. In: *Workshop on Opportunities for Control of Dracunculiasis* (ed. F.R. Ruskin), 1–7. Washington: National Academy Press.

17 Ministry of Health and Medical Education, Islamic Republic of Iran. 1996 Eradication of Dracunculiasis in Islamic Republic of Iran. Country Report. Directorate General of Disease Prevention and Surveillance, Ministry of Health and Medical Education. Report no.: CTD/96.13.

18 World Health Organization (WHO) (1982). Dracunculiasis surveillance. *Wkly Epidemiol. Rec.* 57: 65–72.

19 Macpherson, C.N., Meslin, F.X., and Wandeler, A. (2012). *Dogs, Zoonoses and Public Health*, 237–239. London: CAB International.

20 Department of Health and Human Services (2003). *Detect Every Case, Contain Every Worm! 8th Meeting of National Coordinators Held in Kampala [Guinea*

Worm Wrap-Up no132]. Atlanta: US Centers for Disease Control and Prevention. http://www.cdc.gov/ncidod/dpd/parasites/Guineaworm/wrapup/word132.pdf (accessed 2004 Jan 12).

21 Muller, R. (1985). Life cycle of dracunculus medinensis. In: *Workshop on Opportunities for Control of Dracunculiasis: Contributed Papers*, 13–17. Washington: National Academy Press.

22 Muller, R. (1976). The pathology of experimental *Dracunculus* infection and its relevance to chemotherapy. In: *Pathophysiology of Parasitic Infection* (ed. E.J.L. Soulsby), 133–148. New York: Academic Press.

23 Hours, M. and Cairncross, S. (1994). Long-term disability due to Guinea worm disease. *Trans. R. Soc. Trop. Med. Hyg.* 88: 559–560.

24 Chippaux, J.P. and Massougbodji, A. (1991). Évaluation clinique et épidémiologique de la dracunculose au Bénin. *Méd. Trop.* 51: 269–274.

25 Biswas, G., Sankara, D., Agua-Agum, J., and Maiga, A. (2013). Dracunculiasis (Guinea worm disease): eradication without a drug or a vaccine. *Philos. Trans. R. Soc. Lond. Ser. B Biol. Sci.* 368: 20120146.

26 Egbuna, C. and Ifemeje, J.C. (2015). Biological functions and anti-nutritional effects of phytochemicals in living system. *IOSR Journal of Pharmacy and Biological Sciences (IOSR-JPBS)* 10 (2): 10–19. iosrjournals.org/iosr-jpbs/papers/Vol10-issue2/. . ./C010231019.pdf.

27 Egbuna, C., Mishra, A., and Goyal, M.R. (eds.) (2020). *Preparation of Phytopharmaceuticals for the Management of Disorders*, 1e. US: Academic Press, Elsevier. ISBN: 9780128202845.

28 Egbuna, C., Kumar, S., Ezzat, S.M. et al. (eds.) (2019). *Phytochemicals as Lead Compounds for New Drug Discovery*, 1e. Cambridge, USA: Elsevier. ISBN: 9780128178911. 378 pages.

29 Shashank, K. and Egbuna, C. (eds.) (2019). *Phytochemistry: In vitro and In Silico Updates*, 1e. Singapore: Springer Nature. ISBN: 978-981-13-6919-3. 589 pages.

30 Egbuna, C., Ifemeje, J.C., Udedi, S.C., and Kumar, S. (eds.) (2018). Phytochemistry. In: *Fundamentals, Methods, and Applications*, 1e, vol. 1. New York: Apple Academic Press/CRC Taylor & Francis. ISBN: 978-1771887595. 684 pages, part of a 3 volume set.

31 Egbuna, C., Kumar, S., Ifemeje, J.C., and Kurhekar, J.V. (eds.) (2018). Phytochemistry. In: *Pharmacognosy, Nanomedicine, and Contemporary Issues*, 1e, vol. 2. New York: Apple Academic Press/CRC Taylor & Francis. ISBN: 978-1771887601. 620 pages, part of a 3 volume set.

32 Egbuna, C., Ifemeje, J.C., Kumar, S., and Sharif, N. (eds.) (2018). Phytochemistry. In: *Marine, Industrial, and Advances*, 1e, vol. 3. New York: Apple Academic Press/CRC Taylor & Francis. ISBN: 978-1771887618. 502 pages, part of a 3 volume set.

33 Srivastav, V.K., Egbuna, C., and Tiwari, M. (2019). Plant secondary metabolites as lead compounds for the production of potent drugs. In: *Phytochemicals as Lead*

Compounds for New Drug Discovery., vol. 2019 (eds. C. Egbuna, S. Kumar, S.M. Ezzat, et al.), 3–13. USA: Elsevier, Cambridge.

34 Ezzat, S.M., Jeevanandam, J., Egbuna, C. et al. (2019). Phytochemicals as sources of drugs. In: *Phytochemistry: In vitro and In Silico Updates* (eds. S. Kumar and C. Egbuna), 3–22. Singapore: Springer Nature.

35 Egbuna, C., Ezzat, S.M., Tijjani, H., and Srivastav, V.K. (2019). Synthetic analogs of phytochemicals. In: *Phytochemistry: In vitro and In Silico Updates* (eds. S. Kumar and C. Egbuna), 23–56. Singapore: Springer Nature.

36 Issaka-Tinogah, A., Magnussen, P., Bloch, P., and Yakubu, A. (1994). Lack of effect of ivermectin on prepatent Guinea-worm: a single-blind, placebo-controlled trial. *Trans. R. Soc. Trop. Med.* 88 (3): 346–348.

37 Dalziel, J.M. (1937). The useful plants of west tropical Africa. *London: The Crown Agents for the Colonies*: 612–623.

38 Ibrahim, M.A., Nwude, N., Ogunsusi, R.A., and Aliu, Y.O. (1984). Screening of West African plants for anthelmintic activity. *ILCA Bull.* 17: 19–23.

39 Oliver, B. (1960). *Medicinal Plants in Nigeria*. Ibadan: Nigerian college of Arts, Science and Engineering.

40 Ampofo, O. (1978). Plants that heal. *Helminthol. Abstr.* 47 (6): 247.

41 Ahmad, V.U., Arshad, S., Bader, S. et al. (2008). New terpenoids from *Stachys parviflora* Benth. *Magn. Reson. Chem.* 46 (10): 986–989.

42 Murthy, P.K. and Joseph, S.K. (2011). Plant products in the treatment and control of filariasis and other helminth infections and assay systems for antifilarial/anthelmintic activity. *Planta Med.* 77 (06): 647–661.

43 Haerdi, F. (1964). Die Eingeborenen-Heilpflanzen des Ulanga-Distriktes Tanganjikas (Ostafrika). *Acta Trop.* Suppl 8: 1–278.

44 Stevens, C.G., Ugese, F.D., Otitoju, G.T., and Baiyeri, K.P. (2015). Proximate and anti-nutritional composition of leaves and seeds of *Moringa oleifera* in Nigeria: a comparative study. *Agro-Science* 14 (2): 9–17.

45 Nacoulma OG (1996). Medicinal plants and traditional medicinal practices in Burkina Faso: the case of the Central Plateau. Tome 1 & Tome 2. Doctoral thesis. University of Ouagadougou, Ouagadougou.

46 Kamal, R. and Mathur, N. (2010). Rotenoids from *Lablab purpureus* L. and their bioefficacy against human disease vectors. *Parasitol. Res.* 107 (6): 1481–1488.

47 Belcher, D.W., Wurapa, F.K., Ward, W.B., and Lourie, I.M. (1975). Guinea worm in southern Ghana: its epidemiology and impact on agricultural productivity. *Am. J. Trop. Med. Hyg.* 24 (2): 243–249.

48 Horton, J.A. (1868). *Guinea Worm, or Dracunculus, Its Symptoms and Progress, Causes, Pathological Anatomy, Results and Radical Cure*. London: W. J. Johnson.

49 Spiegler, V., Sendker, J., Petereit, F. et al. (2015). Bioassay-guided fractionation of a leaf extract from *Combretum mucronatum* with anthelmintic activity: oligomeric procyanidins as the active principle. *Molecules* 20: 14810–14832.

50 Famuyiwa, S.O., Oladele, A.T., Adeloye, A.O., and Fakunle, C.O. (2014). Terpenoid compounds from the latex of *Euphorbia Drupifera*. *IFE J. Sci.* 16 (1): 1–4.

51 Bahmani, M., Rafieian-Kopaei, M., Hassanzadazar, H. et al. (2014). A review on most important herbal and synthetic antihelmintic drugs. *Asian Pac. J. Trop. Biomed.* 7 (Suppl 1): S29–S33.

52 Cabardo, D.E. Jr. and Portugaliza, H.P. (2017). Anthelmintic activity of *Moringa oleifera* seed aqueous and ethanolic extracts against *Haemonchus contortus* eggs and third stage larvae. *Int. J. Vet. Sci. Med.* 5: 30–34.

53 Ouattara, M.B., Bationo, J.H., Kiendrebeogo, M., and Nacoulma, O.G. (2020). Phytochemical screening and acute toxicity evaluation of leaves extract of two Fabaceae's species: *Sesbania pachycarpa* DC. And *Indigofera berhautiana* Gillett. *J. Biosci. Med.* 8 (7): 28–34.

11

Medicinal Potentials of Phytochemicals as a Source of Drugs for the Treatment of Onchocerciasis (River Blindness)

Chandrashekar Srinivasa[1,2], Chandan Shivamallu[3], S. R. Santhosh Kumar[1,2], P. Sushma[3], Shiva Prasad Kollur[4], Seema J. Patel[1,2], S. S. Patil[5], R. Suhas[6], Chukwuebuka Egbuna[7,8], and Bui Thanh Tung[9]

[1] *Department of Biotechnology, Davangere University, Davangere, Karnataka, India*
[2] *Department of Food Technology, Davangere University, Davangere, Karnataka, India*
[3] *Department of Biotechnology and Bioinformatics, School of Life Sciences, JSS Academy of Higher Education and Research, Mysore, Karnataka, India*
[4] *Department of Sciences, Amrita School of Arts and Sciences, Mysore, Karnataka, India*
[5] *Virology Laboratory, ICAR – National Institute of Veterinary Epidemiology and Disease Informatics (NIVEDI), Bangalore, Karnataka, India*
[6] *Postgraduate Department of Chemistry, JSS College of Arts, Commerce and Science, Mysore, Karnataka, India*
[7] *Department of Biochemistry, Faculty of Natural Sciences, Chukwuemeka Odumegwu Ojukwu University, Uli, Nigeria*
[8] *Nutritional Biochemistry and Toxicology Unit, World Bank Africa Centre of Excellence, Centre for Public Health and Toxicological Research (PUTOR), University of Port-Harcourt, Port Harcourt, Nigeria*
[9] *VNU School of Medicine and Pharmacy, Vietnam National University, Hanoi, Vietnam*

List of Abbreviations

NTD	Neglected tropical disease
WHO	World Health Organization
DEC	Diethylcarbamazine
PCR	Polymerase chain reaction
ESPEN	Expanded Special Project for the Elimination of Neglected Tropical Diseases in Africa
OEPA	Onchocerciasis Elimination Program for the Americas
PAHO	Pan American Health Organisation
APOC	African Programme for Onchocerciasis Control
OCP	Onchocerciasis Control Programme for West Africa
PTS	Post treatment surveillance
MDA	Mass drug administration

Neglected Tropical Diseases and Phytochemicals in Drug Discovery, First Edition.
Edited by Chukwuebuka Egbuna, Muhammad Akram, and Jonathan Chinenye Ifemeje.

11.1 Introduction

Onchocerciasis is an eye- and skin-related disease that is classified as a neglected tropical disease (NTD). *Onchocerca volvulus* is a parasitic worm that causes onchocerciasis which is transmitted by the repeated bites of blackflies (Figure 11.1), belonging to the genus *Simulium* [1]. The disease also gets the name river blindness as the blackflies responsible for the transmission of the infection breed nearer to the fast-flowing rivers and streams by increasing the risk of individuals living in remote rural villages. The microfilariae which are adult female worms lay thousands of microfilariae (embryonic larvae) which migrate into the skin and eye [2]. Globally, onchocerciasis is the second disease next to trachoma that can cause visual impairment and blindness through the infection. The outcome of this disease includes blindness, skin-related problems like nodules under the skin, intense itching, or rashes [1]. The microfilariae living on the human body is very toxic to the eye and the skin when dead. The death of these worms leads to lesions (eye manifestations) and terrible skin itching. Irreversible blindness and various skin diseases are caused as the result of continuous exposure to lesions. Few skin diseases might be disfiguration and are named "lizard" and "leopard" skin [2].

11.2 Biology and Life Cycle of *Onchocerca volvulus*

Onchocerca volvulus is a nematode or roundworm that causes the disease upon inhibiting the subcutaneous tissues. The life cycle of the nematode involves two types of transmission of the infection and each type again has four stages (Figure 11.2) [3].

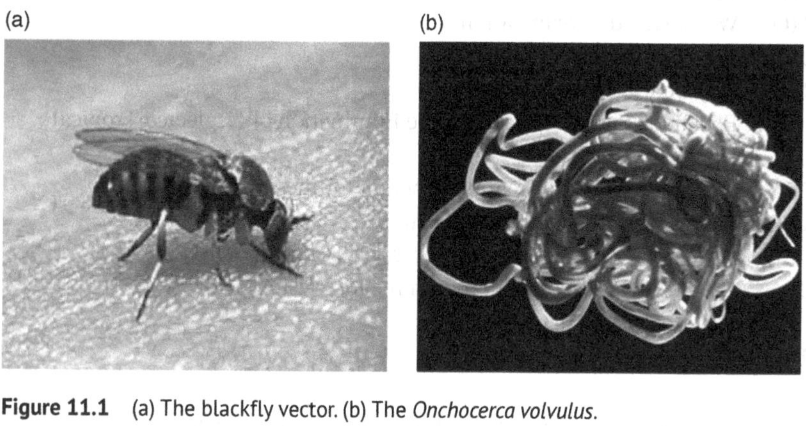

Figure 11.1 (a) The blackfly vector. (b) The *Onchocerca volvulus*.

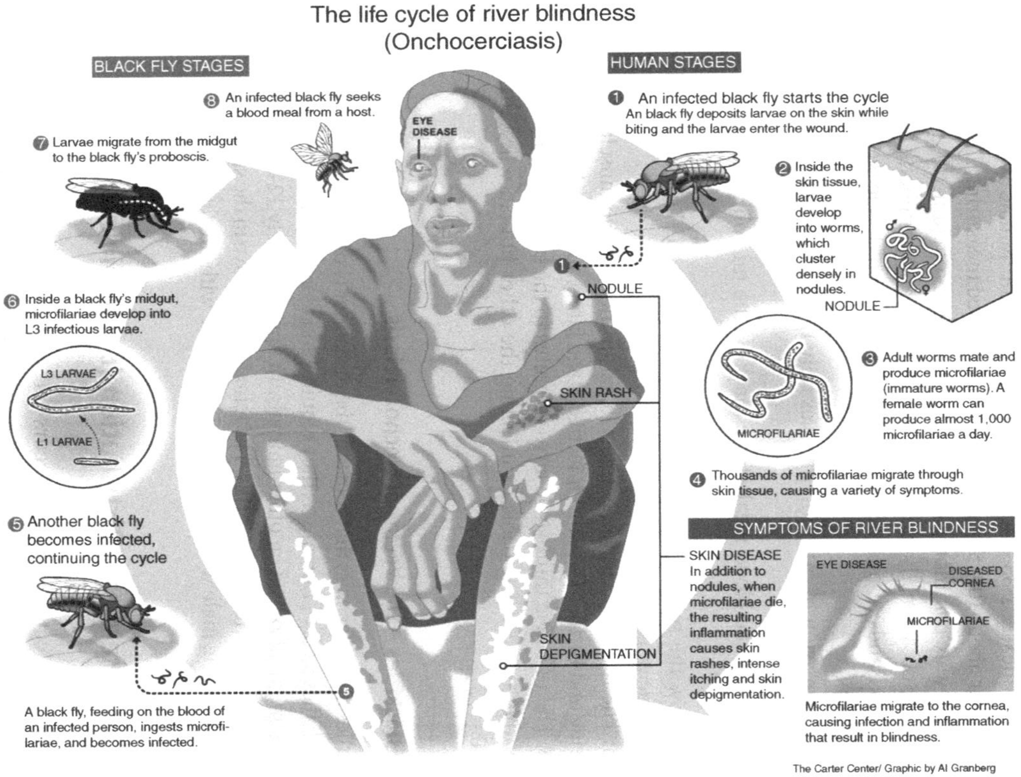

Figure 11.2 The life cycle of onchocerciasis with clinical symptoms. *Source:* Adapted from Parasites – Onchocerciasis (also known as River Blindness) Biology, Centers for Disease Control and Prevention, 2020.

- Type 1 transmission of infection involves only the human-mediated stages of microfilariae.

  Stage 1: This is an initial stage where the third-stage filarial larvae are introduced onto the human skin through the bite wound formed by the infected blackfly.

  Stage 2: Now the larva present in the subcutaneous tissues develops into adult filariae and resides in the nodules of these tissues [4].

  Stage 3: In the nodules, the adult worms can approximately live for 10–15 years. There is the presence of both male and female worms in the nodules. The male worms measure 19–42 mm in length and 130–210 μm in diameter, while the female worms measure 33–50 cm by 270–400 μm. The female worms produce microfilariae for about 10–15 years and are unsheathed measuring 220–360 μm by 5–9 μm [5]. They are commonly found in the lymphatics of connective tissues and the skin but are occasionally found in urine, sputum, and peripheral blood [1, 3].

  Stage 4: Thousands of microfilariae migrate to the skin tissue where it causes many symptoms.

- Type 2 transmission of infection involves only the blackfly-mediated stages of microfilariae.

  Stage 1: During the blood meal of the blackfly, the microfilariae get ingested into its body.

  Stage 2: Now the microfilariae present in the midgut of the blackfly migrate into the thoracic muscles through the hemocoel.

  Stage 3: The first-stage larvae get developed in the thoracic muscles from the microfilariae and further the third infective-stage larvae are also developed [6].

  Stage 4: The infective third-stage larva migrates to the blackfly's proboscis and thus, now the blackfly is capable of infecting the other human when it takes a blood meal [1, 3].

11.2.1 Blackflies – the Vector of Onchocerciasis

The blackflies are the vectors that transmit the disease only during the day time. The female black flies feed on humans as they require blood for ovulation and few species of blackflies might also feed on other animals. In approximately 12–18 months, the microfilariae mature into adult worms. The female worms live in the skin, joints, and muscles whereas the male worms are often seen along with the female worms [7].

11.3 Onchocerciasis Distribution (Epidemiology)

Onchocerciasis infection mainly occurs in tropical climates (Figure 11.3). According to a 2012 report, around 123 million people of 38 endemic countries were at risk of getting onchocerciasis infection and among which 25.7 million

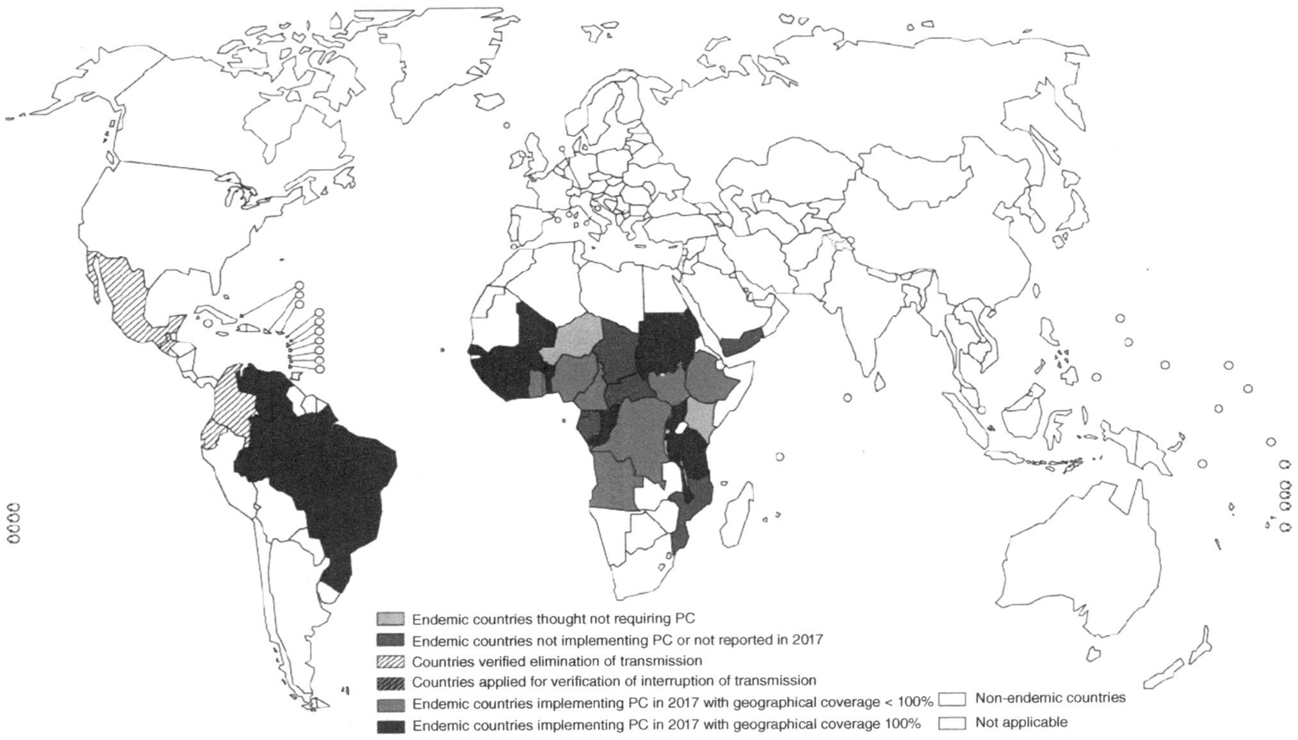

Figure 11.3 The epidemiology of Onchocerciasis infection. *Source:* Onchocerciasis – World Health Organization, 2 Sept 2020.

were infected and 1 million were having a visual impairment or blinded [8]. The African countries were highly affected by this infection whereas the South American countries, such as Brazil and Venezuela, and Yemen, a Middle East country, have shown a limited number of infections [9]. In 2016, the WHO declared effective global guidelines that included certain goals on how to control and eliminate the disease [10]. Several countries of the Americas have been verified and declared as free from onchocerciasis by the WHO [11].

11.3.1 African Perspectives

Sub-Saharan African countries are majorly affected from this infection which includes 31 countries such as Angola, Benin, Burkina Faso, Burundi, Cameroon, Central African Republic, Chad, Republic of Congo, Côte d'Ivoire, Democratic Republic of the Congo, Equatorial Guinea, Ethiopia, Gabon, Ghana, Guinea, Guinea-Bissau, Kenya, Liberia, Malawi, Mali, Mozambique, Niger, Nigeria, Rwanda, Senegal, Sierra Leone, South Sudan, Sudan, Togo, Uganda, and United Republic of Tanzania that were estimated to be at risk [9]. More than 102 million people living in Sudan were at risk due to this infection [12].

11.3.2 American Perspectives

Guatemala is an American country with the highest number, 231 467, of people at risk for this infection in a total of 518 endemic regions that included 42 hyperendemic, 15 mesoendemic, and 38 hypoendemic areas. Mexico is the second country to have 169 869 persons at risk in 39 hyperendemic, 220 mesoendemic, and 411 hypoendemic regions [13]. Venezuela is the third country in America to have around 119 358 people at risk for this infection. Brazil is also one of the contiguous areas that include 12 988 people in 7 hyperendemic, 9 mesoendemic, and 6 hypoendemic (22 endemics) regions. In both Venezuela and Brazil, the affected areas are remote, densely forested, and are migratory across the border. The 2012 report suggested that Brazil was very much close in suppressing the onchocerciasis transmission [8].

11.4 Disease Symptoms

Onchocerciasis infection is caused due to a parasitic worm that leads to several eye- and skin-related diseases. Some individuals do not show any symptoms even if infected with *O. volvulus* [7]. The microfilariae are responsible for the intense inflammatory responses as they move in the subcutaneous tissues of the human body and die [9]. The infected individuals might show severe rashes, itching, changes in skin pigmentation, loss of skin elasticity, and bumps under the skin.

Some individuals even develop eye lesions that can cause visual impairment or even permanent blindness, cataracts, light sensitivity, and itching in the eyes. In rare cases, lymph glands might be swollen [14].

11.5 Disease Diagnosis

Diagnosis of Onchocerciasis refers to the finding of an adult worm or the microfilariae in the human body; also diagnosis is very difficult in light infection stages [15]. The differential diagnosis of acute onchocerciasis includes leprosy, yaws, food allergies, vitamin A deficiency, and syphilis. Certain other parasitic infections might also resemble the symptoms of onchocerciasis [16]. The laboratory diagnosis of this infection includes five different tests.

1) Skin snip test

The first test is to try to identify the nodules present on the skin. After the identification of nodules, the doctor will perform the skin biopsy also called a skin snip test. The procedure includes the removal of 2–5 mg of infected skin (biopsy or snips) and placing it in the saline solution causing the larvae to emerge [14]. Usually, six or multiple snips are taken from different parts of the body [14, 15]. This type of microscopic examination test of snips (biopsies) is referred to as a gold standard in diagnosing the infection as the snips exhibit a high degree of specificity toward *O. volvulus*. The skin snipping test has been recommended as the primary diagnostic test by the WHO [17].

2) Mazzotti test

This test is an alternative for skin snip tests and is also known as a skin patch test as it uses a drug diethylcarbamazine (DEC) patches [14]. DEC is an antihelminthic that upon application to the affected individuals might cause a localized rash within 24–48 hours as it leads to the death of microfilariae [17]. DEC can be used in two ways to test for onchocerciasis. The first way is to give an oral dose of DEC as a medication against the infection. The second way is to put the DEC patch on the skin. If infected, the individual develops severe itching and rashes on the skin [14].

3) Nodulectomy

This test is very rarely used as it involves the examination of worms in the surgically removed nodules and hence got it to name as nodulectomy [14].

4) Polymerase chain reaction

The highly sensitive polymerase chain reaction (PCR) technique requires a small skin sample to perform the test. It is so sensitive that even a low level of infection can be detected. Through the larval DNA amplification, the onchocerciasis infection can be diagnosed in this method [18].

5) Rapid-format antibody card test (serological tests)
It is the most specific and sensitive test used to detect the presence of the infection by detecting the presence of antibodies generated in the human body [14]. A drop of blood is required to perform the card test, as soon as the card detects the antibodies produced due to the infection, there is a change in color. This test is highly recommended as it can be used in the field and does not require any lab access [19].

6) Slit-lamp examination
Slit-lamp examination is used to diagnose eye-related infections. The lesions caused by the larvae in the anterior part of the eye can be examined [14].

11.6 Disease Treatment

People found infected with *O. volvulus* must be treated to prevent skin and eye damage [20]. In 1981, the Mectizan Donation Program led to the discovery of ivermectin that was made to prevent the onchocerciasis-related blindness. Ivermectin is an antiparasitic drug and it does not act directly on the microfilariae wherein it is effective in inhibiting the reproductive process in adult worms [21]. The drug causes the microfilariae to decrease rapidly as it is responsible for all the pathogenic repercussions caused [22]. In 1987, Merck decided to donate the ivermectin as far as the human onchocerciasis is controlled. Through the mass drug administration (MDA) programs, around 2 billion doses of ivermectin were distributed over 29 countries [23]. The endemic populations were mass treated with ivermectin. Microfilariae loads were found to be decreased by 99% after the 14 days of treatment and even 1 year later there was no regain in the number. Regular treatment of ivermectin seemed to reduce the incidence, morbidity, and disability of the infection [22].

Ivermectin has a capacity of a long time or complete blocking the early stage of embryogenesis in the adult female worms. The doxycycline is an alternate treatment used to kill the adult worms that can kill the *Wolbachia* bacteria on which the adult worms depend to survive. Sterilization is the strategy utilized by the drug, as the Wolbachia endobacteria are depleted from the worms [24]. Doxycycline is a novel chemotherapeutic drug given 100 mg/day for 6 weeks which sterilizes the female filariae for 18 months. Due to this effect of doxycycline, there is a reduction in the proportion of male to female worms. The major drawback of ivermectin and doxycycline drugs is that the parasites gain resistance to them and will no longer have their effectiveness in curing or reducing the disease symptoms [25]. This is the reason why alternative medicine is essential in treating onchocerciasis.

11.7 Phytochemicals as a Source of Drugs for Onchocerciasis Treatment

Medicinal plants and herbs have played a significant role in all the fields of health care. A majority of drugs derived from plants are very active against several infectious diseases. For example, the antimalarial drug artemisinin, derived from *Artemisia annua,* was reported to possess filaricidal effects [26].

The fact associated with the drugs ivermectin and doxycycline is that in certain places, the parasite *O. volvulus* develops resistance toward these drugs, and thus, in such situations, there is a need for effective and new pharmaceutical drug interventions to treat onchocerciasis [27]. Utilizing the natural sources to treat this infection has become a novel approach and a good alternative source of antifilarial medicines as they have negligible side effects, cheap, and are readily available [28]. Several plants and few plant families have been investigated for anti-*Onchocerca* properties and their importance has been represented (Table 11.1).

Table 11.1 Plant species and parts active against *Onchocerca* species.

S/no	Family	Plant species	Crude extract	*Onchocerca* species
1	Ochnaceae	*Lophira lanceolata*	Leaves and root bark	*O. ochengi*
2	Anacardiaceae	*Anacardium occidentale*	Bark	*O. volvulus*
3	Euphorbiaceae	*Margaritaria discoidea*	Roots	*O. ochengi*
4	Annonaceae	*Annona senegalensis*	Bark	*O. ochengi*
5	Meliaceae	*Khaya senegalensis*	Aq. bark	*O. volvulus*
6	Apocynaceae	*Rauvolfia vomitaria Afzel*	Leaves and roots	*O. volvulus*
7	Flacourtiaceae	*Homalium africanum Benth*	Roots	*O. ochengi*
8	Combretaceae	*Anogeissus leiocarpus*	Leaves	*O. ochengi*
9	Cucurbitaceae	*Cucurbita pepo ovifera*	Seeds	*O. ochengi*
10	Cyperaceae	*Cyperus articulatus*	Rhizome	*O. volvulus*

11.7.1 Lophira lanceolata

The infusion of *L. lanceolata* plant's bark, roots, and trunk has been widely used to treat various health problems such as menstrual pain in women, diarrhea, constipation, and dysentery. The plant possesses antimicrobial activity against some bacteria and fungi, antipyretic activity, antiplasmodial effects, the potential to cure chronic wound, and antidiarrheal activity. The antifilarial effect of this plant was investigated by Justin et al. in 2017 and reported that the Phyto components of *L. lanceolata* significantly possess the antihelmintic activity against the worms of this infection [29].

11.7.2 Plants Belonging to Anacardiaceae Family

This family of plants constitutes 73 genera and 550 species with rarely subshrubs, shrubs, trees, and lianas. *Anacardium occidentale* is a 15 m long pan tropical tree commonly called "cashew" as it produces inverted heart-shaped fruit. *A. occidentale* is used as a medicine to treat malaria, diabetes, fever, hypertension, and arthritis [30]. Ndouwe et al. [31] demonstrated the *in vitro* antifilarial activity of *A. occidentale* against *O. volvulus* and revealed that the ethanolic bark extract of *A. occidentale* exhibits a strong anti-onchocerca activity [31].

11.7.3 Plants Belonging to Euphorbiaceae Family

The Euphorbiaceae is one of the most diverse and largest flowering plants with about 300 genera most of which are usually subtropical or tropical but rarely temperate. It comprises shrubs, herbs, and trees which are short-lived. *Margaritaria discoidea* is an abundant African tree and its barks are used to treat inflammations, allergies, stomach aches, toothaches, kidney complaints, antioxidant, and antihelmintic activity. *Euphorbia hirta* is the largest genus of Euphorbiaceae family used as a traditional medicine against several health issues like cough, diarrhea, chronic bronchitis, respiratory inflammations, and amoebic dysentery. Both *M. discoidea* and *E. hirta* were reported as anti-*Onchocerca* potent plants [31].

11.7.4 Plants Belonging to Annonaceae Family

It is an important group of angiosperms with about 128 genera and many of which have shown an anti-*Onchocerca* property. *Annona senegalensis* and *Polyalthia suaveolens* are some of the medicinal plants belonging to this family that have a potent antifilarial activity. These two plants have shown a high rate of mortality of *O. volvulus* microfilariae and effective control over the infection when compared with ivermectin [31].

11.7.5 Plants Belonging to Meliaceae Family

The Meliaceae family consists of about 50 genera of flowering plants which usually grow in subtropical and tropical regions and are widely used in traditional medicine for gastrointestinal disorders. *Khaya senegalensis* is a Saharan tree commonly known as African mahogany and is one of the most popular remedies. Including antifungal, anticancer, antimalarial, antidiabetic, and antibacterial activity, the barks and leaves also exhibited the anti-Onchocerca property.

11.7.6 Plants Belonging to Apocynaceae Family

The Apocynaceae family is widely found in subtropical and tropical regions of the world and are angiosperms that constitute about 450 genera of plants. In addition to its antipsychotic, antitumor, antiulcer, aphrodisiac properties, some plant species of this family also showed antifilarial activity such as *Rauvolfia vomitaria Afzel* and *Voacanga Africana Staph* [31].

11.7.7 Plants Belonging to Flacourtiaceae Family

Flacourtiaceae family constitutes about 87 genera and 900 species of hermaphroditic plants which are mostly shrubs or trees. The woody plant *Homalium africanum Benth* species are found in tropics and it is highly diverse in South Asia and Madagascar. These species are usually used as a medicine to treat stomach ulcers, aphrodisiac, skin disease, and malaria. Very few biological activity studies of *H. africanum* have been conducted against parasites [30]. The *in vitro* anti-Onchocerca activity was evaluated by ChoNgwa et al. in 2010 using the roots and bark extracts and thus reported that this species of plants exhibited an effective antifilarial property against the microfilariae with an IC_{50} value of 31.25 μg/ml [32].

11.7.8 Plants Belonging to Combretaceae Family

This family of plants is mostly shrubs, herbs, and trees which are angiosperms consisting of 20 genera that are largely distributed in Africa and Asia in the tropical regions. The plant species belonging to this family have been treated against malaria, diarrhea, inflammations, infections, digestive disorders, diabetes, and bleeding. *Anogeissus leiocarpus* species was studied for its parasitic activity against the *Onchocerca* sp. [30]. Ndjonka et al. (2012) reported that the leaves extract can induce the mortality in adult worms and microfilariae, thus showing the anti-Onchocerca activity with LC_{50} values of 0.06 and 0.09 mg/ml [33].

11.7.9 Plants Belonging to Cucurbitaceae Family

Cucurbitaceae family has huge economic importance worldwide with about 184 genera that have 10 vegetable crops and most of which are distributed in subtropical and tropical regions. *Cucurbita pepo ovifera* is a plant commonly cooked and is used to treat various health issues. The plant has exhibited several properties like antidiabetic, antibacterial, anti-inflammation, antihypertensive as well as antihelmintic properties [30]. Both seeds and leaves of this plant were investigated by Kalmobe et al. in 2017 for their antifilarial activity and have shown remarkable activity against *O. volvulus* adult worms with an LC_{50} value of 17 μg/ml [34].

11.7.10 Plants Belonging to Cyperaceae Family

Cyperus articulatus plant belonging to the Cyperaceae family is found in four continents: Oceania, Latin America, Africa, and Asia. It is a rhizome-bearing plant used as a traditional medicine for migraines, malaria, headaches, and toothaches. *C. articulatus* has shown a wide variety of properties such as anticonvulsant, antifungal, sedative, antibacterial, and trypanocidal [30]. Metuge et al. [28] evaluated the anti-Onchocerca activity of the plants' rhizome against microfilariae and adult worms. The reports suggest that this plant can kill both microfilariae and adults with an IC_{50} value of 31.25 μg/ml [35].

In all, certain families of plants and their various species were evaluated to check their antifilarial activities and most of them were found to show the desired effects against *Onchocerca* sp. [36].

11.8 Future Projection

As mentioned earlier, plants contain relevant and valuable source of anti-infectious properties. Nowadays, various researches have been carried out to show that the new compounds extracted from the plants can be treated on several resistant pathologies. The present study is done to showcase that a few plants used as a medicine against *O. volvulus* are efficient and be used to treat onchocerciasis. In the future, few other plant species can be considered and its *in vitro* studies can be carried out to check their antifilarial activity [37].

11.8.1 Disease Prevention and Elimination

The nongovernment partners with the help of WHO began to distribute the medications to the affected countries to prevent skin diseases and blindness [38]. Various regional onchocerciasis elimination programs such as the present ESPEN

(Expanded Special Project for the Elimination of Neglected Tropical Diseases in Africa), OEPA (the Onchocerciasis Elimination Program for the Americas) that was launched in 1993 in response to the 1991 resolution of PAHO (Pan American Health Organisation), APOC (the African Programme for Onchocerciasis Control) that was launched in 1995, and OCP (the Onchocerciasis Control Programme for West Africa) have supported in controlling this infection [39]. Athree-year PTS (Post Treatment Surveillance) phase was conducted in Sudan in the year 2012 which was led by MOHs (Ministries of Health) and other community partners to educate the people about the transmission and successful eradication of this infection [40, 41].

11.9 Conclusion

The already available drugs were found to gain resistance toward the parasites and hence an alternate treatment is a must to avoid skin diseases and eye complications. The idea of this present study was to focus on the medical potential of phytochemicals as a source of drugs for Onchocerciasis treatment to prevent the growth or to eliminate the adult worms and the microfilariae (parasite *O. volvulus*) which causes Onchocerciasis infection. Thus, the herbs and plants are effective enough to kill the adult worms and to reduce the skin complications related to Onchocerciasis.

References

1 (2020). Parasites – Onchocerciasis (also known as River Blindness). https://www.cdc.gov/parasites/onchocerciasis/index.html.

2 (2020). Blindness and vision impairment Onchocerciasis (river blindness) – disease information. https://www.who.int/blindness/partnerships/onchocerciasis_disease_information/en/.

3 (2020). Parasites – Onchocerciasis (also known as River Blindness) Biology. https://www.cdc.gov/parasites/onchocerciasis/biology.html.

4 Hopkins, A. (2017). Onchocerciasis then and now: achievements, priorities and challenges. *Commun. Eye Health* 30 (100): 92–95.

5 Murdoch, M.E., Asuzu, M.C., Hagan, M. et al. (2002). Onchocerciasis: the clinical and epidemiological burden of skin disease in Africa. *Ann. Trop. Med. Parasitol.* 96 (3): 283–296. https://doi.org/10.1179/000349802125000826.

6 Showler, A.J., Kubofcik, J., Ricciardi, A., and Nutman, T.B. (2019). Differences in the clinical and laboratory features of imported onchocerciasis in endemic individuals and temporary residents. *Am. J. Trop. Med. Hyg.* 100 (5): 1216–1222. https://doi.org/10.4269/ajtmh.18-0757.

7 (2020). Parasites – Onchocerciasis (also known as River Blindness) Disease. https://www.cdc.gov/parasites/onchocerciasis/disease.hl.

8 Sauerbrey, M., Rakers, L.J., and Richards, F.O. Jr. (2018). Progress toward elimination of onchocerciasis in the Americas. *Int. Health.* 10: i71–i78.

9 (2020). *Onchocerciasis – key facts.* https://www.who.int/news-room/fact-sheets/detail/onchocerciasis.

10 Cantey, P.T., Roy, S.L., Boakye, D. et al. (2018). Transitioning from river blindness control to elimination: steps toward stopping treatment. *Int. Health.* 10 (suppl_1): i7–i13. https://doi.org/10.1093/inthealth/ihx049.

11 (2020). Parasites – Onchocerciasis (also known as River Blindness) Epidemiology & Risk Factors. https://www.cdc.gov/parasites/onchocerciasis/index.html.

12 Zarroug, I.M., Hashim, K., El Mubark, W.A. et al. (2016). The first confirmed elimination of an onchocerciasis focus in Africa: Abu Hamed, Sudan. *Am. J. Trop. Med. Hyg.* 95 (5): 1037–1040. https://doi.org/10.4269/ajtmh.16-0274.

13 WHO/Department of Control of Neglected Tropical Diseases (2018). Progress report on the elimination of human onchocerciasis, 2017–2018. *Weekly Epidemiological Record.* 93 (47): 633–648.

14 Seladi-schulman, J. and Stewart, D. (2020). Everything You Should Know About Onchocerciasis (River Blindness), 1–16. https://www.healthline.com/health/onchocerciasis-river-blindness.

15 (2020). Parasites – Onchocerciasis (also known as River Blindness) Diagnosis. https://www.cdc.gov/parasites/onchocerciasis/diagnosis.html.

16 Udall, D.N. (2007). Recent updates on onchocerciasis: diagnosis and treatment. *Clin. Infect. Dis.* 44 (1): 53–60. https://doi.org/10.1086/509325.

17 Unnasch, T.R., Golden, A., Cama, V., and Cantey, P.T. (2018). Diagnostics for onchocerciasis in the era of elimination. *Int. Health.* 10 (suppl_1): i20–i26. https://doi.org/10.1093/inthealth/ihx047.

18 Showler, A.J. and Nutman, T.B. (2018). Imported onchocerciasis in migrants and travelers. *Curr. Opin. Infect. Dis.* 31 (5): 393–398. https://doi.org/10.1097/QCO.0000000000000483.

19 Stingl, P. (2009). Onchocerciasis: developments in diagnosis, treatment and control. *Int. J. Dermatol.* 48 (4): 393–396. https://doi.org/10.1111/j.1365-4632.2009.03843.x.

20 (2020). Parasites – Onchocerciasis (also known as River Blindness) Treatment. https://www.cdc.gov/parasites/onchocerciasis/treatment.html.

21 Abegunde, A.T., Ahuja, R.M., and Okafor, N.J. (2016). Doxycycline plus ivermectin versus ivermectin alone for treatment of patients with onchocerciasis. *Cochrane Database Syst. Rev.* 1: CD011146. https://doi.org/10.1002/14651858.CD011146.pub2. Published 2016 Jan 15.

22 Basáñez, M.G., Pion, S.D., Boakes, E. et al. (2008). Effect of single-dose ivermectin on *Onchocerca volvulus*: a systematic review and meta-analysis. *Lancet. Infect. Dis.* 8 (5): 310–322. https://doi.org/10.1016/S1473-3099(08)70099-9.

23 Walker, M., Pion, S.D.S., Fang, H. et al. (2017). Macrofilaricidal efficacy of repeated doses of ivermectin for the treatment of river blindness. *Clin. Infect. Dis.* 65 (12): 2026–2034. https://doi.org/10.1093/cid/cix616.

24 Hoerauf, A., Mand, S., Adjei, O. et al. (2001). Depletion of wolbachia endobacteria in *Onchocerca volvulus* by doxycycline and microfilaridermia after ivermectin treatment. *Lancet* 357 (9266): 1415–1416. https://doi.org/10.1016/S0140-6736(00)04581-5.

25 Hoerauf, A., Specht, S., Marfo-Debrekyei, Y. et al. (2009). Efficacy of 5-week doxycycline treatment on adult *Onchocerca volvulus*. *Parasitol. Res.* 104: 437–447. https://doi.org/10.1007/s00436-008-1217-8.

26 Cho-Ngwa, F., Monya, E., Azantsa, B.K. et al. (2016). Filaricidal activities on *Onchocerca ochengi* and *Loa loa*, toxicity and phytochemical screening of extracts of *Tragia benthami* and *Piper umbellatum*. *BMC Compl. Altern. Med.* 16 (1): 326. https://doi.org/10.1186/s12906-016-1319-2 Published 2016 Aug 30.

27 Dieudonné, N., Mouraba, A., Abakar, A., and Boursou, D. (2017). *In vivo* toxicity study and antifilarial activity of four plants from Nord-Cameroon. *European Journal of Medicinal Plants* 19: 1–12. https://doi.org/10.9734/EJMP/2017/33290.

28 Metuge, J.A., Nyongbela, K.D., Mbah, J.A. et al. (2014). Anti-Onchocerca activity and phytochemical analysis of an essential oil from *Cyperus articulatus* L. *BMC Compl. Altern. Med.* 14: 223. https://doi.org/10.1186/1472-6882-14-223Published 2014 Jul 7.

29 Kalmobé, J., Ndjonka, D., Boursou, D. et al. (2017). Phytochemical analysis and *in vitro* anthelmintic activity of *Lophira lanceolata* (Ochnaceae) on the bovine parasite *Onchocerca ochengi* and on drug resistant strains of the free-living nematode *Caenorhabditis elegans*. *BMC Compl. Altern. Med.* 17 (1): 404. https://doi.org/10.1186/s12906-017-1904-zPublished 2017 Aug 14.

30 Ndjonka, D., Djafsia, B., and Liebau, E. (2018). Review on medicinal plants and natural compounds as anti-Onchocerca agents. *Parasitol. Res.* 117: 2697–2713. https://doi.org/10.1007/s00436-018-6003-7.

31 Ndouwe, M.H., Ndjonka, D., and Mimpfoundi, R. (2017). Anthelmintic activity, acute toxicity of *Anacardium occidentale* L. (Anacardiaceae) on *Onchocerca ochengi* and *Caenorhabditis elegans*. *Asian J. Med. Health* 5: 1–12. https://doi.org/10.9734/AJMAH/2017/34454.

32 Cho-Ngwa, F., Abongwa, M., Ngemenya, M.N., and Nyongbela, K.D. (2010). Selective activity of extracts of *Margaritaria discoidea* and *Homalium africanum* on *Onchocerca ochengi*. *BMC Compl. Altern. Med.* 10: 62. https://doi.org/10.1186/1472-6882-10-62.

33 Ndjonka, D., Abladam, E.D., Djafsia, B. et al. (2013). Anthelmintic activity of phenolic acids from the axlewood tree *Anogeissus leiocarpus* on the filarial nematode *Onchocerca ochengi* and drug-resistant strains of the free-living nematode *Caenorhabditis elegans*. *J. Helminthol.* 88: 481–488. https://doi.org/10.1017/S0022149X1300045X.

34 Kalmobé, J., Ndjonka, D., Dikti, J., and Liebau, E. (2017). Antifilarial activity of *Cucurbita pepo ovifera* var ovifera (Cucurbitaceae) on *Onchocerca ochengi* adult worms. *Br. J. Pharm. Res.* 17: 1–8. https://doi.org/10.9734/BJPR/2017/33381.

35 Metuge, J.A., Nyongbela, K.D., Mbah, J.A. et al. (2014). Anti-Onchocerca activity and phytochemical analysis of an essential oil from *Cyperus articulatus* L. *BMC Compl. Altern. Med.* 14: 223. https://doi.org/10.1186/1472-6882-14-223.

36 Ndjonka, D., Rapado, L.N., Silber, A.M., and Liebau, E. (2013). Natural products as a source for treating neglected parasitic diseases. *International Journal of Molecular Sciences*: 3395–3439. https://doi.org/10.3390/ijms14023395.

37 Edwige, M.L., Ludovic, M., and Sophie, A. (2018). Are medicinal plants the future of *Loa loa* treatment? *Pharmacognosy Reviews* 12 (23): 133–137. https://doi.org/10.4103/phrev.phrev.

38 (2020). Parasites – Onchocerciasis (also known as River Blindness) Prevention & Control. https://www.cdc.gov/parasites/onchocerciasis/prevent.ml.

39 WHO (1995). Onchocerciasis and its control. Report of a WHO Expert Committee on onchocerciasis control. WHO Technical Report Series 852. https://www.who.int/onchocerciasis/resources/who_trs_852/en/external icon.

40 African Programme for Onchocerciasis Control (2010). Conceptual and operational framework of onchocerciasis elimination with ivermectin treatment. JAF16.6(II). https://www.who.int/apoc/oncho_elimination_report_english.pdfpdf icon.

41 WHO (2016). Onchocerciasis guidelines for stopping mass drug administration and verifying elimination of human onchocerciasis. WHO Criteria and Procedures. https://www.who.int/onchocerciasis/resources/9789241510011/en/external icon.

12

Foodborne Trematodiases and Phytochemicals

Shashank M. Patil[1], Ramith Ramu[1], Prithvi S. Shirahatti[2], and K. Sumana[3]

[1] *Department of Biotechnology and Bioinformatics, School of Life Sciences, JSS Academy of Higher Education and Research, Mysore, India*
[2] *Department of Biotechnology, Teresian College, Mysore, India*
[3] *Department of Microbiology, School of Life Sciences, JSS Academy of Higher Education and Research, Mysore, India*

List of Abbreviations

CAT	Catalase
CCA	Cholangiocarcinoma
***Cs*CaM**	*Clonorchis sinensis* calmodulin
DALYs	Disability-adjusted life years
ELISA	Enzyme linked immuno-sorbent assay
ES	Excretory–secretory
FAO	Food and agriculture association of the United Nations
FLOTAC	Flotation technique
GPx	Glutathione peroxidase
HIV-1/AIDS	Human immuno-deficiency virus/Acquired immuno-deficiency syndrome
HPLC-MS	High pressure liquid chromatography-Mass spectroscopy
IARC	International agency for research on cancer
IEC	Information, education, and communication
iNOS	Inducible nitric oxide synthase
LAMP	Loop-mediated isothermal amplification
LC–MS	Liquid chromatography–mass spectroscopy
MALDI-TOF	Matrix-Assisted Laser Desorption/Ionization-Time of Flight

Neglected Tropical Diseases and Phytochemicals in Drug Discovery, First Edition.
Edited by Chukwuebuka Egbuna, Muhammad Akram, and Jonathan Chinenye Ifemeje.

mtDNA	Mitochondrial DNA
NF-κB	Nuclear Factor kappa-light-chain-enhancer of activated B cells
NGOs	Non-governmental organizations
NGS	Next generation sequencing
NTDs	Neglected tropical diseases
PCR	Polymerase chain reaction
PEG/PC	Polyethylene glycol/propylene carbonate
rDNA	Ribosomal DNA
RNAi	RNA interference
SLE	Systematic lupus erythematosus
SOD	Superoxide dismutase
WHO	World Health Organization

12.1 Introduction

Foodborne trematodiases are one of the precarious diseases among neglected tropical diseases (NTDs), that are caused by trematodes, generally known as flukes. Belonging to the class of digenea under Platyhelminthes, flukes have resulted in a mammoth >56 million infections and >0.6 million disability-adjusted life years (DALYs). Trematodiases have put around 750 million (>10% of the global population) at risk, while 2 million life years are lost to disability and death globally every year [1–3]. In the past, trematodiases predominantly affected Southeast Asian countries. But current epidemiological analyses have marked their prevalence in different parts of the world including USA, Europe, and Africa. In the recent years, it has also been observed that the number of foodborne trematodiases infections have decreased significantly, owing to the factors like urbanization, health education campaigns, socio-economic development, adequate food inspections, regulated use of pesticides, and chemical fertilizers [4, 5].

Around 6000 digenean species have been identified but only 100 of them are known to infect humans. These have been categorized as liver flukes (*Fasciola* spp., *Opisthorchis* spp.), lung flukes (*Paragonimus* spp.), and intestinal flukes (heterophyids, *Echinostoma* spp., *Fasciolopsis buski*). These originated from different sources which predominantly include aquatic animals like fishes, crabs, crayfish, frogs, snails, tadpoles, and snakes and plants like watercress, rice, and vegetables [5, 6]. It is noteworthy that mammals and pet animals could be used as definitive hosts, which makes the foodborne trematodiases as both public and veterinary health threats. These things are sure to affect the society and vulnerable population (Table 12.1). While the public health is largely driven by the clinical aspects of the disease including coinfection, transmission, and secondary

Table 12.1 Summary of taxonomy, medical importance, and epidemiology of foodborne trematodes.

Name of the disease	Family	Genus	Species	Source of human infection	Location in human body	Prevalent countries affected	Total number of infections (global)	Total number of deaths (global)	DALYs (global)	References
Opisthorchiasis	*Opisthorchidae*	*Opisthorchis*	*O. viverrini*, *O. felineus*	Fish, plants	Liver and biliary system	Myanmar, Lao PDR, Cambodia, Southern Vietnam, Thailand, Europe, Russia	8 398 230	1 323	74 367	[1, 2, 7, 8]
Clonorchiasis	*Opisthorchidae*	*Clonorchis*	*C. sinensis*	Fish, plants	Liver and biliary system	China, Vietnam, Lao PDR, Cambodia, and Korea	15 313 219	5 591	275 370	[1, 2, 7, 8]
Fascioliasis	*Fasciolidae*	*Fasciola*	*F. hepatica*, *F. gigantica*	Plants	Liver and biliary system	China, Vietnam, Lao PDR, Cambodia, Korea, Philippines, Central, and South America	2 646 515	0	35 206	[1, 2, 7, 8]
Paragonimiasis	*Troglotrematidae*	*Paragonimus*	*Paragonimus* spp. *complex*	Fish, plants, crustaceans	Pleural cavity and lung; occasional brain invasion	Asia, Africa, South, Central, and North America, Vietnam, China, Thailand, Lao PDR, and Africa	23 155 105	244	196 710	[1, 2, 7, 8]

(Continued)

Table 12.1 (Continued)

Name of the disease	Family	Genus	Species	Source of human infection	Location in human body	Prevalent countries affected	Total number of infections (global)	Total number of deaths (global)	DALYs (global)	References
Fasciolopsiasis	*Fasciolidae*	*Fasciolopsis*	*F. buski*	Plants	Mucosa of small intestine	Taiwan, Cambodia, and China	6 723 551	0	83 699	[1, 2, 7, 8]
Echinostomiasis	*Echinostomatidae*	*Echinostoma*	*E. malayanum, E. revolutum, E. echinatum, E. hortense, E. ilocanum*	Fish, plants, crustaceans	Mucosa of small intestine	Thailand, Taiwan, Malaysia, Philippines, Indonesia, Japan, Korea, Russia, Nepal, India, Hungary, Italy, Romania, Denmark, Egypt, USA, and Brazil				
Heterophyiasis	*Heterophyidae, Microphallidae, Diplostomidae, Lecithodendriidae, Gymnophallidae, Plagiorchiidae*	*Heterophyes, Haplorchis, Metagonimus, Stellantchasmus*	*H. heterophyes, H. taichui, H. pumilio, M. yokogawai, S. falcatus*	Fish, plants	Mucosa of small intestine	China, Korea, Lao PDR, Japan, Taiwan, Thailand, Vietnam, Philippines, Siberia, Turkey, Hawaii, and Balkans				

complications like cholangiocarcinoma (CCA), the veterinary health aspects include a decline in productivity [5]. Despite considerable efforts by international bodies like World Health Organization (WHO) and Food and Agriculture Association of the United Nations (FAO), national and state bodies which organize several control and health promoting campaigns, the diseases continue to affect the global population. Moreover, the research is constantly interfered by difficulties in diagnostics, poor knowledge about the economic consequences of the disease, complexities in racial, ethical, and cultural aspects of human beings from around the globe. Thus, trematodiases remarkably affect low-income countries with aggravated poverty [9]. Trematodiases are positioned fourth in the list of important communicable diseases around the globe, after lower respiratory infections, Human immuno-deficiency virus/Acquired immuno-deficiency syndrome (HIV-1/AIDS), and diarrheal diseases [10].

Treatment of foodborne trematodiases comprises several chemotherapeutic agents. Tribendimidine and mefloquine have shown profound anti-trematode effect *in vivo*, on rodents infected with *Opisthorchis viverrini* and *Clonorchis sinensis*, praziquantel-inhibited *Fasciolopsis buski*, an intestinal fluke. However, they are associated with numerous adverse effects (Table 12.2). Advancements have been found in molecular and immunological diagnostics. Research has been oriented on finding novel targets for drug delivery, based on metabolic profiling evaluations using *Echinostoma caproni* and *Fasciola hepatica* rodents [7]. Also, detection using FLOTAC and mini-FLOTAC flotation technologies have yielded inefficient detection of trematode eggs. Despite the huge amount of work, it becomes difficult for the economically backward communities, which are not capable of affording the medical care comprising chemotherapeutic agents. This is because the NTDs affect the poor countries and has little incentive for pharmaceutical firms from West to invest and manufacture novel trematocidal drugs [8, 12]. Thus, it becomes truly essential to deliberate on the available resources to develop alternatives to chemotherapeutic agents. As medicinal plants have been recognized for comprising several phytochemicals with potential trematocidal properties, advanced research could surely be able to deliver phytotherapeutic agents against trematodiases. Anthelminthic activity of *Momordica charantia* L. against *F. hepatica* eggs has been achieved [13]. Plants like *Albizia anthelmintica*, *Balanites aegyptiaca*, and *Solanum nigrum* have also been reported to possess anthelminthic properties against *F. hepatica* and other parasites [14, 15]. A summary of these plants and their pharmacological effect on foodborne trematodiases has been given in Table 12.3.

Based on these results, advanced studies can be performed with an emphasis on the presence of specific phytochemicals with anti-trematode properties. Special attention needs to be given for the phytochemicals, which bring upon these anti-trematode effects. Identification, isolation, and treatment followed by

Table 12.2 Summary of chemotherapy in foodborne trematodiases.

Name of the disease	Species	Drugs used in human treatment	Dosage	Adverse effects	References
Opisthorchiasis	*O. viverrini*, *O. felineus*	Praziquantel	40 mg/kg single dose, 25 mg/kg three times a day for 2 days, 25 mg/kg single dose	Abdominal pain, nausea, dizziness, and urticaria	[2, 7, 11]
Clonorchiasis	*C. sinensis*	Praziquantel	40 mg/kg single dose, 25 mg/kg three times a day for 2 days, 25 mg/kg single dose	Abdominal pain, nausea, dizziness, and urticaria	[2, 7, 11]
Fascioliasis	*F. hepatica*, *F. gigantica*	Triclabendazole	10 or 20 mg/kg	Fever, nausea, abdominal pain, weakness, pruritus, vomiting, liver enlargement and limited disturbances in liver function	[2, 7, 11]
Paragonimiasis	*Paragonimus* spp. *complex*	Praziquantel, Triclabendazole	Praziquantel: 25 mg/kg oral doses for 3 days after meals. Triclabendazole: 10 mg/kg for 3 days	Abdominal pain, nausea, dizziness, urticaria, weakness, pruritus, vomiting, liver enlargement, and limited disturbances in liver function	[2, 7, 11]
Fasciolopsiasis	*F. buski*	Praziquantel	15 mg/kg	Abdominal pain, nausea, dizziness, and urticaria	[2, 7, 11]
Echinostomiasis	*E. malayanum*, *E. revolutum*, *E. echinatum*, *E. hortense*, *E. ilocanum*	Praziquantel	25 mg/kg, 10–20 mg/kg for minor infections	Abdominal pain, nausea, dizziness, and urticaria	[2, 7, 11]
Heterophyiasis	*H. heterophyes*, *H. taichui*, *H. pumilio*, *M. yokogawai*, *S. falcatus*	Praziquantel	25 mg/kg	Abdominal pain, nausea, dizziness, and urticaria	[2, 7, 11]

Table 12.3 Summary of phytotherapy in foodborne trematodiases.

Name of the disease	Name of the species	Name of the plant	Type of extract	Reported activity	Reported phytochemicals	References
Opisthorchiasis	*O. viverrini*	*Thunbergia laurifolia, Garcinia mangostana, Boesenbergia rotunda*	Not defined	*Thunbergia laurifolia*: Anti-inflammatory, *Garcinia mangostana, Boesenbergia rotunda*: Antioxidant	*Boesenbergia* rotunda: panduratin A, 2′,6′-dihydroxy-4′-methoxychalcone, cardomonin, and flavonoids like alpinetin, pinostrobin, and pinocembrin	[31–33]
Clonorchiasis	*C. sinensis*	*Machilus thunberghii, Prunus mume, Inula helenium, Saussurea lappa*	Not defined	*Machilus thunberghii, Prunus mume*: Lowering seminal receptacle level by widening the bladder, affecting the multiplication, *Inula helenium, Saussurea lappa*: degeneration, necrosis, dilatation, and atrophy of the visceral mass of the worms	*Inula helenium*: santonin-like compounds, isoalantolactone, β-cyclocostunolide, alantolactone, and dihydrolactone	[35]
Fascioliasis	*F. hepatica, F. gigantica*	*Albizia anthelmintica, Balanites aegyptiaca, Atriplex inflata, Solanum nigrum, Abies procera*	Aqueous: *Albizia anthelmintica, Balanites aegyptiaca*, hydromethanol and ethyl acetate: *Atriplex inflata*	Growth inhibition, toxic effects on larvae and eggs, worm deformation	*Atriplex inflata*: triterpenoids and/or sterols possible presence of flavonoids like quercitin. *Solanum nigrum:* glycoalkaloids. *Abies procera*: novel tetracyclic triterpenoids 700015 and 700234	[13–15, 50, 51]
Paragonimiasis	*Paragonimus* spp. *complex*	*Ficus drupacea*	Not defined	Not defined	Not defined	[62]
Fasciolopsiasis	*F. buski*	*Alpinia nigra, Flemingia vestita, Lysimachia ramosa*	Alcohol: *Alpinia nigra*. Ethanol: *Lysimachia ramosa*	*Alpinia* nigra: extreme disorientation in trematode tegument. *Flemingia vestita*: antioxidant, genistein remove the scale-like spines from the body of the worms	Possible presence of genistein	[73–75]

(*Continued*)

Table 12.3 (Continued)

Name of the disease	Name of the species	Name of the plant	Type of extract	Reported activity	Reported phytochemicals	References
Echinostomiasis	*E. caproni*	*Milbraedia paniculate, Securidaca longepedunculata, Xylopia aethiopica, Artemisia annua, Asimina triloba. Garcinia mangostana,* onion, fig, date tree, coconut, ananas, cistrose, garlic, and chicory	PEG/PC: onion, fig, date tree, coconut, ananas, cistrose, garlic, and chicory. Ethanol: *Artemisia annua, Asimina triloba*	*Artemisia annua, Asimina triloba*: antioxidant	Mangostin, mangostin diacetate	[82–84, 109]
Heterophyiasis	*H. taichui, S. falcatus*	*Artocarpus takoocha, Commiphora molmol, Momordica charantia*	Aqueous: *Artocarpus takoocha*	*Commiphora molmol*: Mirazid. *Momordica charantia: tannins*	Tegument amelioration	[94–96]

clinical trials could be extremely helpful in this regard. In this chapter we emphasize on the foodborne trematodiases, elements of infection and transmission, extended pathogenicity, and treatment with phytochemicals. We also highlight the perspective on futuristic approaches that could aid the phytochemical-based drug discovery process.

12.2 Foodborne Trematodiases Affecting Liver

12.2.1 Opisthorchiasis and Clonorchiasis

Opisthorchiasis and clonorchiasis share several common features with respect to pathogenicity and transmission. Opisthorchiasis is caused by liver flukes *Opisthorchis viverrini* and *O. felineus*, whereas clonorchiasis is by *Clonorchis sinensis*. Both *Opisthorchis* spp. *C. sinensis* are known to affect human population at different parts of the world, predominantly Southeast [16, 17]. Human infections occur through the consumption of raw and undercooked freshwater fish (cyprinoid) consisting infective larvae (metacercariae). Around 700 million people are reported to be at risk throughout the world with these two flukes [18, 19]. *O. viverrini* and *C. sinensis* that affect humans are regarded as Group 1 carcinogens by WHO and the International Agency for Research on Cancer (IARC) [20]. These flukes not only induce helminth infections but are also carcinogenic to human beings. Interestingly, the prevalence of *Helicobacter pylori* is associated with both the flukes, causing hepatobiliary diseases, malignancy, and CCA [21].

12.2.1.1 Life Cycle and Transmission

The adult *O. viverrini* and *C. sinensis* trematodes colonize the intra- and extrahepatic regions of human biliary system. These appear as dorsoventrally flattened, bilaterally symmetric organisms (Figure 12.1). The life cycles of all the trematodes appear strikingly similar, with minor differences in the preference of intermediate hosts. While *C. sinensis* flukes measure approximately 10–25 mm by 3–5 mm, *O. viverrini* measures up to 7 by 1.5 mm. The life cycle of these flukes is complicated as it uses digenetic hosts and both modes (sexual and asexual) of reproduction [1, 5]. Embryonated eggs comprising the miracidia larvae get released from the fertile worms into the bile duct, subsequently into the water through feces. These eggs are consumed by Bithynia snails and hatch into sporocysts, which because of asexual reproduction give rise to rediae, in turn become cercariae. Further, the free-living cercaria attach and invade into the skin of around 18 susceptible species of fishes belonging to *Cyprinidae* [22]. Later, inside the skin cercariae encysts themselves as metacercariae, which are infective to the secondary hosts including cats, dogs, and humans after ingesting raw and undercooked fish. With the effect of gastric and intestinal juices in the gut, the metacercariae release

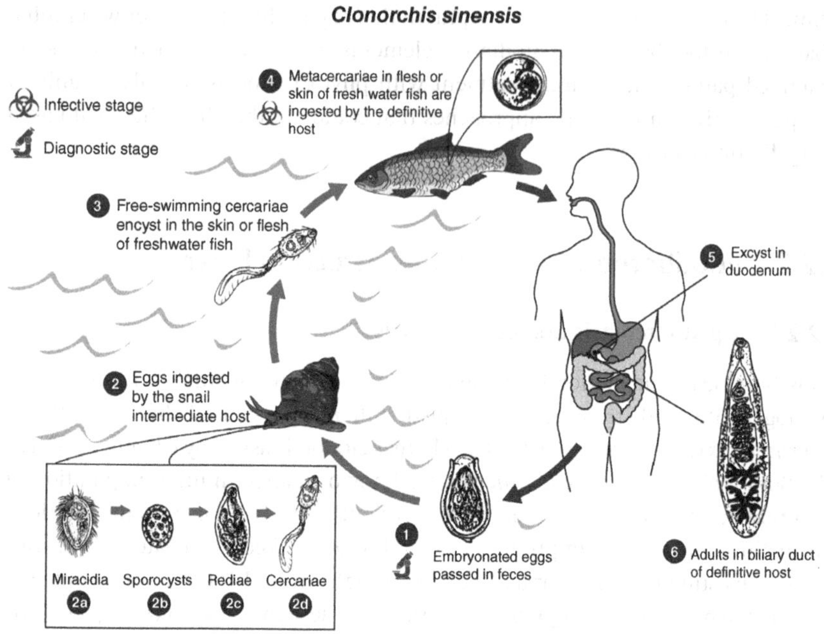

Figure 12.1 Events followed in the life cycle of *Clonorchis sinensis*. *Source:* Centers for Disease Control and Prevention, 15 September 2020. Freely taken from https://www.cdc.gov/dpdx/clonorchiasis/modules/Clonorchis_LifeCycle_lg.jpg (accessed 15 September 2020).

the juvenile flukes, which migrate to the bile duct through the ampulla of Vater, further inhabiting intrahepatic bile ducts to become mature and reproductive flukes. Some mature flukes bind to common bile duct, cystic duct, and gall bladder. Adult liver flukes are hermaphrodites and are capable of living for several years in the gut [2].

12.2.1.2 Epidemiology

Both the flukes *C. sinensis* and *O. viverrini* are reported to be endemic in East and Southeast Asia, predominantly re-emerging in Southeast Asian countries. While *O. viverrini* is said to affect 8–10 million in Lao PDR and Thailand, *O. felineus* further expands to the West, affecting Eurasian and Central European population, causing 1.6 million infections in the Western Siberia and other parts of Russia [2]. Based on the reports, prevalence of *O. viverrini* remains high in Lower Mekong Basin which includes Myanmar, Lao People's Democratic Republic (Lao PDR), Cambodia, Southern Vietnam, and Thailand with 10 million infections. According to the national surveys, Thailand appears to be the origin and the most affected country, with some of the studies reporting it as a hotspot. It is attributed to the intermediate hosts, eating habits, sanitation, and climatic factors [23]. *C. sinensis* was first reported in 1874, where a Chinese carpenter in Calcutta (now Kolkata) died of it. It is also predominant in China, Vietnam, and Korea. Currently, the estimated population at risk of *C. sinensis* is around 200 million worldwide. Over 15 million are said to be infected, with 13 million from China alone [19].

12.2.1.3 Pathogenesis

Infections in both the cases of *O. viverrini* and *C. sinensis* occur with series of pathological events mainly in the liver, intra- and extrahepatic bile ducts, and the gallbladder. Infected animals exhibit hepatomegaly, dilation of bile ducts, and thickened gall bladder. Histopathologic revelations depict the occurrence of inflammation, epithelial desquamation, gallstone, periductal fibrosis, adenomatous, and epithelial hyperplasia, including granuloma formation. In chronic phase, infiltration of the immune cells including plasma cells, monocytes, lymphocytes, and eosinophils are principally observed. Parasite eggs are responsible for granulomatous inflammation, which predominantly occurs in animal models. Goblet cell metaplasia and adenomatous hyperplasia are observed, yet epithelial hyperplasia is more frequently occurring at the early stages through last stage of infection. Periductal fibrosis, along with the invasion of mast cells, eosinophils and gallbladder wall mucosal hyperplasia is also observed [24]. In all the cases of foodborne trematodes, the pathogenesis is highly directed with the secretion of excretory–secretory (ES) products of the flukes. These metabolic products, irrespective of their toxicity potential, interact with biliary epithelium. For example, granulin from *O. viverrini*, is suspected to increase the proliferation of the epithelium cells *in vitro*, which may lead to CCA [25]. ES products from *C. sinensis* are also reported to induce cell proliferation, which was determined *in vitro*. For example, liver inflammation in clonorchiasis is attributed to *C. sinensis* calmodulin (*Cs*CaM), which results in mild to moderate liver fibrosis [19]. As the biliary epithelium harbors antigens, severe inflammation can occur in small bile ducts. This may lead to the ameliorative effects due to the inducible nitric oxide synthase (iNOS) gene expression, which in turn leads to the production of N-nitroso compounds, finally leading to the oxidative DNA damage. Enhanced hepatic activation with these compounds may result in the generation of highly mutagenic conditions in proliferating and inflamed bile duct epithelium [26, 27]. Asymptomatic cases of opisthorchiasis and clonorchiasis usually experience the formation of gallstones and periductal fibrosis, which are not associated with specific external symptoms. These are usually diagnosed with ultrasonic scanning. Symptomatic cases are associated with loss of appetite, diarrhea, indigestion, fullness with pain in the right upper quadrant. In opisthorchiasis, about 5–10% of the infected individuals exhibit flatulence, malaise, and abdominal pain and 5% show hepatomegaly. Severe infections are observed with lassitude, weight loss, ascites, flatulence, and edema. Further complications like obstructive jaundice, cholecystitis, and cholangitis may be present. Notable clinical manifestation of infection is believed to have increased susceptibility to CCA [1, 28].

12.2.1.4 Diagnosis

Diagnosis of both the flukes is conducted through several approaches. The gold standard approaches include determination of presence of eggs in feces, duodenal, and bile fluids. Fecal examination for the parasitological evidence is a widely used diagnostic method. However, accuracy and sensitivity rely on the methods followed during the examination and the expertise of the microscopist. In addition, Stoll's dilution, quantitative formalin ethyl acetate concentration, and KatoKatz thick smear are the frequently used techniques [28]. Due to the morphological similarities, it is difficult to determine the origin of the eggs belonging to different families including *Lecithodendriidae*, *Heterohyidae*, and *Opisthorchiidae*. Modern-day approaches for diagnosis including immunodiagnostics, polymerase chain reaction (PCR), and enzyme linked immuno-sorbent assay (ELISA) have been developed. In the immunodiagnostic methods, non-specific antigens of these flukes are analyzed for binding with antibodies. Fecal and urine antigen detection with ELISA using monoclonal antibodies is also efficiently used. Recent approaches include identification of egg-DNA in fecal specimens, using microRNAs as biomarkers have been used [29, 30].

12.2.1.5 Chemotherapy

Using praziquantel at a single, oral dose of 40 mg/kg concentration is the most widely used regimen for both the flukes. In addition, 25 mg/kg three times a day for two days or single dose of 25 mg/kg are also preferred. However, drugs under development including artemether and artesunate (200 mg/kg) in hamster model infected with *O. viverrini* depict significant reduction in trematodal load. Also, rat model infected with *C. sinensis* showed the effectiveness of artemether, artesunate, and OZ78 at 75 mg/kg doses. However, adverse effects including abdominal pain, nausea, dizziness, and urticaria are observed in case of liver fluke infections including fascioliasis, paragonimiasis, and all intestinal fluke infections [7, 11].

12.2.1.6 Phytotherapy

As the prevalence of *Opisthorchis* spp. induces inflammatory response that could ultimately lead to CCA, some researchers reveal only the efficacy of plant extracts to reduce the inflammation (*Thunbergia laurifolia*) [31] and free radical scavenging (*Garcinia mangostana*) [32], without the information on possible activity of chemical compounds. But further evaluation involving fingerroot (*Boesenbergia rotunda*) reports the reduction of alkaline phosphatase and alanine transaminase enzymes which in turn reduces the number of immune cells. Contrastingly,

N-nitrosodimethylamine, an anti-inflammatory drug, enhances the inflammatory response and is believed to be toxic to the liver. The activity is induced by possible antioxidant compounds including panduratin A, 2′,6′-dihydroxy-4′-methoxychal cone, cardomonin, and flavonoids like alpinetin, pinostrobin, and pinocembrin [33]. In addition, anthocyanin complex prepared from blue butterfly pea petals, purple waxy corn cobs, and turmeric extract downregulates expression of oxidation-related genes like iNOS and nuclear Factor kappa-light-chain-enhancer of activated B cells (NF-κB). It further upregulates the expression of antioxidant genes like glutathione peroxidase (GPx), superoxide dismutase (SOD), and catalase (CAT), thus reducing liver injury and periductal fibrosis in hamsters infected with *O. viverrini* [34]. Therefore, it becomes perceptible that phytochemicals do possess anti-inflammatory and antioxidant activity, which can be utilized against trematodes.

In the case of *C. sinensis*, early reports show that water extracts of *Cassia obutusifolia*, and *Dictamnus dasycarpus* appear to have no effects on trematodal growth in rabbit model. However, *Machilus thunberghii* and *Prunus mume* extracts possess few trematostatic properties like lowering seminal receptacle level by widening the bladder, in turn affecting the multiplication. In case of *Inula helenium* and *Saussurea lappa*, worms are affected through the means of degeneration, necrosis, dilatation, and atrophy of the visceral mass. The root of *I. helenium* is believed to possess santonin-analogues including isoalantolance, β-cyclocostunolide, alantolactone and dihydrolactone with trematocidal properties [35]. Despite several attempts to identify specific phytochemicals from these plants, no lead molecules have been isolated with potential trematocidal properties. Chemotherapeutic drugs including praziquantel, tribendimidine, and mebendazole are currently being used against chlonorchiasis, despite knowing the large potential of medicinal plants to cure the disease [19].

12.2.1.7 Prevention and Control

Opisthorchiasis and clonorchiasis infections can be prevented and controlled by the treatment of animal and human reservoirs, aiming for the reduction in number of eggs as well as preventing them from reaching the source of water. Health promotion with respect to information, education, and communication (IEC) to encourage people to withdraw from extensive consumption of raw and undercooked aquatic sources. Controlling the intermediate host snails is not considered as a feasible option due to the nature of their habitat (water), ubiquitous distribution, low prevalence of infection, and amount of molluscidal agents required [36]. In such scenario, IEC strategies need to be put forward with both beneficial and correctional methodologies in the cultural matters of the communities. Raising awareness, especially among youth and school-aged children could be an effective approach. Involvement of public health agencies has a major impact on reducing the infection rate in the last decade. One health approach involving local governments, nongovernmental organizations (NGOs), private organizations, and citizen could lead to the sustainable control of parasite infections [19].

12.2.2 Fascioliasis

Fascioliasis affects around 2.4–17 million people around the world comprising *F. gigantica* and *F. hepatica* as principal species causing acute and chronic morbidity and has put an estimated population of 80 million at risk. *F. hepatica* is the first parasite to be described, owing to the large burden on sheep farming in 1893 [5]. The disease has originated from Europe, Americas, and Oceania, and two species meet in Africa and Asia. Although fascioliasis is believed to affect distinct geographical areas including Southeast Asia, the liver fluke has transmitted worldwide [37]. They affect domestic ruminants including goats, sheep, and cattle. *Fasciola* spp. also affect humans through contaminating aquatic plants like water hyacinth (*Eichhornia* sp.), water bamboo (*Zizania* sp.), water lily (*Nymphaea* sp.), gankola (*Otelia* sp.), and water chestnut (*Eliocharis* sp.) by existing in the feces of host animals that consume these plants [6].

12.2.2.1 Life Cycle and Transmission

Transmission of *Fasciola* spp. occurs from animals to humans, affecting the liver. The disease is referred to as a zoonosis, where animal reservoirs like goats, sheep, buffaloes, pigs, horses, donkeys, camels, dromedaries, and rodents are involved in disease transmission. Feces from these animals are contaminated with the eggs of parasites that release miracidia into the water body thereby contaminating them. They are in turn known to invade into the body of intermediate hosts such as *Fossaria, Pseudosuccinea*, and *Galba* [37]. Cercariae are formed inside the snails because of morphosis, passing the development stages sporocysts and rediae, which are then released to the water to become encysted metacercariae. These are subsequently consumed by humans and other animals, after inhabitating fresh aquatic plants. Infection may also occur because of floating metacercariae on water used for drinking and washing purposes [6]. Upon ingestion, metacercariae invade the intestinal wall and continue to persist in duodenum to get developed as adults in three to four months. In humans, mature *F. hepatica* appears dorsoventrally flattened and measures 30 mm by 13 mm, whereas *F. gigantica* measures up to 75 mm [2].

12.2.2.2 Epidemiology

Human fascioliasis, currently, is estimated to be reemerging in many countries, including those of its origin and predominance. Trematodal outbreaks occur in communities and households living near water bodies with substandard lifestyle and unhygienic conditions, where poor sanitization is present [38]. Human infection is said to be infrequent in endemic regions like Southeast Asia, where the prevalence is high among animal reservoirs [16]. For example, in Korea predominant infections have been reported are linked with cattle and consumption of contaminated water with *F. hepatica* and *F. gigantica*. In several regions of the Philippines, the disease is still prevalent despite implementing WHO guidelines [39, 40]. Similar scenario has also been reported in Central Europe, with more than 20 helminth species detected regularly in Austria itself among both netizens and tourists [41]. Fascioliasis is an emerging problem with escalating number of cases in Bolivia and other countries of Central and South America. Even Vietnam has faced similar situation in the last decade owing to the climatic changes and/or genetic variations within the *Fasciola* spp [42, 43].

12.2.2.3 Pathogenesis

Infection with these trematodes can cause notable mortality in cattle and sheep, whereas human morbidity depends on the number of worms entering the body and the stage of infection. The acute phase of infection begins with the migration of immature flukes through the liver/hepatic parenchyma [44]. As a result, the parenchymal tissue gets destroyed, as well as parasite death occurs due to inflammatory responses majorly induced by eosinophils. However, these defensive mechanisms can rake up extensive periportal fibrosis. Later in the chronic phase, parasites tend to be less severely present in bile ducts. Due to the inflammatory responses, obstruction of ducts, and enhanced proline levels by flukes, changes in tissues including dilation, fibrosis, and bile duct proliferation occurs. Flukes can be observed as small and flat objects in gallbladder, intra- and extrahepatic bile ducts, with spontaneous movements [44, 45]. As a result of bile duct dilation and lesions, anemia can occur and may further result in death.

12.2.2.4 Diagnosis

Diagnosis can be done during the chronic phase of infection, which is completed with the observation of eggs in feces. The older techniques like sedimentation have been replaced by novel techniques. Eggs strikingly look like those of *Fasciolopsis buski*. It is also difficult to differentiate between eggs of *F. hepatica* and *F. gigantica*. This problem is now resolved with the help of microRNA's, which are used as biomarkers, and can later be detected in isothermal PCR called loop-mediated isothermal amplification (LAMP) [46]. Pseudofascioliasis or false fascioliasis occurs due to the consumption of infected livers containing eggs, which results in the presence of eggs in feces. It is not regarded as an actual infection. For this, liver free diet is suggested prior to the secondary feces examination [47]. Antibody detection technique is the most common procedure followed, which is effective in early infections. These can be detected within two to four weeks of infection in the body, even before the appearance of eggs in feces. This method also helps in case of absence of eggs, or in case of ectopic fascioliasis or during chronic fascioliasis when the number of eggs is low, or to determine the pseudofascioliasis. Modern approaches include ELISA, where ES antigens or purified serum proteins are used [48]. Detection of antigens in infected mice, cattle, and humans has been reported. The novel technique of MM3-COPRO uses CoproGuard, a diluent which preserves the coproantigens for a period of two months and is reliable due its accurate specificity and sensitivity [49].

12.2.2.5 Chemotherapy

Unlike other trematode infections, fascioliasis with *F. hepatica* is found resistant to praziquantel, one of the most common drugs used against trematode infections. Thus, triclabendazole with bithionol is being used as an alternative. Triclabendazole at a dose of 10 or 20 mg/kg is found to be effective, without adverse effects [10]. Several compounds have been found to cure the fascioliasis in rodent models including artemether and artesunate and (semisynthetic derivatives of artemisinin), 1,2,4-trioxolane OZ78, and tribendimidine. However, treatment with triclabendazole can result in adverse effects like fever, nausea, abdominal pain, weakness, pruritus, vomiting, liver enlargement, and limited disturbances in liver function [7, 11].

12.2.2.6 Phytotherapy

Treatment of fascioliasis with herbal applications has been achieved with different fractions of plant extracts, which are attributed to several chemical compounds existing in those plants. Though pioneering efforts in late 1990s resulted only in proving the efficacy of plant extracts to comprise trematocidal properties, advanced studies in the twenty-first century have proved the trematocidal potential of phytochemicals upon their identification. For example, in the beginning, aqueous extracts of *Albizia anthelmintica* and *Balanites aegyptiaca* plants were reported with the reduction of fluke numbers in goats upon consumption [14]. But advanced studies performed with *Atriplex inflata* extracts attributed the trematocidal activity to the presence of triterpenoids and/or sterols, along with the prediction of possible presence of flavonoids like quercitin. The hydromethanol and ethyl acetate extracts of *A. inflata* inhibit different stages of *F. hepatica*, including the sporocysts and rediae that are present in intermediate host snail [50]. Also, glycoalkaloids present in unripe fruit of *Solanum nigrum* are believed to reduce the trematodal load along with leaf extracts. Interestingly, the extracts induce toxic effects on target eggs and larvae, without aberrating host tissues [15]. The modern-day evaluation of plant extracts with high pressure liquid chromatography–mass spectroscopy (HPLC–MS) technology confirmed the presence of quercitin and prominent presence of other flavonoids in the extracts and fractions of *Momordica charantia* leaves, which inhibit *F. hepatica* larval growth in eggs, inhibiting their hatching [13]. Further, novel tetracyclic triterpenoids 700015 and 700234 from *Abies procera* possess anti-fasciolicide properties, which inhibit the growth of *F. hepatica*, by disrupting their surface membrane [51]. These results make evident that phytochemicals could be used to treat fascioliasis with appropriate doses for both humans and animals.

12.2.2.7 Prevention and Control

Prevention of fascioliasis must include strategic immunization of livestock and other reservoir animals. Though advances in vaccine development process are encouraging, it takes years to develop functional and commercially available vaccines after conducting clinical trials [37, 52]. Controlling the infection with avoiding raw and undercooked food from aquatic origin may sound practically good but it is difficult to avoid traditional methods of nutrition. Also, controlling the intermediate hosts (genera *Pseudosuccinea, Galba*, and *Lymnaea*) using molluscidal drugs could be done, though not a feasible approach. Health education and awareness camps comprising detailed analysis on disease, including diagnostic difficulties, and results of community-based studies in endemic regions could be useful [4, 6]. In the wake of possible resistance of flukes to the available chemotherapeutic drugs and delayed vaccines development process, drug development from plants and extracted phytochemicals could be extremely useful for the treatment.

12.3 Foodborne Trematodiases Affecting Lungs

12.3.1 Paragonimiasis

Paragonimus spp., or generally known as the lung flukes, are regarded as one of the most damaging species among foodborne trematodes. Paragonimiasis occurs in different parts of the world, including Asia, Africa, South, and Central America [10]. About 15 species of *Paragonimus* are known to affect humans, whereas *P. westermani* is said be the most common parasite. *P. heterotremus* is regarded as an etiologic agent of human infection in Vietnam, China, Thailand, and Lao PDR. Paragonimiasis is also reported in Africa (*P. africanus*) and North America (*P. kellicotti*). It follows the same digenetic hosts including freshwater molluscs, crabs, shrimps, and crayfishes. The disease also spreads through definitive hosts like toddy cats, dogs, tigers, civet cats, and mongooses [53, 54].

12.3.1.1 Life Cycle and Transmission

Infective metacercariae present in crustacean hosts are consumed by the final host in either raw or undercooked manner, excyst in the small intestine followed by penetrating through intestinal wall into the abdominal cavity. The final hosts normally include the members of Canidae, Felidae, man, and mammals [55]. These larvae further move into the muscles, liver, diaphragm followed by the lungs, which are regarded as the sites of maturation. Matured larvae lay eggs, which are transferred to the water through the fecal stream. Eggs are also transferred through coughing and spitting. These eggs become matured and develop into miracidia, free swimming larvae that swim freely and invade freshwater snails. In case of crabs and crayfish, the infection occurs through consuming mature microcercus cercariae. The cercariae further develop into metacercariae, that are to be ingested by the definitive hosts. Adult worms are hermaphrodites, appear as flattened, coffee-bean shaped, and measure up to 7.5–12 mm by 4–6 mm [56].

12.3.1.2 Epidemiology

Prevalence of paragonimiasis is expanded in three endemic focal regions including Asia (Thailand, China, Taiwan, Lao PDR, Japan, Vietnam, Korea, and Philippines), Central and South America (Columbia, Ecuador, Costa Rica, and Peru), and Africa (Nigeria, Cameroon, and Gambia) [9, 57]. Around 20 million people are affected with these lung flukes worldwide, while an estimated 293 million people are at risk. These endemic regions have been determined as such because of the eating habits and condition of food components that chiefly include host animals like crabs, snails, and crayfishes. Considering both the genders, the prevalence of the disease is more in age groups of 25–29 years [9].

12.3.1.3 Pathogenesis

Pathological symptoms are the results of mechanical damage caused by the migration of the worms from the gut to the lungs, and subsequent release of toxic metabolites. In case of long-term infections, ectopic migrations occur to the subcutaneous sites at the extremities and aberrant sites like brain [58]. The flukes also result in lung hemorrhage, inflammation, parenchymal necrosis, and fibrotic encapsulation. Acute infections are normally associated with chronic cough and brownish sputum with blood-streaks. In this regard, there is a chance of misunderstanding paragonimiasis with bronchial asthma and chronic bronchitis. In case of chronic infections, abdominal and cerebral paragonimiasis may occur, based on the location of the flukes [2, 58].

12.3.1.4 Diagnosis

Diagnosing paragonimiasis is based on clinical aspects as well as laboratory examinations. Microscopic examination of eggs in feces or sputum is done to determine the origin of the eggs. The microscopic evaluation of the eggs is not considered as a feasible approach due to the small size of the eggs. However, supporting approaches like X-ray and immunodiagnosing could be relied on for better results [59].

Serological tests including ELISA are conducted using antigen–antibody reaction in humans. In addition, PCR-based DNA detection of *Paragonimus* spp. eggs and worms from the infected could reveal accurate results [60]. Reliable serological tests have been applied for immunodiagnosis of human paragonimiasis; these include ELISA and immunoblotting-based methods. Immunoblotting for the detection of IgG4 antibody to ES products of adult *P. heterotremus* provides a sensitive and specific test. Recently, a recombinant protein of a *P. westermani* egg antigen was tested as an ELISA antigen, which offers a highly sensitive and specific assay for serodiagnosis. A PCR for detection of DNA from *Paragonimus* eggs in feces or sputa has potential in clinical epidemiological studies [61].

12.3.1.5 Chemotherapy

Triclabendazole and praziquantel have been recommended by WHO for treating paragonimiasis. Treatment of praziquantel involves 25 mg/kg oral doses for three days after meals [10]. Triclabendazole at a dose of 10 mg/kg daily for three days has resulted in 98.5% cure of the disease. Though bithionol is used, the limiting success and more adverse effects have made the drug unfeasible for the human consumption [7, 60].

12.3.1.6 Phytotherapy

Paragonimiasis is observed with infection in lungs caused by *Paragonimus* spp. To this date, there have been no updates on phytochemical-based treatment for the disease. Though chemotherapeutic agents like praziquantel, and triclabendazole have been used since 1970s, no major research has been deliberated in the development of plant-based or phytochemical-based treatment. Available literature cites the adequacy of *Ficus drupacea* leaves, without mentioning the type of extract, and its accurate effect [62]. Although several phytochemical compounds are present in *F. drupacea* root and aerial parts, no specifications are available on their isolation and subsequent effects on *Paragonimus* spp. However, it can be deduced that molluscidal properties of medicinal plants could be used to kill the *Paragonimus* spp. that exist in intermediate host snail in different developmental stages [63]. In the vista, methanolic extract of seeds, β-solamarine, glycoalkaloids and saponins from *Solanum elaeagnifolium* possess molluscidal activity against freshwater snail (*Galba truncatula*) [64]. Though with this approach one can treat the symptoms of paragonimiasis, *in vivo* models infected with *Paragonimus* spp. could be analyzed for trematocidal activity, prior to conducting dose-dependent studies.

12.3.1.7 Prevention and Control

Prevention and control of the paragonimiasis requires the effective approaches to interfere the life cycle. The eventful elimination of the disease can be achieved with usage of molluscidal agents to kill the intermediate hosts like snails. But this is not regarded as an effective approach, as they are the ubiquitous part of the environment. Apart from mass utilization of chemotherapeutic agents, health promotion and awareness camps can be conducted in endemic areas regarding environmental sanitation, clean and hygienic food preparation, and utilization of sterilized food products. One health approach using different sectors of the health control could be a promising approach [2].

12.4 Foodborne Trematodiases Affecting Intestine

12.4.1 Fasciolopsiasis

Fasciolopsiasis is a resultant intestinal infection of a giant fluke known as *Fasciolopsis buski*. It shares the common family (*Fasciolidae*) with the liver flukes *F. hepatica* and *F. gigantica* that are categorized under the genus *Fasciola*, thus resembling the latter in morphology and developmental cycle. Like all the foodborne trematodes, *F. buski* infection predominantly occurs in Southeast Asian countries, including India [65, 66]. About 10 million people are infected with *F. buski* in South Asia. It is believed to affect the humans through infecting snails, followed by the attachment of metacercariae to the aquatic plants. Consumption of such plants in either raw or undercooked conditions, eating the raw nuts, and fruits could lead to the infection. Rabbits, pigs, and dogs serve as natural definitive hosts [67].

12.4.1.1 Life Cycle and Transmission

Definitive hosts including humans get infected post-consumption of bulbs, stems, roots, or the seed pots of aquatic plants attached with metacercariae. Plants include water chestnut, water hyacinth, water bamboo, lotus or caltrops, etc. [68]. Upon consumption, *F. buski* metacercariae excyst with the aid of gastric and intestinal juices in duodenum, and the resultant juvenile flukes attach to the epithelial cells of duodenal and jejunal wall. Larvae develop to become mature worms within three months and produce an estimated 10 000–25 000 eggs per worm/day. The eggs appear as large, ellipsoidal, yellowish in color, with operculum single end (diameter 27 mm). They measure 120–140 mm by 70–90 mm. Adult worms are usually elongated, oval, and measure up to 2–10 cm by 0.8–3 cm. Embryonation and development of the eggs may occur over 3–7 weeks, in case of freshwater sources, followed by hatching and penetrating into the snails belonging to genera *Hippeutis*, *Gyraulus*, and *Segmentina*. The miracidia then hatch and transform into encysted metacercariae in snails, through asexual multiplication as sporocysts, rediae, and cercariae [65].

12.4.1.2 Epidemiology

Extensive infections occur in Thailand, which is regarded as a hotspot due to heavy rainfall and floods, ultimately leading to fecal contamination of water. The consumption of feces in pigs appears to be one of the mainstream sources of infection in Thailand. Infection by *F. buski* is most prevalent among school-aged children, with highest number of infections reported within the age group 10–14 [2]. This is because of the extensive consumption of water caltrops by school-aged children, where the number of worms could exceed by 800 per child. However, the highest infection rates were found in children in the same age group in Taiwan. Meanwhile, the disease also occurs in few endemic countries like China with 85% prevalence, Taiwan (25–61%), Bangladesh (8.6–50%), and Cambodia (0.04%) [65].

12.4.1.3 Pathogenesis

Adhering of adult worms on the duodenal and jejunal wall can result in the feeble eosinophilia that is <35%, in which there is a risk of developing inflammation and ulcer formation from mechanical injuries. The ulcers thus formed sometimes bleed with capillary damages, and may further lead to petechial bleeding, forming intestinal abscesses. In case of mild infection, lower levels in hematocrit and serum levels of vitamin B_{12} are reported. In this context, no other alterations in other biomolecules or the nutrients are observed. This condition may arise from either the competitive utilization of vitamin B_{12} or the impaired absorption of the same in damaged intestinal mucosa [2]. Higher number of flukes in the hosts can result in severe pathology and acute intestinal obstruction in children. Excessive ulceration may lead to interference in digestion, causing malabsorption which in turn leads into malnutrition. In chronic and long-term infections, edema can also occur due to the toxic ES metabolites from the worms. Clinical manifestations occur at both mild and chronic infections, and may include mild abdominal colic, flatulence, loss of appetite, chronic diarrhea, nausea, vomiting, eosinophilia, ulceration, and fever. Due to the duodenal ulceration, abdominal pain is observed. Long-term effects on patients can result in the development of edema of the face, eyelid, abdomen, legs; weakness; anemia; and anorexia. *F. buski* infection in case of pregnant women is reported with risk of systematic lupus erythematosus (SLE) [69]. Death is inevitable in case of untreated severe infections [2, 70].

12.4.1.4 Diagnosis

Fecal examination is a commonly used approach for diagnosis in laboratories. Other diagnostic tools include Stoll's dilution, formalin ethyl acetate centrifugation, direct smears, and KatoKatz techniques. Although these approaches have been used efficiently, the differentiation between the eggs of *Fasciola* spp., *Fasciolopsis* spp., and *Echinostoma* spp. has been difficult due to the small size and morphological similarity between the eggs [28]. However, modern-day approaches like nucleic acid sequencing, where nuclear ribosomal DNA (rDNA) and mitochondrial (mtDNA) can reveal the differences. Also, approaches like next generation sequencing (NGS) primed with PCR can distinguish between different strains of the parasites available [71, 72].

12.4.1.5 Chemotherapy

Praziquantel has been a choice of many researchers, being able to show anti-trematodal effect on diverse class of foodborne trematodes. It is normally given in different dosages according to the age of the infected. Oral and single dose of 15, 25, and 40 mg/kg have been given to school-aged children, who were found with severe infection in Thailand. Praziquantel has shown its efficiency even at the low dosage of 15 mg/kg. In case of other drugs, *in vivo* studies using pigs have resulted in the effective inhibition of trematodal growth for oxyclozanide, triclabendazole, and rafoxanide, where reduced fecal egg count and lowering of clinical symptoms were reported [2, 65].

12.4.1.6 Phytotherapy

Treatment of intestinal flukes causing fasciolopsiasis with isolated phytochemicals and plant extracts has revealed the efficacy of phytochemicals on the intestinal flukes. The alcoholic extract of *Alpinia nigra* results in extreme disorientation in trematode tegument. As a result, vacuoles form up to the basal lamina separating syncytium from the tegument at several regions. Loss of parenchyma material and connecting tubules result in the syncytium result in deprivation of trematode from protective layers [73]. Also, root-tuber peel of *Flemingia vestita* and genistein induce and promote the NO radicals, that in turn are believed to increase neurotoxicity, neural degradation, tissue dysfunction, and DNA damage in the worms [74]. Further treatment with ethanolic extract of *Lysimachia ramosa* leaves and genistein removes the scale-like spines from the body of the worms [75].

12.4.1.7 Prevention and Control

Considering the effect of praziquantel, community-based treatment programs were carried out to control the escalating infections in Indonesia. But this effort was confounded by the recurring infections and subsequent migration of the patients. It becomes evident that it is difficult to control the disease by chemotherapeutic approach alone. In the absence of vaccine, it becomes essential to put forward nonclinical methods of prevention like following hygienic practices while preparing and consuming food, avoiding consumption of aquatic plants, and vegetables without sterilization, use of boiled or fermented vegetables. Sterilization of human and pig feces before using them in agricultural practices would surely help to interrupt the developmental cycle of the fluke. Filtering and

UV treatment of drinking water also could be beneficial in endemic areas. Along with this, health awareness among children and youth by public health agencies can be effective [2, 67].

12.4.2 Echinostomiasis

Unlike other foodborne trematodiases, human echinostomiasis is a resultant disease of at least 17 species of the genera *Echinostoma*, under the family *Echinostomatidae*. The most common including *E. malayanum*, *E. revolutum*, *E. echinatum*, *E. Hortense*, and *E. ilocanum*. Though associated with minor medical importance, infections linked with these flukes are said to be prevalent in endemic regions [5]. Adult trematodes affect diverse classes of mammals and aquatic birds through digenetic life cycle and spread through the consumption of aquatic products inhabited by the pathogenic metacercariae. In addition to snails, *Echinostoma* spp. use crustaceans, fish, amphibians, as intermediate hosts. The resultant infection occurs in animals and birds in the premises of water resources, subsequently appears in humans consuming the aquatic products [76].

12.4.2.1 Life Cycle and Transmission

It is interesting to note that echinostomes possess feeble host specificity with diverse host range, the property which is not seen in the other trematodes. Though events in the digenetic (human and sylvatic) life cycle resemble the other trematodes, the sylvatic hosts include fishes, brackish and freshwater molluscs (opisthobranch or pulmonated and bivalves), amphibians (frogs and tadpoles), and crustaceans [76]. As these animals constitute a colossal amount of diet in endemic areas. Adult and infective flukes are found in mammals and birds of these regions. The metacercariae inhabit the aquatic plants and animals to be consumed by humans. They excyst in the gut, releasing the juvenile flukes. The juvenile flukes attach to the intestinal wall and grow. The eggs thus formed are released into the water through fecal stream, which are consumed by the intermediate hosts that facilitate the asexual development of larvae to metacercariae through developmental stages. Aquatic birds, dogs, and rats are regarded as the major reservoirs of echinostomiasis. Notable morphological variations exist in case echinostomes, varying from 5 to >10 mm in length [77].

12.4.2.2 Epidemiology

Predominant human infections occur in the areas that practice ingestion of raw and undercooked molluscs and fish. Infection is common in Asian countries including Thailand, Taiwan, Malaysia, Philippines, Indonesia, Korea, Russia, Nepal, and India [76, 78, 79]. The prevalence ranges from 1 to 50% in these countries. Cases of *E. liliputanus* were first reported in China, with a prevalence of 13.4% among 2426 people in a province. In addition, infections with *E. japonicus* have also been found with 4.9% prevalence. It has also been found in Lao PDR and Korea. *E. cinetorchis* was first reported in Japan and Korea. Also, a high prevalence of *E. echinatum* with 43% reaching 96% in some parts of Korea. Prevalence of *E. hortense* has been found in a province, among residents of Korea (22.4%), where China recorded six people getting infected with it. *E. ilocanum* was first detected in Philippines with the prevalence of 7–17% in a local community. The prevalence has also occurred in India, Java, and China. Further, *E. revolutum* occurs largely in Africa and Europe, most of it can be found in Asia. Echinostomiasis was first detected in Taiwan and an estimated prevalence in Taiwan ranges between 2.8 and 6.5%. In addition, a prevalence of 2.4–7.5% was reported in a province of Cambodia. In addition to these, echinostomiasis has also been reported in Hungary, Italy, Romania, Denmark, Egypt, USA, and Brazil [77].

12.4.2.3 Pathogenesis

The disease usually affects families from poor economic background that live in rural and unhygienic areas and prefer raw, undercooked and unsterilized food. Drinking untreated water also plays a major role in elevated incidence [9]. Major pathological incidences associated with the occurrence of lesions in the villi of the jejunal wall, where the parasites attach. This may further lead to the inflammation followed by ulceration of the mucosa. Recent evaluations report these lesions as adenocarcinoma. Interestingly, echinostomes are not highly pathogenic to humans with only few reported cases of clinical features. Light infections are usually asymptomatic, while heavy infections result in diarrhea, weight loss, fatigue, eosinophilia, vomiting, anorexia, and abdominal pain. These symptoms are not considered as lethal, compared to the other trematodes [77].

12.4.2.4 Diagnosis

Since the eggs appear as large, operculated and embryonated in fecal examination, and share common morphology with those of *Fasciola* and Fasciolopsis, it is difficult to determine the actual source. However, exclusion chemotherapy can help in the recovery of adult worms from feces, which helps in the determination of worms based on morphology [80]. Collar spines and circumoral disc present around the oral sucker are believed to play a significant role in the identification. Also, it is noteworthy that morphological disorientations can occur in case of chemotherapy. Apart from fecal examination, the disease can be diagnosed with gastroduodenal endoscopy. Till date, immunological techniques for the diagnosis have not been developed. However, recent updates show that using PCR technique, two separate DNAs of the trematodes can be compared with identifying distinct markers [81].

12.4.2.5 Chemotherapy

Due to its broad spectrum anti-trematodal effects, praziquantel has been the drug of choice echinostome infections. A single dose of 25 mg/kg being used for infection of other intestinal fluke infections, echinostome infections being less infective, are given with a slightly lower, oral, single dose of 10–20 mg/kg concentration [7, 11]. Apart from praziquantel, several other drugs have been reported with potential for treating the disease. OZ78 has been successful in treating the infected rodent model. Echinostomicidal effect of benzimidazoles, where the inhibition of glycoconjugate secretion on the surface of *E. caproni* has been reported. Furthermore, multiple treatments with mebendazole showed complete reduction in trematodal load. Also, single doses of clorsulon (100 mg/kg) and rafoxanide (50 mg/kg) was able to kill 100% of flukes in a mice model within 24 hours [77].

12.4.2.6 Phytotherapy

Pioneer studies conducted on the *Echinostoma* spp. using different plant extract did not reveal any of the specific compound attributable for anti-trematodal effects. For example, ethanolic crude extracts from 50 medicinal plants were evaluated for their anti-trematodal activity *in vitro*, where crude extract of *Milbraedia paniculate* whole plant, root extract of *Securidaca longepedunculata*, and bark extract of *Xylopia aethiopica* were found to be effective against *E. caproni* [82]. In addition, aqueous, chloroform, polyethylene glycol/propylene carbonate (PEG/PC) extracts made from onion, fig, date palm, coconut, ananas, cistrose, garlic, and chicory plants were evaluated for the anti-trematodal activity. However, no specific phytoconstituent with anti-trematodal property was observed [83]. Furthermore, ethanolic extracts of *Artemisia annua* and *Asimina triloba* killed *Echinostoma caproni*, attributing antioxidant flavonoids [84]. *E. caproni* is also sensitive to mangostin, a xanthone isolated from *Garcinia mangostana* and its synthetic derivative known as Mangostin diacetate. But the mechanisms of these phytochemicals are yet to be deciphered [84].

12.4.2.7 Prevention and Control

Prevention of echinostomiasis could be completed with the usage of sterilized vegetables and aquatic plants, as the consumption of these products is associated with high prevalence of the disease. Also, consumption of aquatic animals like fish and mollusk should be practiced after proper sterilization. Proper examination of these aquatic animals for eggs could be a useful approach. Using the ingredients like vinegar which possess anti-trematodal properties also could help in controlling the disease. As the disease is associated with low host specificity, health promotion and awareness programs along with the medical support could surely aid in the eradication of echinostomiasis [77].

12.4.3 Heterophyiasis

Heterophyiasis is caused by several minute intestinal flukes belonging to diverse family of intestinal flukes. They are extremely minute in size (<2.5 mm length) and can affect fish-eating birds and mammals including human. They belong to the families including *Heterophyidae*, *Microphallidae*, *Diplostomidae*, *Lecithodendriidae*, *Gymnophallidae*, and *Plagiorchiidae* [5, 9]. As the geographic distribution of flukes is found only in family *Heterophyidae*, these flukes are categorized as heterophyids. Heterophyiasis is principally caused by different species of genus *Heterophyes*, including *H. nocens*, *H. dispar*, *H. heterophyes* [85]. In addition, other genera like *Haplorchis*, *Carneophallus*, and *Metagonimus* may also cause the disease. There are no major differences between heterophyiasis and other trematodes in terms of infection and transmission, which occur through digenetic life cycle using aquatic plants and fish [68].

12.4.3.1 Life Cycle and Transmission

Heterophyes heterophyes, *Haplorchis taichui*, *Haplorchis pumilio*, *Metagonimus yokogawai*, and *Stellantchasmus falcatus* are known to be a few of the many heterophyids known to parasitize humans [2]. Adult worms adhere and inhabit in the intestinal mucosal wall of birds and mammals, where they produce embryonated eggs to be excreted through feces. Thereupon they get absorbed by freshwater snails (*Semisulcospira* and *Melanoides*, etc.) to get hatched out to release miracidia. These miracidia develop into sporocysts after penetrating intestinal wall and further develop into rediae and cercariae through asexual reproduction. The cercariae become encysted to become metacercariae and attach to the plants and fish, later to be consumed by either humans or other host animals, which get infected [86].

12.4.3.2 Epidemiology

Heterophyiasis is predominantly found in Asian countries including China, Korea, Lao PDR, Japan, Taiwan, Thailand, Vietnam, and Philippines. It is also found in Siberia, Turkey, Hawaii, and the Balkans. The discovery of rich diversity of the heterophyids is attributed to the chemotherapy, which revealed the maximum number of worms with notable variety that a man can get infected with [85]. *M. yokogawai* affects around 1.2% of the population in Korea and is prevalent in those areas that consume raw sweetfish (5–20%). In addition, *H. nocens* and *H. dispar* are known to be found in Japan, Korea, and China, and *H. dispar* and *H. heterophyes* exist in Egypt, Sudan, the Middle East, and Southeast Europe. *S. falcatus,* along with other heterophyids is reported to have occurred in Vietnam [87, 88].

12.4.3.3 Pathogenesis

Infections with heterophyids are generally asymptomatic or associated with only mild intestinal discomforts including abdominal pain, comprise mucous diarrhea, intermittent that occur because of adherence of adult worms to the intestinal walls. The attachment of the worms to the wall results in mild inflammation, mainly eosinophilia, hyper secretion of mucin, superficial necrosis, and bleeding at the site of attachment. Histopathological examination of tissues affected by *H. taichui* reveal hemorrhages in mucosal wall, inflammation, villi deformation, fibrosis of the submucosa, followed by ulceration [89]. Though symptoms appear frequently in severe infections, they fade in due time, with flukes remaining in the gut, and they may recur in case of further infections. This leads to occasional diarrhea, as reported in endemic regions. Infections are sometimes considered lethal because the presence of eggs in cardiac muscles can lead into cardiac dilatation and of flukes in cardiac beriberi. There is a chance of entering of adult flukes and eggs to other body organs including brain, lungs, spleen, and heart through bloodstream [86]. Trapped eggs in the walls of the gut may elicit granulomatous fibrosis and lesions. Heterophyid myocarditis is a severe condition where cardiac enlargement, embolism of the capillaries, subepicardial edema, epicardium hemorrhage, fragmentation of muscle fiberes, cough, fatigue, dyspnea, ascites, palpitation, cyanosis, loss of reflexes, and abnormal heart sounds are observed. Eggs or worms present in the brain or spinal cord may lead into loss of motor and sensory functions, transverse myelitis, and other neurological diseases [90].

12.4.3.4 Diagnosis

Diagnosis primarily relies on fecal examination for eggs. This method is not feasible due to the morphological features between the eggs of heterophyids and other species. Low output of eggs (34–45) and low intensity of infection (<100 worms) are responsible for this. Eggs look strikingly like those of *Clonorchis* spp. and *Opisthorchis* spp. However, several differential diagnosis methods have been developed. Extraintestinal cases of the disease can be dealt with autopsy or surgery, followed by careful searching [2]. Modern-day approaches like immunodiagnosis (antigen- and antibody-based detection), PCR, DNA sequencing have also been developed for different species causing heterophyiasis [91, 92].

12.4.3.5 Chemotherapy

Praziquantel at a single dose of 25 mg/kg is recommended since it is highly efficient and a broad spectrum anti-trematodal drug [93]. No recommendations of other drugs are reported.

12.4.3.6 Phytotherapy

In phytotherapy, aqueous extract of *Artocarpus takoocha* kills *Haplorchis taichui*, a causative agent of heterophyiasis. Though the extract is able to dissociate the tegument of the fluke, specific phytoconstituent(s) responsible for this action is/are yet to be elucidated [94]. Although mirazid, a plant-based drug from *Commiphora molmol* has been proved to possess anti-heterophyiasis activity, further and advanced phytochemical- and plant extract-based studies need to be conducted against the rest of the intestinal flukes [95]. In addition, *Momordica charantia* (bitter gourd) extracts at different concentrations were able to kill *Stellantchasmus falcatus*. This anthelminthic activity of the extract was attributed to tannins, which ameliorates the tegumental surface of the trematode. Due to the rupturing of the surface, bleeding and curving of the spines were observed. However, other metabolites like saponins, alkaloids, and flavonoids were also reported with anti-trematodal effects on different worms [96].

12.4.3.7 Prevention and Control

Since the pathological effects of the disease are not effective on humans, at least not as of other trematodes, treatment with praziquantel can be encouraging. In the absence of vaccine, health awareness camps, promotion of health education on safe consumption of aquatic products can surely be able to lift off the endemic areas from disease burden. With the limited information available on few species, it is also advised to conduct surveys and advanced studies to determine the clinical consequences and effect of therapeutic agents [2].

12.5 Perspectives and Projections

Foodborne trematodes have been infecting many countries of the world with their extensive pathogenicity, and this is supported equally by the eating habits of people who are unaware of the possible infection with trematodes. The chemotherapeutic agents used so far have given good results; but the risk of adverse effects and possible drug resistance prevail the efforts of public health agencies around the world [7, 8, 97]. In the vista, phytochemicals from various plants, irrespective of their total medicinal potential, could be useful in treating the foodborne trematodiases. It becomes noteworthy that, despite several health promoting efforts like awareness camps, mass chemotherapy, and health education, the diseases continue to spread in both endemic and non-endemic areas. For example, FAO has issued control guidelines to be implemented at various levels of administration including for citizens who could practice successful preventive measures [98]. Food security plays an important role in the maintenance of good health. Thus, every individual must be taught to practice hygienic modes of food preparation and consumption, without casting a negative effect on the traditional practices [99]. This could be further modified with contributions from different sectors of the society to practice "One-Health Approach" [100]. In addition to this, environmental sanitation is also essential, as the trematodes spread through fecal matter deposited near water sources. In the perspective of medicine, significant improvisations have been done in the last decade. Development of chemotherapeutic models for the drug analysis, targeting the virulent protein for the inhibition of the associated pathogenic mechanisms, designing efficient and accurate diagnostic tools and techniques like LC–MS, NGS, MALDI-TOF (omics approach) have been done [26, 101, 102]. Advanced methods like RNAi technology, proteomic analysis for possible target search, and DNA sequencing have also been done. Even in this regard, there have been some gaps in research and followed methodologies, which needs to be taken care of. For example, people may go untreated in mass drug administration, polyparasitism, uneven distribution of worms, technical limitations in diagnosis and research, increasing population, political factors, economic burden, etc. Considering these multifactorial options, treatment of the infected becomes difficult despite colossal amount of research and funding [103].

As many of the endemic countries belong to the substandard levels in the global economy, it becomes essential to utilize available natural resources as source of medicine [104–108]. Thus, usage of cost-effective measures like phytotherapy becomes a feasible approach. Several studies have been conducted and the anti-trematodal activity of plant extracts has been reported. Though many of them have restricted the research to the extract level [50, 62], few advanced studies have reported the activity of individual phytochemical compounds that can be utilized for plant-based drug discovery process [13, 51]. The reported phytochemicals can be easily extracted from the plants. These phytochemical compounds need to be characterized, evaluated for their cytotoxic effects and possibilities of their synthesis in *in vitro* conditions need to be studied. In addition, qualitative and quantitative assays using individual compounds in a dose-dependent and combination approaches could reduce the burden of multifactorial results. These isolated compounds could be further modified according to requirement based on cytotoxic evaluations done in primary screening. These compounds could be analyzed *in silico* for effective binding and inhibitory effect on specific component or any part of the identified target [104, 105, 108]. Further, the bioavailability of the compounds could be dealt to modify the compound to match the physiological conditions of the target organ. Usage of *in silico* methods like molecular modeling and docking can save a lot of animals, which are used in laboratories. Following these protocols would surely help the treatment of parasites through plant-based drugs.

12.6 Conclusion

Foodborne trematodiases are known for their extensive capability of transmission and pathogenesis, resulting in the escalation of infections all over the world. The infective nature of the disease is attributed to both the causative agents and the humans due to the repeated unhygienic practices followed during consumption of aquatic food. The chapter focuses on the various factors of the disease like life cycle, transmission, epidemiology, pathogenesis, diagnosis, available therapeutic agents, phytotherapy, and preventive measures. The aim of the chapter is to highlight the burden of the foodborne trematodes with respect to latest advancements in disease factors listed above as well as the phytochemical-based approach to cure the disease. Although approaches to develop the plant-based methods have been practiced in the past 20 years, noticeable research has not been completed in this aspect, leaving behind much of the facts unknown about the efficacy of the plant extracts and their reported constituents. This chapter has highlighted the work done so far, in the phytotherapeutic approach and further deliberates on the need of the focus in research and development. It can thus be deduced that with the following of protocol given, it could be easier to cure food-borne trematodiases with phytochemical-based drugs.

References

1 Fürst, T., Keiser, J., and Utzinger, J. (2012). Global burden of human food-borne trematodiases: a systematic review and meta-analysis. *The Lancet Infectious Diseases* [Internet] 12 (3): 210–221.

2 Sripa, B., Kaewkes, S., Intapan, P.M. et al. (2010). Food-borne trematodiases in Southeast Asia. Epidemiology, pathology, clinical manifestation and control [Internet]. *Advances in Parasitology* 72: 305–350.

3 WHO (2020). Foodborne trematode infections [Internet]. https://www.who.int/news-room/fact-sheets/detail/foodborne-trematode-infections (accessed 11 August 2020).

4 Fried, B. and Abruzzi, A. (2010). Food-borne trematode infections of humans in the United States of America. *Parasitology Research* 106 (6): 1263–1280.

5 Keiser, J. and Utzinger, J. (2009). Food-borne trematodiases. *Clinical Microbiology Reviews* 22 (3): 466–483.

6 Fried, B., Graczyk, T.K., and Tamang, L. (2004). Food-borne intestinal trematodiases in humans. *Parasitology Research* 93 (2): 159–170.

7 Keiser, J., Duthaler, U., and Utzinger, J. (2010). Update on the diagnosis and treatment of food-borne trematode infections. *Current Opinion on Infective Diseases* 23 (5): 513–520.

8 Keiser, J. and Utzinger, J. (2007). Advances in the discovery and development of trematocidal drugs. *Expert Opinion on Drug Discovery* 2 (sup1): S9–S23.

9 Toledo, R., Esteban, J.G., and Fried, B. (2012). Current status of food-borne trematode infections. *European Journal of Clinical Microbiology & Infectious Diseases* 31 (8): 1705–1718.

10 Keiser, J. and Utzinger, J. (2005). Emerging foodborne trematodiases. *Emerging Infectious Diseases* 11 (10): 1507–1514.

11 Keiser, J. and Utzinger, J. (2007). Food-borne trematodiasis: current chemotherapy and advances with artemisinins and synthetic trioxolanes. *Trends in Parasitology* 23 (11): 555–562.

12 Catalano, S., Symeou, A., Marsh, K.J. et al. (2019). Mini-FLOTAC as an alternative, non-invasive diagnostic tool for *Schistosoma mansoni* and other trematode infections in wildlife reservoirs. *Parasites & Vectors* [Internet] 12 (1): 1–9.

13 Pereira, C.A.J., Oliveira, L.L.S., Coaglio, A.L. et al. (2016). Veterinary parasitology anti-helminthic activity of *Momordica charantia* L. against *Fasciola hepatica* eggs after twelve days of incubation in vitro. *Veterinary Parasitology* [Internet] 228: 160–166.

14 Koko, W.S., Galal, M., and Khalid, H.S. (2000). Fasciolicidal efficacy of *Albizia anthelmintica* and *Balanitesa egyptiaca* compared with albendazole. *Journal of Ethnopharmacology* 71: 247–252.

15 Hammami, H. and Ayadi, A. (2008). Molluscicidal and antiparasitic activity of *Solanum nigrumvillosum* against *Galba truncatula* infected or uninfected with *Fasciola hepatica*. *Journal of Helminthology* 82 (3): 235.

16 Hotez, P.J. and Ehrenberg, J.P. (2010). Escalating the global fight against neglected tropical diseases through interventions in the Asia Pacific region. *Advances in Parasitology* 72: 31–53.

17 Araki, H., Ong, K.I., Lorphachan, L. et al. (2018). Mothers' *Opisthorchis viverrini* infection status and raw fish dish consumption in Lao People's Democratic Republic: determinants of child infection status. *Tropical Medicine and Health* 46 (1): 29.

18 Sithithaworn, P., Andrews, R.H., Van De, N. et al. (2012). The current status of opisthorchiasis and clonorchiasis in the Mekong Basin. *Parasitology International* 61 (1): 10–16.

19 Tang, Z.L., Huang, Y., and Yu, X.B. (2016). Current status and perspectives of *Clonorchis sinensis* and clonorchiasis: epidemiology, pathogenesis, omics, prevention and control. *Infectious Diseases of Poverty* 5 (1): 71.

20 Vale, N., Gouveia, M.J., Gärtner, F., and Brindley, P.J. (2020). Oxysterols of helminth parasites and pathogenesis of foodborne hepatic trematodiasis caused by *Opisthorchis* and *Fasciola* species. *Parasitology Research* 23: 1–1.

21 Sripa, B., Deenonpoe, R., and Brindley, P.J. (2017). Co-infections with liver fluke and Helicobacter species: a paradigm change in pathogenesis of opisthorchiasis and cholangiocarcinoma? *Parasitology International* 66 (4): 383–389.

22 Pitaksakulrat, O., Sithithaworn, P., Laoprom, N. et al. (2013). A cross-sectional study on the potential transmission of the carcinogenic liver fluke *Opisthorchis viverrini* and other fishborne zoonotic trematodes by aquaculture fish. *Foodborne Pathogens and Disease* 10 (1): 35–41.

23 Suwannatrai, A., Saichua, P., and Haswell, M. (2018). Epidemiology of *Opisthorchis viverrini* infection. *Advances in Parasitology* 101: 41–67.

24 Lvova, M.N., Tangkawattana, S., Balthaisong, S. et al. (2012). Comparative histopathology of *Opisthorchis felineus* and *Opisthorchis viverrini* in a hamster model: an implication of high pathogenicity of the European liver fluke. *Parasitology International* 61 (1): 167–172.

25 Smout, M.J., Laha, T., Mulvenna, J. et al. (2009). A granulin-like growth factor secreted by the carcinogenic liver fluke, *Opisthorchis viverrini*, promotes proliferation of host cells. *PLoS Pathogens* 5 (10): e1000611.

26 Pakharukova, M.Y., Zaparina, O.G., Kapushchak, Y.K. et al. (2019). *Opisthorchis felineus* infection provokes time-dependent accumulation of oxidative hepatobiliary lesions in the injured hamster liver. *PLoS One* 14 (5): e0216757.

27 Kim, T.S., Pak, J.H., Kim, J.B., and Bahk, Y.Y. (2016). Clonorchis sinensis, an oriental liver fluke, as a human biological agent of cholangiocarcinoma: a brief review. *BMB Reports* 49 (11): 590.

28 Marcos, L.A., Terashima, A., and Gotuzzo, E. (2008). Update on hepatobiliary flukes: fascioliasis, opisthorchiasis and clonorchiasis. *Current Opinion in Infectious Diseases* 21 (5): 523–530.

29 Upatham, E.S. and Viyanant, V. (2003). *Opisthorchis viverrini* and opisthorchiasis: a historical review and future perspective. *Actatropica* 88 (3): 171–176.

30 Worasith, C., Kamamia, C., Yakovleva, A. et al. (2015). Advances in the diagnosis of human opisthorchiasis: development of *Opisthorchis viverrini* antigen detection in urine. *PLoS Neglected Tropical Diseases* 9 (10): e0004157.

31 Wonkchalee, N., Boonmars, T., Laummaunwai, P. et al. (2013). A combination of praziquantel and the traditional medicinal plant *Thunbergia laurifolia* on *Opisthorchis viverrini* infection and cholangiocarcinoma in a hamster model. *Parasitology Research* 112 (12): 4211–4219.

32 Aukkanimart, R., Boonmars, T., Sriraj, P. et al. (2015). Anthelmintic, anti-inflammatory and antioxidant effects of *Garcinia mangostana* extract in hamster opisthorchiasis. *Experimental Parasitology* [Internet] 154: 5–13.

33 Boonjaraspinyo, S., Boonmars, T., Aromdee, C., and Kaewsamut, B. (2010). Effect of fingerroot on reducing inflammatory cells in hamster infected with *Opisthorchis viverrini* and *N*-nitrosodimethylamine administration. *Parasitology Research* 106 (6): 1485–1489.

34 Intuyod, K., Priprem, A., Limphirat, W. et al. (2014). Anti-inflammatory and anti-periductal fibrosis effects of an anthocyanin complex in *Opisthorchis viverrini*-infected hamsters. *Food Chemistry and Toxicology* [Internet] 74: 206–215.

35 Rhee, J.K., Baek, B.K., and Ahn, B.Z. (1985). Structural investigation on the effects of the herbs on *Clonorchis sinensis* in rabbits. *The American Journal of Chinese Medicine* 13 (1–4): 119–125.

36 Jongsuksuntigul, P. and Imsomboon, T. (2003). Opisthorchiasis control in Thailand. *Actatropica* 88 (3): 229–232.

37 Mas-Coma, S., Bargues, M.D., and Valero, M.A. (2005). Fascioliasis and other plant-borne trematode zoonoses. *International Journal of Parasitology* 35 (11–12): 1255–1278.

38 Mas-Coma, S., Valero, M.A., and Bargues, M.D. (2009). Fasciola, lymnaeids and human fascioliasis, with a global overview on disease transmission, epidemiology, evolutionary genetics, molecular epidemiology and control. *Advances in Parasitology* 69: 41–146.

39 Leonardo, L., Hernandez, L., Magturo, T.C. et al. (2020). Current status of neglected tropical diseases (NTDs) in the Philippines. *Acta Tropica* 203: 105284.

40 Shin, E.H., Guk, S.M., Kim, H.J. et al. (2008). Trends in parasitic diseases in the Republic of Korea. *Trends in Parasitology* 24 (3): 143–150.

41 Auer, H. and Aspöck, H. (2014). Helminths and helminthoses in Central Europe: general overview and diseases caused by trematodes (flukes). *Wiener Medizinische Wochenschrift* 164 (19–20): 405–413.

42 Mas-Coma, S., Valero, M.A., and Bargues, M.D. (2009). Climate change effects on trematodiases, with emphasis on zoonotic fascioliasis and schistosomiasis. *Veterinary Parasitology* 163 (4): 264–280.

43 Le, T.H., Van De, N., Agatsuma, T. et al. (2008). Human fascioliasis and the presence of hybrid/introgressed forms of *Fasciola hepatica* and *Fasciola gigantica* in Vietnam. *International Journal for Parasitology* 38 (6): 725–730.

44 Lim, J.H., Mairiang, E., and Ahn, G.H. (2008). Biliary parasitic diseases including clonorchiasis, opisthorchiasis and fascioliasis. *Abdominal Imaging* 33 (2): 157.

45 Yesildag, A., Yıldız, H., Demirci, M. et al. (2009). Biliary fascioliasis: sonographic appearance patterns. *Journal of Clinical Ultrasound* 37 (1): 26–30.

46 Tran, D.H. and Phung, H.T. (2020). Detecting *Fasciola hepatica* and *Fasciola gigantica* microRNAs with loop-mediated isothermal amplification (LAMP). *Journal of Parasitic Diseases* 44 (2): 364–373.

47 Mazeri, S., Sargison, N., Kelly, R.F. et al. (2016). Evaluation of the performance of five diagnostic tests for *Fasciola hepatica* infection in naturally infected cattle using a Bayesian no gold standard approach. *PLoS One* 11 (8): e0161621.

48 Sarkari, B. and Khabisi, S.A. (2017). Immunodiagnosis of human fascioliasis: an update of concepts and performances of the serological assays. *Journal of Clinical and Diagnostic Research: JCDR* 11 (6): OE05.

49 Ubeira, F.M., Muiño, L., Valero, M.A. et al. (2009). MM3-ELISA detection of *Fasciola hepatica* coproantigens in preserved human stool samples. *The American Journal of Tropical Medicine and Hygiene.* 81 (1): 156–162.

50 Hamed, N., Njeh, F., Damak, M. et al. (2015). Atividademoluscicida e larvicida das partesaéreas de Atriplexinflata contra o molusco Galba truncatula, hospedeirointermediário da *Fasciola hepatica*. *Revista do Instituto de Medicina Tropical de São Paulo.* 57 (6): 473–479.

51 Whiteland, H.L., Chakroborty, A., Forde-Thomas, J.E. et al. (2018). An Abeisprocera-derived tetracyclic triterpene containing a steroid-like nucleus core and a lactone side chain attenuates in vitro survival of both *Fasciola hepatica* and *Schistosoma mansoni*. *International Journal of Parasitology – Drugs and Drug Resistance* [Internet] 8 (3): 465–474.

52 Hotez, P.J., Strych, U., Lustigman, S., and Bottazzi, M.E. (2016). Human anthelminthic vaccines: rationale and challenges. *Vaccine* 34 (30): 3549–3555.

53 Fischer, P.U. and Weil, G.J. (2015). North American paragonimiasis: epidemiology and diagnostic strategies. *Expert Review of Anti-Infective Therapy* 13 (6): 779–786.

54 Singh, T.S., Sugiyama, H., and Rangsiruji, A. (2012). Paragonimus & paragonimiasis in India. *The Indian Journal of Medical Research.* 136 (2): 192.

55 Calvopiña, M., Romero, D., Castañeda, B. et al. (2014). Current status of Paragonimus and paragonimiasis in Ecuador. *Memórias do Instituto Oswaldo Cruz* 109 (7): 849–855.

56 Diaz, J.H. (2013). Paragonimiasis acquired in the United States: native and nonnative species. *Clinical Microbiology Reviews* 26 (3): 493–504.

57 Saleri, N. and Ryan, E.T. (2019). Respiratory Infections. In: *Travel Medicine*, 4e (eds. J.S. Keystone, P.E. Kozarsky, B.A. Connor, et al.), 527–537. Elsevier.

58 Blair, D., Agatsuma, T., and Wang, W. (2007). Paragonimiasis. In: *Food-Borne Parasitic Zoonoses: Fish and Plantborne Parasites*, World Class Parasites, vol. 11 (eds. K.D. Murrell and B. Fried), 117–150. New York: Springer.

59 Kalhan, S., Sharma, P., Sharma, S. et al. (2015). *Paragonimus westermani* infection in lung: a confounding diagnostic entity. *Lung India: Official Organ of Indian Chest Society* 32 (3): 265.

60 Rathore, M.H. (2006). Chapter 38 – Toxocariasis, hydatid disease of the lung, strongyloidiasis, and pulmonary paragonimiasis. In: *Kendig's Disorders of the Respiratory Tract in Children*, 7e (eds. V. Chernick, T.F. Boat, R.W. Wilmott and A. Bush), 566–577. W.B. Saunders.

61 Intapan, P.M., Sanpool, O., Thanchomnang, T. et al. (2012). Molecular identification of a case of *Paragonimus pseudoheterotremus* infection in Thailand. *The American Journal of Tropical Medicine and Hygiene.* 87 (4): 706–709.

62 Giang, P.M. and Otsuka, H. (2018). New compounds and potential candidates for drug discovery from medicinal plants of Vietnam. *Chemical & Pharmaceutical Bulletin* 66 (5): 493–505.

63 Procop, G.W. (2009). North American paragonimiasis (caused by *Paragonimus kellicotti*) in the context of global paragonimiasis. *Clinical Microbiology Reviews* 22 (3): 415–446.

64 Njeh, F., Feki, H., Koubaa, I. et al. (2016). Molluscicidal activity of *Solanum elaeagnifolium* seeds against *Galba truncatula* intermediate host of *Fasciola hepatica*: identification of β-solamarine. *Pharmaceutical Biology* 54 (4): 726–731.

65 Mas-Coma, S. (2014). Helminth-Trematode: *Fasciolopsis buski*. In: (ed. Y. Motarjemi), 146–157. Encyclopedia of Food Safety: Academic Press.

66 Chai, J.Y. and Jung, B.K. (2019). Epidemiology of trematode infections: an update. In: *Digenetic Trematodes*, 359–409. Springer Champions.

67 Fox, L.M. (2012). Intestinal trematodes. In: *Principles and Practice of Pediatric Infectious Diseases*, 1363–1364. Elsevier.

68 Slifko, T.R., Smith, H.V., and Rose, J.B. (2000). Emerging parasite zoonoses associated with water and food. *International Journal for Parasitology* 30 (12-13): 1379–1393.

69 Fiamma, M., Longoni, S.S., Ngo, T.M. et al. (2015). *Fasciolopsis buski* infection in a Vietnamese pregnant woman with systemic lupus erythematosus. *The Journal of Infection in Developing Countries* 9 (6): 670–673.

70 Biswal, D.K., Roychowdhury, T., Pandey, P., and Tandon, V. (2018). De novo genome and transcriptome analyses provide insights into the biology of the trematode human parasite *Fasciolopsis buski*. *PLoS One* 13 (10): e0205570.

71 Ma, J., Sun, M.M., He, J.J. et al. (2017). *Fasciolopsis buski* (Digenea: *Fasciolidae*) from China and India may represent distinct taxa based on mitochondrial and nuclear ribosomal DNA sequences. *Parasites & Vectors* 10 (1): 101.

72 Biswal, D.K., Ghatani, S., Shylla, J.A. et al. (2013). An integrated pipeline for next generation sequencing and annotation of the complete mitochondrial genome of the giant intestinal fluke, *Fasciolopsis buski* (Lankester, 1857) Looss, 1899. *PeerJ* 1: e207.

73 Roy, B., Dasgupta, S., and Tandon, V. (2009). Ultrastructural observations on *Fasciolopsis buski* and its alterations caused by shoot extract of *Alpinia nigra*. *Microscopy Research and Technique* 72 (2): 61–66.

74 Kar, P.K., Tandon, V., and Saha, N. (2002). Anthelmintic efficacy of *Flemingia vestita*: genistein-induced effect on the activity of nitric oxide synthase and nitric oxide in the trematode parasite, *Fasciolopsis buski*. *Parasitology International* 51 (3): 249–257.

75 Challam, M., Roy, B., and Tandon, V. (2010). Effect of *Lysimachiaramosa* (*Primulaceae*) on helminth parasites: motility, mortality and scanning electron microscopic observations on surface topography. *Veterinary Parasitology* 169 (1-2): 214–218.

76 Graczyk, T.K. and Fried, B. (1998). Echinostomiasis: a common but forgotten food-borne disease. *The American Journal of Tropical Medicine and Hygiene.* 58 (4): 501–504.

77 Toledo, R. and Esteban, J.G. (2016). An update on human echinostomiasis. *Transactions of the Royal Society of Tropical Medicine and Hygiene* 110 (1): 37–45.

78 Sohn, W.M., Chai, J.Y., Yong, T.S. et al. (2011). *Echinostoma revolutum* infection in children, Pursat Province, Cambodia. *Emerging Infectious Diseases* 17 (1): 117.

79 Sah R, Khadka S, Hamal R, Poudyal S. Human echinostomiasis: a case report. *BMC Research Notes* 2018;11(1):1-6.

80 Toledo, R. and Fried, B. (2014). *Helminth-Trematode: Echinostoma* (ed. Y. Motarjemi), 134–139. Encyclopedia of Food Safety, Academic Press.

81 Leles, D., Cascardo, P., dos Santos, F.A. et al. (2014). Insights about echinostomiasis by paleomolecular diagnosis. *Parasitology International* 63 (4): 646–649.

82 Koné, W.M., Vargas, M., and Keiser, J. (2012). Anthelmintic activity of medicinal plants used in Côte d'Ivoire for treating parasitic diseases. *Parasitology Research* 110 (6): 2351–2362.

83 Abdel-Ghaffar, F., Semmler, M., Al-Rasheid, K.A. et al. (2011). The effects of different plant extracts on intestinal cestodes and on trematodes. *Parasitology Research* 108 (4): 979–984.

84 Ferreira, J.F., Peaden, P., and Keiser, J. (2011). In vitro trematocidal effects of crude alcoholic extracts of *Artemisia annua, A. absinthium, Asiminatriloba,* and *Fumaria officinalis*. *Parasitology Research* 109 (6): 1585–1592.

85 Chai, J.Y. (2014). Helminth-Trematode: *Heterophyes heterophyes*. In: *Encyclopedia of Food Safety* (ed. Y. Motarjemi), 158–163. Academic Press.

86 Bruschi, F. and Gomez-Morales, M.A. (2017). Parasites. In: *Foodborne Diseases (3)* (eds. C.E.R. Dodd, T. Aldsworth, R.A. Stein, et al.), 305–324. Academic Press.

87 Chontananarth, T. and Wongsawad, C. (2010). Prevalence of *Haplorchis taichui* in field-collected snails: a molecular approach. *The Korean Journal of Parasitology* 48 (4): 343.

88 Thien, P.C., Dalsgaard, A., Thanh, B.N. et al. (2007). Prevalence of fishborne zoonotic parasites in important cultured fish species in the Mekong Delta, Vietnam. *Parasitology Research* 101 (5): 1277–1284.

89 Ashour, D.S., Othman, A.A., and Radi, D.A. (2014). Insights into regulatory molecules of intestinal epithelial cell turnover during experimental infection by *Heterophyes heterophyes*. *Experimental Parasitology* 143: 48–54.

90 Blauwet, L.A. and Cooper, L.T. (2010). Myocarditis. *Progress in Cardiovascular Diseases* 52 (4): 274–288.

91 Lovis, L., Mak, T.K., Phongluxa, K. et al. (2009). PCR diagnosis of *Opisthorchis viverrini* and *Haplorchis taichui* infections in a Lao community in an area of endemicity and comparison of diagnostic methods for parasitological field surveys. *Journal of Clinical Microbiology* 47 (5): 1517–1523.

92 Han, E.T., Yang, H.J., Park, Y.J. et al. (2014). *Metagonimus yokogawai*: a 100-kDa somatic antigen commonly reacting with other trematodes. *The Korean Journal of Parasitology*. 52 (2): 201.

93 Chai, J.Y., Han, E.T., Park, Y.K. et al. (2000). High endemicity of *Metagonimus yokogawai* infection among residents of Samchok-shi, Kangwon-do. *The Korean Journal of Parasitology* 38 (1): 33.

94 Wongsawad, C., Wongsawad, P., Luangphai, P., and Kumchoo, K. (2005). In vitro effects of aqueous extract from *Artocarpus takoocha* Roxb on tegumental surface of *Haplorchis taichui*. *The Southeast Asian Journal of Tropical Medicine and Public Health* 36: 112.

95 Massoud, A.M., El-Shazly, A.M., and Morsy, T.A. (2007). Mirazid (*Commiphora molmol*) in treatment of human heterophyiasis. *Journal of the Egyptian Society of Parasitology* 37 (2): 395–410.

96 Poolperm, S. and Jiraungkoorskul, W. (2017). An update review on the anthelmintic activity of bitter gourd, Momordica charantia. *Pharmacognosy Reviews* 11 (21): 31.

97 Fairweather, I., Brennan, G.P., Hanna, R.E. et al. (2020). Drug resistance in liver flukes. *International Journal for Parasitology: Drugs and Drug Resistance* 12: 39–59.

98 Food and Agriculture Organization of the United Nations (FAO) (2018). Summary of the tripartite meeting to accelerate prevention and control of neglected foodborne parasitic zoonoses in selected Asian countries [Internet]. http://www.fao.org/3/ca6886en/ca6886en.pdf (accessed 11 August 2020; updated 16–18 October 2018).

99 Abdussalam, M., Käferstein, F.K., and Mott, K.E. (1995). Food safety measures for the control of foodborne trematode infections. *Food Control* 6 (2): 71–79.

100 Johansen, M.V., Lier, T., and Sithithaworn, P. (2015). Towards improved diagnosis of neglected zoonotic trematodes using a One Health approach. *Acta Tropica* 141: 161–169.

101 Huang, X.H., Qian, M.B., Zhu, G.H. et al. (2020). Assessment of control strategies against *Clonorchis sinensis* infection based on a multi-group dynamic transmission model. *PLoS Neglected Tropical Diseases* 14 (3): e0008152.

102 Toledo, R., Bernal, M.D., and Marcilla, A. (2011). Proteomics of foodborne trematodes. *Journal of Proteomics* 74 (9): 1485–1503.

103 Haçarız, O. and Sayers, G.P. (2016). The omic approach to parasitic trematode research – a review of techniques and developments within the past 5 years. *Parasitology Research* 115 (7): 2523–2524.

104 Egbuna, C., Kumar, S., Ezzat, S.M. et al. (eds.) (2019). *Phytochemicals as Lead Compounds for New Drug Discovery*, 1e. Cambridge, USA: Elsevier.

105 Shashank, K. and Egbuna, C. (eds.) (2019). *Phytochemistry: In vitro and In Silico Updates*, 1e. Singapore: Springer Nature.

106 Tijjani, H., Zangoma, M.H., Mohammed, Z.S. et al. (2020). Polyphenols: classifications, biosynthesis and bioactivities. In: *Functional Foods and*

Nutraceuticals: Bioactive Components, Formulations and Innovations (eds. C. Egbuna and G.-D. Tupas), 389–414. Switzerland AG: Springer Nature.

107 Walag, A.M.P., Ahmed, O., Jeevanandam, J. et al. (2020). Health benefits of Organosulfur compounds. In: *Functional Foods and Nutraceuticals: Bioactive Components, Formulations and Innovations* (eds. C. Egbuna and G.-D. Tupas), 445–472. Switzerland AG: Springer Nature.

108 Ezzat, S.M., Jeevanandam, J., Egbuna, C. et al. (2019). Phytochemicals as sources of drugs. In: *Phytochemistry: In vitro and In Silico Updates* (eds. S. Kumar and C. Egbuna), 3–22. Singapore: Springer Nature.

109 Keiser, J., Vargas, M., and Winter, R. (2012). Anthelminthic properties of mangostin and mangostin diacetate. *Parasitology International* 61 (2): 369–371.

Part IV

Bacterial Infections

13

Bacteriocidal Effects of Phytochemicals on *Mycobacterium ulcerans*, the Causative Agents of Buruli Ulcer

Alloysius Chibuike Ogodo[1], Francis Ushie Ebuara[1], and Chidera Favour Ogodo[2]

[1] Department of Microbiology, Faculty of Pure and Applied Sciences, Federal University Wukari, Wukari, Nigeria
[2] Department of Medical Services, Federal University Wukari, Wukari, Nigeria

List of Abbreviations

BU	Buruli Ulcer
MIC	Minimum Inhibitory Concentration
NTD	Neglected Tropical Disease
WHO	World Health Organization

13.1 Introduction

Mycobacterium ulcerans infection is one of the most common infections of mycobacteria. It is ranked third worldwide after *Mycobacterium tuberculosis* and *Mycobacterium leprae* infections especially in immunocompromised individuals as well as one of the most neglected tropical diseases (NTD) [1–3]. This infection is reported to affect people living in tropical, sub-tropical and temperate regions and has been found and identified in Africa, South America, Southeast Asia, and the Western pacific region (Figure 13.1) where it affects people of all ages. However, infection has been found more in children below 15 years of age with more disease burden on people living in rural areas where there are limited access to good health services [4, 5]. The infection caused by *Mycobacterium ulcerans* is referred to as buruli ulcer (BU).

Neglected Tropical Diseases and Phytochemicals in Drug Discovery, First Edition.
Edited by Chukwuebuka Egbuna, Muhammad Akram, and Jonathan Chinenye Ifemeje.

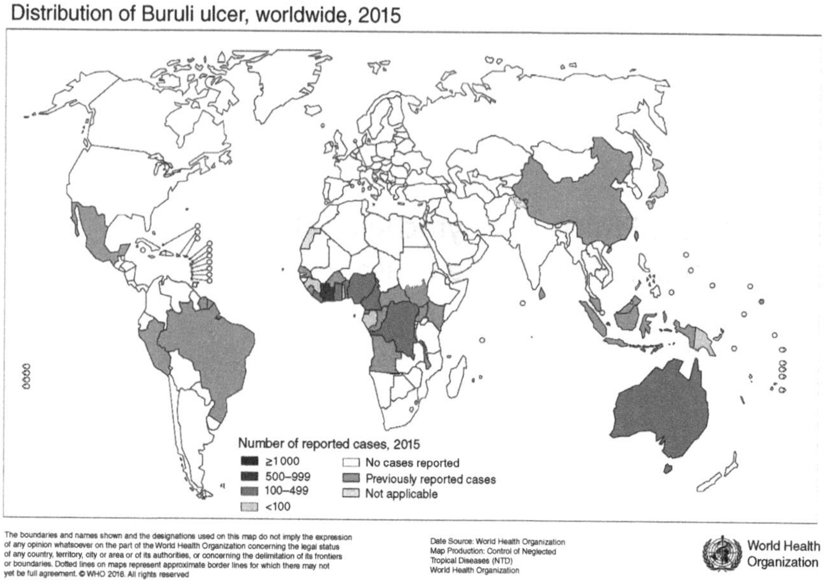

Figure 13.1 Global distribution of buruli ulcer disease as of 2015. *Source:* Distribution of Buruli ulcer, worldwide, 2015. World Health Organization. Retrieved from http://gamapserver.who.int/mapLibrary/Files/Maps/Buruli_2015.png?ua=1 (accessed 15 September 2020).

Mycobacterium ulcerans infection (buruli ulcer) (Figure 13.2) mostly affects the lower extremities. It may also affect the face, chest wall, abdomen, and back which could present in both active and inactive forms. The active form is characterized by lesions that are non-ulcerative, ulcers, nodules, plaques and edematous indurations. The onset of the infection appears as small nodule that later turns ulcerative, giving rise to an ulcer with a cotton wool appearance. The skin surrounding the lesion thickens and darkens with subsequent development of scar. The inactive form is characterized by evidence of previous infection with depressed stellate scar with or without sequellae [2, 6].

Buruli ulcer are commonly associated with a wet trauma especially in a stagnant water body or wetlands. Although the exact mechanisms or mode of transmission of *Mycobacterium ulcerans* in not fully understood, several methods have been suggested. For instance, *Aedes camptorhynchus*, a mosquito species in Australia and *Naucoridae,* an aquatic organism in West Africa have been implicated in the transmission where exposure to wetlands were reported. Transmission through animal bite, gunshot injury, trauma through superficial or deep layers of skin has been reported and it is likely that *Mycobacterium ulcerans* was on the surface of the skin of the victims before the trauma and was then inoculated in the bite or trauma [2, 7, 8].

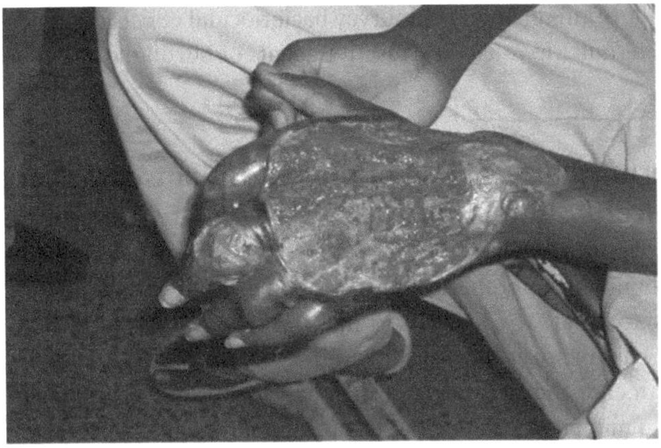

Figure 13.2 A typical buruli ulcer in a 17-year-old boy identified during the assessment. *Source:* Adapted from Centers for Disease Control and Prevention – DPDx; Global Health, Division of Parasitic Diseases. Freely taken from: https://wwwnc.cdc.gov/eid/article/13/5/07-0065-f1 (accessed 15 September 2020).

Recommended treatment for *Mycobacterium ulcerans* infection consists of surgical excision and/or skin grafting. However, spontaneous healing may occur [2]. Antibiotics using clofazimine has been reported earlier [8]. The success of a pilot study in 2005 from Ghana [9], confirming that human lesion can be sterilized with antibiotics led to the introduction of new treatment guideline by WHO using antibiotic combination therapy for buruli ulcer. This protocol is a combination treatment involving oral intake of rifampicin, 10 mg/kg body weight daily for eight weeks, combined with streptomycin 15 mg/kg body weight for eight weeks. However, this treatment has some limitations such as long treatment duration (about 56 days), toxicity, streptomycin administration by parenteral means and drug–drug interactions caused by rifampicin as well as burden on families due to long stay during treatment. On the other hand, surgery which is inevitable for some forms of the disease is neither accessible nor affordable to a large population [1, 10–12]. These limitations opened way for the search for more effective treatment options which plant-based products or phytochemicals may hold great potential.

Moreover, traditional plant-based treatment is the first approach by people living in remote areas where it is believed that the modern treatments are expensive and with side effects [13, 14]. Therefore, plant products, based on traditional knowledge may present a good source of early treatment and prevention of buruli ulcer. Although, there are limited knowledge on the real efficacy of medicinal plants and active principles as well as the standardization of the plant products [15]. The scope of this chapter is to survey bacteriocidal effects of phytochemicals on *Mycobacterium ulcerans*, the causative agents of buruli ulcer.

13.2 Ethnopharmacological Use of Plants to Control *Mycobacterium ulcerans* Infection

There are wide variations in the proportion of the population using traditional plant-based products to treat *Mycobacterium ulcerans* infections worldwide. However, in the rural African areas, the use of plants for medicine plays an important role in daily health care and well-being of the people who sometimes prefer plant-based products to modern medicine. In this process mostly practiced by traditional healers, parts or whole plants are prepared and administered as oral suspension and/or as pomade and bandages. Most times the remedies involved two or more parts of different plants [15]. In many rural settlements people usually rely on traditional medicine for treatment of various diseases caused by different microorganisms including mycobacteria. This could be due to cultural belief, cost, convenience, or inaccessibility of modern treatment.

Some plants or medicinal plants are prescribed or bought as over-the-counter therapeutics for the prevention and treatment of diseases [16]. In Africa, the vast majority (70–80%) of people consult traditional medical practitioners for their healthcare needs [17], including the management of buruli ulcer disease [15]. Ethnopharmacological studies have revealed several assay methods employed to estimate antimycobacterial activity of plant extracts, pure compounds, and plant fractions. These are described by Fokou et al. [15] as the proportion method and the rasazurin microtiter assay. The proportion method applied Löwenstaein–Jensen medium. This uses a drug containing media associated with or without dilution of the drug and drug-free media. Then an inoculum that has been standardized is applied to the drug containing media and drug-free media and incubated. The concentration that inhibited not lower than 99% of *Mycobacterium ulcerans* was regarded as minimum inhibitory concentration (MIC). On the other hand, the *Resazurin Microtiter Assays (REMA)* or alamar Blue-based assays are sensitive and powerful tools used to detect inhibition of growth by a substance at various concentration ranges on mammalian cell lines and microorganism like the mycobacteria. This method is performed in the 96-well microplates and can be read with a spectrophotometer or spectrofluorometer or visually. In the latter case, the MIC was the lowest concentration of extract that prevents a color change of resazurin from blue to pink. Viable mycobacteria will reduce the blue resazurin into the pink-colored resorufin.

13.3 Activities of Selected Plants and Bioactive Compounds Against *Mycobacterium ulcerans*

Several authors have reported medicinal species with biological activity against *M. ulcerans in vitro* including *Aloe vera, Alstonia boonei* De Wild, *Pupalia lappacea* Juss, *Lannea nigritana* Keay, *Aglaonema commutatum* Schott, *Ageratum conyzoides, Senna occidentalis* (L.) (Table 13.1).

13.3.1 Aloe vera

Aloe vera is a perennial medicinal plant which is commonly used for beauty and treatment of skin infections. Moreover, it has been associated with wound healing [50]. Addo et al. [20] have demonstrated its *in vitro* activities against seven different *Mycobacterium ulcerans* strains. Phenolic compounds such as chromone, anthrone, anthraquinone, aloin, and emodin have been reported in *Aloe vera* and are known for their antibacterial effects. Also, its anti-inflammatory action mediated through cyclooxygenase pathway inhibition and reduction of prostaglandin E_2 production from arachidonic acid was reported. An anti-inflammatory compound, C-glucosyl chromone has been identified from *Aloe vera* gel extracts. These properties (antibacterial, wound healing, and anti-inflammatory) support the use of *Aloe vera* as an antimycobacterial agent. In addition, *A. vera* contains glucomannan and gibberellin. These compounds lead to increase in collagen and proteoglycan synthesis upon their interactions with receptors of growth factors as well as macrophages. Moreover, oral administration of *A. vera* extract results in accelerated tissue repair due to collagen cross-linking. Furthermore, acemannan, saponin, and genin isolated from *Aloe* leaves have been demonstrated to accelerate wound-healing process [22, 50, 51].

13.3.2 *Jatropha curcas* L.

Jatropha curcas leaf extract has been demonstrated to inhibit *Mycobacterium ulcerans* growth (MIC = 250 μg/ml) *in vitro* and used traditionally to cure buruli ulcer. *J. curcas* plant parts contain several phytochemicals and bioactive compounds such as apigenin, alkaloids, flavanoids, isovitexin, sapogenins, sterol, vitexin, stigmasterol, β-D-sitosterol, β-D-glucoside, and alcohol as well as 1-triacontanol [15, 25]. Proteolytic enzymes from the plant latex have shown wound-healing capacity [26].

Table 13.1 Some plants and their reported bioactive compounds against *Mycobacterium ulcerans*.

S/N	Plant species	Plant parts	Isolated phytochemical/ bioactive compound	MIC (μg/ml)	References
1	*Pupalia lappacea* Juss	Leaves	Alkaloids, terpenoids, glycosides, saponins, flavonoids, tannins, coumarins, and steroids	40	[14, 18, 19]
2	*Aglaonema commutatum* Schott.	Leaves	Calcium oxalate	40	[20, 21]
3	*Aloe vera* (L.) Burm. f.	Leaves	Carbohydrate (glucomannans, manose 6-phosphate), phenolic compounds (chromone, anthraquinone)	40	[15, 20]
4	*Alstonia boonei* De Wild.	Leaves, bark	Alkaloids (*N*-formylechitamidine, echitamidine, echitamine, voacangine, akuammidine), iridoids (boonein, loganin), tannins, steroids, glycosides, saponins, flavonoids, and terpenoids and triterpenoids	40	[20, 22, 23]
5	*Zanthoxylum zanthoxyloides* (Lam.)	Leaves, root-bark	Essential oils, alkaloids, coumarins, sterols	12.5–25	[20, 24]
6	*Jatropha curcas* L.	All plant parts	Pigenin, alkaloids, flavanoids, isovitexin, sapogenins, sterol, vitexin, stigmasterol, β-D-sitosterol, β-D-glucoside, alcohol, 1-triacontanol	250	[15, 25, 26]

Table 13.1 (Continued)

S/N	Plant species	Plant parts	Isolated phytochemical/ bioactive compound	MIC (μg/ml)	References
7	*Syzygium aromaticum* (L.) Merr. & L.M. Perry	Seeds, clove buds	Essential oil, eugenol, eugenyl acetate, β-caryophyllene, vanillin, crategolicacid, tannins, gallotannic acid, methyl salicylate, flavonoids, kaempferol, rhamnetin, etc.	25	[20, 27]
8	*Senna occidentalis* (L.)	Leaves	Acrosin, aloe-emodin, emodin, anthraquinones, anthrones, sitosterols, tannins, and xanthosine	40	[20, 28, 29]
9	*Bridelia ferruginea* Benth	Stem barks	Alkaloids, tannins, polyphenols, saponins, and other secondary metabolites	40	[20, 30]
10	*Cleome viscosa*	Leaf, whole plant	Alkaloids, flavonoids, tannins, saponins, terpenes, carbohydrates	40	[22, 31–33]
11	*Spathodea campanulata*	Leaves, root, bark, flower	Contains carbohydrates, proteins, glycosides, steroids, tannins, alkaloids, anthraquinone, flavonoids, and saponins	25	[15, 34–37]
12	*Phyllanthus fraternus*	Whole plant	Aminoglycosides, alkaloids, flavonoids, saponins, steroids, tannins, estradiol, corilagin, ellagic acid, gallic acid, rutinoside, lupa, and methyl salicylate	40	[20, 22, 38]

(Continued)

Table 13.1 (Continued)

S/N	Plant species	Plant parts	Isolated phytochemical/ bioactive compound	MIC (μg/ml)	References
13	*Capsicum annum*	Fruits	Capsaicin Saponins, alkaloids, quaternary bases, anthracenosides, flavanosides, flavonds, coumarin derivatives, steroid glycosides, anthocyanosides, essential oils, waxes, colored materials (mainly capsanthin, capsorubin, zeaxanthin, cryptoxanthin, and lutein), and several capsaicinoids	40	[20, 39, 40]
14	*Ageratum conyzoides*	Leaves	Methoxylated flavonoids, chromenes, dihydrolycopsamine, pyrrolizidine alkaloids, acetyl-lycopsamine	40	[20, 41, 42]
15	*Lannea nigritana* Keay	Leaves	Tannins, phenolic compounds (Lanneanol)	40	[20, 43]
16	*Psidium guajava* (guava)	Leaves	Phenolics, flavonoid, carotenoid, volatile oil, tannins, terpenoid	40	[20, 44]
17	*Allium sativum* L.	Cloves	Alliin, γ-glutamylcysteine, ajoenes, and vinyldithiins, allicin	0.78–6.25	[20, 22]
18	*Solanum torvum* Sw.	Leaves	Stigmasterol, isoflavonoid sulfate, chlorogenone, 22-β-*O*-spirostanol oligoglycosides, steroidal glycosides, 26-*O*-β-glucosidase, C-22 steroidal lactone saponins, tetratriacontanoic acid, sitosterol, campesterol, stigmasterol	40	[20, 22, 45, 46]

Table 13.1 (Continued)

S/N	Plant species	Plant parts	Isolated phytochemical/ bioactive compound	MIC (μg/ml)	References
19	*Hydrastis canadensis*	Root	Organic acids, fatty acids, alkaloids, volatile oil, flavonoids, sterols (*β*-sitosterol 3-*O*-*β*-D-glucoside), resin	0.39–6.25	[20, 22, 24]
20	*Annickia chlorantha*	Stem	Berberine and protoberberine alkaloids	1.95–7.81	[47]
21	*Greenwayodendron suaveolens*	Various plant parts	Polysin, greenwayodendrin-3-1, *N*-acetylpolyveoline, 3-*O*-acetyl greenwayodendrin, polyveoline	3	[48, 49]

13.3.3 Cleome viscosa

Cleome viscosa (tickweed, spider or wild mustard) plant occurs in grassland and woodland. Traditionally, the leaves of *Cleome viscosa* and the whole plant are used to treat skin infections, wounds, ulcers, and inflammations. The aerial parts are reported to have wound-healing, analgesic, antimicrobial, and anti-ulcer properties [31, 32]. The plant contains alkaloids, flavonoids, tannins, saponins, and terpenes in addition to carbohydrates which could be responsible for the antibacterial and wound-healing capacities [33].

13.3.4 *Alstonia boonei* De Wild

Alstonia boonei (devil tree) has been associated with the treatment of diseases in many parts of Africa with the bark reported to be useful in wound and ulcer treatment [52]. The leaves show *in vitro anti-Mycobacterium ulcerans* activity. Also, many phytochemicals have been isolated from the plant, including alkaloids (*N*-formylechitamidine, echitamidine, echitamine, voacangine, akuammidine), iridoids (boonein, loganin), tannins, steroids, glycosides, saponins, flavonoids, and terpenoids and triterpenoids (ursolic acid, lupeol) which might support its reported *Mycobacterium ulcerans* inhibitory effects [20, 22].

13.3.5 *Spathodea campanulata* P. Beauv.

Spathodea campanulata leaves are used in the treatment of ulcers and skin conditions as well as buruli ulcers and swollen cheeks [15, 34, 35]. According to the report of Addo et al. [20], decoctions from the root of *S. campanulata* inhibited the growth of seven isolates of *Mycobacterium ulcerans* (MIC = 25 μg/ml). A range of phytochemicals have been identified from the different parts of the plant. For instance, the bark contains carbohydrates, proteins, glycosides, steroids, alkaloids, and tannins while the leave and flower contains phytochemicals such as tannins, alkaloids, glycosides, carbohydrates, anthraquinone, flavonoids, and saponins [36, 37]. The potency of this plant against *M. ulcerans* could be attributed to these phytochemicals.

13.3.6 Phyllanthus fraternus

Phyllanthus fraternus is a small, erect, annual herb that grows 30–40 cm in height and indigenous to the rainforests of the Amazon and tropical Africa [53]. *Phyllanthus fraternus* is used to treat various diseases including blennorrhagia, diabetes, fever, flu, tumors, dysentery, jaundice, and vaginitis [54]. *P. fraternus* has *in vitro* antibacterial activity against *M. ulcerans* strains [20]. The plant contains a variety of phytochemicals such as aminoglycosides, alkaloids, flavonoids, saponins, steroids, tannins and other secondary metabolites including estradiol, corilagin, ellagic acid, gallic acid, rutinoside, lupa, and methyl salicylate [22, 38]. These phytochemicals and metabolites may be responsible for their antimycobacterial activity.

13.3.7 *Pupalia lappacea* Juss

Pupalia lappacea, commonly found in forests and savannah regions of tropical Africa and Asia, is a medicinal plant used for the treatment of wound infections [18, 55]. *P. lappacea* leaf extract contains alkaloids, glycosides, flavonoids, saponins, tannins, coumarins, terpenoids, and steroids such as 1-docosanol, stearicacid, stigmasterol, sitosterol, *N*-benzoyl-L-phenylalaninolacetate, setosterol-3-*O*-Dglucopyranoside, stigmasterol-3-*O*-D-glucopyranoside, and 20-hydroxyl ecdysone. These phytochemicals promote *M. ulcerans* wound infections healing process in mammals [18].

13.3.8 Capsicum annum

Capsicum annum has an active compound called capsaicin used in the management of a variety of medical conditions. This plant has been reported to have antibacterial and wound-healing properties especially when combined with

Allium sativum and *Pothos scandens* L. Saponins, alkaloids, quaternary bases, anthracenosides, flavanosides, flavonds, coumarin derivatives, steroid glycosides, anthocyanosides, essential oils, waxes, colored materials (mainly capsanthin, capsorubin, zeaxanthin, cryptoxanthin, and lutein), and several capsaicinoids are the major phytochemicals found in this plant which could be responsible for their bioactivity [39, 40]. The macerated fruits are used to treat buruli ulcer [20].

13.3.9 Ageratum conyzoides

Ageratum conyzoides is used in traditional medicine across African. It has been reported to be effective in the treatment of ulcers, wounds, and burns. *Ageratum conyzoides* contains phenolic phytochemicals including methoxylated flavonoids and chromenes and dihydrolycopsamine as well as pyrrolizidine alkaloids and acetyl-lycopsamine with other derivatives [41, 42].

13.3.10 *Lannea nigritana* Keay

Lannea nigritana is a medicinal plant used traditionally for the treatment and management of various disease conditions including wounds. Phytochemical studies on this plant are scanty. However, studies show that the plant contains tannins and phenolic compounds (Lanneanol) [20, 43]. These phytochemicals could be responsible for their wound-healing and antimycobacterial properties.

13.3.11 Psidium guajava

Psidium guajava (guava) is indigenous to tropical America which has been used traditionally to treat various diseases of human including ulcers, cholera, and wounds. Reports indicated that *Psidium guajava* has anti-inflammatory, antibacterial, and antimycobacterial properties. Phytochemicals such as phenolics, flavonoid, carotenoid, volatile oil, tannins, terpenoid, and triterpene have been isolated, which are believed to contribute to its biological activities [20, 44].

13.3.12 *Bridelia ferruginea* Benth

Bridelia ferruginea Benth is a medicinal tree plant whose stem barks are used to treat secondary bacterial wound infections. It has also been reported to be a potent inhibitor of *Mycobacterium ulcerans* strains. The plant contains alkaloids, tannins, polyphenols, saponins and other secondary metabolites which are believed to be responsible for their reported antimycobacterial activity [20, 30].

13.3.13 *Senna occidentalis* (L.)

Senna occidentalis is widely used in tropical and rainforest regions of the world as cholagogue, laxative, vermifuge, analgesic, hepatoprotective, febrifuge, and diuretic. The paste prepared from the leaves facilitates healing of wounds by external application. Similarly, the leaf extract has been found to be effective against many microorganisms including *Mycobacterium ulcerans* strains. The presence of compounds such as acrosin, aloe-emodin, emodin, anthraquinones, anthrones, sitosterols, tannins, and xanthosine can justify its reported antimycobacterial activity [20, 28, 29].

13.3.14 *Aglaonema commutatum* Schott

Aglaonema commutatum is a common ornamental plant used for the treatment of BU. Leaf washings of *Aglaonema commutatum* are inhibitory to *Escherichia coli*, *Pseudomonas aeruginosa*, and *Staphylococcus aureus* which are commonly involved in wound infection [20, 56].

Other common medicinal plants reported to have antimycobacterial activity, at least *in vitro* include *Solanum torvum* Sw. Commonly known as Turkey berry, *Allium sativum* L., *Syzygium aromaticum*, *Hydrastis canadensis*, *Zanthoxylum zanthoxyloides*, *Annickia chlorantha* commonly known as Yellow Wood, etc. [22].

13.4 Compounds with Potency Against *Mycobacterium ulcerans*

There is limited research that attempts to isolate compounds from plants to treat BU. However, two studies conducted separately in 2014 and 2015 respectively, isolated 2,3,6-trihydroxymethyl benzoate from *Sorindeia juglandifolia* [15, 48] and holadyamine, holaphyllonol, holamine/holaphyllamine, and then compound C (structure yet to be confirmed) (Figure 13.3) from *Holarrhena floribunda*. These compounds demonstrated MICs ranging from 50 to 125 mg/ml [15, 57].

13.5 Future Prospects of Phytochemicals Against *M. ulcerans*

Phytochemicals occur naturally in plants (fruits, vegetables, legumes, and grains) which have been reported to be effective in disease management and in alleviating the risk of diseases [58]. The information on the various phytochemicals of most plants with anti-*M. ulcerans* activity is scanty [59], however, available literature (partly discussed in this chapter), shows that the prospects of phytochemicals

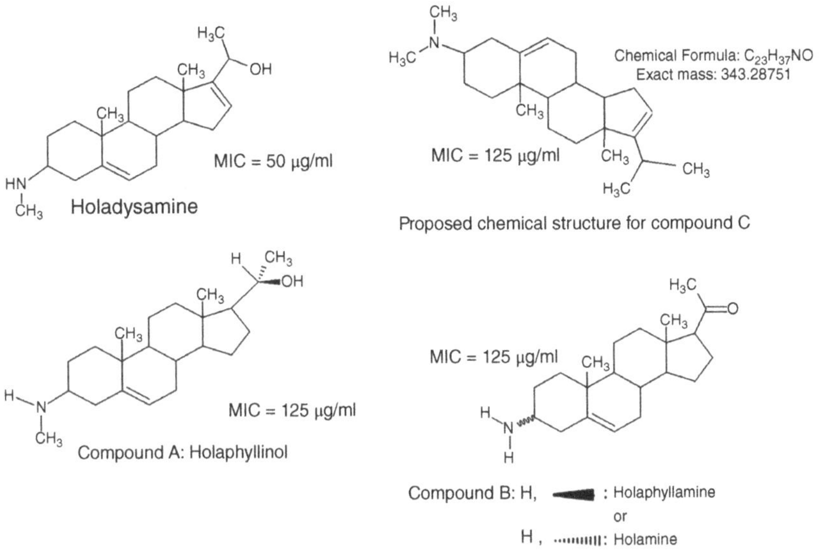

Figure 13.3 Isolated compounds with activity against *Mycobacterium ulcerans*. *Source:* Fokou et al. [22].

against *Mycobacterium ulcerans* is high. This is because the medicinal values of plants are believed to be dictated by their phytochemical constituents [60].

Many people in rural areas within the African continent depend on medicinal plants to treat their ailments and since prehistoric times, various parts of plants are used in the prevention and treatment of various diseases [61–69], hence, the need to study their phytochemical constituents against buruli ulcer. Moreover, different plants have been used as a source of raw materials for the development of new drugs either in a pure compound form or as extract. Therefore, validation studies on the plants with anti-buruli ulcer should be done. Further studies (especially *in vivo*) of preparations from identified plant species are required. Phytochemical studies which will open way for isolation, identification and characterization of bioactive compounds with anti-*Mycobacterium ulcerans* properties should be a major focus. Determination of the dose regimen and toxicity levels of these plant-based products should be the focus of further studies.

13.6 Conclusion

This chapter discussed the effect and potentials of phytochemicals against *Mycobacterium ulcerans*, the causative agent of buruli ulcer (a neglected tropical disease) and highlighted some plants as well as isolated plant components with potentials to combat the ulcers associated with BU. Many plant species and phytochemical have been linked to the treatment of BU and showed inhibition of the etiologic agent (*Mycobacterium ulcerans*). However, there is limited information in literature on the isolated active principle from the various plants with claims of anti-BU properties as well as limited *in vivo* studies of the plant preparations/phytochemicals against BU. Hence, the need for more researches in this area.

References

1 Fokou, P.V.T., Kissi-Twum, A.A., Yeboah-Manu, D. et al. (2016). In vitro activity of selected West African medicinal plants against *Mycobacterium ulcerans* Disease. *Molecules* 21: 445. https://doi.org/10.3390/molecules21040445.

2 Ramahia, J.W., Annabb, H., Al Karmic, M. et al. (2017). Chronic cutaneous mycobacterial ulcers due to *Mycobacterium ulcerans* (Buruli ulcer): the first indigenous case report from Jordan and a literature review. *International Journal of Infectious Diseases* 58 (2017): 77–81.

3 World Health Organization (2013). Buruli ulcer (*Mycobacterium ulcerans* Infection). Fact Sheet N 199, 2013. http://www.who.int/mediacentre/factsheets/fs199/en (accessed 22 August 2020).

4 Pieter, S., Anthony, A., Bayonne, L.M. et al. (2006). Heterogeneity among *Mycobacterium ulcerans* isolates from Africa. *Emerging Infectious Diseases* 12: 844–847.

5 World Health Organization (2017). *Buruli Ulcer Disease (Mycobacterium ulcerans Infection)*. Geneva: WHO http://www.who.int/mediacentre/factsheets/fs199/en/index.html (accessed 20 April 2020).

6 Chukwuekezie, O., Ampadu, E., Sopoh, G. et al. (2007). Buruli Ulcer, Nigeria. *Emerging Infectious Diseases* 13 (5): 782–783.

7 Johnson, P.D., Azuolas, J., Lavender, C.J. et al. (2007). *Mycobacterium ulcerans* in mosquitoes captured during outbreak of Buruli ulcer, Southeastern Australia. *Emerging Infectious Diseases* 13: 1653–1660.

8 Walsh, D.S., Portaels, F., and Meyers, W.M. (2008). Buruli ulcer (*Mycobacterium ulcerans* infection). *Transactions of the Royal Society of Tropical Medicine and Hygiene* 102: 969–978.

9 Etuaful, S., Carbonnelle, B., Grosset, J. et al. (2005). Efficacy of the combination Rifampin–Streptomycin in preventing growth of *Mycobacterium ulcerans* in early lesions of Buruli ulcer in humans. *Antimicrobial Agents and Chemotherapy* 49: 3182–3186.

10 Johnson, P.D. (2010). Should antibiotics be given for Buruli ulcer? *Lancet* 2010 (20): 664–672.

11 Kibadi, K., Boelaert, M., Fraga, A.G. et al. (2010). Response to treatment in a prospective cohort of patients with large ulcerated lesions suspected to be Buruli ulcer (*Mycobacterium ulcerans* disease). *PLoS Neglected Tropical Diseases* 4: e736.

12 Zhang, T., Li, S.Y., Converse, P.J. et al. (2013). Rapid, serial, non-invasive assessment of drug efficacy in mice with autoluminescent *Mycobacterium ulcerans* infection. *PLoS Neglected Tropical Disease* 7: e2598.

13 Stienstra, Y., van der Graaf, W.T.A., Asamoa, K., and van der Werf, T.S. (2002). Beliefs and attitudes toward Buruli ulcer in Ghana. *The American Journal of Tropical Medicine and Hygiene* 67 (2): 207–213.

14 Johnson, R.C., Makoutode, M., and Hougnihin, R. (2004). Traditional treatment for Buruli ulcer in Benin. *Médecine Tropicale* 64 (2): 145–150.

15 Fokou, P.V.T., Nyarko, A.K., Appiah-Opong, R. et al. (2015). Ethnopharmacological reports on anti-Buruli ulcer medicinal plants in three West African countries. *Journal of Ethnopharmacology* 172: 297–311.

16 Dean, G., Wan-Ying, W., Min, Y. et al. (2015). A holistic approach to the quality control of traditional Chinese medicines. *Science* 347 (Suppl. 6219): S29–S31.

17 Cunningham, A.B. (1993). *African Medicinal Plants: Setting Priorities at the Interface Between Conservation and Primary Health Care.* People and Plant Working Paper 1, 1–49. Paris, France: United Nations Educational, Scientific and Cultural Organization (UNESCO) Press.

18 Udegbunam, S.O., Udegbunam, R.I., Muogbo, C.C. et al. (2014). Wound healing and antibacterial properties of methanolic extract of *Pupalia lappacea* Juss in rats. *BMC Complementary and Alternative Medicine* 14 (157) https://doi.org/10.1186/1472-6882-14-157.

19 Naidu, P.P., Madakka, M., and Rajesh, B. (2014). *Pupalia lappacea* Juss [L]: a review of phytochemistry and therapeutic application. *Asian Journal of Pharmaceutical and Clinical Research* 7 (1): 15–18.

20 Addo, P., Quartey, M., Abbas, M. et al. (2007). In-vitro susceptibility of *Mycobacterium ulcerans* to herbal preparations. *The Internet Journal of Tropical Medicine* 4: 1–11.

21 Fuller, T.C. and McClintock, E.M. (1986). *Poisoneous Plant of California*, 269–270. Los Angeles: University of California Press.

22 Fokou, P.V.T., Nyarko, A.K., Appiah-Opong, R. et al. (2015). Update on medicinal plants with potency on *Mycobacterium ulcerans*. *BioMed Research International* 2015: 917086. (pp. 1–16). http://dx.doi.org/10.1155/2015/917086.

23 Chime, S.A., Ugwuoke, E.C., OnyishiIV, E.C. et al. (2013). Formulation and evaluation of *Alstonia boonei* stem bark powder tablets. *Indian Journal of Pharmaceutical Sciences.* 75: 226–230.

24 Matu, E.N. (2011). *Zanthoxylum zanthoxyloides* (Lam.) Zepern. & Timler. Record from PROTA4U [Internet] (ed. G.H. Schmelzer and A. Gurib-Fakim). Wageningen, Netherlands: PROTA (Plant Resources of Tropical Africa/

Ressources végétales de l'Afrique tropicale). http://www.prota4u.org/search.asp (accessed 6 September 2020).

25 Prasad, D.M.R., Izam, A., and Khan, M.M.R. (2012). *Jatropha curcas*: plant of medical benefits. *Journal of Medicinal Plant Research* 6: 2691–2699.

26 Lans, C., Harper, T., Georges, K., and Bridgewater, E. (2001). Medicinal and ethnoveterinary remedies of hunters in Trinidad. *BMC Complementary and Alternative Medicine* 1: 10. https://doi.org/10.1186/1472-6882-1-10.

27 Mittal, M., Gupta, V., Parashar, P. et al. (2014). Phytochemical evaluation and pharmacological activity of *Syzygium aromaticum*: a comprehensive review. *International Journal of Pharmacy and Pharmaceutical Sciences* 6 (8): 67–72.

28 Yadav, J.P., Arya, V., Yadav, S. et al. (2010). *Cassia occidentalis* L.: a review on its ethnobotany, phytochemical and pharmacological profile. *Fitoterapia* 81 (4): 223–230.

29 Silva, M.G.B., Aragao, T.P., and Vasconcelos, C.F.B. (2011). Acute and subacute toxicity of *Cassia occidentalis* L. stem and leaf in Wistar rats. *Journal of Ethnopharmacology* 136 (2): 341–346.

30 Kolawole, O., Oguntoye, S.O., Agbede, O., and Olayemi, A.B. (2006). Studies on the efficacy of *Bridelia ferruginea* Benth. Bark extract in reducing coliform load and BOD of domestic wastewater. *Ethnobotanical Leaflets* 10: 228–238.

31 Panduraju, T., Parvathi, B., Rammohan, M., and Reddy, C.S. (2011). Wound healing properties of *Cleome viscosa* Linn. *Hygeia: Journal for Drugs and Medicines* 3 (1): 41–45.

32 Mohammad, F. and Avijit, M. (2013). Wound healing potentiality of methanolic extract of aerial parts of *Cleome viscose*. *International Journal of PharmTech Research* 5 (3): 978–982.

33 Upadhyay, A., Chattopadhyay, P., Goyary, D. et al. (2014). Topical application of *Cleome viscosa* increases the expression of basic fibroblast growth factor and type III collagen in rat cutaneous wound. *BioMed Research International* 2014: 680879.

34 Adjet, A., Fokou, G., Kouame, D., and Mosi, L. (2013). Soins de l'ulcere de buruli dans le district sanitaire de Yamoussoukro. In *Conference Internationale Africa. sur l'Ecosant'e*, 2013.

35 Sam, G.H., Adu, F., and Annan, K. (2013). Ethnobotanical survey of medicinal plants used for the treatment of diarrhoea and skin ulcer in the Brong Ahafo region of Ghana. *Journal of Medicinal Plant Research* 7: 3280–3285.

36 Ilodigwe, E.E., Akah, P.A., and Nworu, C.S. (2010). Anticonvulsant activity of ethanol leaf extract of *Spathodea campanulata* P. Beauv (Bignoniaceae). *Journal of Medicinal Food* 13 (4): 827–833.

37 Ilodigwe, E.E., Akah, P.A., and Nworu, C.S. (2010). Evaluation of the acute and subchronic toxicities of ethanol leaf extract of *Spathodea campanulata* P. Beauv. *International Journal of Applied Research in Natural Products* 3 (2): 17–21.

38 Christian, M. (2013). Steroids – chemical constituents of *Phyllanthus fraternus* Webster through TLC and HPTLC. *International Research Journal of Chemistry* 3: 39–49.

39 Ayyanar, M. and Ignacimuthu, S. (2009). Herbal medicines for wound healing among tribal people in Southern India: ethnobotanical and scientific evidences. *International Journal of Applied Research in Natural Products* 2 (3): 29–42.

40 Lagu, C. and Kayanja, F.B.I. (2013). Acute toxicity profiles of aqueous and ethanolic extracts of *Capsicum annum* seeds from South Western Uganda. In: *Insights from Veterinary Medicine* (ed. R. Payan-Carreira), 257–268. London: InTech Open http://dx.doi.org/10.5772/53599.

41 Diallo, A.S., Eklu-Gadegkeku, K., Agbonon, A. et al. (2010). Acute and sub-chronic (28-day) oral toxicity studies of hydroalcohol leaf extract of *Ageratum conyzoides* L (Asteraceae). *Tropical Journal of Pharmaceutical Research* 9 (1): 463–467.

42 Bosi, C.F., Rosa, D.W., and Grougnet, R. (2013). Pyrrolizidine alkaloids in medicinal tea of *Ageratum conyzoides*. *Brazilian Journal of Pharmacognosy* 23 (3): 425–432.

43 Kapche, G.D.W.F., Laatsch, H., and Fotso, S. (2007). Lanneanol: a new cytotoxic dihydroalkylcyclohexenol and phenolic compounds from *Lannea nigritana* (Sc. Ell.) Keay. *Biochemical Systematics and Ecology* 35 (8): 539–543.

44 Metwally, A.M., Omar, A.A., Ghazy, N.M. et al. (2011). Monograph of *Psidium guajava* L. leaves. *Pharmacognosy Journal* 3: 89–104.

45 Mohan, M. and Bhandare, S. (2012). Protective effect of *Solanum torvum* against testicular toxicity in male wistar rats. *International Journal of Pharmacy and Pharmaceutical Sciences* 4 (3): 188–192.

46 Agrawal, A.D., Bajpei, P.S., Patil, A.A., and Bavaskar, S.R. (2010). *Solanum torvum* Sw. – a phytopharmacological review. *Der Pharmacia Lettre* 2 (4): 403–407.

47 Kimbi, H.K. and Fagbenro-Beyioku, A.F. (1996). Efficacy of *Cymbopogon giganteus* and *Enantia chrantha* against chloroquine resistant *Plasmodium yoelii nigeriensis*. *East African Medical Journal* 73 (10): 636–637.

48 Donfack, V.F.D., Roque, S., Trigo, G. et al. (2014). Antimycobacterial activity of selected medicinal plants extracts from Cameroon. *International Journal of Biological and Chemical Sciences* 8 (1): 273–288.

49 Ngantchou, I., Nyasse, B., Denier, C. et al. (2010). Antitrypanosomal alkaloids from *Polyalthia suaveolens* (Annonaceae): their effects on three selected glycolytic enzymes of *Trypanosoma brucei*. *Bioorganic and Medicinal Chemistry Letters* 20 (12): 3495–3498.

50 Surjushe, A., Vasani, R., and Saple, D.G. (2008). *Aloe vera*: a short review. *Indian Journal of Dermatology* 53 (4): 163–166.

51 Mukherjee, P.K., Nema, N.K., Maity, N. et al. (2014). Phytochemical and therapeutic profile of *Aloe vera*. *Journal of Natural Remedies* 14 (1): 1–26.

52 Igbinaduwa, P.O. and Obasuyi, O. (2012). Phytochemical screening and antibacterial activity of stem bark extract of *Alstonia Boonei* Dewild (Apocynaceae). *Journal of Pharmaceutical and Allied Sciences* 9 (3): 1574–1579.

53 Matur, B.M., Matthew, T., and Ifeanyi, C.I.C. (2009). Analysis of the phytochemical and in vivo antimalaria properties of *Phyllanthus fraternus* Webster extract. *New York Science Journal* 2 (5): 12–19.

54 Leslie, T. (2003). *Herbal Secrets of the Rainforest, Traditional Medicine*, 2e. Geneva, Switzerland: Sage Press Incorporated, WHO.

55 Akindele, A.J., Salako, O.A., and Ohonbamu, U.V. (2014). Evaluation of the antidiarrhoeal activity of the hydroethanolic leaf extract of *Pupalia lappacea* Linn. Juss. (Amaranthaceae). *Journal of Ethnopharmacology* 151 (2): 984–989.

56 Kamel, F.H., Ismael, H.M., and Ahmed, A.A. (2012). Effect of natural surface secretes of some common ornamental plants leaves on pathogenic micro-organisms. *Journal of Life Sciences* 6: 1387–1390.

57 Yemoa, A., Gbenou, J., and Afolabi, D. (2015). Beninese medicinal plants as a source of antimycobacterial agents: bioguided fractionation and in vitro activity of alkaloids isolated from *Holarrhena floribunda* used in traditional treatment of buruli ulcer. *BioMed Research International* 2015: 835767.

58 Onyeka, E.U. and Nwambekwe, I.O. (2007). Phytochemical profile of some green leafy vegetables in south east Nigeria. *Nigerian Food Journal* 25: 67–76.

59 Oduse, A., Idowu, M.A., and Adegbite, A. (2012). Chemical and phytochemical profile of some uncommon green leafy vegetables consumed in South-West, Nigeria. *Journal of Environmental Science, Toxicology and Food Technology* 1 (3): 22–26.

60 Fallah, H.S.M., Alavian, H.R., Heydari, M.R., and Abolmaali, K. (2005). The efficacy of liv-52 on liver cirrhotic patients: a randomized, double blind, placebo-controlled first approach. *Phytomedicine* 9: 619–624.

61 Egbuna, C. and Ifemeje, J.C. (2015). Biological functions and anti-nutritional effects of phytochemicals in living system. *IOSR Journal of Pharmacy and Biological Sciences (IOSR-JPBS)* 10 (2): 10–19. iosrjournals.org/iosr-jpbs/papers/Vol10-issue2/.../C010231019.pdf (accessed 17 March 2020).

62 Ifemeje, J.C., Egbuna, C., Udedi, S.C., and Iheukwumere, H.I. (2014). Phytochemical and in vitro antibacterial evaluation of the ethanolic extract of the stem bark of *Entada africana* Guill. & Perr and *Sarcocephalus latifolus*. *International Journal of Biochemistry Research & Review* 4 (6): 584–592. https://doi.org/10.9734/IJBCRR/2014/11554.

63 Iheukwumere, I.H., Dimejesi, S.A., Iheukwumere, C.M. et al. (2020). Plasmid curing potentials of some medicinal plants against citrate negative motile salmonella species. *European Journal of Biomedical and Pharmaceutical Sciences* 7 (5): 40–47.

64 Egbuna, C., Kumar, S., Ezzat, S.M. et al. (eds.) (2019). *Phytochemicals as Lead Compounds for New Drug Discovery*, 1e. Cambridge, USA: Elsevier.

65 Shashank, K. and Egbuna, C. (eds.) (2019). *Phytochemistry: In vitro and In Silico Updates*, 1e. Singapore: *Springer Nature*.

66 Egbuna, C., Ifemeje, J.C., Udedi, S.C., and Kumar, S. (eds.) (2018). *Phytochemistry, Volume 1: Fundamentals, Methods, and Applications*, 1e. New York: Apple Academic Press/CRC Taylor & Francis.

67 Egbuna, C., Kumar, S., Ifemeje, J.C., and Kurhekar, J.V. (2018). *Phytochemistry, Volume 2: Pharmacognosy, Nanomedicine, and Contemporary Issues*, 1e. New York: Apple Academic Press/CRC Taylor & Francis.

68 Egbuna, C., Ifemeje, J.C., Kumar, S., and Sharif, N. (eds.) (2018). *Phytochemistry, Volume 3: Marine, Industrial, and Advances*, 1e. New York: Apple Academic Press/CRC Taylor & Francis.

69 Egbuna, C., Sharif, N., and Jabeen, S. (2018). Pharmacognosy and prehistoric uses of medicinal plants. In: *Phytochemistry, Volume 2: Pharmacognosy, Nanomedicine, and Contemporary Issues* (eds. C. Egbuna, S. Kumar, J.C. Ifemeje and J.V. Kurhekar), 3–15. New York: Apple Academic Press.

14

Bactericidal Effects of Phytochemicals on *Mycobacterium leprae*, the Causative Agent of Leprosy

Chukwuebuka Egbuna[1,2], Muhammad Akram[3], Surajudeen Abiola Abdulrahman[4], Andrew G. Mtewa[5,6], Mithun Rudrapal[7], Kingsley C. Patrick-Iwuanyanwu[2,8], Saher Rahat[3], Iram Ghaffar[3], Amna Siddique[3], Jonathan Chinenye Ifemeje[1], Michael C. Olisah[9], Mihnea-Alexandru Găman[10,11], Bui Thanh Tung[12], Chandan Shivamallu[13], Chukwuemelie Zedech Uche[14], and Johra Khan[15,16]

[1] Department of Biochemistry, Faculty of Natural Sciences, Chukwuemeka Odumegwu Ojukwu University, Uli, Nigeria
[2] Nutritional Biochemistry and Toxicology Unit, World Bank Africa Centre of Excellence, Centre for Public Health and Toxicological Research (PUTOR), University of Port Harcourt, Port Harcourt, Nigeria
[3] Department of Eastern Medicine and Surgery, Government College University Faisalabad, Faisalabad, Pakistan
[4] Health Education England, Leicester, United Kingdom
[5] Chemistry Section, Malawi Institute of Technology, Malawi University of Science and Technology, Thyolo, Malawi
[6] Pharm-Biotechnology and Traditional Medicine Center (PHARMBIOTRAC), Mbarara University of Science and Technology, Mbarara, Uganda
[7] Department of Pharmaceutical Chemistry, Rasiklal M. Dhariwal Institute of Pharmaceutical Education and Research, Chinchwad, Pune, Maharashtra, India
[8] Department of Biochemistry, Faculty of Science, University of Port Harcourt, Port Harcourt, Nigeria
[9] Department of Medical Biochemistry, Faculty of Basic Medical Sciences, Chukwuemeka Odumegwu Ojukwu University, Uli, Nigeria
[10] "Carol Davila" University of Medicine and Pharmacy, Bucharest, Romania
[11] Department of Hematology, Center of Hematology and Bone Marrow Transplantation, Fundeni Clinical Institute, Bucharest, Romania
[12] VNU School of Medicine and Pharmacy, Vietnam National University, Hanoi, Vietnam
[13] Department of Biotechnology and Bioinformatics, School of Life Sciences, JSS Academy of Higher Education and Research, Mysore, Karnataka, India
[14] Department of Medical Biochemistry and Molecular Biology, Faculty of Basic Medical Sciences, University of Nigeria, Enugu Campus, Nigeria
[15] Department of Medical Laboratory Sciences, College of Applied Medical Sciences, Majmaah University, Saudi Arabia
[16] Health and Basic Sciences Research Center, Majmaah University, Majmaah, Saudi Arabia

Neglected Tropical Diseases and Phytochemicals in Drug Discovery, First Edition.
Edited by Chukwuebuka Egbuna, Muhammad Akram, and Jonathan Chinenye Ifemeje.

List of Abbreviations

BB	Borderline Borderline leprosy
BL	Borderline lepromatous leprosy
BT	Tuberculoid leprosy
CDC	Centers for Disease Control and Prevention
DNA	Deoxyribonucleic acid
ENL	Erythema nodosum leprosum
IL	Interleukin
LL	Lepromatous leprosy
MB	Multibacillary leprosy
MDT	Multidrug therapy
PB	Paucibacillary leprosy
PCR	Polymerase chain reaction
PDB	Protein Data Bank
RNA	Ribonucleic acid
WHO	World Health Organization

14.1 Introduction

Leprosy (Hansen's disease) is an infectious disease of the skin, peripheral nerves, and mucous membranes of the upper respiratory tract and eyes. It is caused by *Mycobacterium leprae* and is prevalent in both low- and middle-income countries. Over 200 000 new cases of leprosy were reported from 159 countries in 2018 [1]. The causative agent of leprosy was first discovered in 1873 by the Norwegian Physician Gerhard Armauer, who named it Hansen's bacillus. In the early days when knowledge about leprosy treatment was still very limited, infected people faced many problems with the morbidity associated with the chronicity of the disease and slow recovery. Leprosy remains a neglected infectious disease, absent from the mainstream of studies and research, and still requires scientific attention in order to achieve eradication.

Mycobacterium leprae is a large, curved, rod-shaped bacteria measuring 1.5–8 microns in length and 0.2–0.5 microns in diameter. Microscopically, *M. leprae* stains red with Ziehl Neelsen (ZN) stain due to its high lipid content. The *M. leprae* bacterium is believed to multiply in number through a binary fission method at a temperature of 27–30 °C which is favorable for its growth. The central feature of leprosy is the impairment of the peripheral nerves and the associated sensory and motor deprivation underlying many of the definitive features of this ailment, including plantar ulcers, cracks, clawed hands, skin wounds, and incomplete closure of eyelids. The individuals suffering from leprosy are reported to feel no pain due to damage of peripheral nerves that conduct pain, a noteworthy feature of the disease. However, neuropathic pain is increasingly being reported as a feature of leprosy in some patients. Tissue inflammation causes nociceptive receptors to trigger off, and this process generally takes place during the activation of the immune system [2].

Although the exact mode of transmission of *M. leprae* is not completely understood, it is believed that when the respiratory system is exposed to airborne particles for prolonged periods, then the pathogen can enter the nasal pathway [3]. However, other ways of entry into the body include direct contact with skin eruptions, blood transmission, and breastfeeding by an infected mother. *M. leprae* is not very virulent; the bacilli reside inside the infected cells and thus most leprosy patients are noninfectious.

Multibacillary leprosy (MB) patients (Figure 14.1), nevertheless, excrete out bacteria from the mucosa of the nasal cavity and skin [4]. The Schwann cells of the peripheral nerves and histiocytes (tissue macrophages) present in the dermis are mainly affected by *M. leprae*. Although *M. leprae* is transmitted via frequent and close contact with infected patients, these patients immediately lose infectivity when receiving multidrug therapy (MDT). While the infective agent is completely treated with MDT, permanent deformity may arise due to particular tropism for the peripheral nerves of *M. leprae*.

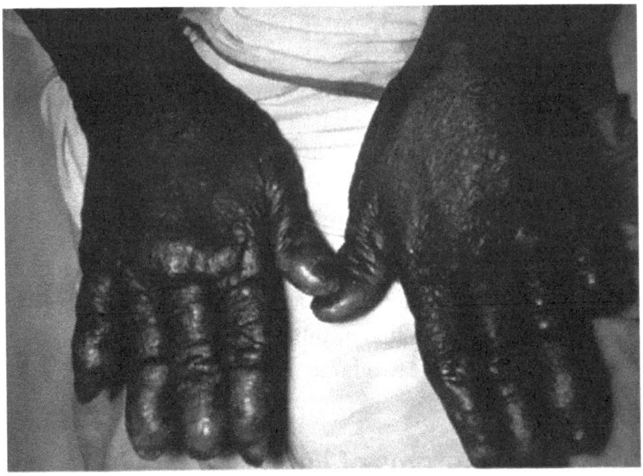

Figure 14.1 Dorsal view of the hands of a patient suffering from nodular lepromatous leprosy, which under the newer WHO standards, is classified as multibacillary leprosy. *Source:* Adapted from Centers for Disease Control and Prevention. Courtesy of the WHO. Freely taken from: https://wwwnc.cdc.gov/eid/article/21/12/et-2112-f1.

14.2 Epidemiology of Leprosy

Leprosy is prevalent in developing countries, especially in the tropical regions (Figure 14.2). MDT was introduced in the early 1980s and has remarkably reduced the prevalence of this disorder. The majority of the cases worldwide were reported in about 105 countries mostly located in Southeast Asia, Africa, Americas, Western Mediterranean, and Eastern Pacific. According to the WHO, countries like India, Indonesia, and Brazil recorded over 10 000 new cases of leprosy between 2011 and 2015 (Figure 14.2), an indication that the fight against leprosy is far from over. Again, some countries that did not report new cases were found to harbor persons suffering from leprosy.

Globally, the prevalence rate of leprosy in 2012 was 0.34 per 10 000 individuals [5]. India and Brazil accounted for the largest number of cases worldwide (Table 14.1). In 2011, the prevalence rate of leprosy in Brazil was 1.54 cases per 10 000 individuals, with further 33 955 new cases detected, 61% of which were multibacillary [6]. Similar to India, Brazil has yet to accomplish its target to become a leprosy-free country (determined by prevalence which is lower than one case per 10 000 inhabitants) as it remains a disease of public health concern in these regions. Looking randomly at Brazil, there is a random distribution of leprosy in different areas of the country with a prevalence rate per 10 000 population. In Brazil, the following epidemiological indicators are used to track the public health response against leprosy: new cases detection rate, the number of new cases in children with age less than 15 years, and the rate of new cases with grade-2 impairment.

To improve global comparability, the interpretation of epidemiological data collected from different countries, including India and Brazil, requires careful harmonization of the data and health indicators. For instance, countries need to:

- be in a position to define cases in new terms.
- exclude cases of treatment for a short duration.
- monitor paucibacillary leprosy (PB) patients, on single-dose treatment.

Following the adoption of these recommendations, the rate of new cases detected was remarkably reduced [7]. Since 2000, not only has the prevalence of leprosy dropped significantly in Brazil, but also the number of new cases has been declining because primary care is available to a large number of patients [8].

14.3 Clinical Sign and Symptoms of Leprosy

Following a mean incubation period of about five years after exposure to the infective agent, clinical symptoms usually develop when the cellular immunity of the host is weak. Cellular immunity is the primary defense against the *M. leprae*

Figure 14.2 Geographical distribution of new cases of Hansen's disease reported to the WHO in 2015. *Source:* Courtesy of the WHO.

Table 14.1 Global new cases of leprosy between 2011 and 2015.

Reported cases	Africa	Asia	America
>10 000 new cases	—	India (127,326 new cases) – 60% global prevalence. Indonesia (17 202 new cases) – 8% global prevalence	Brazil (26 395 new cases) – 13% global prevalence
Between 1 000 and 10 000	Democratic Republic of Congo, Ethiopia, Madagascar, Mozambique, Nigeria, and United Republic of Tanzania	**Southeast Asia**: Bangladesh, Myanmar, Nepal, and Sri Lanka; and from **Western Pacific**, the Philippines. Collectively, these countries reported 19 069 new cases, 14% of all new cases globally	—
The remaining 10 286 new cases (5%) were reported by 92 countries			
30 countries reported 0 new cases			
92 countries did not report, several of which are known to have cases of leprosy			

Source: Data from Centers for Disease Control and Prevention. Available at https://www.cdc.gov/leprosy/transmission/index.html.

infection. Cytokines are present within skin lesions and are directly linked to the development of clinical manifestations. In response to PB, T-helper cells (Th)1 cytokines, for instance, interleukin (IL)-2 and interferon-γ, play a major role in cellular immunity. Th-2 cytokines, e.g. IL-4, IL-5, and IL-10, stimulate the humoral immune response and also influence MB. Thus, cytokine profiles are a fundamental part of the body's response to PB and MB and are inversely related.

Because the incubation period of the disease can vary, symptoms may occur as early as one year or up to twenty years after the exposure to the bacteria. Red and painful skin lesions are the most common clinical manifestation of leprosy. A large number of asymmetrical lesions with raised external borders and hypochromic centers are found on the affected person's body. This causes skin denervation and ulceration in the compression areas. All of these changes are the first phase in the interruption of nerve root of transmission. Lesions are usually brown with defined borders and mostly found on the cold parts of the body. Sometimes, there are no other symptoms of this infectious skin disease except for dry skin. Hand and feet skin edema may also be present along with other symptoms. When leprosy reaches an advanced stage, the patient's face may exhibit a peculiar appearance (leonine facies) and the loss of eyelashes may also occur. The mucous membranes of the following body parts are affected: eyes, bones, joints, lymph nodes, and upper airway systems. If the disease progresses and remains untreated, it may affect various body systems such as the cardiovascular system, renal system, lymphatic system, endocrine system, respiratory system, etc., leading to complications.

14.4 Classification of Leprosy

According to the Ridley–Jopling classification I, Tuberculoid leprosy (BT) is classified under PB, while Borderline Borderline leprosy (BB), Borderline lepromatous leprosy (BL), and Lepromatous leprosy (LL) are classified under MB [9]. Treatment regimens were recommended based on this classification. According to this classification, a BI value of ≥2 at any skin site suggests the requirement of therapy for MB, while a BI value <2 suggests therapy for PB.

Until 1988, if a skin smear at any site was found positive, it was enough to recommend treatment for MB. The new mechanism of counting skin lesions has been recently recommended by the WHO as it is field-friendly and used to decide whether the patients should be treated against PB or MB (PB ≤5 lesions; MB >5 lesions).

There are two main types of immune reactions that are involved in leprosy which influence the immunomodulation of the body. They are:

1) **Leprosy Type 1 reactions (T1Rs)**, also known as reversal reactions: This reaction is also called delayed hypersensitivity reaction. In T1Rs, antigens of *M. leprae* are found in skins and nerves with localization in Schwann cells and macrophages. This localization is attributed to the damages in nerve cells and cause of neuritis, skin lesions, and edema in hands, feet, and face. This affects about 30% of patients with leprosy.
2) **Type 2 reactions (T2Rs)**, also known as erythema nodosum leprosum (ENL): This type of immune reaction is found in patients with BL and LL leprosy and to some degree in individuals with BB. T2Rs causes tender, erythematous, subcutaneous nodules on skin which can cause fever, anorexia, enlarged lymph nodes, edema, and so on. High levels of circulating TNF are found in blood plasma of patients with T2Rs. ENL is induced by the activation of the complement cascade and deposition of immune complexes which results in type-III hypersensitivity reaction or vasculitis.

14.5 Diagnosis of Leprosy

The diagnosis is mainly clinical. In asymptomatic patients, the diagnosis of leprosy is a difficult task for physicians because the bacillus is not easily found in the nerve biopsy. In 89% of cases, patients lose their sensory control and in 81% of cases, motor control is lost. An important symptom of leprosy is nerve thickness with or without any prominent inflammation in patients with a previous history of skin lesions. Other common symptoms of leprosy are nerve pain, nerve tenderness, numbness, and loss of sensation in affected parts of the relevant nerve. Beyond the clinical symptoms, polymerase chain reaction (PCR) is one of the most authentic and reliable tests in the diagnosis of leprosy, particularly in asymptomatic patients [10]. Two to three million people become permanently disabled due to leprosy [11].

14.6 Management of Leprosy

MDT, usually a combination of different classes of antibiotics, is required to eliminate the bacteria. Dapsone in combination with rifampicin is usually recommended for patients suffering from PB whereas a combination of dapsone, clofazimine, and rifampicin is recommended for MB [12]. Dapsone acts by inhibiting the synthesis of dihydrofolic acid. It competes with para-aminobenzoate by binding itself to the active site of dihydropteroate. The main target of dapsone is the proteins inactive dihydropteroate synthase 1 (UniProt ID: P0C0X1) and dihydropteroate synthase 2 (UniPort ID: P0C0X2) [12]. However, dihydropteroate 2 lacks dihydropteroate action and does not play a part in the metabolism of folate. Rifampicin inhibits the enzyme DNA-dependent RNA polymerase which results in the suppression of RNA synthesis and leads to death of the cell. Clofazimine disrupts the cell cycle by binding to microbial DNA and kills the bacteria.

14.7 Computational Drug Development Aspects of Leprosy

Drug development against leprosy can be approached in several ways. These include the nervous system and its other affiliated systems as targets or specific proteins responsible for the growth and/or exacerbation of the disease. All these targets will be required to be validated before the subjection of various drug tests. Since phytocompounds are a large pool of compounds with many properties that work in synergy and sometimes as antagonists, it is important to understand specifically isolated and purified compounds that could interact with the selected biological targets. Considering natural compounds as possible sources of drugs against leprosy, it is also possible to subject these molecules to semi-synthesis to form more suitable compounds by modifying and optimizing various functional groups and bonds to improve one or more drug-like properties for optimal results. This section looks into a protein from *M. leprae* which can be obtained freely from the protein databank with a PDB ID 2NTV, where it is already co-crystallized with a ligand with PDB ID P1H, a PTH-NAD adduct. Figure 14.3 shows the protein and Figure 14.4 shows the co-crystalized ligand, both of which were visualized independently in the discovery studio, with water molecules removed for a clearer view. The advantage of studying the ligand and the active sites of receptor proteins is that it gives a picture and direction toward which a chemist working with a phytocompound database or their allied semi-synthetics can easily see which ones would fit best into the active pockets of the proteins. In this regard, the size, shape, functional groups, bonding types, and flexibility of the ligand will be essential to decide which molecules are best to be developed into drugs. Further,

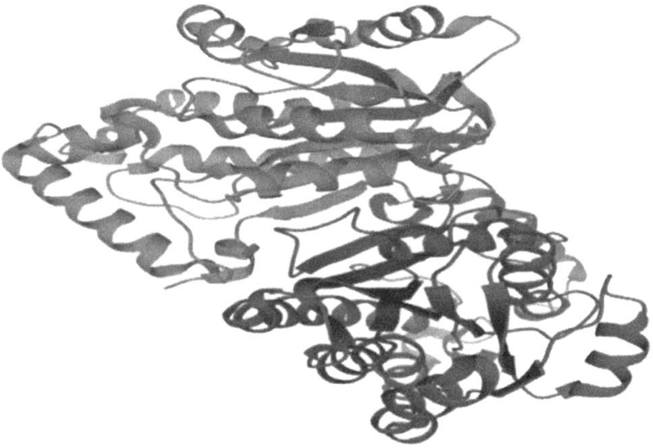

Figure 14.3 2NTV protein from *Mycobacterium leprae* as a potential drug target in leprosy.

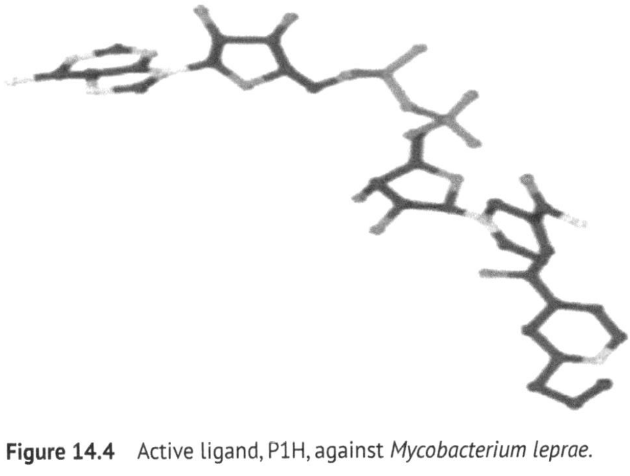

Figure 14.4 Active ligand, P1H, against *Mycobacterium leprae.*

into the work, the visualization shows Figure 14.5 which is a simple figure showing the ligand seated in its rightful region of interaction with the various amino acids in the protein. Figure 14.6 shows specific types of interactions with specific amino acids, data which are crucial as they show the chemist whether the drug being developed, active or not, would fit long enough onto the binding pockets of the proteins for effective results or would be there for too long, and possibly end up in some toxic results or else, would be there for too short a time and have little or no impact onto the proteins. This information is obtained from such data as binding categories, binding energy, and ligand efficiency which should be further explored by the computational chemists for the discovery and development of new and effective drugs.

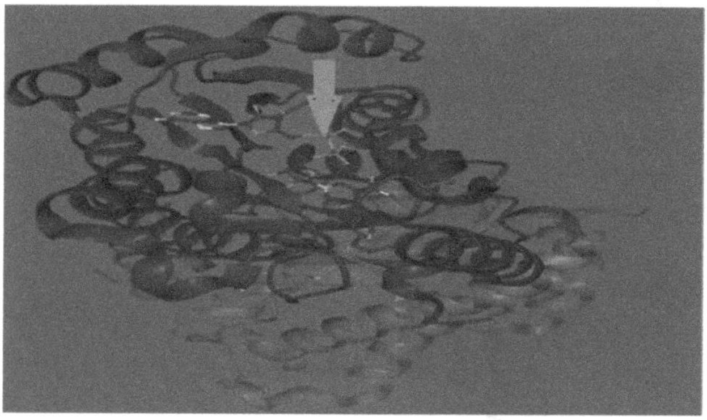

Figure 14.5 Co-crystalized ligand with protein of *Mycobacterium leprae*.

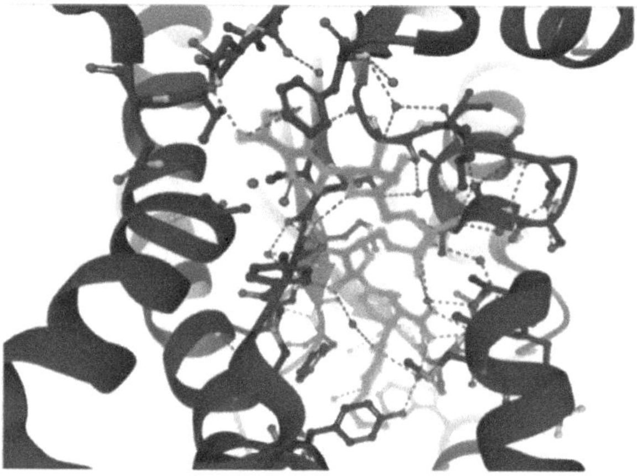

Figure 14.6 Ligand interactions with specific amino acids in *Mycobacterium leprae*.

14.8 Bactericidal Activity of Different Phytochemicals in Leprosy

Since ancient times, a large number of drugs have been derived from plants. Their source may be the plant as a whole or any specific part of a plant, e.g. fruit, flower, leaves, roots, stem, etc. Gums, resins, and waxes are excretory forms of drugs that are obtained from plants. In the following sections below, we briefly discuss several common plants rich in phytochemicals that have shown bactericidal activity in the treatment of leprosy.

A list of some important medicinal plants/herbs along with their phytochemicals effective against leprosy is presented in Table 14.2.

14.8.1 *Tinospora cordifolia* Linn.

Tinospora cordifolia L. is a shrub that belongs to the family Menispermaceae. This large climbing shrub is found throughout India and in some parts of China. Its English name is moonseed. In Hindi, it is called Giloy. It contains the following chemical constituents: alkaloids, steroids, glycosides, steroids, and polysaccharides. Roots, flowers, leave, fruit, and stems are parts of the plant used in obtaining effective remedies against several microbes. The methanolic extract of *T. cordiflora* shows strong antibacterial activity. It decreases the bacterial growth and improves the phagocytic activity of neutrophils. The methanolic extracts of leaves and flowers increase the activity of antioxidant enzymes, activate T and B cells, and increase the production of antibodies against microbes. Its active constituents, which play major roles in its antibacterial activity, are 11-hydroxymustakone, *N*-methyl-2-pyrollidone, *N*-Formylannonain, Cordifolioside, and syringing [13, 14].

14.8.2 *Khaya senegalensis* Ders. A. Juss.

The plant *Khaya senegalensis* belongs to the family Meliaceae. This herb is locally used in Nigeria for wound healing, mucous diarrhea, and dysentery. It grows as a medium to a full-size tree in its natural habitat. The leaves and bark of this plant show the main role in the antibacterial activity. The chemical constituents of this plant are saponins, glycosides, tannins, alkaloids, terpenoids, and flavonoids. The active constituents supporting its antioxidant and antimicrobial activity are polyphenols which are abundantly found in the bark. Polyphenols show their antimicrobial activity by inhibiting the hydrolyzing enzyme which is released from bacteria, therefore inhibiting its entrance into a normal body cell. This herb also protects the body from oxidative destruction which is caused by microbes through the production of free radicals. Polyphenols exhibit bactericidal action by increasing the activity of lymphocytes, antioxidant enzymes, and the production of antibodies against the microbes [15, 16].

Table 14.2 Some important medicinal plants/herbs and their phytochemicals against leprosy.

No.	Name of plant	Phytochemical constituents
1	*Acacia catechu* Willd.	Saponins, Tannins, Phenolic compounds, Flavonoids, Alkaloids, Glycosides, Carbohydrates
2	*Adhatoda vasica* Linn.	Alkaloids, Flavonoids, Polyphenols, Glycosides, Carbohydrates
3	*Aegle marmelos* Linn.	Flavonoids, Phenolic compounds, Phytosterols, Saponins
4	*Allium sativum* Linn.	Saponins, Tannins, Steroids, Cardiac glycosides, Carbohydrates
5	*Amaranthus spinosus* Linn.	Fixed oils and fats, Carbohydrates, Glycosides, Gum, Mucilage, Phenolic substances, Protein/amino acids, Tannins, Saponins
6	*Andrographis paniculata* Wall.	Steroids, Phenols, Terpenoids, Alkaloids, Saponins, Flavonoids
7	*Asparagus racemosus* Linn.	Phytosterols, Triterpenoids, Saponins, Alkaloids, Carbohydrates, Flavonoids, Tannins and phenolic compounds, Proteins, Glycosides, Fixed oils and fatty acids
8	*Azadirachta indica* Linn.	Alkaloids, Carbohydrates, Reducing sugars, Flavonoids, Glycosides, Tannins, Phenols, Saponins, Triterpenoids, Steroids
9	*Brassica nigra* Linn.	Saponins, Alkaloids, Flavonoids, Glycosides, Reducing sugar, Phlobatannins, Terpenoids
10	*Butea monosperma* (Lam.) Taub.	Carbohydrate, Flavonoid, Glycosides, Saponins, Terpenoids, Steroids
11	*Calotropis procera* Linn.	Proteins, Amino acid, Alkaloid, Phenols, Saponins, Flavonoids, Tannin, Fatty acids, Triterpenoids, Steroids, Reducing sugars, Glycosides
12	*Carpotroche brasiliensis* (Raddi)	Fatty acids, Carbohydrates, Flavonoids
13	*Cassia fistula* Linn.	Anthraquinones, Flavonoids, Alkaloids, Terpenoids, Reducing sugars, Saponins, Tannins, Phlobatannins, Steroids
14	*Centella asiatica* Linn.	Alkaloids, Protein/Amino acid, Reducing sugars, Saponins, Tannins, Triterpenoids, Starch, Fatty acids
15	*Cordia dichotoma* Forsk.	Alkaloids, Flavonoids, Saponins, Tannins, Proteins, Terpenoids, Reducing sugars
16	*Coriandrum sativum* Linn.	Carbohydrates, Reducing sugars, Glycosides, Triterpenes, Proteins, Terpenoids
17	*Curcuma longa* Linn.	Alkaloids, Tannins, Steroids, Saponins, Flavonoids, Polyphenols, Amino acids, Glycosides

Table 14.2 (Continued)

No.	Name of plant	Phytochemical constituents
18	*Dioscorea alata* Linn.	Tannins, Flavonoids, Phenolic compounds, Alkaloids, Saponins
19	*Eclipta alba* (Linn.) Hassk.	Alkaloids, Saponin, Flavonoids, Tannins, Steroids
20	*Euphorbia tirucalli* Linn.	Flavonoids, Phenol, Proanthocyanidins
21	*Gmelina arborea* Roxb.	Alkaloids, Flavonoids, Tannins, Lignin, Glycosides, Terpenoids, Saponins, Carbohydrates, Fatty acids
22	*Gynocardia odorata* Robert Brown	Flavonoids, Proteins, Triterpenoids, Tannins, Glycosides, Saponins, Fixed oils, Proteins, Alkaloids, Carbohydrates
23	*Holarrhena antidysenterica* Foxh.	Alkaloids, Flavonoids, Sterols, Quinine, Glycosides
24	*Hydnocarpus kurzii* (King) Warb.	Fatty acids, Flavonoids, Carbohydrates
25	*Ipomoea digitata* Linn.	Flavonoids, Fatty acids, Saponins, Glycosides, Alkaloids, Terpenoids, Coumarins
26	*Jatropha gossypifolia* Linn.	Phenolic compounds, Tannins, Starch, Saponin, Lignin
27	*Khaya senegalensis* Ders. A. Juss.	Saponins, Glycosides, Tannins, Alkaloids, Terpenoids, Polyphenols, Flavonoids
28	*Mimosa pudica* Linn.	Flavonoids, Saponins, Glycosides, Alkaloids, Terpenoids, Coumarins
29	*Ocimum sanctum* Linn.	Saponin, Sterols, Anthroquinones, Flavonoids, Phenolic compounds, Tannins, Terpenoids, Alkaloids, Carbohydrates
30	*Phyllanthus emblica* Linn.	Flavonoids, Tannins, Glycosides, Steroids, Saponins, Carbohydrates
31	*Plumeria rubra* Linn.	Tannins, Phlobatannins, Saponins, Flavonoids, Steroids, Terpenoids, Cardiac glycosides, Reducing sugars
32	*Psoralea corylifolia* Linn.	Coumarins, Flavonoids, Saponins, Glycosides, Alkaloids, Steroids
33	*Ricinus communis Linn.*	Flavonoids, Saponins, Glycosides, Alkaloids, Steroids
34	*Sesamum indicum* Linn.	Polyphenols, Alkaloids, Flavonoids, Terpenoids, Glycosides
35	*Sphaeranthus indicus* Linn.	Phenols, Tannins, Steroids, Terpenoids, Alkaloids, Flavones
36	*Terminalia bellirica* Roxb.	Alkaloids, Flavonoids, Saponins, Phenolic compounds, Steroids, Tannins, Glycosides

(Continued)

Table 14.2 (Continued)

No.	Name of plant	Phytochemical constituents
37	*Terminalia catappa*	Tannins, Flavonoids, Steroids, Xanthones, Saponins
38	*Terminalia chebula* Retz.	Alkaloids, Flavonoids, Saponins, Phenolic compounds, Steroids, Tannin, Glycosides
39	*Theobroma cacao* Linn.	Tannins, Saponins, Cardiac glycosides, Terpenoids, Flavonoids
40	*Tinospora cordifolia* Linn.	Alkaloids, Steroids, Glycosides, Flavonoids, Polysaccharides
41	*Trichilia emetica* Vahl.	Anthraquinones, Xanthines, Cardioactive glycosides, Flavonoids, Terpenoids, Lignans, Saponins
42	*Withania somnifera* Dunal.	Glycosides, Condensed tannins, Flavonoids, Alkaloids, Amino acids, Steroids, Terpenoids, Starch, Reducing sugars

Source: Based on [30–33].

14.8.3 *Azadirachta indica* Linn.

Azadirachta indica L. (Neem) is a member of the Meliaceae family that is native to India and neighboring countries. It is an important medicinal plant found in Chinese, Ayurvedic, and Unani medicines. All parts of the tree have been used traditionally for the treatment and prevention of numerous human ailments. For instance, the leaves, bark, and seeds are extensively used for skin diseases such as leprosy, scabies, intestinal worms, wounds, ulcers, and many more. Various phytochemical constituents, such as alkaloids, triterpenoids, and their glycosides, limonoids, tannins, flavonoids, polyphenols, fatty acids, saponins, and steroids have been reported to exhibit a wide range of biological activities (antimicrobial, antileprotic, antiviral, anti-inflammatory, antiulcer, anticarcinogenic, and so on) of the neem tree. The chief active constituent is azadirachtin and the others are nimbolinin, nimbin, nimbidin, nimbidol, sodium nimbinate, gedunin, salannin, β-sitosterol, polyphenolic flavonoids, and quercetin. Fresh leaves containing these phytochemicals have been claimed to possess antimicrobial activities. The health-promoting effect is mainly attributed to its antimicrobial and antioxidant properties [17, 18].

14.8.4 *Gynocardia odorata* Roxb.

Gynocardia odorata, Robert Brown (*Chaulmoogra odorata*, Roxburgh; *Hydnocarpus odorata*, Lindley) is grown as a very large tree belonging to the family Achariaceae, which is native to East India. The seeds and oil (*Chaulmoogra* oil) of this plant have been traditionally used in certain skin diseases. *Chaulmoogra* oil is mixed with fresh butter, and used as a remedy for cutaneous diseases such as leprosy, skin diseases, scrofula, syphilis, and also in rheumatism. The seeds are rich in various phytoconstituents including flavonoids, fixed oils, proteins, triterpenoids, tannins, glycosides, and alkaloids. Chaulmoogra oil has been reported to be effective against leprosy and tuberculosis. Furthermore, important bioactive constituents that are present in seeds include triglycerides of fatty acids, sterols, flavonoids, and flavonolignans. Hydnocarpin, a flavonolignan, has been reported to potentiate antimicrobial and anticancer activities [19, 20].

14.8.5 *Cassia fistula* Linn.

Cassia fistula Linn. (Family: Caesalpinaceae), commonly known as the "Golden Shower Tree," has been used in different traditional medicines for various human ailments since ancient times. In Ayurveda, this plant is known as "disease killer." It is widely grown in many Asian countries like Bangladesh, India, China, Hong Kong, Philippines, Malaysia, Indonesia, and Thailand. Different parts of the plant are used for different medicinal purposes. The bark and leaves are mainly used for skin diseases like leprosy. The fruit pulp is used as a mild laxative. Flowers are used as antipyretic, while the root is used as a diuretic, antileprotic, and antisyphilitic. Seeds are used as carminative, analgesic, and laxative. The fruits, stem bark, seeds, and leaves of this plant contain a variety of biologically active compounds such as anthraquinones, flavonoids, alkaloids, glycosides, tannins, saponins, steroids, terpenoids, and reducing sugar those have various medicinal properties. Some important phytochemicals include fatty acids (linoleic, oleic, and stearic acid), oxalic acids, terpenoids, oxyanthra-quinones, anthraquinones derivatives. Fruits of *C. fistula* contain Rhein Glycosides fistulic acids, sennosides A/B, rhein glycosides, anthraquinones, and flavanoid-3-ol-derivatives, ceryl lcohol, kaempferol, bianthraquinone glycosides, fistulin, and phytol. These phytochemicals are believed to be responsible for the biological potential (antileprotic, anti-inflammatory, antioxidant, antimicrobial, antiulcer, antitumor, laxative, analgesic-antipyretic, antipruritic, etc.) of *C. fistula* [21–23].

14.8.6 *Butea monosperma* (Lam.) Taub.

Butea monosperma (Lam.) Taub belonging to the family Fabaceae is a deciduous tree and is found in the dry deciduous forests of Central and Western India. This plant is commonly known as "Flame of the forest." This tree grows to 50 ft high, with stunning flower clusters. This plant is extensively used in Ayurveda, Unani, Homeopathy, and Traditional systems of medicine. Flowers of *B. monosperma* are used as in the treatment of leprosy and other skin diseases along with other ailments such as convulsions, gouty arthritis, ulcer, diarrhea, and liver disorders. In phytochemical studies of the flower extract, triterpenes, flavonoids, and glycosides like butein, butin, isobutrin, coreopsin, isocoreopsin, sulphurein, monospermoside, isomonospermoside, chalcones, aurones, and steroids have been reported to be present. The antileprotic activity is believed to be due to antioxidant, antimicrobial, and anti-inflammatory activities of this plant [24, 25].

14.8.7 *Terminalia bellirica* Roxb.

Terminalia belerica Roxb. (Combretaceae) is also referred to as Beleric Myrobalan in English and is locally known as *Bahera* in India. It has been used for centuries in Ayurveda. It is a large (10–20 m) deciduous tree with characteristic thick brownish-gray bark and is found throughout the Indian forests and plains, and in other Asian countries. The dried ripe fruit of *T. bellirica* has traditionally been used in the treatment of skin diseases like leprosy, inflammatory disorders, piles, dysenteric-diarrhea, cough, dyspepsia, asthma, bronchitis, liver disease, and eye diseases. It is believed that the triterpenoid present in the fruits possesses antimicrobial activity. Some predominant phytoconstituents include β-sitosterol, flavonoids, tannins, catechol, gallic acid, ellagic acid, ethyl gallate, galloyl glucose, terpenoids (chebulinic acid, chebulagic acid, belleric acid), lignans (termilignan and thannilignan), saponins (bellericoside and bellericanin), coumarins, and alkaloids. These phytochemicals possess significant antioxidant activity [26, 27].

14.8.8 *Ricinus communis* Linn.

Ricinus communis L., commonly called castor and belonging to the Euphorbiaceae, is widely cultivated in India, Africa, China, Brazil, and Thailand. The castor oil plant is a soft wooden small tree grown throughout tropical regions. The castor oil has been used traditionally in the treatment of leprosy, rheumatic arthritis, paralysis, epilepsy, and constipation. The castor oil obtained from castor seeds contains mainly ricinoleic acid along with stearic, oleic, and linoleic acids. The leaves have been reported to possess bioactive components like epicatechin and isoquercetin. The plant has been reported to exhibit antileprotic, antibacterial, antioxidant, analgesic, hepatoprotective, anti-inflammatory, purgative, and anti-tuberculate activities [28, 29].

14.9 Future Prospects

Natural resources, especially plants, have been a valuable source of primary metabolites (carbohydrates, proteins and amino acids, fats/lipids and fatty acids, vitamins and minerals) and secondary metabolites (also known as phytochemicals, e.g.

flavonoids, alkaloids, terpenoids, saponins, tannins, polyphenols, organosulfur compounds, and so on) which are essential for man's survival as sources of nutrients (foods) and medicines [34–52]. The numerous bioactive compounds, now running in tens of thousands which are freely at man's disposal, have paved way for drug discovery opportunities against life-threatening diseases [34, 36, 37, 43, 47], including neglected tropical diseases. On record, some of these bioactive compounds have proven to be effective against pathogenic microorganisms. What remains a mystery are the mechanisms behind which some of these phytochemicals works, in which if known could provide more insight to medicinal chemists and drug developers on a specific chemical entities to adopt as a model, then optimize and predict the activities of similar chemical entities [51, 52].

The incidence of leprosy has persisted for so long with the new emergence of drug-resistant (DR) and multidrug-resistant (MDR) *M. leprae* now an issue of major concern [53–55]. This resistance is attributed to missense mutations in *M. leprae* genome which implies that the use of known antibiotics might be counterproductive for effective treatment. This revelation calls for seriousness and joint collaborative research in finding new drugs from chemical reservoirs available in plants and other natural resources while employing sophisticated modern methods and techniques.

14.10 Conclusion

Leprosy is a chronic infectious disease of the skin, peripheral nerves, and mucous membranes of the upper respiratory tract and eyes, and is caused by *M. leprae*. Infected patients show a variety of signs and symptoms but most common skin lesions and nerve dysfunction. PCR is one of the most important diagnostic tests when patients do not show any symptoms and *M. leprae* is absent in the biopsy samples. MDT has been proven to be a highly effective treatment of this infection. Beyond synthetic drugs, many valuable natural compounds are also used and are highly effective against *M. leprae*. Of these, we discussed the bactericidal effect of two, *T. Cordifolia* and *K. Senegalensis*. Both of these have shown strong antioxidant and antibacterial activity. Today, leprosy's impact on health worldwide has been reduced, yet it remains a burden in countries such as India and Brazil. Based on the activity of various extracts from natural molecules, drug developers must go the extra mile and study-specific interactions of an individual or multiple phytocompounds with various biological targets to determine the level of intended activity, mode of action, and possibly help to shape and standardize possible dosing later on just as it has been described in the various works of the authors. Computational tools help to guide before a lot of investment is made in the laboratory, thereby providing direction as well as cutting costs.

Acknowledgments

Authors would like to thank the Deanship of Scientific Research at Majmaah University for supporting this work. Authors are also thankful to World Bank, University of Port Harcourt, Majmaah University, National Universities Commission, the Association of African Universities and other research institutes for their support.

References

1 World Health Organization (WHO) (2020). Fact sheets: leprosy. https://www.who.int/news-room/fact-sheets/detail/leprosy (accessed 2 September 2020).
2 Stump, P.R., Baccarelli, R., Marciano, L.H. et al. (2004). Neuropathic pain in leprosy patients. *Int. J. Lepr. Other Mycobact. Dis.* 72 (2): 134–138.
3 Davey, T.F. and Rees, R.J. (1974). The nasal discharge in leprosy: clinical and bacteriological aspects. *Lepr. Rev.* 45: 121–134.
4 Job, C.K., Jayakumar, J., Kearney, M., and Gillis, T.P. (2008). Transmission of leprosy: a study of skin and nasal secretions of household contacts of leprosy patients using PCR. *Am. J. Trop. Med. Hyg.* 78: 518–521.
5 World Health Organization (WHO) (2012). Global leprosy situation, 2012. *Wkly Epidemiol. Rec.* 87: 317–328.
6 Lastória, J.C. and Abreu, M.A. (2014). Leprosy: review of the epidemiological, clinical, and etiopathogenic aspects-part 1. *An. Bras. Dermatol.* 89 (2): 205–218.
7 Talhari, S., Grossi, M.A., Oliveira, M.L. et al. (2012). Hansen's disease: a vanishing disease? *Mem. Inst. Oswaldo Cruz.* 107: 13–16.
8 Penna, M.L.F., Oliveira, M.L.W., Carmo, E.H. et al. (2008). The influence of increased access to basic healthcare on the trends in Hansen's disease detection rate in Brazil from 1980 to 2006. *Rev. Soc. Bras. Med. Trop.* 41: 6–10.
9 Ridley, D.S. and Jopling, W.H. (1966). Classfication of leprosy according to immunity: a five-group system. *Int. J. Lepr. Other Mycobact. Dis.* 34: 255–273.
10 Jardim, M.R., Antunes, S.L., Santos, A.R. et al. (2003). Criteria for diagnosis of pure neural leprosy. *J. Neurol.* 250 (7): 806–809.
11 Sundar, R.D., Settu, S., Shankar, S. et al. (2018). Potential medicinal plants to treat leprosy—a review. *Res. J. Pharm. Technol.* 11 (2): 813–821.
12 Halder, S.T., Dhorajiwala, T.M., and Samant, L.R. (2019). Multiple docking analysis and In silico absorption, distribution, metabolism, excretion, and toxicity screening of anti-leprosy phytochemicals and dapsone against dihydropteroate synthase of *Mycobacterium leprae. Int. J. Mycobacteriol.* 8 (3): 229.
13 Mittal, J., Sharma, M.M., and Batra, A. (2014). *Tinospora cordifolia*: a multipurpose medicinal plant-A. *J. Med. Plants* 2 (2): 32–47.

14 Reddy, N.M. and Reddy, R.N. (2015). *Tinospora cordifolia* chemical constituents and medicinal properties: a review. *Scholars Acad. J. Pharm.* 4 (8): 364–369.

15 Karou, D., Dicko, M.H., Simpore, J., and Traore, A.S. (2005). Antioxidant and antibacterial activities of polyphenols from ethnomedicinal plants of Burkina Faso. *Afr. J. Biotechnol.* 4 (8): 823–828.

16 Makut, M.D., Gyar, S.D., Pennap, G.R., and Anthony, P. (2008). Phytochemical screening and antimicrobial activity of the ethanolic and methanolic extracts of the leaf and bark of Khaya senegalensis. *Afr. J. Biotechnol.* 7 (9): 1216–1219.

17 Alzohairy, M.A. (2016). Therapeutics role of *Azadirachta indica* (Neem) and their active constituents in diseases prevention and treatment. *Evid. Based Complement. Alternat. Med.* 2016: 7382506.

18 Kumar, V.S. and Navaratnam, V. (2013). Neem (*Azadirachta indica*): prehistory to contemporary medicinal uses to humankind. *Asian Pac. J. Trop. Biomed.* 3 (7): 505–514.

19 Rogers, L. (1921). Chaulmoogra oil in leprosy and tuberculosis: the successful treatment of leprosy by injections of soluble preparations of the fatty acids of chaulmoogra and other oils and its bearing on the tuberculosis problem. *Lancet* 197 (5101): 1178–1180.

20 Sahoo, M.R., Dhanabal, S.P., Jadhav, A.N. et al. (2014). Hydnocarpus: an ethnopharmacological, phytochemical and pharmacological review. *J. Ethnopharmacol.* 154 (1): 17–25.

21 Ali, M.A. (2014). *Cassia fistula* Linn: a review of phytochemical and pharmacological studies. *Int. J. Pharm. Sci. Res.* 5 (6): 2125–2130.

22 Sharma, D.K. (2017). Enumerations on phytochemical, pharmacological and ethnobotanical properties of *Cassia fistula* Linn: yellow shower. *J. Phytopharmacol.* 6 (5): 300–306.

23 Bhalodia, N.R., Nariya, P.B., Acharya, R.N., and Shukla, V.J. (2012). *In vitro* antibacterial and antifungal activities of *Cassia fistula* Linn. fruit pulp extracts. *Ayu* 33 (1): 123–129.

24 Shruti, V., Hegde, G.R., Shruti, M., and Shreedevi, S. (2014). Pharmacognostical studies on *Butea monosperma* (Lam.) Taub (Faboideae) flower. *Int. J. Pharm. Phytopharmacol. Res.* 4 (1): 34–36.

25 Gupta, A., Pandey, S., Maithili, V., and Yadav, J.S. (2008). Anti-inflammatory activity of *Butea monosperma* (Lam.) Taub. Bark extract. *Asian J. Pharm. Clin. Res.* 1 (1): 95–97.

26 Dharmaratne, M.P.J., Manoraj, A., Thevanesam, V. et al. (2018). *Terminalia bellirica* fruit extracts: *In-vitro* antibacterial activity against selected multidrug-resistant bacteria, radical scavenging activity and cytotoxicity study on BHK-21 cells. *BMC Complement. Altern. Med.* 18: 325.

27 Kumar, N. and Paul Khurana, S.M. (2018). Phytochemistry and medicinal potential of the *Terminalia bellirica* Roxb. (*Bahera*). *Indian J. Nat. Prod. Resour.* 9 (2): 97–107.

28 Ramanjaneyulu, A.V., Anudradha, G., Ramana, M.V. et al. (2017). Multifarious uses of castor (*Ricinus communis* L.). *Int. J. Econ. Plants* 04 (04): 170–176.

29 Ladda, P.L. and Kamthanr, R.B. (2014). *Ricinus communis* (Castor): an overview. *Int. J. Res. Pharmacol. Pharmacother.* 3 (2): 136–144.

30 Ghosh, G. (2017). Traditional use of plants against leprosy in India: a review of the recent literature. *J. Innov. Pharm. Biol. Sci.* 4 (4): 55–64.

31 Mukesh, S.K., Nagori, K., Badwaik, H. et al. (2011). Potential antileprotic herbal drugs: a comparative review of marketed products. *J. Pharm. Res.* 4 (6): 1875–1876.

32 Solanki, R. (2015). In-vitro antileprotic activity of ethanolic extracts of *Mimosa Pudica* L. and *Prosopis Cineraria* L. (Druce). *Int. J. Sci. Res.* 4 (12): 508–509.

33 Khare, C.P. (2007). *Indian Medicinal Plants*. Berlin/Heidelberg: Springer-Verlag.

34 Egbuna, C. and Ifemeje, J.C. (2015). Biological functions and anti-nutritional effects of phytochemicals in living system. *IOSR J. Pharm. Biol. Sci. (IOSR-JPBS)* 10 (2): 10–19. iosrjournals.org/iosr-jpbs/papers/Vol10-issue2/…/C010231019.pdf.

35 Egbuna, C., Mishra, A., and Goyal, M.R. (eds.) (2020). *Preparation of Phytopharmaceuticals for the Management of Disorders*, 1e. US: Academic Press, Elsevier. ISBN: 9780128202845.

36 Egbuna, C., Kumar, S., Ezzat, S.M. et al. (eds.) (2019). *Phytochemicals as Lead Compounds for New Drug Discovery*, 1e. USA: Elsevier, Cambridge. ISBN: 9780128178911. 378 pages.

37 Shashank, K. and Egbuna, C. (eds.) (2019). *Phytochemistry: In vitro and In Silico Updates*, 1e. Singapore: Springer Nature. ISBN: 978-981-13-6919-3. 589 pages.

38 Egbuna, C., Ifemeje, J.C., Udedi, S.C., and Kumar, S. (eds.) (2018). Phytochemistry. In: *Fundamentals, Methods, and Applications*, 1e, vol. 1. New York: Apple Academic Press/CRC Taylor & Francis. ISBN: 978-1771887595. 684 pages, part of a 3 volume set.

39 Egbuna, C., Kumar, S., Ifemeje, J.C., and Kurhekar, J.V. (eds.) (2018). Phytochemistry. In: *Pharmacognosy, Nanomedicine, and Contemporary Issues*, 1e, vol. 2. New York: Apple Academic Press/CRC Taylor & Francis. ISBN: 978-1771887601. 620 pages, part of a 3 volume set.

40 Egbuna, C., Ifemeje, J.C., Kumar, S., and Sharif, N. (eds.) (2018). Phytochemistry. In: *Marine, Industrial, and Advances*, 1e, vol. 3. New York: Apple Academic Press/CRC Taylor & Francis. ISBN: 978-1771887618. 502 pages, part of a 3 volume set.

41 Islam, M.T., Ali, E.S., Uddin, S.J. et al. (2018). Andrographolide, a diterpene lactone from *Andrographis paniculata* and its therapeutic promises in cancer. *Cancer Lett.* 420: 129–145. https://doi.org/10.1016/j.canlet.2018.01.074.

42 Gheorghe, G., Stoian, A.P., Gaman, M. et al. (2019). The benefits and risks of antioxidant treatment in liver diseases. *Rev. Chim.* 70 (2): 651–655. https://doi.org/10.37358/RC.19.2.6977.

43 Găman, A.M., Egbuna, C., and Găman, M.A. (2020). Natural bioactive lead compounds effective against haematological malignancies. In: *Phytochemicals as Lead Compounds for New Drug Discovery* (eds. C. Egbuna, S. Kumar, J. Ifemeje, et al.), 95–115. Elsevier.

44 Hassan, S., Faiza, N., Farooq, M.A. et al. (2020). Nutritional and health benefits of seafoods. In: *Functional Foods and Nutraceuticals* (eds. C. Egbuna and T.G. Dable). Cham: Springer. https://doi.org/10.1007/978-3-030-42319-3_13.

45 Salehi-Sahlabadi, A., Varkaneh, H.K., Shahdadian, F. et al. (2020). Effects of phytosterols supplementation on blood glucose, glycosylated hemoglobin (HbA1c) and insulin levels in humans: a systematic review and meta-analysis of randomized controlled trials. *J. Diabet. Metab. Disord.* 19: 625–632. https://doi.org/10.1007/s40200-020-00526-z.

46 Akram, M. et al. (2020). Vitamins and minerals: types, sources and their functions. In: *Functional Foods and Nutraceuticals* (eds. C. Egbuna and G. Dable Tupas). Cham: Springer. https://doi.org/10.1007/978-3-030-42319-3_9.

47 Xiong, P., Niu, L., Talaei, S. et al. (2020). The effect of berberine supplementation on obesity indices: a dose- response meta-analysis and systematic review of randomized controlled trials. *Complement. Ther. Clin. Pract.* 39: 101113. https://doi.org/10.1016/j.ctcp.2020.101113.

48 Gaman, A.M., Buga, A.M., Gaman, M.A., and Popa-Wagner, A. (2014). The role of oxidative stress and the effects of antioxidants on the incidence of infectious complications of chronic lymphocytic leukemia. *Oxidative Med. Cell. Longev.* 2014: 158135. https://doi.org/10.1155/2014/158135.

49 Júnior, L.G.A., Dimitrova Tchekalarova, J., Atanasova, M. et al. (2018). Anticonvulsant effect of anacardic acid in murine models: putative role of GABAergic and antioxidant mechanisms. *Biomed. Pharmacother.* 106: 1686–1695. https://doi.org/10.1016/j.biopha.2018.07.121.

50 Lin, Y., Shi, D., Su, B. et al. (2020). The effect of green tea supplementation on obesity: a systematic review and dose–response meta-analysis of randomized controlled trials [published online ahead of print, 2020 May 5]. *Phytother. Res.* 34: 2459–2470. https://doi.org/10.1002/ptr.6697.

51 Ezzat, S.M., Jeevanandam, J., Egbuna, C. et al. (2019). Phytochemicals as sources of drugs. In: *Phytochemistry: In vitro and In Silico Updates* (eds. S. Kumar and C. Egbuna), 3–22. Singapore: Springer Nature.

52 Egbuna, C., Ezzat, S.M., Tijjani, H., and Srivastav, V.K. (2019). Synthetic analogs of phytochemicals. In: *Phytochemistry: In vitro and In Silico Updates* (eds. S. Kumar and C. Egbuna), 23–56. Singapore: Springer Nature.

53 Avanzi, C., Busso, P., Benjak, A. et al. (2016). Transmission of drug-resistant leprosy in Guinea-Conakry detected using molecular epidemiological approaches. *Clin. Infect. Dis.* 63 (11): 1482–1484.

54 You, E.Y., Kang, T.J., Kim, S.K. et al. (2005). Mutations in genes related to drug resistance in *Mycobacterium leprae* isolates from leprosy patients in Korea. *J. Infect.* 50 (1): 6–11.

55 Benjak, A., Avanzi, C., Singh, P. et al. (2018). Phylogenomics and antimicrobial resistance of the leprosy bacillus *Mycobacterium leprae. Nat. Commun.* 9 (1): 352. https://doi.org/10.1038/s41467-017-02576-z.

15

Medicinal Plants and Phytochemicals Effective Against *Chlamydia trachomatis*, the Causative Agent of Trachoma

Yusuf Abubakar[1], Chukwuebuka Egbuna[2,3], Ahmed Olatunde[4], Habibu Tijjani[1], and Uchenna Estella Odoh[5]

[1] Department of Biochemistry, Natural Product Research Laboratory, Bauchi State University, Gadau, Nigeria
[2] Department of Biochemistry, Faculty of Natural Sciences, Chukwuemeka Odumegwu Ojukwu University, Uli, Nigeria
[3] Nutritional Biochemistry and Toxicology Unit, World Bank Africa Centre of Excellence, Centre for Public Health and Toxicological Research (PUTOR), University of Port-Harcourt, Port Harcourt, Nigeria
[4] Department of Biochemistry, Abubakar Tafawa Balewa University, Bauchi, Nigeria
[5] Department of Pharmacognosy and Environmental Medicines, Faculty of Pharmaceutical Sciences, University of Nigeria, Nsukka, Nigeria

List of Abbreviations

BLTR	Bilamellar tarsal rotation
CPAF	*Chlamydia* protease-like activity factor
GET 2020	Global Elimination of Blinding Trachoma by 2020
SAFE strategy	**S** – Surgical treatment, **A** – Antibiotic treatment for acute infection, **F** – Face washing, and **E** – Environmental changes to improve sanitation

15.1 Introduction

Trachoma, one of the neglected tropical diseases, is the leading cause of eye disease worldwide. Trachoma is the third most common cause of blindness globally, besides cataract and glaucoma [1]. The first and precise description of trachoma was from fifteenth century BC in Ebers Papyrus, a collection of medical

Neglected Tropical Diseases and Phytochemicals in Drug Discovery, First Edition.
Edited by Chukwuebuka Egbuna, Muhammad Akram, and Jonathan Chinenye Ifemeje.

information found in Egypt [2]. Trachoma is a chronic disease that causes kerato-conjunctivitis resulting from a gram-negative, obligate intracellular bacterium called *Chlamydia trachomatis*. This infection is primarily found in childhood and progresses to a complication resulting in vision loss and blindness [3]. Based on the clinical perspective, trachoma is divided into two categories, namely active (early) and cicatricial (late-stage) disease. The active disease, mostly found in childhood, is categorized by chronic recurrent follicular conjunctivitis. Later in life, chronic sequelae of trachoma, including conjunctival scarring, trichiasis, and corneal scarring then all lead to visual impairment [4, 5]. Poverty (inadequate personal hygiene), lack of potable water supply, and overpopulation are the significant facilitators of the disease transmission [6]. Trachoma is still endemic in several countries (Figure 15.1). The endemic countries are majorly found in Africa, low rural areas of Latin America, the Western Pacific, Asia, and the Middle East [7, 8]. Trachoma has been of global health concern over the years, but recently, it has been eradicated in most developing countries [9]. Additionally, to the economy, some widespread implementation of trachoma control programs and measures have played an essential role in the eradication of the disease and reducing its level in some underdeveloped areas worldwide.

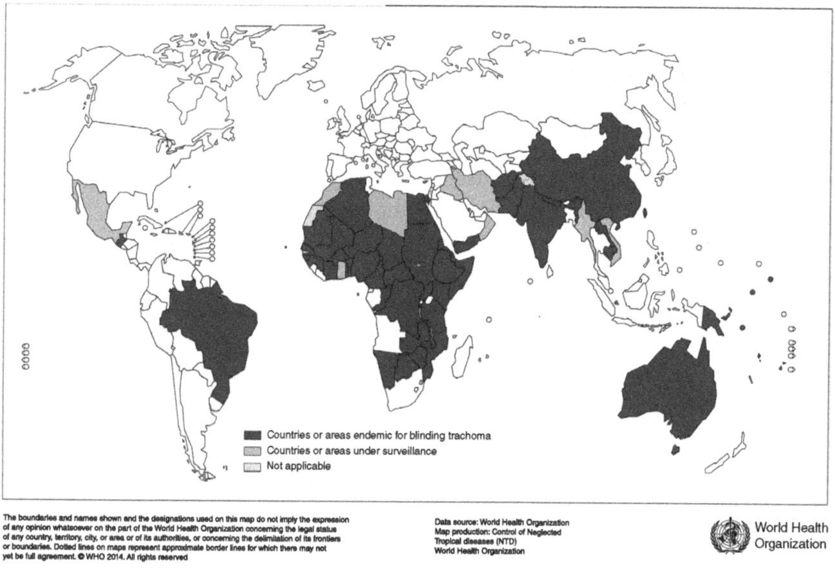

Figure 15.1 Global distribution of trachoma, 2012. *Source:* Distribution of trachoma, worldwide, 2012. World Health Organization. Retrieved from https://www.who.int/trachoma/Global_Trachoma_hr.png?ua=1.

15.2 The Causative Organism of Trachoma

Trachoma is caused by infection with a gram-negative, obligate bacterium *C. trachomatis* [10]. The active form of the disease is characterized by an inflammation of the conjunctiva leading to subsequent scarring, and the repetition of this infection and scarring causes the inward turning of the eyelid, which is a condition called trichiasis. The disease is characterized by an extreme painful rubbing of the eyelashes against the eyeball, leading to corneal scarring, light intolerance, and blindness. A 2011 report from 53 countries by the World Health Organization has found that approximately 21.4 million people are affected by active trachoma, while 7.2 million have their blindness caused by trichiasis [11]. *C. trachomatis* has a single chromosome and multiple plasmids copies which function as a virulence factor [12]. *C. trachomatis* serovars A, B, Ba, and C are the causative agents for endemic trachoma. The genital tract infection is mainly caused by serovars D–K, which affect mostly the eye leading to ophthalmia neonatorum in newly born infants or adult conjunctivitis. The serotype of the basis of the tissue tropism has not been fully elucidated [13]. However, the main form of transmission of this infection can be through direct (such as eye to eye, person to person) or indirect (through contaminated objects) routes. There is a possibility that some insect-like houseflies and eye slickers may serve as mechanical vectors responsible for the spread of the infection [14]. The incubation period lasts for 5–12 days after infection by *C. trachomatis*. It is noteworthy that every individual is susceptible to trachoma, while the most sensitive are children [15] and those in endemic regions. The infection involves a complex interaction of the immune response with *C. trachomatis* which may go on for years [16]. The disease begins in the form of active follicular trachoma, with papillary hypertrophy and inflammatory cells infiltrating lymphocytes, macrophages, neutrophils, and plasma cells (conjunctiva) punctuated by lymphoid follicles [17], which extend through the conjunctiva, more common in the upper tarsal conjunctiva. The microscopic studies of the cornea could reveal epithelial keratitis, the subepithelial infiltrates, and the extent of the limbic vessels of the cornea. The inflammatory response to primary infection to the conjunctiva by *C. trachomatis* leads to mild, self-limited conjunctivitis which is referred to as inclusion conjunctivitis. Some studies reveal that there is resistance after the first contact with the infection, but the resistance is partial because, after inoculation, a new condition is developed [18]. The manifestation of trachoma is successive in conjunctival reinfection due to individuals living where the infection is endemic, and this favors the possibilities of getting infected again. Reinfection with the disease leads to hypersensitive immune response that becomes increasingly resistant, resulting in a succession of pathophysiological phenomena that categorize trachoma [13, 19]. Matrix metalloproteinases are an integral part of tissue homeostasis involved in the destruction of tissue and fibrosis. The inflammatory cells produce proteolytic enzymes, epithelium, and fibroblast. Active trachoma and scarring caused by trachoma increases the number of matrix metalloproteinases [20–22]. The fibrosis that is formed due to the agent's response to exerting traction on the upper eyelid, resulting in distortion, causes the eyelashes to touch the eyeball (trichiasis) which can cause an ulcer. Nevertheless, scaring and corneal pacification can lead to varying degrees of decreased visual acuity to blindness [13]. Loss of vision usually occurs within a year and about 30% of untreated individuals with trichiasis [23]. If vision is not lost, trichiasis could cause substantial disability and reduces life quality [24, 25]. Additionally, follicles may appear in the limbus region while keratitis occurs in the upper part of the eye, and neovascularization [26]. Tearing, foreign body sensation, photophobia, and mild purulent discharge in small quantities are inflammatory trachoma symptoms, and purulent will only be seen when another bacterial conjunctivitis associated with trachoma is involved. A severe infection or older conjunctiva can become thickened and edematous. In extreme cases of trachoma, specifically in younger children, they are asymptomatic. An individual suffering from entropion, trichiasis, and corneal ulceration experience constant pain and intense photophobia [27].

15.3 Treatment of Trachoma

The treatment of trachoma is essentially clinical. This is achieved by external eye examination by the use of binocular loupe 2.5-fold increase (Figure 15.2). In examining the eye for disease treatment, eyelid and cornea should be first observed to verify the presence or absence of entropion, trichiasis, and corneal opacities. The upper eyelid should then be everted to examine the central area of tarsal conjunctiva, despising the eyelid and the corners' edges. Conjunctiva is normally smooth, thin, and transparent with pink coloration [28].

The main aim of treating trachoma is to cure the infection and subsequent interruption of the transmission chain. WHO founded the Alliance for the Global Elimination of Blinding Trachoma by 2020 (GET 2020) [29]. The year after, a World Health Assembly resolution called for the elimination of trachoma by 2020 through the use of a strategy referred to as SAFE strategy (**S** – Surgical treatment, **A** – Antibiotic treatment for acute infection, **F** – Face washing, and **E** – Environmental changes to improve sanitation) (Figure 15.2). It is believed that the combined health and development approach would rapidly eliminate blinding (endemic or hyperendemic) trachoma. Like in Australia, which is the only developed country where trachoma is endemic, the entire SAFE program's implementation has led to less than complete success [30].

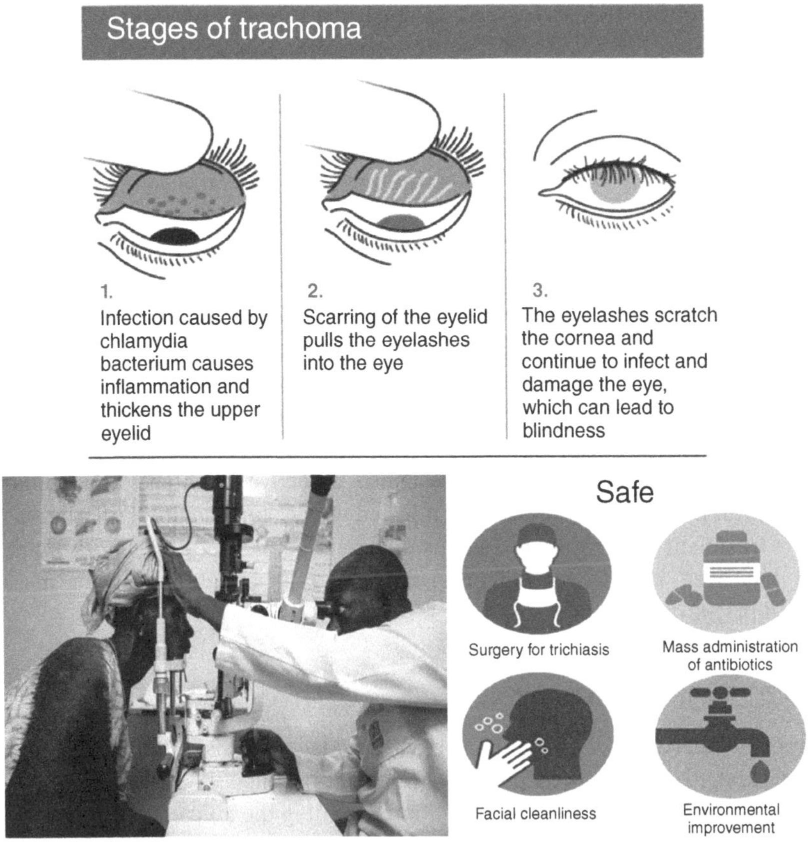

Figure 15.2 Stages of trachoma, need for early diagnosis and safe practices. *Source:* Adapted from Nicola Pearson, Mapping trachoma to eliminate blindness, SciDev.Net. Supported by UK Department for International Development.

15.3.1 Conventional Treatment

Antibiotics Azithromycin oral or topical tetracycline is used to manage *Chlamydia* in countries where it is endemic. Surgical treatment of trichiasis and several numbers of different surgical procedures can be used to treat trichiasis [31]. The one recommended by the WHO is bilamellar tarsal rotation (BLTR) or lid rotation surgery. All patients suffering from trichiasis, but still not clear if surgery is required due to less severe trichiasis, may tend to epilate the affected lashes and wait until the disease progresses before undertaking surgery [32, 33]. In standard BLTR surgery, a partial-thickness incision is made through the skin and orbicularis, followed by another incision through the conjunctiva and tarsus [34]. However, there is no proof of trichiasis surgery having better visual effects than epilation in patients with less severe trichiasis, which suggests that epilation may be appropriate for minor trichiasis where surgery is not available or accessible [32–38]. However, high recurrence rates have been reported, but research has shown that recurrence may be reduced by a single dose of Azithromycin at the time of surgery [20, 23, 37, 39, 40]. Given that Azithromycin reduces recurrence after surgery, there may be a continued presence of *C. trachomatis* or that other pathogens are contributing to the process. Again, poor surgical outcomes have been a barrier to some trachoma programs [34].

In the antibiotic treatment for acute infection, Azithromycin is the drug of choice for trachoma control. In part, this is because one annual dose over several years, depending on baseline prevalence, appears to eliminate the infection. More so, it is simpler to be administered orally than the use of non-recommended tetracycline ointment used previously for a period ranging from four to six weeks. Antibiotics reduce the risk and prevalence of active trachoma and ocular *Chlamydial* infection in communities infected with *C. trachomatis*. Still, the size of the treatment in individuals remains uncertain because *C. trachomatis* infection has been shown to re-emerge in communities that have been mass treated [41, 42].

15.3.2 Traditional Medicine

Ethnobotanical and herbal medicine have provided a means of identifying new anti-*Chlamydial* natural compounds. Moreover, it is now known that various resources often do not provide information regarding the efficacy of these treatments but provide only insights into species, which could be of interest to further research. In a study conducted in 2011, 34 Bapedi traditional healers from Limpopo Province in South Africa were interviewed concerning treatments, especially for sexually transmitted diseases, and *Chlamydia* was among the four most common conditions that were identified. Regarding *Chlamydia*, about eight plant species were used in treating this disease. They include *Aloe marlothii, Eucomis pallidiflora, Gethyllis namaquensis, Hypoxis obtuse, Kleinia longiflora, Protea caffra, Tribulus terrestris,* and *Ziziphus mucronata* [43]. Generally, using the root type, extractions in water-based through oral route were found to be the commonest, and the preparations usually involved more than one plant [43]. A survey of herbal medicines used in the treatment of sexually transmitted diseases found that an active compound called berberine played a vital role in treating Ocular *C. trachomatis* [44]. Berberine (Figure 15.3) can be obtained in a variety of plant species [45] and can be found in the powdered dried root, tinctures, or fluid and extracts of *Berberis aquifolium, Berberis vulgaris,* and *Hydrastis canadensis.*

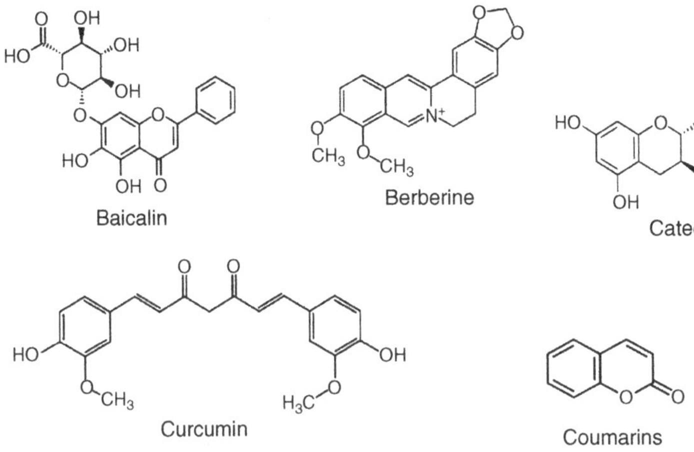

Figure 15.3 Structures of some phytochemicals active against trachoma. *Source:* Modified from [45].

15.3.3 Phytochemical Lead Compounds Effective Against Trachoma

There are diverse biochemical compounds present in natural materials that have found wide usefulness in folkloric and modern medicine. The medicinal effects of plants can be attributed to the presence of phytochemicals or secondary plant metabolites [46, 47]. They include the major phytochemical groups of quinones, flavonoids, polyphenols, tannins, coumarins, terpenoids, essential oils, alkaloids, lectins, polypeptides, glycosides, and saponins [48–52]. Each of these groups may exhibit the same or similar mechanism action which are usually defined by their chemical structure [47].

15.3.3.1 Catechins

Catechins (Figure 15.3) are ubiquitous constituents of plants. They are synthesized through the flavonoids' biosynthetic pathway [53]. Catechins being the main component of tea polyphenol extract as studied by Yamazaki et al. [54], was found to exhibit an *in vitro* inhibitory effect on *C. trachomatis* and other bacteria such as *Chlamydia pneumonia*. However, the required inhibitory concentrations are relatively high compared with the MICs of antibiotics, such as tetracyclines. Catechins damage or disrupt the permeability of lipid bilayers, and therefore it is likely that the antibacterial properties observed for this secondary metabolite may be due to the mechanism of cytoplasmic membrane damage [55]. The tea extract was found to constitute a wide range of tea polyphenols, including epicatechin gallate, epicatechin, epigallocatechin gallate, epigallocatechin, and gallocatechin gallate [54].

15.3.3.2 Baicalin

Baicalin (Figure 15.3) is a flavone derived from natural sources, including *Radix scutellaria* and *Scutellariae baicalensis*. The baicalin activities include their antimicrobial and anti-inflammatory activities and potent inhibitors of several strains of *Chlamydia*, including *C. trachomatis* and *C. pneumonia in vitro* and *in vivo* studies [54, 56, 57]. Baicalin is a potent blocker of *C. trachomatis* infection *in vitro* [58]. Baicalin successfully blocked the infection of Hep-2 cells and is observed to possess the ability of downregulating the production of a *Chlamydia* secreted protein, *Chlamydia* protease-like activity factor (CPAF) [59]. CPAF was shown to decrease host transcription factors such as RFX5, which are required for significant histocompatibility complex antigen expression, thus allowing *Chlamydia* to bypass efficient immune detection [60, 61], which serves as an essential defense mechanism for *Chlamydial* bacteria. Therefore, by baicalin targeting and downregulating the production of CPAF, the immune system may more effectively detect *Chlamydial* infections [58].

15.3.3.3 Lipids

Lipids are the essential components of the cell membrane. Studies have shown that lipids exhibit some antimicrobial activity and possess *anti-Chlamydial* activity [62, 63]. Bergsson et al. [62] showed that *C. trachomatis* could be inactivated when exposed to lauric acid, a 12-carbon saturated fatty acid (Figure 15.4) for 10 minutes at 10 mM. Similar activity was obtained with capric acid (Figure 15.4), a 10 carbon saturated fatty acid [62]. However, not all tested fatty acids gave the same result as lauric and capric acids, such as caprylic acid (8 : 0), monocaprylin (8 : 0), monolaurin (12 : 0), myristic acid (14 : 0), palmitoleic acid (16 : 1), monopalmitolein (16 : 1), oleic acid (18 : 1), and monoolein (18 : 1) (Figure 15.4) at concentrations of 20 mM.

Compounds shown in Figure 15.5 are supported by a previous study, which proposes that linoleic acid disrupted both the viral envelope of vesicular stomatitis virus and the cellular membrane on Vero cells [64]. Evidence in support of the potential for the production of specific anti-*Chlamydial* formulations incorporating bioactive compounds, several pharmaceutical formulations containing monocaprin (Figure 15.4), has already been developed. They can be potent inactivators of *C. trachomatis*, HSV-2, and human immunodeficiency virus *in vitro* [62]. Large-scale application of such compounds have been reported, and the detergent nonoxynol-9 (Figure 15.5) has also been widely studied and applied for the formulation of drug for the prevention of STDs such as *C. trachomatis* infection even though it is known to be toxic in cell cultures and cause irritation of mucosal tissues with frequent usage [62]. Monocaprin is believed to exhibit a more significant inhibitory effect against *Chlamydia* with low toxicity in *in-vivo* model at significant concentrations [62]. Synthetic lipids have demonstrated membrane disruption ability and have also been shown to exhibit the characteristics of being used in anti-*Chlamydial* formulations [63].

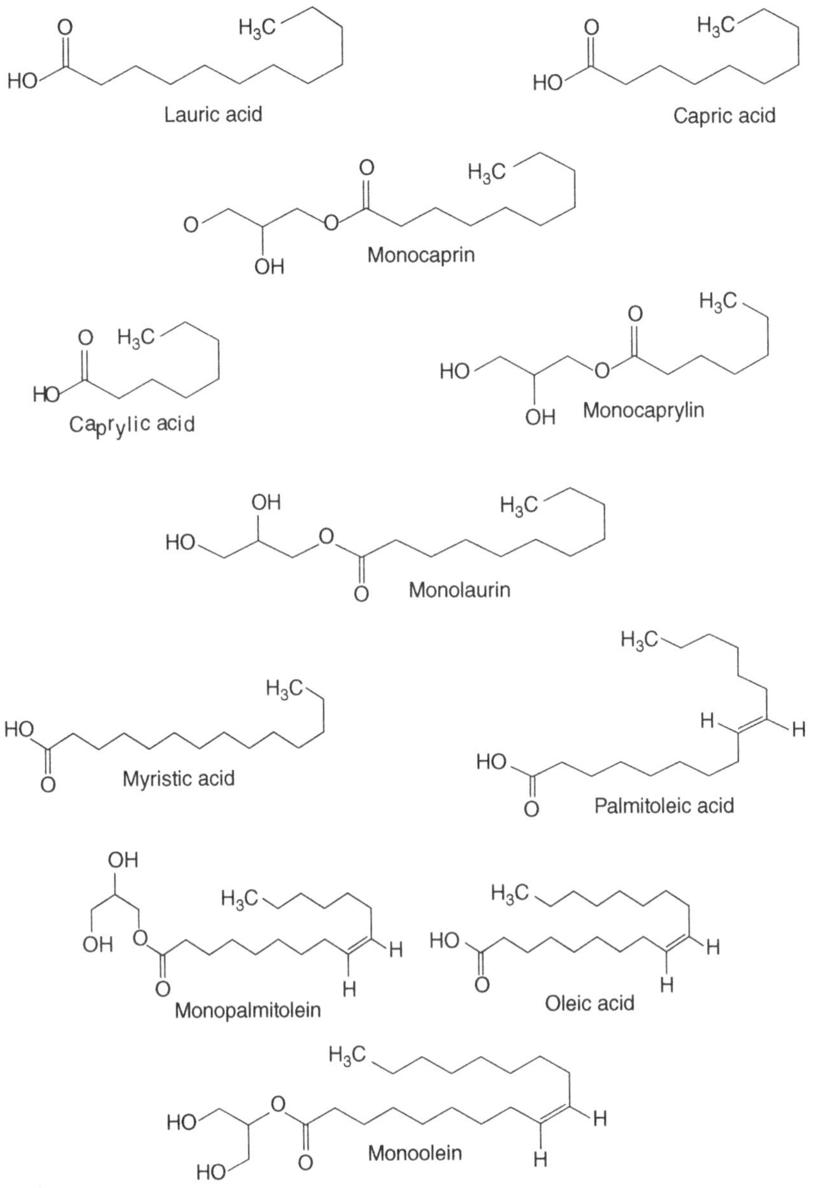

Figure 15.4 Structures of some lipids against trachoma. *Source:* Based on Bergsson [62].

Lampe et al. [63] studied the activity of five synthetic lipids derived from human breast milk. Natural compounds were isolated and modified to maintain antimicrobial activity by increasing their stability and solubility. The most active of these, 2-*O*-octyl-*sn*-glycerol (Figure 15.6), significantly inhibited the growth of

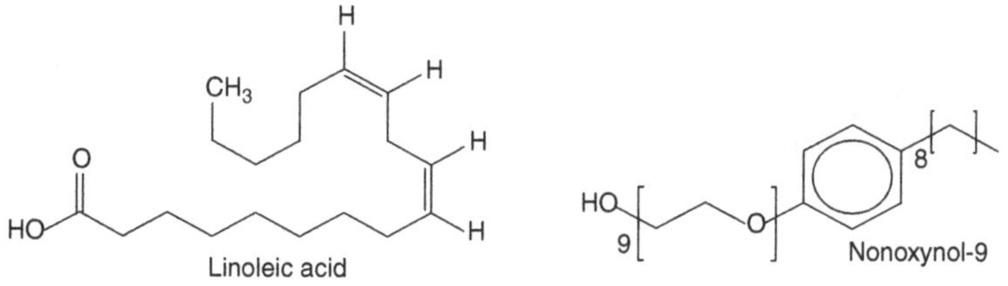

Figure 15.5 Structure of linoleic acid and nonoxynol-9 used against trachoma. *Source:* Based on [64].

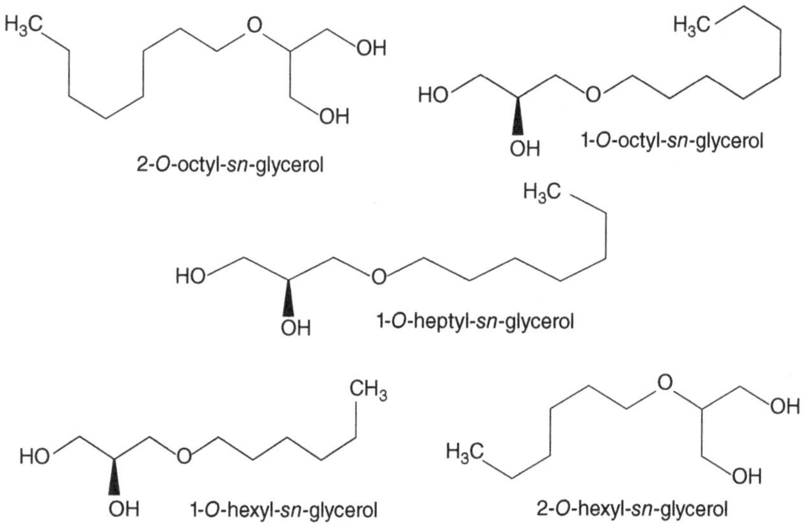

Figure 15.6 Structure of some synthetic lipids derived from human breast milk (2-*O*-octyl-*sn*-glycerol, 1-*O*-octyl-*sn*-glycerol, 1-*O*-heptyl-*sn*-glycerol, 1-*O*-hexyl-*sn*-glycerol, and 2-*O*-hexyl-*sn*-glycerol) used against trachoma. *Source:* Based on [63].

C. trachomatis (7.5 mM) for 120 minutes [63]. The remaining four compounds, 1-*O*-octyl-*sn*-glycerol, 1-*O*-heptyl-*sn*-glycerol, 2-*O*-hexyl-*sn*-glycerol, and 1-*O*-hexyl-*sn*-glycerol (Figure 15.6), exhibited less activity. The observed anti--*Chlamydial* activity of 2-*O*-octyl-*sn*-glycerol indicates the possibility that it also possesses some specific anti-*Chlamydial* effects, as reported in previous studies. These synthetic lipids have shown significant inhibitory activity against other bacteria such as *Escherichia coli*, *Salmonella enteritidis*, and *Staphylococcus epidermidis* [65].

15.3.3.4 Peptides

A wide variety of peptides exists in both plants and animals that exhibits broad-spectrum antimicrobial activities [66]. The main mechanism of antimicrobial action of many peptides is by membrane disruption [66]. The investigation into the antibacterial activity of two cecropin peptides (D2A21 and D4E1) found that they inhibit *C. trachomatis* and could be formulated into effective topical microbicides. Also, cecropin peptides exhibited broad-spectrum activity *in vitro* against human immunodeficiency virus, *Neisseria gonorrhoea*, *Gardnerella vaginalis*, *Trichomonas vaginalis*, and *Candida albicans* [67].

15.4 Future Prospects

Several different formulations of commonly used medicinal plant species have exhibited anti-*Chlamydial* activity in both *in vitro* and *in vivo* models and clinical studies. However, the specific compounds and mechanisms of action are speculative and are subject to further studies. Knowing that medicinal plants are chemical factories, it is recommended that funding opportunities be made available for rigorous research to discover new and effective drugs against trachoma.

15.5 Conclusion

Neglected tropical diseases have been the cause of death, especially in developing countries due to poor sanitary conditions and lack of adequate healthcare facilities. Trachoma, a disease caused by a gram-negative, obligate bacterium, *C. trachomatis,* is a neglected tropical disease classified as the third most common cause of blindness and glaucoma worldwide. Due to the threat of trachoma to human health and the need to address it, the World Health Assembly resolution summon for the elimination of trachoma by 2020 through the use of a strategy referred to as SAFE strategy: **S** – Surgical treatment, **A** – Antibiotic treatment for acute infection, **F** – Face washing, and **E** – Environmental changes to improve sanitation. Under the category of plant-based therapy, natural lead compounds such as baicalin, berberine, catechins, curcumin, coumarins, lauric acid, capric acid, monocaprin, caprylic acid, monocaprylin, monolaurin, myristic acid, palmitoleic acid, monopalmitolein, oleic acid, and monoolein among others have been reported to be active against *C. trachomatis*. These lead compounds need to be fully explored for the development of new drugs against trachoma.

References

1 Burton, M.J. (2007). Trachoma: an overview. *Br. Med. Bull.* 84: 99–116.

2 Wright, H.R., Turner, A., and Taylor, H.R. (2008). Trachoma. *Lancet* 371 (9628): 1945–1954.

3 Hu, V.H., Harding-Esch, E.M., and Burton, M.J. (2010). Epidemiology and control of trachoma: systematic review. *Tropical Med. Int. Health* 15 (6): 673–691.

4 Bird, M., Dawson, C.R., and Schachter, J.S. (2003). Does the diagnosis of trachoma adequately identify ocular chlamydial infection in trachoma-endemic areas? *J. Infect. Dis.* 187 (10): 1669–1673.

5 Javaloy, J., Ferrer, C., Vidal, M.T., and Alió, J.L. (2003). Follicular conjunctivitis caused by *Chlamydia trachomatis* in an infant Saharan population: molecular and clinical diagnosis. *Br. J. Ophthalmol.* 87: 142–146.

6 Modabber, M., Arora, S., and Sher, J. (2012). Trachoma: a disease of poverty. *MUMJ* 9 (1): 21–25.

7 Wright, H.R., Turner, A., and Taylor, H.R. (2007). Trachoma and poverty: unnecessary blindness further disadvantages the poorest people in the poorest countries. *Clin. Exp. Optom.* 90 (6): 422–428.

8 Alp, A. (2010). Trachoma in Turkey and Australia – a tale of two countries. *Turk. J. Ophthalmol.* 40 (5): 300–303.

9 Taylor, H.R. (2009). Doyne lecture: trachoma, is it history. *Eye (Lond)* 23 (11): 2007–2022.

10 Smith, J.L., Haddad, D., and Polack, S. (2011). Mapping the global distribution of trachoma: why an updated Atlas is needed. *PLoS Negl.Trop. Dis.* 5 (6): 973.

11 World Health Organization (2012). Global WHO alliance for the elimination of blinding trachoma by 2020. In: *Weekly Epidemiological Record 87*, 161–168. Geneva, Switzerland: WHO.

12 Clarke, I.N. (2011). Evolution of *Chlamydia trachomatis. Ann. N. Y. Acad. Sci.* 1230: E11–E18.

13 Carlson, J.H., Porcella, S.F., McClarty, G., and Caldwell, H.D. (2005). Comparative genomic analysis of *Chlamydia trachomatis* oculotropic and genitotropic strains. *Infect. Immun.* 73: 6407–6418.

14 Brasil, Ministério da Saúde, and Secretaria de Vigilância em Saúde (2005). Guia de vigilância epidemiológica/Ministério da Saúde. In: *Secretaria de Vigilância em Saúde*, 6e. Brasília: Ministério da Saúde.

15 Hu, V.H., Holland, M.J., and Burton, M.J. (2013). Trachoma: protective and pathogenic ocular immune responses to *Chlamydia trachomatis. PLoS Negl. Trop. Dis.* 7: e2020.

16 Ngondi, J., Onsarigo, A., and Adamu, L. (2005). The epidemiology of trachoma in Eastern Equatoria and Upper Nile States, southern Sudan. *Bull. World Health Organ* 83: 904–912.

17 El-Asrar, A.M., Van den Oord, J.J., Geboes, K. et al. (1989). Immunopathology of trachomatous conjunctivitis. *Br. J. Ophthalmol.* 73: 276–282.

18 Lopes, M.F.C., Luna, E.J.A., Medina, N.H. et al. (2013). Prevalência de tracoma entre escolares brasileiros. *Rev Saúde Pública* 47 (3): 451–459.

19 WHO. *Report of the 3rd Global Scientific Meeting on Trachoma Elimination, 19–20 July 2010*. Baltimore, MA: Johns Hopkins University. http://www.who.int/blindness/publications/3RDGLOBALSCIENTIFICMEETINGONTRACHOMA.pdf (accessed 26 Febraury 2016).

20 Burton, M.J., Kinteh, F., and Jallow, O. (2005). A randomised controlled trial of azithromycin following surgery for trachomatous trichiasis in the Gambia. *Brit. J. Ophthalmol.* 89 (10): 1282–1288.

21 Natividad, A., Freeman, T.C., and Jeffries, D. (2010). Human conjunctival transcriptome analysis reveals the prominence of innate defense in *Chlamydia trachomatis* infection. *Infect. Immun.* 78: 4895–4911.

22 Burton, M.J., Rajak, S.N., and Bauer, J. (2011). Conjunctival transcriptome in scarring trachoma. *Infect. Immun.* 79: 499–511.

23 Bowman, R.J.C., Faal, H., and Myatt, M. (2002). Longitudinal study of trachomatous trichiasis in the Gambia. *Br. J. Ophthalmol.* 86 (3): 339–343.

24 Frick, K.D., Melia, B.M., Buhrmann, R.R., and West, S.K. (2001). Trichiasis and disability in a trachoma-endemic area of Tanzania. *Arch. Ophthalmol.* 119: 1839–1844.

25 Dhaliwal, U., Nagpal, G., and Bhatia, M.S. (2006). Health-related quality of life in patients with trachomatous trichiasis or entropion. *Ophthalmic Epidemiol.* 13: 59–66.

26 BRASIL Ministério da Saúde (2001). Fundação Nacional de Saúde. Manual de Controle do Tracoma Brasília, versão.

27 Kasper, D. and Fauci, A. (2015). Infecções Bacterianas. Doenças Infecciosas de Harrison. *AMGH Editora* 2: 626–628.

28 Lucieni, C.B.F., Silvana, A.S., Carlos, R.P. et al. (2010). Tracoma em crianças do ensino fundamental no município de Bauru – Estado de São Paulo, Brasil. *ArqBras Oftalmol.* 73 (5): 4337.

29 WHO (2020). Prevention of blindness and visual impairment: trachoma. https://www.who.int/blindness/causes/trachoma/en/ (accessed 2 August 2020).

30 Lansingh, V.C., Mukesh, B.N., Keeffe, J.E., and Taylor, H.R. (2010). Trachoma control in two Central Australian Aboriginal communities: a case study. *Int. Ophthalmol.* 30 (4): 367–375.

31 Ferreira, I.S., Bernardes, T.F., and Bonfioli, A.A. (2010). Trichiasis. *Semin. Ophthalmol.* 25 (3): 66–71.

32 Rajak, S.N., Habtamu, E., and Weiss, H.A. (2011). Surgery versus epilation for the treatment of minor trichiasis in Ethiopia: a randomised controlled noninferiority trial. *PLOS Med.* 12: e1001136.

33 Rajak, S.N., Collin, J.R.O., and Burton, M.J. (2012). Major review: trachomatous trichiasis and its management in endemic countries. *Surv. Ophthalmol.* 57: 105–135.

34 Gower, E.W., West, S.K., and Harding, J.C. (2013). Trachomatous trichiasis clamp vs standard bilamellar tarsal rotation instrumentation for trichiasis surgery. Results from a randomized clinical trial. *JAMA Ophthalmol.* 131 (3): 294–301.

35 Woreta, T.A., Munoz, B.E., Gower, E.W. et al. (2009). Effect of trichiasis surgery on visual acuity outcomes in Ethiopia. *Arch. Ophthalmol.* 127 (11): 1505–1510.

36 Burton, M.J. (2012). Global trichiasis scientific meeting report. In: *Proceedings of the 16th Meeting of GET*, 2020. Tanzania: Moshi.

37 Woreta, F., Munoz, B., Gower, E. et al. (2012). Three-year outcomes of the surgery for trichiasis, antibiotics top revent recurrence trial. *Arch. Ophthalmol.* 130 (4): 427–431.

38 Khandekar, R.B., Al Harby, S.S., and Vora, U.P. (2012). Lid surgery for trachomatous trichiasis is negatively associated with visual disabilities and visual impairments in Oman. *East. Mediterr. Health J.* 18 (11): 1107–1113.

39 West, S.K., West, E.S., and Alemayehuetal, E. (2006). Single-doseazithromycin prevents trichiasis recurrence following surgery: randomized trial in Ethiopia. *Arch. Ophthalmol.* 124 (3): 309–314.

40 West, S., Alemayehu, W., Munoz, B., and Gower, E.W. (2007). Azithromycin prevents recurrence of severe trichiasis following trichiasis surgery: STAR trial. *Ophthal. Epidemiol.* 14 (5): 273–277.

41 West, S.K., Munoz, B., and Mkocha, H. (2005). Infection with *Chlamydia trachomatis* after mass treatment of a trachoma hyperendemic community in Tanzania: a longitudinal study. *Lancet* 366 (9493): 1296–1300.

42 West, S.K., Munoz, B., Mkocha, H. et al. (2011). Number of years of annual mass treatment with azithromycin needed to control trachoma in hyper-endemic communities in Tanzania. *J. Infect. Dis.* 204 (2): 268–273.

43 Semenya, S.S., Potgieter, M.J., and Erasmus, L.J.C. (2013). Indigenous plant species used by Bapedi healers to treat sexually transmitted infections: their distribution, harvesting conservation and threats. *South Afr. J. Bot.* 87: 66–75.

44 Vermani, K. and Garg, S. (2002). Herbal medicines for sexually transmitted diseases and AIDS. *J. Ethnopharmacol.* 80: 49–66.

45 Wang, M.W., Hao, X., and Chen, K. (2007). Biological screening of natural products and drug innovation in China. *Philos. Trans. R. Soc. B. Biol. Sci.* 362: 1093–1105.

46 Egbuna, C., Ifemeje, J.C., Kryeziu, T.L. et al. (2018). Introduction to phytochemistry. In: *Phytochemistry: Fundamentals, Methods, and Applications*, vol. 1 (eds. C. Egbuna, J.C. Ifemeje, S.C. Udedi and S. Kumar), 3–35. New York: Apple Academic Press.

47 Egbuna, C., Ezzat, S.M., Tijjani, H., and Srivastav, V.K. (2019). Synthetic analogs of phytochemicals. In: *Phytochemistry: In vitro and In Silico Updates* (eds. S. Kumar and C. Egbuna), 23–56. Singapore: Springer Nature.

48 Egbuna, C., Kumar, S., Ezzat, S.M. et al. (eds.) (2019). *Phytochemicals as Lead Compounds for New Drug Discovery*, 1e. Cambridge, USA: Elsevier. 378 pages.

49 Shashank, K. and Egbuna, C. (eds.) (2019). *Phytochemistry: In vitro and In Silico Updates*, 1e. Singapore: Springer Nature. ISBN: 978-981-13-6919-3. https://link.springer.com/book/10.1007/978-981-13-6920-9. 589 pages.

50 Egbuna, C., Ifemeje, J.C., Udedi, S.C., and Kumar, S. (eds.) (2018). Phytochemistry. In: *Fundamentals, Methods, and Applications*, 1e, vol. 1. New York: Apple Academic Press/CRC Taylor & Francis. ISBN: 978-1771887595. https://www.taylorfrancis.com/books/e/9780429426223. 684 pages, part of a 3 volume set.

51 Egbuna, C., Kumar, S., Ifemeje, J.C., and Kurhekar, J.V. (eds.) (2018). Phytochemistry. In: *Pharmacognosy, Nanomedicine, and Contemporary Issues*, 1e, vol. 2. New York: Apple Academic Press/CRC Taylor & Francis. ISBN: 978-1771887601. https://www.taylorfrancis.com/books/e/9780429426193. 620 pages, part of a 3 volume set.

52 Egbuna, C., Ifemeje, J.C., Kumar, S., and Sharif, N. (eds.) (2018). Phytochemistry. In: *Marine, Industrial, and Advances*, 1e, vol. 3. New York: Apple Academic Press/CRC Taylor & Francis. ISBN: 978-1771887618. https://www.taylorfrancis.com/books/e/9780429426155. 502 pages, part of a 3 volume set.

53 Tijjani, H., Egbuna, C., and Luka, C.D. (2018). Biosynthesis of phytochemicals, chapter 2. In: *Phytochemistry, Fundamentals, Modern Techniques, and Applications*, vol. 1, 37–78. Canada: Apple Academic Press Inc.

54 Yamazaki, T., Inoue, M., Sasaki, N. et al. (2003). *In vitro* inhibitory effects of tea polyphenols on the proliferation of Chlamydia trachomatis and *Chlamydia pneumoniae*. *Jpn. J. Infect. Dis.* 56: 143–145.

55 Ikigai, H., Nakae, T., Hara, Y., and Shimamura, T. (1993). Bactericidal catechins damage the lipid bilayer. *Biochim. Biophys. Acta BBA Biomembr.* 1147: 132–136.

56 Kuang, Z., Fu, L., Wo, H., and Zhang, S. (2005). Exploring the mechanism of baicalin in influencing cells infected with *Chlamydia pneumoniae* by observing its effect on receptors. *J. Guangzhou Univers. Tradit. Chin. Med.* 3: 014.

57 Liu, Y.D., Wu, W., Wang, S. et al. (2006). The influence of Baicalin on the level of Serum TNF-α, IL-6, IL-10 in diet-induced hyperlipidemic and CPn infected mice. *Sichuan Zhong Yi* 24: 19–20.

58 Hao, H., Aixia, Y., Nancai, Y., and Wen, S. (2009). Baicalin suppresses expression of *Chlamydia protease*-like activity factor in Hep-2 cells infected by *Chlamydia trachomatis*. *Fitoterapia* 80: 448–452.

59 Mandeel, Q.A. and Al-Laith, A.A.A. (2007). Ethnomycological aspects of the desert truffle among native Bahraini and non-Bahraini peoples of the Kingdom of Bahrain. *J. Ethnopharmacol.* 110: 118–129.

60 Zhong, G., Fan, P., Ji, H. et al. (2001). Identification of a chlamydial protease-like activity factor responsible for the degradation of host transcription factors. *J. Exp. Med.* 193: 935–942.

61 Fan, P., Dong, F., Huang, Y., and Zhong, G. (2002). *Chlamydia pneumoniae* secretion of a protease-like activity factor for degrading host cell transcription factors is required for major histocompatibility complex antigen expression. *Infect. Immun.* 70: 345–349.

62 Bergsson, G., Arnfinnsson, J., Karlsson, S.M. et al. (1998). *In vitro* inactivation of *Chlamydia trachomatis* by fatty acids and monoglycerides. *Antimicrob. Agents Chemother.* 42: 2290–2294.

63 Lampe, M.F., Ballweber, L.M., Isaacs, C.E. et al. (1998). Killing of *Chlamydia trachomatis* by novel antimicrobial lipids adapted from compounds in human breast milk. *Antimicrob. Agents Chemother.* 42: 1239–1244.

64 Thormar, H., Isaacs, C.E., Brown, H.R. et al. (1987). Inactivation of enveloped viruses and killing of cells by fatty acids and monoglycerides. *Antimicrob. Agents Chemother.* 31: 27–31.

65 Isaacs, C.E., Kashyap, S., Heird, W.C., and Thormar, H. (1990). Antiviral and antibacterial lipids in human milk and infant formula feeds. *Arch. Dis. Child.* 65: 861–864.

66 Zasloff, M. (2002). Antimicrobial peptides of multicellular organisms. *Nature* 415: 389–395.

67 Ballweber, L.M., Jaynes, J.E., Stamm, W.E., and Lampe, M.F. (2002). *In vitro* microbicidal activities of cecropin peptides D2A21 and D4E1 and gel formulations containing 0.1% to 2% D2A21 against *Chlamydia trachomatis*. *Antimicrob. Agents Chemother.* 46: 34–41.

16

Yaws (Endemic Treponematoses) Drug Discovery from Phytochemicals

An Informatics Protocol for Drug Target Identification to Phytochemical Inhibitor Screening and Validation

Zarrin Basharat[1] *and Arisha Khoso*[2]

[1] *Jamil-ur-Rahman Center for Genome Research, Dr. Panjwani Center for Molecular Medicine and Drug Research, International Center for Chemical and Biological Sciences, University of Karachi, Karachi, Pakistan*
[2] *HEJ Research Institute of Chemistry, International Center for Chemical and Biological Sciences, University of Karachi, Karachi, Pakistan*

List of Abbreviations

ADMET	Absorption, distribution, metabolism, and excretion
CDSs	Coding DNA sequences
DEG	Database of essential genes
DNA	Deoxyribonucleic acid
GMQE	Global model quality estimation
NCBI	National Center for Biotechnology Information
NTDs	Neglected tropical diseases
QMEAN	Qualitative model energy analysis
Rg	Radius of gyration
RMSD	Root mean square deviation
RMSF	Root mean square fluctuation
WHO	World Health Organization

Neglected Tropical Diseases and Phytochemicals in Drug Discovery, First Edition.
Edited by Chukwuebuka Egbuna, Muhammad Akram, and Jonathan Chinenye Ifemeje.

16.1 Introduction

Neglected tropical diseases (NTDs) are a group of chronic, debilitating, and disabling diseases that are caused by different pathogens like bacteria, protozoa, parasites, viruses, and fungi. They are much prevalent in tropical and subtropical developing countries, where poverty is rife [1]. Treponematosis is a medical word used to refer to any of the diseases caused by four members of the bacterial genus *Treponema* which includes Syphilis (*Treponema pallidum pallidum*), Yaws (*Treponema pallidum pertenue*) [2], Bejel (*Treponema pallidum endemicum*), and Pinta (*Treponema carateum*). Yaws (also known as framboesia or pian), an NTD mostly affecting poverty-stricken populace (Figure 16.1), closely resembles the bacterium *T. palladium palladium*, causative agent of venereal syphilis. It is similar serologically and morphologically but not identical [3]. Its length ranges from 10 to 15 μm and typically, it has a diameter of about 0.2 μm and is visible in dark field illumination light microscopy [4]. It is surrounded by a cytoplasmic and outer membrane cover, although clear space has not been observed during examination, except in few cases [5]. They have specific crock screw motility because of endoflagellum that enables it to swim in gel-like environment much efficiently as

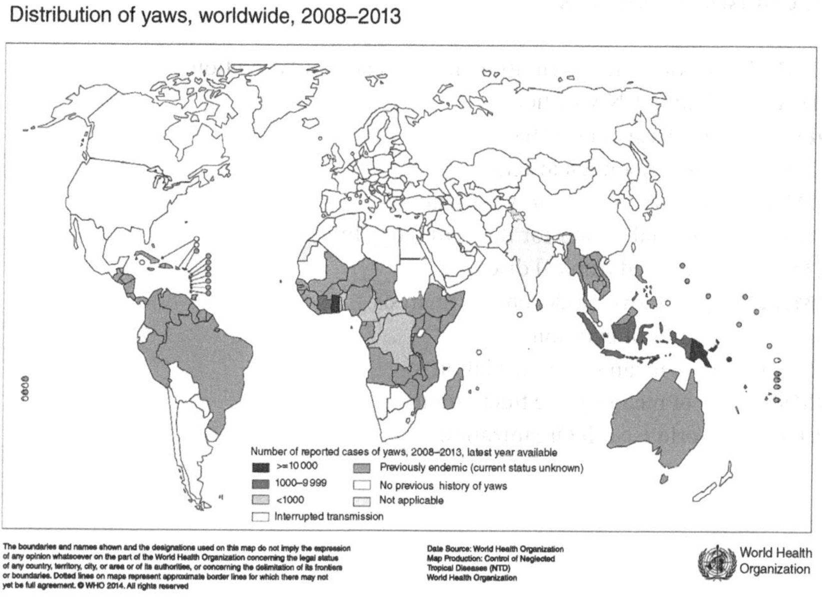

Figure 16.1 Yaws global distribution. *Source:* Distribution of Yaws, worldwide, 2008–2013. World Health Organisation. Retrieved from http://gamapserver.who.int/mapLibrary/Files/Maps/Yaws_2013.png.

in connective tissues, thus, motility plays a vital role in their virulence and pathogenesis [6]. It is more prevalent in humid tropical areas and usually gets spread through skin to skin contact. Children living in poor densely populated areas are primarily the victims of yaws [7].

As far as disease manifestation of yaws is concerned, early stage of disease progression has similarities with syphilis but late stage is limited to skin, joints, and bones. On the contrary, syphilis in its late stages can affect organs, organ systems, and tissues badly but rarely visceral, congenital, and central nervous system are involved [8]. Progression of yaws appears to be a multistage disease process that predominantly affects skin, bones, and cartilage. The primary yaws are also known as "mother yaws" and start to progress with an unusual redness (erythromatous) of the skin at the inoculation site, following the incubation period of usually about 9–90 days [9]. It manifests as ulcers that are typically painless with raised edges and soft base. Moreover, it can also appear as papilloma [10]. Later, conditions become much worse and lead to secondary yaws if the initial stage remains untreated. It starts predominantly afflicting skin and bones. Secondary yaws causes hyperkeratotic papules on palms and soles, osteoperiostitis that results in swelling and severe bone pain. Secondary yaws can also be considered as polyostotic as it affects many bones including radius ulna, phalanges causing polydactylitis [11, 12]. In the case of prolonged untreated early yaws, patients go into a latent phase where there are no clinical symptoms but have only reactive serological tests. Relapse from latent to active yaws might develop later on and can serve as a source of transmission of disease [13]. Disease progression can further lead to tertiary yaws, a much severe stage which commonly manifests skin lesions, hyperkeratosis in palm areas, nodules in the area of joints (Figure 16.2), and ulcerate that gradually develops into tissue necrosis [14]. Disease throughout progression becomes much more complicated and can cause medical complication as choric osteoperiostitis, destructive ulcer gangosa, and goundou exostosis of maxillary bones.

Since the earliest time, many plants are used for the cure of several human infectious diseases. It is because of the presence of compounds in them that can stop biological activity associated with disease progression. Phytochemical inhibition of human pathogens is much sought-after since it helps prevent the phenomenon of antibiotic resistance and add to the pool of drugs. In this study, we have used a subtractive genomics approach to identify core drug targets of the causative agent of yaws, i.e. *T. pallidum* subsp. pertenue and applied virtual screening approach to find out potent inhibitors from traditional Tibetan medicinal plants. This ancient system of medicine is also known as Amchi or Sowa-Rigpa. Along with other aspects of this medicine system, it is centered on aboriginal Tibetan understanding and use of local plants for the benefit of the people's health. Sowa Rigpa is now being sold globally as it has produced a significant industry-valued

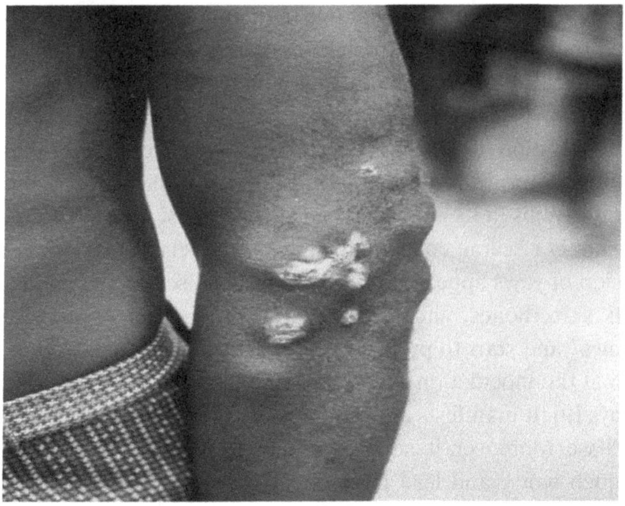

Figure 16.2 Yaws infection caused by *Treponema pertenue* bacteria: Juxta-articular nodules on the elbow. *Source:* Adapted from Wikipedia. Courtesy: CDC/Dr. Peter Perine. Freely taken from https://en.wikipedia.org/wiki/Yaws#/media/File:Yaws_01.jpg.

over half a billion US dollars [15]. ADMET profiling and dynamics simulation were carried out to validate the usability and binding of this compound to the target protein.

16.2 Methods

The genome sequence of 14 strains of *T. pallidum* subsp. pertenue were obtained from NCBI (Table 16.1). Pan-genomics was conducted as described previously [16], with sequence identity clustering of 70% in USearch clustering. Pan-genome study was used to identify the core region coding DNA sequences (CDSs). These CDSs were subjected to blast against DEG and CEG database, to find out essential genes with a cutoff E-value $\leq 10^{-10}$ and bit-score ≥100, after CD-Hit analysis (~60% criteria). Human and gut flora nonhomologous sequences (E-value >0.005 and $>10^{-4}$, respectively) were then sifted out [17]. Druggable targets were filtered and subjected to structural modeling using SWISS-Model [18]. Molecular docking analysis against in-house database of 52 Sowa-Rigpa compounds (available from authors on request) from Yeshi et al. [19] was carried out. Docking was carried out using Molecular operating environment software v. 2015. Energy minimization and preparation of ligands for docking was followed by setting parameters as: receptor = lipoprotein; pharmacophore = none; docking placement = triangle

Table 16.1 Accessory and unique gene statistics for *Treponema palladium* subsp. pertenue.

Serial no.	Name	No. of accessory genes	No. of unique genes	No. of exclusively absent genes
1	CDC 2575	835	0	0
2	CDC-1	841	0	0
3	CDC2	848	8	0
4	Gauthier	853	3	0
5	Ghana051	835	0	0
6	Kampung Dalan K363	841	1	0
7	LMNP-1	835	2	0
8	R2-64919 SamoaD	486	10	90
9	R3-SAL-007 substr. TD6	498	6	92
10	R7-KUN2-001 substr. SE7	503	15	81
11	R7-PUT1-002 substr. JG8	484	5	102
12	R8-KOS-007 substr. JG8	813	22	2
13	SamoaD	856	4	0
14	Sei Geringging K403	836	1	1

matching; scoring = London dG and GBVI/WSA dG [20]. ADMET profiling was done using admetSAR 1.0 (http://lmmd.ecust.edu.cn/admetsar1).

Dynamics simulation of 3000 picoseconds (ps) (or 3 nanoseconds [ns]) each for receptor protein without ligand, receptor protein binding, and unbinding with ligand was conducted using AMBER16 program, according to instructions in reference manual (https://ambermd.org/doc12/Amber16.pdf). First of all, coordinate and topology file construction of the complex was done with Tleap module. Other specifications were: protein forcefield = ff14SB and ligand forcefield = gaff; Water = explicit, System counter ions = Na^+ or Cl^-; Molecule solvation = TIP3P water in octahedron box extended 10 Å in each direction from the solute; Minimization = Sander module (steepest descent = 2000 steps; conjugated gradient = 3000 steps); Backbone atoms of protein = fixed; All atoms = fixed; Water, ions, and hydrogens = unfixed; MD simulation = Pmemd module; System heating 10–300 K in 30 ps; pressure = 1 atm; Conditions = periodic boundary; Equilibration time = 3000 ps. Restrain file during unbinding was made using

CAVER3.0 [21]. The final result was the average value of work profiles from 10 independent trajectories. Trajectory analysis (RMSD, Rg, RMSF, B-factor and Hydrogen Bond) was done using Cpptraj module [22]. Binding energy decomposition was analyzed using decomposition module of AMBER16 software. Free energy calculation was calculated by the molecular mechanics/Poisson–Boltzmann (Generalized Born) surface area or MM/PB(GB)SA and empirical method [23].

16.3 Results and Discussion

16.3.1 Genome Analysis

Pan-genome statistics of the 14 studied species disclosed that parameter "*b*" was 0.07. This means that pan-genome is almost closed. This depicts a clonal population structure of the species. Previously, Jaiswal et al. [24] conducted a pan-genome analysis of the genus *Treponema* and corroborated pan-genome of this genus is open for some species, increasing at a very slow pace for others, while it was almost closed for subsp. pertenue. Although they conducted analysis with only 7 genomes of the specie compared with availability of larger sample size, i.e. 14 genomes at the time of this study. However, the findings complement their analysis that even doubling the number of genomes had no effect on the pan-genome size. They mined ~1000 core genes for the species while with addition of more genomes, this number has been reduced to 183 in this study. 10 species had more than 800 accessory genes while other four genomes had a significantly reduced gene count (Table 16.1). These were the same genome showing largest number of exclusively absent genes while their unique gene count ranged from 5 to 15. Highest number of unique gene count was observed for the genome *T. palladium* subsp. pertenue R8-KOS-007 substr. JG8.

16.3.2 Therapeutic Candidate Mining

Genome subtraction-based drug designing aims to subtract smaller pieces from the larger ones to identify the unique set of genes that are typical for specific activities. In this study, we attempted to identify unique and essential genes for organism's survival from core regions of the 14 organisms of *T. palladium* subsp. pertenue. Essential genes might belong to adaptability, growth, pathogenesis, and metabolism function category while their deletion or inhibition can become fatal for the survival of the organism [25, 26]. These genes were further subtracted from human genome so that only nonhost homolog genes were further analyzed (Figure 16.3). Following this approach and using the obtained 183 core genes for

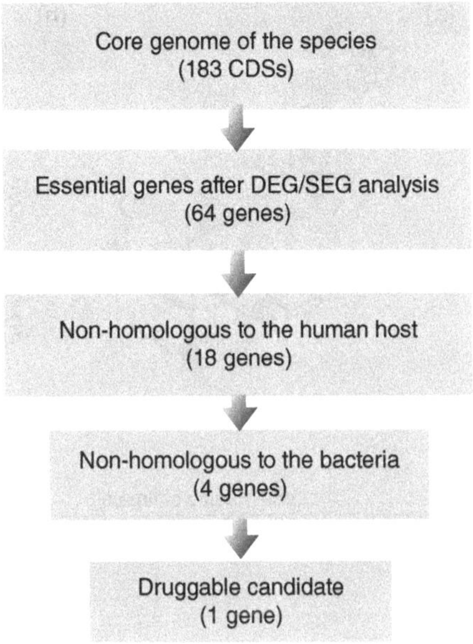

Figure 16.3 Flowchart showing steps and relative number of mined gene sequences.

therapeutic candidate mining analysis, only one drug candidate, i.e. a periplasmic binding lipoprotein, was mined that had potential of targeting for inhibition of all studied *T. palladium* subsp. pertenue genomes.

16.3.3 Structure Modeling

Protein structure provides valuable insights into drug discovery, and its functional characterization is one of the main problems in biological sciences. The 3D structure of proteins facilitates its functional characterization low-resolution protein structure gives insight for fold recognition. Furthermore, high-resolution structures help in inferring and understanding their general role, ligand specificities, and inhibitor designing [27]. Experimentally deduced protein structures are still time-consuming, so computational approaches are needed to be used in order to fill the gap in the number of known sequences and 3D structures [28].

The translated protein product for the mined candidate gene was obtained. The template coding for a variant D-glucose-binding protein with PDB ID: 6BGD, from *T. pallidum* subsp. pallidum str. Nichols was used for homology modeling. It had 61% similarity and 91% coverage. ProMod 3.0 [29] at the backend calculated the monomer structure with several quality parameters (Figure 16.4a–c) with a calcium ion ligand. GMQE and QMEAN scores were 0.97 and 0.46, respectively. GMQE and QMEAN are assessment of model quality, with respect to native experimental structure. Higher value (between 0 and 1) of GMQE and near-zero

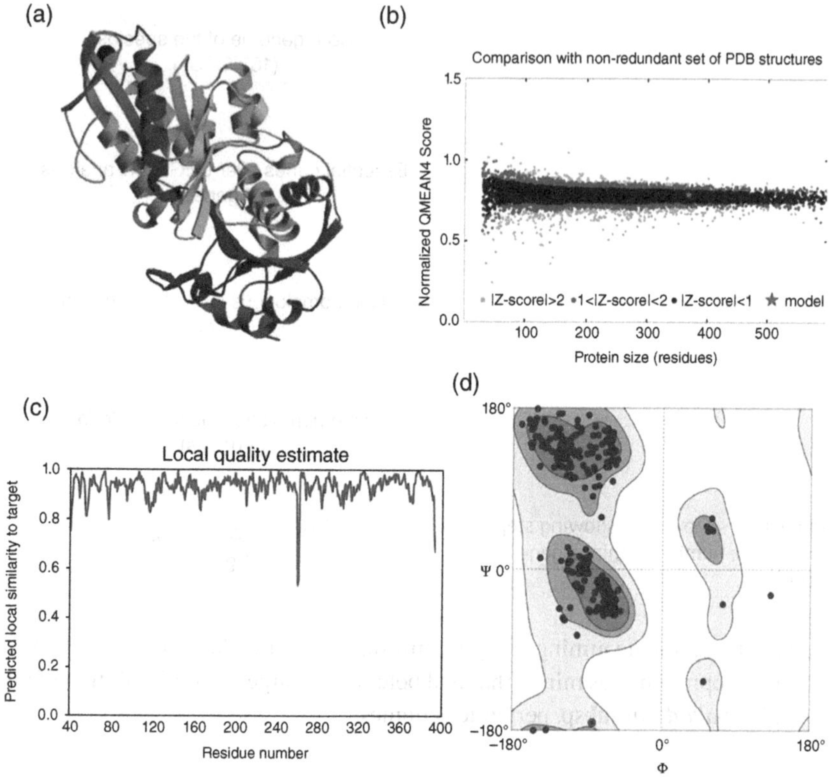

Figure 16.4 (a) 3D Structure model of the target lipoprotein. (b) Comparison of 3D model with nonredundant structures in PDB. (c) Local similarity of variant D-glucose-binding protein from *Treponema pallidum* subsp. pallidum str. Nichols to the target lipoprotein of *Treponema pallidum* subsp. pertenue. (d) Ramachandran plot of modeled structure with a large number of residues in the favored regions. *Source:* Based on [29].

value of QMEAN implies good structure [30]. Ramachandran favored residues were 98.36%, which also renders the model as very good quality (Figure 16.4d).

16.3.4 Docking/Virtual Screening

Among various high-throughput techniques, virtual screening is the most popular and powerful tool for searching chemical entities from thousands of chemical compound databases and then select a limited number of new entities that can bind and fit in the particular binding site of the protein and have certain biological activities [31]. The term "virtual screening" was coined in the late 1990s that describes the use of computational algorithms for models and identification of novel biologically active compounds [32]. Virtual screen has been proved as breakthrough technique in many cases of drug discovery but still this has been remained questioned as several case studies have published and that has become evident that the virtual screening method mainly relies on the macromolecular target being investigated; in other words, this technique seems to be problem-centric rather than method-centric [33]. Structural-based virtual screening has mounted a number of success stories among which docking-based virtual screening is most applicable in drug discovery [34].

The chosen compounds for screening belonged to the traditional Tibetan medicine. Practice of healing with plants originated in central Tibet between seventh and twelfth century CE and get integrated with medical knowledge of Indian, Chinese, Persian, and Central Asian traditions. Slowly, this folk medicine knowledge got transmitted to Mongolia, Nepal, Bhutan, Indian Himalayas, and subsequently Russia [35]. It is used for curing a wide range of conditions, including chronic metabolic disorders, autoimmune disorders, inflammatory, and other acute and chronic diseases [15].

Energy-minimized 3D structure was screened virtually against Sowa Rigpa compounds, using a molecular docking approach. An alkaloid, Isorhoeadine (IUPAC name: (1*S*,14*R*,24*S*)-24-methoxy-13-methyl-5,7,19,21,25-pentaoxa-13-azahexacyclo[12.11.0.02,10.04,8.015,23.018,22]pentacosa-2,4(8),9,15(23),16,18(22)-hexaene) from *Meconopsis horridula,* with least binding energy value of −9.569 was the top-scoring inhibitor of target lipoprotein (Table 16.2). Several hydrogen and other interactions were observed (Figure 16.5). Receptor lipoprotein was exposed more at residues Asp220, Asn222, Arg310, Asp337, Phe338, and Ser341 compared with some exposure at Gln240, Ala342, Ala221, Val242, and Tyr308. Asp337, Phe338, and Ser341 interacted with solvent by forming a triad. Ala221, Ala312, and Asp220 donated hydrogens to Isorhoeadine while Arg120 and Gln240 accepted hydrogen atoms. Asp337 formed an ionic interaction with the Isorhoeadine. Distance of interactions ranged between 2 and 3 Å. Energy values for hydrogen-donating bonds were lower than acceptor bonds and almost equaled ionic interaction energy, i.e. −4.2 kcal/mol.

Isorhoeadine has also been extracted from *Papaver rhoeas* (red poppy) [36], *Papaver rupifragum* [37], and more recently from *P. rhoeas* [38]. Slavik and Slavikova [39] also characterized it from Himalayan plant of the genus *Meconopsis* belonging to the Papaveraceae family. Tibetan traditional medicinal plant *M. horridula* is a herb from the same family [40] that grows at an altitude of 3000–5000 m in the Qinghai-Tibet plateau area. It is used as Tibetan folk medicine used mainly for the treatment of headaches and fractures [41]. For hundreds of years, *M. horridula* has been used to clear away heat, relief in pain, mobilization of static blood, and used to treat bruises. It contains alkaloids, flavonoids, and phenylpropanoids. Apart from use as sedative, it is used for the treatment of cancer [42], osteomyelitis, and liver diseases.

Table 16.2 Top 10 inhibitor compounds from Tibetan traditional medicinal plants.

S. no.	Compound name	S-value	Weight	Hydrogen donors	Hydrogen acceptors	Source plant
1	Isorhoeadine	−9.569	829.69	14	20	*Meconopsis horridula*
2	Cyanidin 3-*O*-[(6-*O*-malonyl-2-*O*-beta-D-xylopyranosyl)-beta-D-glucopyranoside]-7-*O*-beta-D-glucopyranoside	−7.565	610.521	10	15	*Meconopsis horridula*
3	Kaempferol 3-(6″-*p*-coumarylglucoside)-7-glucoside	−7.081	756.666	10	16	*Anaphalis contorta*
4	Procyanidin B3	−6.815	578.526	10	12	*Bergenia purpurascens*
5	Tiliroside	−6.793	594.525	7	11	*Anaphalis contorta*
6	Bellidifolin	−6.501	730.631	12	15	*Gentiana algida*
7	Norstictic acid	−6.270	798.791	11	16	*Usnea articulata*
8	Herbacetin 3-glucoside	−6.002	302.238	5	6	*Meconopsis paniculata*
9	Fecosterol	−5.825	396.659	1	1	*Usnea longissima*
10	Leptopinine	−5.777	367.357	0	4	*Hypecoum leptocarpum*

ADMET profiling of Isorhoeadine showed non-AMES toxic, noncarcinogenic, and category III acute oral toxicity. This category consists of compounds with LD50 >500 and <5000 mg/kg. CYP450 and P-glycoprotein inhibition was negative, along with low blood–brain barrier passage and intestinal absorption. Aqueous solubility (Log *S*) was found to be −2.5614, fish toxicity (pLC_{50}) was 1.3511 mg/l, *Tetrahymena Pyriformis* toxicity (pIGC50) was 0.2642 μg/l, rat acute toxicity (LD_{50}) was 2.4472 mol/kg while Caco-2 permeability (Log P_{app}) was predicted as −1.0982 cm/s. ADMET showed it was safe to consume this drug. However, fish and *T. Pyriformis* toxicity was high and dose needs to be verified in mice and other important model organisms before proceeding for human administration.

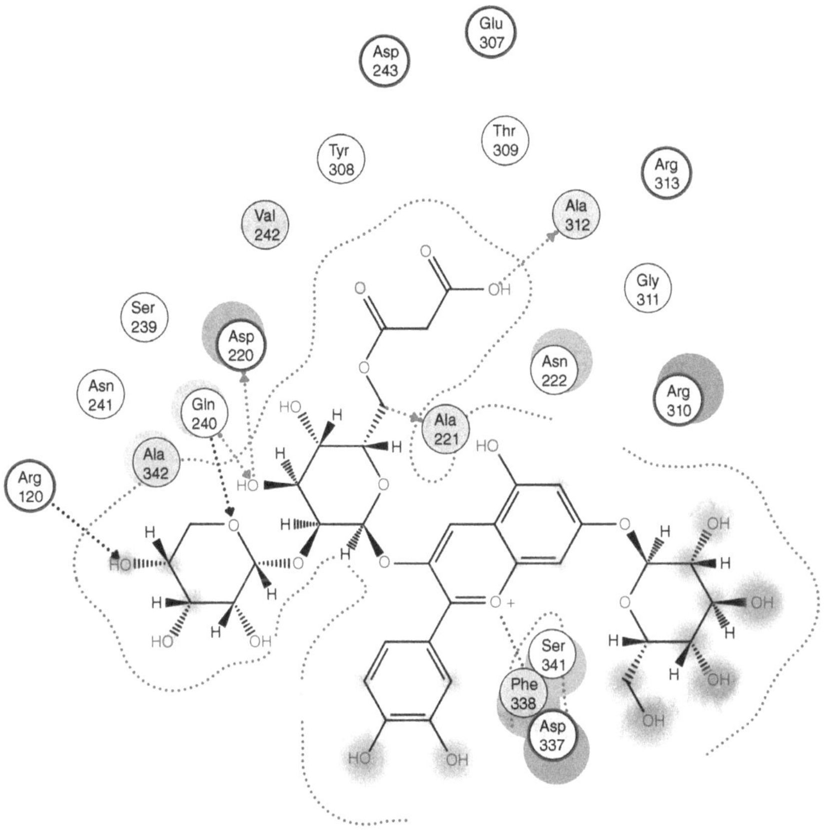

Figure 16.5 A 2D plot of the lipoprotein–Isorhoeadine complex. Arrowheads indicate electron transfer, contour line shows solvent exposure for the three residues Ser341, Phe338, and Asp337.

16.3.5 Dynamics Simulation

Dynamics simulation results were obtained to elucidate various aspects of ligand–receptor binding. Several parameters were calculated like root mean square deviation (RMSD), root mean square fluctuation (RMSF), radius of gyration (Rg), native contact fraction, etc. (Figure 16.6a–j). These are explained below, along with values and their possible implications.

RMSD is the measure of the average distance between the atoms of superimposed structures. RMSD plot indicates change in equilibrium of system in ligand bound and unbound state (Figure 16.6a). It depicts more stability in ligand-bound state as the RMSD is between 1 and 2 Å while in unbound state, it increases more than 4 Å after 2 ns (Figure 16.6b). RMSF was analyzed to record changes in the

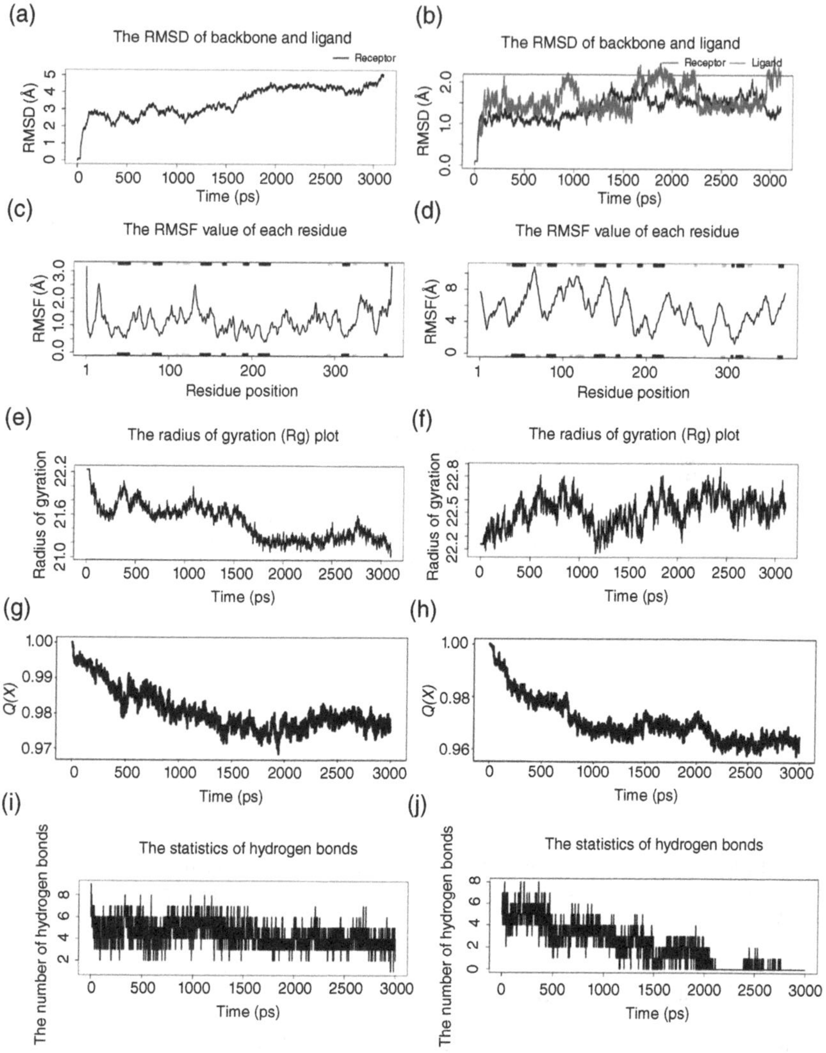

Figure 16.6 (a) RMSD of receptor over 3000 ps timescale. (b) RMSD of ligand and receptor over 3000 ps timescale. (c) RMSF of receptor. (d) RMSF of receptor and ligand. (e) Rg of receptor over 3000 ps timescale. (f) Rg of receptor and ligand over 3000 ps timescale. (g) Fraction of native contacts of receptor over 3000 ps timescale. (h) Fraction of native contacts of receptor and ligand over 3000 ps timescale. (i) H-bond formation in ligand association/binding. (j) H-bond formation in ligand dissociation/unbinding.

atomic positions based on the CA atoms. Larger fluctuations were predicted for loop regions while least for helices (Figure 16.6c). Similarly, less fluctuations for lipoprotein receptor (deviation below 3 Å) while increased for complex (up to 9 Å) (Figure 16.6d). Here, the fluctuation of each residue was deliberated for the

20 ligand-binding residues (shown in Figure 16.5). RMSF for Arg120 varied from 1.16 to 8.82 Å, Asp220 from 0.72 to 5.97 Å, Ala221 from 0.74 to 6.45 Å, Asn222 from 0.95 to 6.57 Å, Ser239 from 1.30 to 5.64 Å, Gln240 from 1.63 to 6.13 Å, Asn241 from 1.49 to 5.50 Å, Val242 from 1.55 to 5.23 Å, Asp243 from 1.19 to 4.57 Å, Glu307 from 1.03 to 1.29 Å, Tyr308 from 0.81 to 1.74 Å, Thr309 from 0.73 to 2.28 Å, Arg310 from 0.68 to 2.12 Å, Gly311 from 0.64 to 2.03 Å, Ala312 from 0.56 to 2.71 Å, Arg313 from 0.68 to 3.02 Å, Asp337 from 1.90 to 5.28 Å, Phe338 from 1.97 to 5.74 Å, Ser341 from 1.72 to 6.32 Å, and Ala342 from 1.82 to 6.89 Å. Comparative RMSF for ligand bound and unbound protein was observed for Glu307 and Tyr308.

MM/PB(GB)SA result consists of electrostatic, van der Waals energy, total gas phase energy, nonpolar and polar contributions to solvation, etc. Sharp decrease in electrostatic molecular mechanic force field energy was observed at and around residue 280, while final estimated binding free energy (ΔPB/ΔGB) and van der Waals contribution from molecular mechanics was lowest in the region around the residues at positions 183, 280, and 300. Overall, ΔGB was −24.43, ΔPB was −17.19, solvation energy was 14.88 while van der Waal's energy was −54.83. Apart from interacting residues Ala221, Tyr308, Arg310, Gly311, Arg313, Phe338, Ser341, and Ala342, additional two residues Ala223 and Pro345 were identified as having important from interactive point of view, based on energy profiling. Electrostatic, gas phase, sum of polar and nonpolar energy was least for Ala223 and Pro345.

Rg is a substantial gauge for predicting the macromolecular structural activity due to protein fold change. Thus, it provides a vital review of the equilibrium disturbed by protein unfolding. Rg of receptor was between 21.0 and 22.2 Å (Figure 16.6e) while it ranged from 22.2 to 22.8 Å in ligand-bound state (Figure 16.6f). Low Rg and, accordingly, tight packing is characteristic of α/β-proteins and since target lipoprotein consisted of both α-helices and β-sheets, results showed the protein was in compact form. The values were slightly higher in ligand-bound state, compared with native globular state, where it achieved compaction at around 2000 ps. Nonnative contact measurement is inapt to protein folding analysis, so energetically favorable fraction of native contacts $Q(x)$ was analyzed. After 500 ps, a sharp descent was observed in fraction of native contact analysis for both ligand unbound (Figure 16.6g) and bound form (Figure 16.6h). It was lower than native state in complex state, after 2500 ps. It depicts a transition state when ligand is bound and unfolding of protein is bit high. H-bond numbers ranged between 4–6 and then 2–4 after 1600 ps of simulation in ligand-bound state of the lipoprotein (Figure 16.6i). During unbinding, it is clearly visible that the number reached less than 2 after 2000 ps and got zero around 3000 ps (Figure 16.6j). Reaction coordinate started at 15.8 and ended at 60 Å.

Molecular dynamics has aided drug discovery, with a whooping number of publications reporting this technique in the recent years. Simulations not only capture the behavior of receptor proteins but ligand binding and movement of ligand across the receptor, at atom scale. We utilized the same approach to analyze protein stability, compactness, and energy parameters in the native and ligand-bound state. Simulations proved that the functional mechanism is not impaired by unfolding. Only a slight change in compactness is observed after binding while RMSF is considerably changed at pocket binding residues, which directly make bonds with ligand. However, this does not lead to a large RMSD and hydrogen bonds along with fraction of native contacts remains in a suitable range.

16.4 Future Projection

Pan-genomics is a useful technique which is increasingly being utilized for core genome identification, with availability of more and more genome sequencing data. Entire set of genes within a species, class, or other defined category is taken [43] and categorized as accessory, core, or unique. Apart from information about evolution and clonality of a species [44], it can be used for identification of common set of genes in species, which may be targeted for drug design against all the included species. These core genes usually have essential function which can be targeted for stopping replication, translation, or other important process of survival. Inhibition can, thus, lead to demise of all studied organisms in the species, instead of just targeting a single bacterium. We are hopeful that this technique will proliferate even more in future and its combination with other techniques such as docking and dynamics simulation will bear more good results while targeting pathogens.

Molecular docking and dynamics simulation have been utilized previously for identifying potent and novel inhibitors, providing useful results without lab work [45–51]. In this chapter, we have defined a simple workflow which couples pan-genomics with docking and dynamics simulation to identify and validate the inhibition of a vital lipoprotein of the *T. pallidum* subsp. pertenue. Use of compounds from Sowa Rigpa is a gaze into the untapped potential of medicinal phytocompounds from understudied medicinal plants. This helps dig new compounds against pathogens and aid in our quest against neglected tropical diseases. We are confident that coupling of this workflow with machine learning techniques and better scoring methods will lead to quicker and improved information. Better ADMET profiling can also achieve superior projection of identified compounds and use in clinical trials. It is estimated that machine learning-based future ADMET analysis will continue to improve. We also feel that benchmarking of various pan-genomic, structure prediction, docking, and dynamics simulation approaches could further add to our knowledge and provide a standardized framework for inhibitor screening against pathogens. Clonal and recombinant bacteria could benefit from specialized approaches, suited to their population structures. Knowledge of different evolutionary models and association studies could be incorporated to further enhance the findings. A suitable timescale and parameters for dynamics simulation could also benefit the validation module.

16.5 Conclusion

In this study, we utilized an informatics approach to find a therapeutic candidate in causative agent of a disfiguring disease yaws. It can spread quickly from one individual to the other. Infectious nature makes it important to study and find cure against it. However, it has been understudied and WHO has pledged to eradicate it. We have assessed pan, core, and unique genes of its causative agent, *T. pallidum* subsp. pertenue. The identified drug target through a subtractive genomics approach, a lipoprotein, was modeled using SWISS-model and subjected to molecular docking-based virtual screening against Sowa Rigpa compounds. Toxicity parameters showed the top-score compound Isorhoeadine from the genus *Meconopsis* was safe to use while molecular docking and dynamics simulation showed a good binding affinity for the receptor lipoprotein. This compound could curb *T. pallidum* subsp. pertenue and help eradicate yaws. It should be tested *in vitro* and *in vivo* to endorse our findings. Our protocol can be used for other diseases caused by bacteria, with availability of genome sequence data.

References

1 Molyneux, D.H., Savioli, L., and Engels, D. (2017). Neglected tropical diseases: progress towards addressing the chronic pandemic. *The Lancet* 389 (10066): 312–325.

2 Cejkova, D., Zobanikova, M., Chen, L. et al. (2012). Whole genome sequences of three *Treponema pallidum* ssp. pertenue strains: yaws and syphilis treponemes differ in less than 0.2% of the genome sequence. *PLoS Neglected Tropical Diseases* 6 (1): e1471.

3 Singh, A.E. and Romanowski, B. (1999). Syphilis: review with emphasis on clinical, epidemiologic, and some biologic features. *Clinical Microbiology Reviews* 12 (2): 187–209.

4 Ovcinnikov, N.M. and Delektorskij, V.V. (1970). *Treponema pertenue* under the electron microscope. *British Journal of Venereal Diseases* 46 (5): 349.

5 Engelkens, H.J.H., Vuzevski, V.D., Ten Kate, F.J.W. et al. (1991). Ultrastructural aspects of infection with *Treponema pallidum* subspecies pertenue (Pariaman strain). *Genitourinary Medicine* 67 (5): 403–407.

6 Charon, N.W. and Goldstein, S.F. (2002). Genetics of motility and chemotaxis of a fascinating group of bacteria: the spirochetes. *Annual Review of Genetics* 36: 47–73.

7 Mabey, D. (2012). Oral azithromycin for treatment of yaws. *The Lancet* 379 (9813): 295–297.

8 Román, G.C. and Román, L.N. (1986). Occurrence of congenital, cardiovascular, visceral, neurologic, and neuro-ophthalmologic complications in late yaws: a theme for future research. *Reviews of Infectious Diseases* 8 (5): 760–770.

9 Jung, R.C. (1985). Handbook of endemic treponematoses. *The American Journal of Tropical Medicine and Hygiene* 34: 1234–1234.

10 Kositz, C., Butcher, R., and Marks, M. (2017). Images in clinical tropical medicine: new diagnostics for yaws. *The American Journal of Tropical Medicine and Hygiene* 96 (1): 3–4.

11 Engelkens, H.J.H., Ginai, A.Z., Judanarso, J. et al. (1990). Radiological and dermatological findings in two patients suffering from early yaws in Indonesia. *Genitourinary Medicine* 66 (4): 259–263.

12 Mitjà, O., Hays, R., Lelngei, F. et al. (2011). Challenges in recognition and diagnosis of yaws in children in Papua New Guinea. *The American Journal of Tropical Medicine and Hygiene* 85 (1): 113–116.

13 Engelkens, H.J.H., Vuzevski, V.D., and Stolz, E. (1999). Nonvenereal treponematoses in tropical countries. *Clinics in Dermatology* 17 (2): 143–152.

14 Hackett, C.J. (1953). Extent and nature of the yaws problem in Africa. *Bulletin of the World Health Organization* 8 ((1–3)): 127–182.

15 Dhondrup, W., Tidwell, T., Wang, X. et al. (2020). Tibetan medical Informatics: an emerging field in Sowa Rigpa pharmacological and clinical research. *Journal of Ethnopharmacology* 250: 112481.

16 Basharat, Z., Yasmin, A., He, T., and Tong, Y. (2018). Genome sequencing and analysis of *Alcaligenes faecalis* subsp. phenolicus MB207. *Scientific Reports* 8 (1): 1–0.

17 Basharat, Z., Zaib, S., Yasmin, A., and Tong, Y. (2019). Drug target identification and virtual screening in pursuit of phytochemical intervention of *Mycobacterium chelonae*. *bioRxiv*: 315408.

18 Guex, N., Peitsch, M.C., and Schwede, T. (2009). Automated comparative protein structure modeling with SWISS-MODEL and Swiss-PdbViewer: a historical perspective. *Electrophoresis* 30 (S1): S162–S173.

19 Yeshi, K., Gyal, Y., Sabernig, K. et al. (2019). An integrated medicine of Bhutan: Sowa Rigpa concepts, botanical identification, and the recorded phytochemical and pharmacological properties of the eastern Himalayan medicinal plants. *European Journal of Integrative Medicine* 29: 100927.

20 Rahman, N., Basharat, Z., Yousuf, M. et al. (2020). Virtual screening of natural products against Type II transmembrane serine protease (TMPRSS2), the priming agent of coronavirus 2 (SARS-CoV-2). *Molecules* 25 (10): 2271.

21 Chovancova, E., Pavelka, A., Benes, P. et al. (2012). CAVER 3.0: a tool for the analysis of transport pathways in dynamic protein structures. *PLoS Computational Biology* 8 (10): e1002708.

22 Roe, D.R. and Cheatham, T.E. III (2013). PTRAJ and CPPTRAJ: software for processing and analysis of molecular dynamics trajectory data. *Journal of Chemical Theory and Computation* 9 (7): 3084–3095.

23 Genheden, S. and Ryde, U. (2015). The MM/PBSA and MM/GBSA methods to estimate ligand-binding affinities. *Expert Opinion on Drug Discovery* 10 (5): 449–461.

24 Jaiswal, A.K., Tiwari, S., Jamal, S.B. et al. (2020). The pan-genome of *Treponema pallidum* reveals differences in genome plasticity between subspecies related to venereal and non-venereal syphilis. *BMC Genomics* 21 (1): 33.

25 Traut, T. (1998). A minimal gene set for cellular life derived by comparison of complete bacterial genomes. *Chemtracts* 11 (3): 234.

26 Sakharkar, K.R., Sakharkar, M.K., and Chow, V.T.K. (2004). A novel genomics approach for the identification of drug targets in pathogens, with special reference to *Pseudomonas aeruginosa*. *In Silico Biology* 4 (3): 355–360.

27 Evers, A., Gohlke, H., and Klebe, G. (2003). Ligand-supported homology modelling of protein binding-sites using knowledge-based potentials. *Journal of Molecular Biology* 334 (2): 327–345.

28 Fiser, A. (2004). Protein structure modeling in the proteomics era. *Expert Review of Proteomics* 1 (1): 97–110.

29 Peitsch, M.C. (1996). ProMod and Swiss-Model: internet-based tools for automated comparative protein modelling. *Biochemical Society Transactions* 24 (1): 274–279.

30 Studer, G., Rempfer, C., Waterhouse, A.M. et al. (2020). QMEANDisCo—distance constraints applied on model quality estimation. *Bioinformatics* 36 (6): 1765–1771.

31 Bajorath, J. (2001). Selected concepts and investigations in compound classification, molecular descriptor analysis, and virtual screening. *Journal of Chemical Information and Computer Sciences* 41 (2): 233–245.

32 Klebe, G. (2006). Virtual ligand screening: strategies, perspectives and limitations. *Drug Discovery Today* 11 (13–14): 580–594.

33 Schneider, G. (2010). Virtual screening: an endless staircase? *Nature Reviews: Drug Discovery.* 9 (4): 273–276.

34 Villoutreix, B., Eudes, R., and Miteva, M. (2009). Structure-based virtual ligand screening: recent success stories. *Combinatorial Chemistry and High Throughput Screening* 12 (10): 1000–1016.

35 Kloos, S., Madhavan, H., Tidwell, T. et al. (2020). The transnational Sowa Rigpa industry in Asia: new perspectives on an emerging economy. *Social Science and Medicine* 245: 112617.

36 Rey, J.P., Levesque, J., Pousset, J.L., and Roblot, F. (1992). Analytical studies of isorhoeadine and rhoeagenine in petal extracts of *Papaver rhoeas* L. using high-performance liquid chromatography. *Journal of Chromatography A* 596 (2): 276–280.

37 Slavikova, L. and Slavik, J. (1980). Alkaloids from *Papaver rupifragum* BOISS. et REUT. *Collection of Czechoslovak Chemical Communications* 45 (3): 761–763.

38 Coban, I., Toplan, G.G., Özbek, B. et al. (2017). Variation of alkaloid contents and antimicrobial activities of *Papaver rhoeas* L. growing in Turkey and northern Cyprus. *Pharmaceutical Biology* 55 (1): 1894–1898.

39 Slavik, J. and Slavikova, L. (1977). Alkaloids of some Himalayan species of *Meconopsis* genus. *Collection of Czechoslovak Chemical Communications* 42 (1): 132–139.

40 Sulaiman, I.M. and Hasnain, S.E. (1996). Random amplified polymorphic DNA (RAPD) markers reveal genetic homogeneity in the endangered Himalayan species *Meconopsis paniculata* and *M. simplicifolia*. *Theor etical and Applied Genetics* 93 (1–2): 91–96.

41 Wu, H., Ding, L., Shen, J. et al. (2009). A new proaporphine alkaloid from *Meconopsis horridula*. *Fitoterapia* 80 (4): 252–254.

42 Fan, J., Wang, Y., Wang, X. et al. (2015). The antitumor activity of *Meconopsis horridula* Hook, a traditional Tibetan medical plant, in murine leukemia L1210 cells. *Cellular Physiology and Biochemistry* 37: 1055–1065.

43 Lapierre, P. and Gogarten, J.P. (2009). Estimating the size of the bacterial pan-genome. *Trends in Genetics* 25 (3): 107–110.

44 Muzzi, A. and Donati, C. (2011). Population genetics and evolution of the pan-genome of *Streptococcus pneumoniae*. *International Journal of Medical Microbiology* 301 (8): 619–622.

45 Singh, N., Tiwari, S., Srivastava, K.K., and Siddiqi, M.I. (2015). Identification of novel inhibitors of *Mycobacterium tuberculosis* PknG using pharmacophore based virtual screening, docking, molecular dynamics simulation, and their biological evaluation. *Journal of Chemical Information and Modeling* 55 (6): 1120–1129.

46 Agnihotri, P., Mishra, A.K., Mishra, S. et al. (2017). Identification of novel inhibitors of *Leishmania donovani* γ-glutamylcysteine synthetase using structure-based virtual screening, docking, molecular dynamics simulation, and *in vitro* studies. *Journal of Chemical Information and Modeling* 57 (4): 815–825.

47 Kashyap, K. and Kakkar, R. (2020). Pharmacophore-enabled virtual screening, molecular docking and molecular dynamics studies for identification of potent and selective histone deacetylase 8 inhibitors. *Computers in Biology and Medicine*: 103850.

48 Li, W., Tang, Y., Liu, H. et al. (2008). Probing ligand binding modes of human cytochrome P450 2J2 by homology modeling, molecular dynamics simulation, and flexible molecular docking. *Proteins: Structure, Function, and Bioinformatics* 71 (2): 938–949.

49 Raj, U., Kumar, H., and Varadwaj, P.K. (2017). Molecular docking and dynamics simulation study of flavonoids as BET bromodomain inhibitors. *Journal of Biomolecular Structure and Dynamics* 35 (11): 2351–2362.

50 Rampogu, S., Baek, A., Zeb, A., and Lee, K.W. (2018). Exploration for novel inhibitors showing back-to-front approach against VEGFR-2 kinase domain (4AG8) employing molecular docking mechanism and molecular dynamics simulations. *BMC Cancer* 18 (1): 264.

51 Yuan, M., Luo, M., Song, Y. et al. (2011). Identification of curcumin derivatives as human glyoxalase I inhibitors: a combination of biological evaluation, molecular docking, 3D-QSAR and molecular dynamics simulation studies. *Bioorganic and Medicinal Chemistry* 19 (3): 1189–1196.

Part V

Viral Infections

17

Dengue Fever: A Brief Overview and Insights into the Potential Applicability of Phytochemicals in Its Management

Muhammad Akram[1], Charles Oluwaseun Adetunji[2], Chukwuebuka Egbuna[3,4], Shaista Jabeen[5], Olugbemi Tope Olaniyan[6], Nebechi Jane Ezeofor[7], Osikemekha Anthony Anani[8], Umme Laila[1], Mihnea-Alexandru Găman[9,10], Kingsley C. Patrick-Iwuanyanwu[4,11], Jonathan Chinenye Ifemeje[3], Chukwudi Jude Chikwendu[3], Michael C. Olisah[12], and Mithun Rudrapal[13]

[1] Department of Eastern Medicine and Surgery, Government College University Faisalabad, Faisalabad, Pakistan
[2] Applied Microbiology, Biotechnology and Nanotechnology Laboratory, Department of Microbiology, Edo State University Uzairue, Iyamho, Edo State, Nigeria
[3] Department of Biochemistry, Faculty of Natural Sciences, Chukwuemeka Odumegwu Ojukwu University, Uli, Nigeria
[4] Nutritional Biochemistry and Toxicology Unit, World Bank Africa Centre of Excellence, Centre for Public Health and Toxicological Research (PUTOR), University of Port Harcourt, Port Harcourt, Nigeria
[5] Department of Zoology, DKM College for Women (Autoonomous), Vellore, India
[6] Laboratory for Reproductive Biology and Developmental Programming, Department of Physiology, Edo University Iyamho, Iyamho, Nigeria
[7] Department of Food Technology, School of Applied Science and Technology, Federal Polytechnic, Oko, Nigeria
[8] Laboratory of Ecotoxicology and Forensic Biology, Department of Biological Science, Faculty of Science, Edo University Iyamho, Iyamho, Nigeria
[9] "Carol Davila" University of Medicine and Pharmacy, Bucharest, Romania
[10] Department of Hematology, Center of Hematology and Bone Marrow Transplantation, Fundeni Clinical Institute, Bucharest, Romania
[11] Department of Biochemistry, Faculty of Science, University of Port Harcourt, Port Harcourt, Nigeria
[12] Department of Medical Biochemistry, Faculty of Basic Medical Sciences, Chukwuemeka Odumegwu Ojukwu University, Uli, Nigeria
[13] Department of Pharmaceutical Chemistry, Rasiklal M. Dhariwal Institute of Pharmaceutical Education and Research, Chinchwad, Pune, Maharashtra, India

Neglected Tropical Diseases and Phytochemicals in Drug Discovery, First Edition.
Edited by Chukwuebuka Egbuna, Muhammad Akram, and Jonathan Chinenye Ifemeje.

List of Abbreviations

DENV-2	Dengue virus serotype-2
DF	dengue fever
DHF	Dengue hemorrhagic fever
DSS	Dengue shock syndrome
RNA	Ribonucleic acid
WHO	World Health Organization

17.1 Introduction

The dengue virus is a viral agent carried by a specific species of Aedes mosquito: *Aedes aegypti*. Dengue fever (DF) is common in Latin America and in Asia (Figure 17.1). Therefore, the World Health Organization (WHO) has established several models that could safeguard human life in accordance with the objectives, aspirations and sustainable goals of the millennium. In addition, WHO [1] has developed a detailed report on the global policy and sustainable management needed to control and eradicate the spread of the dengue virus. According to this white paper, it was estimated that, in the next 50 years, the incidence of DF will gradually increase 30 times in the subtropical and tropical regions of the world.

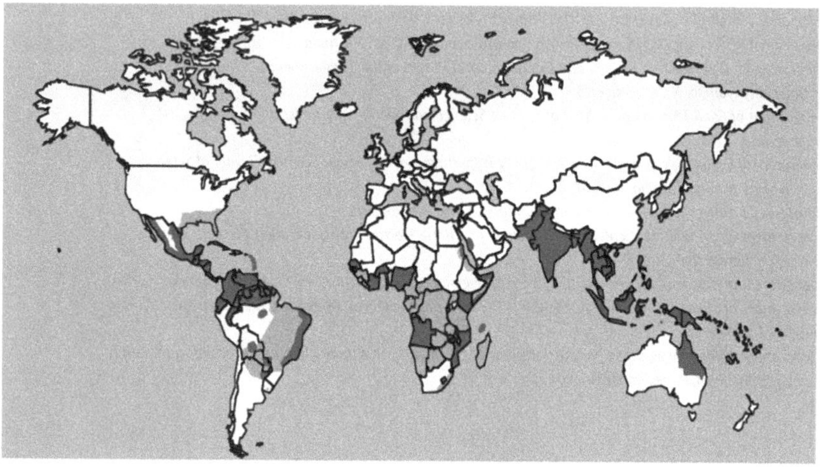

Figure 17.1 *A. aegypti* and dengue distribution in 2006. (in Red) *A. aegypti* distribution with history of epidemic dengue. (Light blue) *A. aegypti* distribution without history of epidemic dengue. *Source:* Map produced by USDA, Wikipedia. Available at https://en.wikipedia.org/wiki/Dengue_fever#/media/File:Dengue06.png. (accessed 18 September 2020).

Globally, DF constitutes a great public concern which has caused enormous financial loss, stress to healthcare services, and human anguish. It has been emphasized that there is a need to forestall and reduce the global burden of this infection, since its associated morbidity and mortality has led to tremendous losses to the global population. In order to avert all the aforementioned challenges, the WHO has suggested several ways in which the dengue infection can be regulated: improving health maintenance levels, recommendation systems, educating health workers, efficient management of possible epidemics, and repositioning the health system to spot out new cases. In addition, morbidity from dengue can also be controlled by domestic water supervision, promoting and integrating the management of the vector, entomological and epidemiological surveillance coordination, and detection and prediction of possible outbreaks with the use of appropriate computer models [31].

The dengue virus is a single-stranded positive polarity ribonucleic acid (RNA) virus which belongs to the Flaviviridae family. It is a viral agent enveloped by a membrane [2]. DF is also known as bone fever because during the course of the infection, the pain is more severe in the muscles and joints. It has been reported that 2.5 billion people are infected worldwide, with children being the most affected category of the population. The chronic form of the infection is called dengue haemorrhagic fever and is responsible for nearly 21 000 deaths each year [3].

It has been discovered that the classic DF or breakbone fever resembles the symptoms of influenza. It primarily affects infants, young people and adults, and infected patients can be symptomatic or asymptomatic. Asymptomatic carriers can transmit the virus. The *Aedes aegypti* mosquito is the primary vector of the infection, whereas *Aedes albopictus* is the secondary vector of DF. The span of high fever is of about 3–14 days after the bite of this mosquito. The most common symptoms in the infected patients are anorexia, nausea, headache, retro-orbital pain, myalgias, arthralgias and rashes. Also, febrile seizures and dehydration are the two main complications found in dengue virus fever patients. Treatment includes: controlling the temperature, relieving pain, avoiding aspirin, anti-inflammatory drugs, and a significant intake of fluids.

Numerous plant extracts and their isolated phytochemicals have been used in the management of infectious diseases. This chapter explores the potential applicability of plants and phytochemicals in the management of DF.

17.1.1 Dengue Hemorrhagic Fever

The dengue hemorrhagic fever (DHF) is spread to humans through the bites of *Aedes aegypti aedes* and *Aedes albopictus* mosquitoes. This epidemic is more common in tropical African regions, America, Western Pacific and South East Asia [4].

The eggs of *Aegypti aedes* can survive in the presence of small amounts of water and transform into larvae before they mature [5]. This infection is more common in children within the age of 10. Moreover, it has been discovered that the symptoms of DF include high fever, headache, as well as respiratory and intestinal involvement. Infants are more vulnerable to this infection which can be responsible for their death [6]. Moreover, an early diagnosis and an increased awareness regarding the DF (DF) has been reported to reduce complications if patients are managed properly. Also, it is important to understand all the aspects of the disease: transmission, clinical presentation, diagnosis, management, and prevention [7].

17.1.2 Dengue Shock Syndrome

The dengue shock syndrome (DSS) is a severe form of hemorrhagic DF. It often occurs between two and six days usually with symptoms which includes sudden collapse or prostration, cold and clammy extremities, perioral cyanosis, weak and thread-like pulse, in addition to hemorrhagic manifestations. Also, the DSS is recognized as the immediate occurrence of shock and hemorrhage into the ears and skin, or as epistaxis, melena and (or) hematemesis [8].

17.2 Dengue Epidemiology

The first record of DF was documented in the Chinese medicipedia from the Jin dynasty (265–420 CE) and was referred to as "water poisoning" linked with flying insects. Benjamine Rush was the man who first coined the term "breakbone fever." The very first case was confirmed in 1789 based on the symptoms of myalgia and arthralgia. The conveyance of DF by mosquitoes and its viral etiology were elucidated in the twentieth century. The first outbreak of DHF was observed in Manila in the year 1953 but since then the rate of the epidemic of DF has increased, in the 1950s many countries reporting an outbreak of DF. From 1980 to 1990, DF became a priority to public health worldwide along with malaria. DF epidemics have been reported in Asia and Latin America, whereas areas free of dengue are Europe and Nepal. Various kinds of intervention, programs and remedies are available to combat dengue [9–12].

17.3 Dengue Fever Symptoms

There are four different serotypes of the dengue virus: DENV-1, DENV-2, DENV-3, and DENV-4 [13]. These serotypes are found in different regions of the world. It has been discovered that being infected with one serotype could lead to the

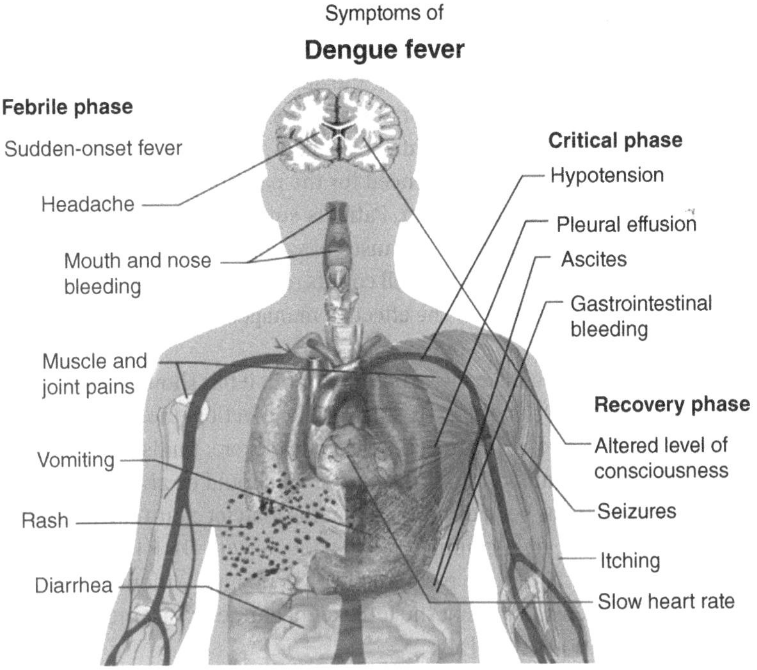

Figure 17.2 Symptoms of dengue fever (schematic depiction). *Source:* Mikael Häggström, wikipedia. Available at https://commons.wikimedia.org/wiki/File:Dengue_fever_symptoms.svg (accessed 18 September 2020).

development of life-long immunity for the same serotype, but only short-term immunity to other serotypes [14]. The common symptoms of DF (Figure 17.2) include high grade fever, headache, pain behind the eyes, arthritis. Also, infants and young children affected by DF develop a rash on their body together with all the above aforementioned symptoms. Platelet activity might also be affected, resulting in bleeding from different sites [9]. Furthermore, if DF becomes severe, then it may evolve into DHF which is associated with a high mortality rate. Before showing the above symptoms, a patient may be asymptomatic or may present mild symptoms. Early hospitalization of patients with the above symptoms prevents further complications and inflammatory response which could be fatal [15].

17.4 Treatment and Management of Dengue

There is no specific drug available for the treatment of dengue to date. Moreover, there are different options that could be used for the effective management of this viral infection. This infection is more common in travelers, therefore there is a need for proper vaccination against it before travelling. The application of painkillers might relieve the pain, while adequate consumption of fluids and proper rest as well as early diagnosis could prevent the adverse outcomes of this infection. Travelers must use insect repellent lotions and wear shirts with long sleeves and trousers and must try to sleep inside, especially if traveling to any of the affected areas by DF [5]. Also, there is a need for the patient to be hospitalized if severe dehydration and bleeding develop. Patients suffering from this infection must avoid the use of aspirin because it causes bleeding. Also, infected persons needs to check their red and white blood cell counts regularly. Moreover, there are several vaccines that are available for the effective management of the four different serotypes of DF [9].

Ribavirin has been effective in *in vivo* experiments, but it has shown less efficacy against mosquito viruses [10]. In the early stage of the infection, the use of antipyretic drugs might be effective. If the infection becomes severe, fluids such as albumin 5%, ringer lactate, normal saline, and dextrose 5% should be given to prevent dehydration. Moreover, it has been discovered that the platelet count can decrease, and thus blood transfusion are recommended in patients with DHF [11]. Codeine and acetaminophen have also been recommended to dengue patients to relieve muscular pain. It has been shown that the tetravalent chimeric vaccine is a suitable option in the prevention of DF [12].

17.5 Therapeutic Effects of Medicinal Plants Against Dengue

It has been observed that the process involved in the preparation of anti-dengue vaccine is one of the most complicated processes. Therefore, the utilization of plant-based formulations has been found to be effective for the management of DF. Some of the medicinal plants that showed pragmatic effect on DF are *Androphis paniculata* (Family Acanthaceae), *Memordica charantia* (Family Cucurbiteceae), *Citrus limon* (Family Rutaceae), *Ocimum sanctum* (Family Lamiaceae), *Cymbopogan citratus* (Family Poaceae), *Pelargonium citrosum* (Family Geramiun), *Carica papaya* (Family Caricaceae), *Psidium guajava* (Family Myrtaceae) [30, 41, 42].

Phytochemicals of medicinal plants works in several ways depending on the type of phytochemical (Table 17.1). Based on some evidence, phytochemicals works by restoring the RBCs and WBCs count to normal level by inhibiting the multiplication of dengue viruses that are present in the infected patient's body. For example, it was found that the methanolic extract of *Andrographis paniculata* exhibits suppressive effect against dengue virus 1 [16]. The leaf extract of neem inhibits the multiplication process of dengue 2 virus. Also, the extract from the

Table 17.1 Some medicinal plants active against dengue.

S/no	Plants	Chemical composition	Biological activity
1)	*Androphis paniculata* 	Andrographolide ($C_{20}H_{30}O_5$)	Cyto-toxic activity against cell line
2)	*Memordica charantia* 	Momordicin and Charantin, Charine, Cryptoxanthin, Cucurbitins, Cucurbitacins, Cucurbitanes, Cycloartenols, Diosgenin	Antioxidant, antidiabetic, immune enhancing, neuroprotective, antitumor and antimicrobial, antiviral, antilypolitic

(Continued)

Table 17.1 (Continued)

S/no	Plants	Chemical composition	Biological activity
3.	*Citrus limon* 	Limonene β-pinene, 85–90% volatile oil (400 components) 1–15% non-volatile oil components	Antibacterial, antifungal, antimoulds, antiviral, pest control, insect repellents
4.	*Ocimum sanctum* 	Eugenol (1-hydroxy-2-methoxy-4-allylbenzene)	Bronchitis, bronchial asthma, malaria, diarrhea, dysentery, skin diseases, arthritis, painful eye diseases, chronic fever, insect bite

5.	*Cymbopogan citratus* 	93–69% Oil Geranial (27.04%), neral (19.93%) and myrcene (27.04%) *E. citriodora* essential oil citronellal (83.50%)	Anti-inflammatory, insecticide, insect repellent, antipyretic, analgesic
6.	*Pelargonium citrosum* 	Citronellal	Mosquito repellent, anti-inflammatory, anti-tumor, anti-nociceptive, hepato-protective, wound healing, antimicrobial, antiviral

Rotunda boesenbergia herb is effective against dengue 2 and 3 virus with some protein digesting enzymes. The leaf extract from the shrub *Carica papaya* which is normally administered to patient gives positive results by increasing the number of plasma cells and showed a greater efficacy against this infections [17]. In a study performed in Philippines, the aqueous extracts from the leaves of *Hirta euphorbia* were given to patients orally. The result shows positive effects against bleeding due to DHF and internal wounds within 24 hours after treatment. The oils from the flower extract of *Alba lippia* is effective against the multiplication process of dengue 2 virus, while it also exhibited anti-viral effect against dengue 1 and 3 but least effect on dengue 4 virus. Moreover, the ethanolic extracts from the leaf of *Apiculata rhizophora* showed antiviral activity against dengue 2 virus while a higher level of inhibition was reported against the virus [10, 18]. Also, antipyretics and sponging are required to keep the body temperature below 40 °C. The treatment of DHF is entirely symptomatic and aims at controlling the clinical manifestations of shock and hemorrhage [19].

17.6 Scientific Evidence of Ameliorative Effects of Phytochemicals on Dengue Fever

Medicinal plants and phytochemicals unequivocally remain the savior of man as source of food and drugs for survival [20–29]. The authors of this chapter have worked extensively in the area of Phytochemistry where they described the various bioactive compounds with antiviral properties. This section discusses some works of previous scientist who investigated the effects of medicinal plants and phytochemicals against dengue virus. Table 17.2 is a summary of some work previously reported against dengue virus [30].

In the study by Eng-Chong et al. [49], authors demonstrated that *Boesenbergia rotunda* belonging to the family of *Zingiberaceae* possess several medical properties particular inhibitory effect against dengue virus protease in dengue virus serotype-2. Many bioactive molecules such as cyclohexenyl chalcone, 4-hydroxypanduratin A and panduratin A can inhibit DENV-2 NS3 protease enzyme. Kadir et al. [10] discovered that *Gymnogongrus torulosus* belonging to the family *Phyllophoraceae*, notable for DL-galactan has anti-dengue effects when tested *in vitro* against dengue virus serotype-2. They also reported that the physiological mechanism of action is brought through the interference and binding with cell surface glycoprotein receptor. Teixeira et al. [50] found that *Cladosiphon okamuranus* belonging to the family *Chordariaceae* possessing high amount of polysaccharide fucoidans and galactans are capable of inhibiting DENV-2 in a dose-dependent manner.

Table 17.2 Some medicinal plants active against dengue.

Plant	Family	Constituents	Roles	References
Cladosiphon okamuranus	Chordariaceae	Fucoidan (sulfated polysaccharide)	Inhibition of the dengue virus (DENV-2)	[32]
Salvia. madrensis	Lamiaceae	7-*O*-methyl-glabranine and glabranine (Flavonoids)	Inhibitory properties against dengue virus (DENV-2) proliferation and replication processes	[33]
Houttuynia cordata	Saururaceae	Hyperoside	Effective against dengue virus (DENV-2)	[34]
Myrtopsis corymbosa	Rutaceae	Myrsellin, ramosin, myrsellinol, haplopin, skimmianine, γ-fagarin	Inhibits dengue virus (DENV) polymerase	[35]
Nilavembu kudineer	—	Comprises of eight herbal mixtures	Used in India against dengue virus	[37]
Eugenia uniflora	Myrtaceae	Flavonoids	Inhibits viral replication of dengue virus in Huh7it-1 cell line	[38]
Ceiba pentandra	Malvaceae	Flavonoids	Inhibits viral replication of dengue virus in Huh7it-1 cell line	[38]
Melia azadirachta	Meliaceae	Tannins, limonoids, polyphenols	Larvicidal action of ethanolic fruit extracts (along with *Azadirachta indica*) against the dengue mosquito *Aedes aegypti*	[39]
Toona ciliate	Meliaceae	Tannins, limonoids, polyphenols	Effective against dengue mosquito vector *Aedes aegypti*	[39]
Psidium guajava	Myrtaceae	Flavonoids (quercetin)	Leaf extract inhibits dengue (DENV) infection (anti-dengue activity)	[40, 41]

(*Continued*)

Table 17.2 (Continued)

Plant	Family	Constituents	Roles	References
Pipe spp. (Aristolochia acuminate, Monotropa uniflora)	Aristolochiaceae Monotropaceae	Chromenes, alkaloids, amides, terpenoids, and flavonoids	Anti-dengue activity against DENV-2	[43]
Piper nigrum	Piperaceae	Limonoids, grandisin, amides, lignan, stilbenes, sesquiterpenes, quinones, coumarins, amides	Effective against *Aedes aegypti* with larvicidal activity	[44]
Carica papaya	Caricaceae	Phenolic compounds, flavonoids (quercetin), tannins, alkaloids, saponins, vanillic acid, vitamin C, terpenoids, enzymes papain, carbohydrates, chymopapain, lycopene, Danielone, a phytoalexin, saponins, karpain, carpaine, glycosides plus steroids	Anti-dengue activity through increase in platelets number due to high expression of platelet-activating factor receptor gene plus arachidonate 12-lipooxygenase gene, thrombocytopenic effects in dengue infection	[42, 45, 46]
Euphorbia hirta	Euphorbiaceae	Triterpenes, phytosterols, tannins, polyphenols, flavonoids	increases platelet counts	[46]
Ocimum sanctum	Lamiaceae	Oleanolic acid, rosmarinic acid, ursolic acid, eugenol, linalool, carvacrol, β-caryophyllene, etc.	Antiviral, immunomodulatory effect	[47]
Cissampelos pariera	Menispermaceae	Bisbenzylisoquinolines, aporphines alkaloids	Antiviral, anti-inflammatory and antipyretic	[48]

Sithisarn et al. [51] revealed that *Cladogynos orientalis* with the family *Euphorbiaceae* promotes anti-dengue activity by effectively inhibiting human hepatocarcinoma. Shah et al. [52] reported that *Cymbopogon citratus* from *Poaceae* family is known to be very rich in bioactive molecules like alcohols, terpenes, ketones, esters plus aldehyde, thus capable of inhibiting DENV-1 infected Vero E6 cells. The oil extracted from *Cymbopogon citratus* has been reported to be rich in citral β, citral α, nerol geraniol, terpinolene, citronellal, geranyl acetate, terpinol methyl heptenone, myrecene, flavonoids, luteolin, quercetin, isoorientin 2′-*O*-rhamnoside, apiginin, kaempferol which revealed various pharmacological, and biological activities. Teixeira et al. [50] demonstrated that *Cryptonemia crenulata* belonging to the family *Halymeniaceae* possesses high amount of sulfated polysaccharides galactan which can selectively inhibit dengue virus serotype-2 replication in Vero cells. Kadir et al. [10] investigated that *Flagellaria indica* belonging to the family *Flagellariaceae* exhibits anti-dengue activities in Vero cells through MTT assays.

Kadir et al. [10] demonstrated that *Gymnogongrus griffithsiae* from the family *Phyllophoraceae* are rich in sulfated polysaccharide kappa carrageenan biomolecule with inhibitory properties against Vero cells in dengue virus serotype-2. In another experiment conducted by the same authors, it has been reported that *Mimosa scabrella* of *Fabaceae* family contains galactomannans which showed inhibitory property against dengue virus serotype-1 and yellow fever virus. *Momordica charantia* belonging to the family *Cucurbitaceae* showed pronounced inhibitory activity against dengue virus serotype-1 in an experiment conducted utilizing antiviral assay. Cho et al. [53] demonstrated that *Hippophae rhamnoides* of *Elaeagnaceae* family is a rich plant with promising medicinal value containing linoleic, palmitoleic, linolenic acids, and oleic acids. The anti-dengue activity has been reported by the authors against dengue type-2 virus infected blood-derived macrophages with decreased level of proinflammatory cytokines such as IFN-γ plus TNF-α determined by traditional plaque assay methods.

Meenadevi et al. [54] demonstrated that *Leucaena leucocephala* of *Fabaceae* family promotes inhibitory effects against dengue virus serotype-1 plus yellow fever virus due to the presence of galactomannans polysaccharides biomolecules. Tahir et al. [55] found that Kushenol K, Canthin-6-one 9-*O*-beta-glucopyranoside and Kushenol W possess high potential against dengue virus with possible development as anti-dengue molecules.

Rosmalena et al. [56] evaluated some medicinal plants and phytochemicals with potential anti-dengue activities *in vitro* and *in silico*. The plants include *Acorus calamus, Myristica fatua,* and *Cymbopogon citratus*. Authors found that these Indonesian herbs possess activity against DENV-2 [36, 40]. Klawikkan et al. [57] revealed that *Piper retrofractum* from Piperaceae family possess anti-dengue activity in dengue virus serotype-2 infected cells in MTT assay. Authors isolated galactomannans polysaccharides from the plant and reported its inhibitory activities against dengue virus serotype-1 plus yellow fever virus. Sekhar and Debasmita [58] investigated that *Meristiella gelidium* from the family of *Solieriaceae* has great potential as antiviral agent due to the presence of kappa carrageenan. They revealed that *Lippia citriodora* plus *Lippia alba* oil belonging to *Verbenaceae* family possesses diverse pharmacological properties with considerable inhibitory potential against dengue virus serotype cell replication. *Houttuynia cordata* has widely been utilized for several pharmacological activities with anti-dengue activity in dengue virus serotype-2 in Vero cells due to the high presence of hyperoside biomolecule. Kadir et al. [10] reported that *Quercus lusitanica* from *Fagaceae* family is a potential inhibitory agent against dengue-2 in C6/36 cell replication suggesting the downregulation of downstream protein NS1 expression. They also reported that *Rhizophora apiculate* from Rhizophoraceae family might offer great potential for anti-dengue properties in dengue virus serotype-2 (DENV-2).

Rees et al. [59] revealed that *Zostera marina* belonging to Zosteraceae family possesses many active ingredients such as zosteric acid, *p*-sulfoxy-cinnamic acid with huge antiviral activity against dengue virus DENV-2 evaluated by quantitative RT-PCR assay methods. *Uncaria tomentosa* of *Rubiaceae* family possesses notable pharmacological activity due to the presence of active biomolecules such as pentacyclic oxindole alkaloids and flavonoids capable of inhibiting proliferation and growth of several types of viral strains particularly viral antigen (DENV-Ag). Again, Sanchez et al. [60] demonstrated that *Tephrosia crassifolia, Tephrosia viridiflora, and Tephrosia madrensis* from the family of *Fabaceae* contains high level of bioflavonoids such as 7-*O*-methyl-glabranine, hildgargtol A, Methyl-hildgardtol A, and glabranine with antiviral property against dengue virus. Bourjot et al. [61] revealed that *Anacolosa pervilleana* from Olacaceae family possesses acetylenic compounds like β-amyrone, cyanogenic glycoside, terpenoids, octadeca-9,11,13-triynoic acid, (*S*)-sambunigrin, lupenone with anti-dengue virus activity. Kataria et al. [62] investigated that *Glycyrrhiza glabra* from *Leguminosae* family is widely grown across different parts of the globe with high presence of glycyrrhizic acid, glycyrrhizin, glycyrrhetinic acid, sugars, asparagine, resin, and starch as bioactive constituents. Authors revealed that the plant possesses antiviral activity against dengue virus mainly due to the presence of glycyrrhizin along with ribavirin, interferon, and 6-azauridine.

Teixeira et al. [50] revealed that *Flacourtia ramontchi* from Flacourtiaceae family contain ample amount of phenolic glycosides possessing antiviral DENV NS5 polymerase activity. Authors also revealed that *Trigonostemon cherrieri* from *Euphorbiaceae* family possess numerous terpenes biomolecules with anti-dengue potential. It was revealed that all these phytochemicals displayed promising inhibitory results against dengue virus. Brandão et al. [63] revealed that *Arrabidaea pulchra* from *Bignoniaceae* family possesses several pharmacological activity which could offer a great potential against anti-dengue activity. It has been reported that *Arrabidaea pulchra* possess anti-DENV-2 activity conducted by MTT colorimetric bioassay. Simões et al. [64] revealed that acacetin-7-Orutinoside and petcolinarin from *Distictella elongate* exhibits anti-dengue property. Another group of scientists, Zandi et al. [65, 66] revealed that *Scutellaria baicalensis* from *Lamiaceae* family contains the flavonoid baicalein which has anti-viral activity. It was observed that baicalein showed strong inhibitory anti-DENV-2 serotype and virucidal action. *Cryptocarya chartacea* Kostern from *Lauraceae* family has high amount of flavonoid pinocembrin and chartaceones as bioactive constituents which could be developed for antiviral activity against dengue virus. *Boerhaavia diffusia* and *Tinospora cardifolia* belonging to the family *Nyctaginaceae* have been known to possess important biological properties including anti-dengue activity. Sarangi and Padhi [67] revealed that *Chondrus crispus* possesses polysaccharide carrageen bioactive molecule with strong anti-viral properties particularly against dengue virus type-2 infections. *Gastrodia elata* from *Orchidaceae* family possess significant amount of phenolic compounds such as D-glucans with inhibitory property against dengue virus type-2. In a study, they revealed that *Kaempferia parviflora* from *Zingiberaceae* family contains flavonoids and borneol capable of inhibiting dengue type-2 virus strains. *Phyllanthus urinaria* from *Phyllanthaceae* family possesses strong biomolecules important with many pharmacological activities including anti dengue property.

Udom et al. [68] revealed that *Piper sarmentosum* from *Piperaceae* family possess pharmacologically active and larvicidal biomolecules such as α-ascarone, ascaricin, βsitosterols which could be harnessed for their development as drug molecules against DF. Patil and Khan [69] found that *Catunagegam spinose* belonging to the *Rubiaceae* family possess potent bioflavonoids with pharmacological, insecticidal and pesticidal activities particularly against dengue mosquito vector *Aedes aegypti* due to the presence of flavonoids, alkaloids, phytosterols, glycosides, triterpenoidal saponins, veleric acid, essential oil, resin, and tannins. Rajalakshmi et al. [70] reported that *Adhatoda vasica* extracts could be utilized for the management of DF due to the presence of sulfated polysaccharides fucoidans, phenolics, terpenoids, flavonoids, quercetin, alkaloids, chalcone compounds, and polycyclic quinones.

17.7 Conclusion

The efforts for drug discovery against dengue virus needs to be intensified owing to the fact that medicinal plants and its chemical constituents have been found to possess anti-dengue activities. Plants from the following families

(Chordariaceae, Lamiaceae, Saururaceae, Rutaceae, Myrtaceae, Malvaceae, Meliaceae, Aristolochiaceae, Monotropaceae, Piperaceae, Caricaceae, Cucurbitacea, Acanthaceae, Euphorbiaceae, Menispermaceae and Labiatae) need to be further investigated and validated through in vivo studies. The various health agencies, pharmaceutical companies (R&D), sponsors and everyone in the field of drug discovery need to channel energy to drug discovery and development against dengue virus so that lives lost annually as a result of dengue infection can at best be minimized.

References

1 WHO (1996). *Report of the Consultation on Key Issues in Dengue Vector Control Towards the Operationalization of a Global Strategy, WHO, Geneva, 6–10 June 1995*. Geneva: World Health Organization (CTD/FIL(DEN)/IC/96.1).

2 Low, J.G., Sung, C., Wijaya, L. et al. (2014). Efficacy and safety of celgosivir in patients with dengue fever (CELADEN): a phase 1b, randomised, double-blind, placebo-controlled, proof-of-concept trial. *The Lancet Infectious Diseases* 14 (8): 706–715.

3 Whitehead, S.S., Durbin, A.P., Pierce, K.K. et al. (2017). In a randomized trial, the live attenuate tetravalent dengue vaccine TV003 is well-tolerated and highly immunogenic in subjects with flavivirus exposure prior to vaccination. *PLoS Neglected Tropical Diseases* 11 (5): e0005584.

4 Parks, W. and Lloyd, L. (2004). Planning social mobilization and communication for dengue fever prevention and control: a step-by-step guide. https://www.who.int/denguecontrol/resources/9241591072/en (accessed 21 September 2020).

5 World Health Organization (2017). Weekly bulletin on outbreaks and other emergencies: week 44: 28 October–3 November. *Weekly Bulletin on Outbreaks and Other Emergencies*: 1–15.

6 Nimmannitya, S. (2017). Dengue hemorrhagic fever: problems and progress. *Paediatrica Indonesiana* 15 (3–4): 93–104.

7 Gubler, D.J. (1998). Dengue and dengue hemorrhagic fever. *Clinical Microbiology Reviews* 11 (3): 480–496.

8 Nagaram, P.P., Piduru, P., Munagala, V.K., and Matli, V.V. (2017). Clinical and laboratory profile and outcome of dengue cases among children attending a tertiary care hospital of South India. *International Journal of Contemporary Pediatrics* 4 (3): 1074–1080.

9 Jahan, F. (2011). Dengue fever (DF) in Pakistan. *Asia Pacific Family Medicine* 10 (1): 1–4.

10 Kadir, S.L.A., Yaakob, H., and Zulkifli, R.M. (2013). Potential anti-dengue medicinal plants: a review. *Journal of Natural Medicines* 67: 677–689.

11 Rajapakse, S., Rodrigo, C., and Rajapakse, A. (2012). Treatment of dengue fever. *Infection and Drug Resistance* 5: 103.

12 Rather, I.A., Parray, H.A., Lone, J.B. et al. (2017). prevention and control strategies to counter dengue virus infection. *Frontiers in Cellular and Infection Microbiology* 7: 336. https://doi.org/10.3389/fcimb.2017.00336.

13 Biernat, B., Stańczak, J., Szostakowska, B. et al. (2015). Different serotypes of dengue virus (DENV) imported by Polish travellers from dengue endemic areas to Poland. *International Maritime Health* 66 (2): 72–76.

14 Chen, J., Ong, A., and Low, S. (2008). An unusual complication of dengue infection. *Singapore Medical Journal* 49 (12): e340–e342.

15 Lum, L.C., Goh, A.Y., Chan, P.W. et al. (2002). Risk factors for hemorrhage in severe dengue infections. *The Journal of Pediatrics* 5: 629–631.

16 Frederico, E., Cardoso, A., Moreira-Marconi, E. et al. (2017). Anti-viral effects of medicinal plants in the management of dengue: a systematic review. *African Journal of Traditional, Complementary, and Alternative Medicines* 14 (S): 33–40.

17 Singh, P.K. and Rawat, P. (2017). Evolving herbal formulations in management of dengue fever. *Journal of Ayurveda and Integrative Medicine* 8: 207–210.

18 Tantawichien, T. (2015). Dengue fever and dengue hemorrhagic fever in adults. *The Southeast Asian Journal of Tropical Medicine and Public Health* 46 (suppl 1): 79–98.

19 Fujimoto, D.E. and Koifman, S. (2014). Clinical and laboratory characteristics of patients with dengue hemorrhagic fever manifestations and their transfusion profile. *Revista Brasileira de Hematologia e Hemoterapia* 36 (2): 115–120.

20 Egbuna, C. and Tupas, G.D. (eds.) (2020). *Functional Foods and Nutraceuticals – Bioactive Components, Formulations and Innovations*, 1e. Cham, Switzerland: Springer Nature.

21 Srivastav, V.K., Egbuna, C., and Tiwari, M. (2019). Plant secondary metabolites as lead compounds for the production of potent drugs. In: *Phytochemicals as Lead Compounds for New Drug Discovery* (eds. C. Egbuna, S. Kumar, S.M. Ezzat, et al.), 3–13. Cambridge, USA: Elsevier.

22 Kaliyaperumal, S., Periyasamy, K., Balakrishnan, U. et al. (2019). Antiviral phytocompounds for drug development: a data mining studies. In: *Phytochemicals as Lead Compounds for New Drug Discovery* (eds. C. Egbuna, S. Kumar, S.M. Ezzat, et al.), 239–243. Cambridge, USA: Elsevier.

23 Egbuna, C. and Ifemeje, J.C. (2015). Biological functions and anti-nutritional effects of phytochemicals in living system. *IOSR Journal of Pharmacy and Biological Sciences (IOSR-JPBS)* 10 (2): 10–19. https://doi.org/10.9790/3008-10231019.

24 Egbuna, C., Mishra, A., and Goyal, M.R. (eds.) (2020). *Preparation of Phytopharmaceuticals for the Management of Disorders*, 1e. US: Academic Press, Elsevier.

25 Egbuna, C., Kumar, S., Ezzat, S.M. et al. (eds.) (2019). *Phytochemicals as Lead Compounds for New Drug Discovery*, 1e. Cambridge, USA: Elsevier.

26 Shashank, K. and Egbuna, C. (eds.) (2019). *Phytochemistry: In vitro and In Silico Updates*, 1e. Singapore: *Springer Nature.*

27 Egbuna, C., Ifemeje, J.C., Udedi, S.C., and Kumar, S. (eds.) (2018). *Phytochemistry, Volume 1: Fundamentals, Methods, and Applications*, 1e. New York: Apple Academic Press/CRC Taylor & Francis.

28 Egbuna, C., Kumar, S., Ifemeje, J.C., and Kurhekar, J.V. (eds.) (2018). *Phytochemistry, Volume 2: Pharmacognosy, Nanomedicine, and Contemporary Issues*, 1e. New York: Apple Academic Press/CRC Taylor & Francis.

29 Egbuna, C., Ifemeje, J.C., Kumar, S., and Sharif, N. (eds.) (2018). *Phytochemistry, Volume 3: Marine, Industrial, and Advances*, 1e. New York: Apple Academic Press/CRC Taylor & Francis.

30 Kaushik, S., Kaushik, S., Sharma, V., and Yadav, J.P. (2018). Antiviral and therapeutic uses of medicinal plants and their derivatives against dengue viruses. *Pharmacognosy Reviews* 12: 177–185.

31 World Health Organization (1999). *Strengthening Implementation of the Global Strategy for Dengue Fever and Dengue Haemorrhagic Fever, Prevention and Control*. Report of the Informal Consultation. Geneva: WHO.

32 Srivastava, M. and Kapoor, V.P. (2005). Seed galactomannans: an overview. *Chemistry & Biodiversity* 2: 295–317.

33 Talarico, L.B., Zibetti, R.G.M., Noseda, M.D. et al. (2005). The antiviral activity of sulfated polysaccharides against dengue virus is dependent on virus serotype and host cells. *Antiviral Research* 66: 103–110.

34 Leardkamolkarn, V., Srigulpanit, W., Phurimsak, C. et al. (2012). The inhibitory actions of *Houttuynia cordata* aqueous extract on dengue virus and dengue-infected cells. *Journal of Food Biochemistry* 26: 86–92.

35 Kumar, S., Kumar, S., Rehman, I. et al. (2012). Molecular herbal inhibitors of dengue virus, an update. *IJMAP* 2: 1–21.

36 Yarnell, E. (2016). Herbs for emerging viral infectious diseases. *Alternative and Complementary Therapies* 22 (4): 164–174. http://doi.org/10.1089/act.2016.29062.eya.

37 Mattummal, R., Gopi, D.K., Parameswaran, S.R. et al. (2018). Bioactive molecules in siddha polyherbal nilavembu kudineer alleviating symptoms of dengue/chikugunya. *Traditional Medicine Research* 3 (5): 215–229. https://doi.org/10.12032/TMR201813080.

38 Dewi, B.E., Angelina, M., Ardiantara, S. et al. (2019). Antiviral activity of *Ceiba pentandra* and *Eugenia uniflora* leaf extracts to dengue virus serotype-2 in Huh 7it-1 cell line. In: *The 4th Biomedical Engineering's Recent Progress in Biomaterials, Drugs Development, Health, and Medical Devices AIP Conference*

Proceedings, vol. 2193, 030003-1–030003-7. AIP Publishing https://doi.org/10.1063/1.5139340.

39 Mubeen, M., Jayakumari, S., Dinesh Kumar, P., and Yogeshwaran, V. (2018). A review on anti-dengue activity of selected plant species from Meliaceae family. *Drug Invention Today* 10 (4): 460–463.

40 Saptawati, L., Febrinasari, R.P., Yudhani, R.D. et al. (2017). In vitro study of eight Indonesian plants extracts as anti-Dengue virus. *Health Science Journal of Indonesia* 8 (1): 12–18.

41 Trujillo-Correa, A.I., Quintero-Gil, D.C., Diaz-Castillo, F. et al. (2019). In vitro and in silico anti-dengue activity of compounds obtained from *Psidium guajava* through bioprospecting. *Complementary and Alternative Medicine* 19: 298. https://doi.org/10.1186/s12906-019-2695-1.

42 Pentewar, R., Sharma, S., Kore, P. et al. (2017). Papaya leaf extract to treat dengue: a review. *Scholars Academic Journal of Pharmacy (SAJP)* 6 (4): 113–119.

43 Marques, A.M. and Kaplan, M.A.C. (2015). Active metabolites of the genus *Piper* against *Aedes aegypti*: natural alternative sources for dengue vector control. *Universitas Scientiarum* 20 (1): 61–82. https://doi.org/10.11144/Javeriana.SC20-1.amgp.

44 Ahmad, N., Fazal, H., Abbasi, B.H. et al. (2012). Biological role of *Piper nigrum* L. (black pepper): a review. *Asian Pacific Journal of Tropical Biomedicine* 2012: S1945–S1953. https://doi.org/10.1016/S2221-1691(12)60524-3.

45 Mrinal Saraf, L. and Kavimandan, B. (2018). Dengue fever: role of *Carica papaya*. *International Journal of Health Sciences & Research* 8 (1): 249–258.

46 Kumar, N. (2010). Re: Papaya leaves for speedy rise of platelet count in Dengue. *British Medical Journal* http://www.bmj.com/rapidresponse/2011/11/03/papaya-leaves-speedy-rise-platelet-countdengue (accessed 21 October 2013).

47 Solanki, H.A., Shankar, S.S., and Pandya, H.A. (2017). Dengue as a druggable disease: a review. *Journal of Fever* 1 (1): 1001. (1–19).

48 Shah, K., Qureshi, S.S., Gupta, J.K. et al. (2017). Ethnomedicinal uses, phytochemistry and pharmacology of *Cissampelos pareira*: a review. *Omics E-Books Group*: 1–19.

49 Eng-Chong, T., Yean-Kee, L., Chin-Fei, C. et al. (2012). *Boesenbergia rotunda*: from ethnomedicine to drug discovery. *Evidence-based Complementary and Alternative Medicine* 2012: 473637. https://doi.org/10.1155/2012/473637.

50 Teixeira, R.R., Pereira, W.L., Oliveira, A.F.C.D.S. et al. (2014). Natural products as source of potential dengue antivirals. *Molecules* 19: 8151–8176. https://doi.org/10.3390/molecules19068151.

51 Sithisarn, P., Rojsanga, P., Sithisarn, P., and Kongkiatpaiboon, S. (2015). Antioxidant activity and antibacterial effects on clinical isolated *Streptococcus suis* and *Staphylococcus intermedius* of extracts from several parts of *Cladogynos* *orientalis* and their phytochemical screenings. *Evidence-Based Complementary and Alternative Medicine* 2015: 908242. http://dx.doi.org/10.1155/2015/908242.

52 Shah, G., Shri, R., Panchal, V. et al. (2011). Scientific basis for the therapeutic use of *Cymbopogon citratus*, stapf (Lemon grass). *Journal of Advanced Pharmaceutical Technology & Research* 2 (1): 3–8.

53 Cho, H., Cho, E., Jung, H. et al. (2014). Antioxidant activities of Sea buckthorn leaf tea extracts compared with green tea extracts. *Food Science and Biotechnology* 23 (4): 1295–1303.

54 Meenadevi, V.N., Ariharan, V.N., and Prasad, N.P. (2013). Nutritive value and potential uses of *Leucaena leucocephala* as biofuel – a mini review. *Research Journal of Pharmaceutical, Biological and Chemical Sciences* 4 (1): 515–521.

55 Qamar, M.T., Maryam, A., Muneer, I. et al. (2019). Computational screening of medicinal plant phytochemicals to discover potent pan-serotype inhibitors against dengue virus. *Scientific Reports* 9: 1433. https://doi.org/10.1038/s41598-018-38450-1.

56 Rosmalena, R., Elya, B., Dewi, B.E. et al. (2019). The antiviral effect of Indonesian medicinal plant extracts against dengue virus in vitro and in silico. *Pathogens* 8: 85. https://doi.org/10.3390/pathogens8020085.

57 Klawikkan, N., Nukoolkarn, V., Jirakanjanakit, N. et al. (2011). Effect of Thai medicinal plant extracts against dengue virus in vitro. *Mahidol University Journal of Pharmaceutical Science* 38 (1–2): 13–18.

58 Sekhar, S.S. and Debasmita, D. (2013). Anti-dengue medicinal plants: a mini review. *Research and Reviews: Journal of Pharmacognosy and Phytochemistry* 1 (2): 5–9.

59 Rees, C.R., Costin, J.M., Fink, R.C. et al. (2008). In vitro inhibition of dengue virus entry by *p*-sulfoxy-cinnamic acid and structurally related combinatorial chemistries. *Antiviral Research* 80: 135–142.

60 Sanchez, I., Garibay, F.G., Taboada, J., and Ruiz, B.H. (2000). Antiviral effect of flavonoids on the dengue virus. *Phytotherapy Research* 14: 89–92.

61 Bourjot, M., Leyssen, P., Eydoux, C. et al. (2012a). Chemical constituents of *Anacolosa pervilleana* and their antiviral activities. *Fitoterapia* 83: 1076–1080.

62 Kataria, R., Hemraj, Singh, G. et al. (2013). Pharmacological activities on *Glycyrrhiza glabra* – a review. *Asian Journal of Pharmaceutical and Clinical Research* 6 (1): 5–7.

63 Brandão, G.C., Kroon, E.G., Souza, D.E.R. et al. (2013). Chemistry and antiviral activity of *Arrabidaea pulchra* (Bignoniaceae). *Molecules* 18: 9919–9932.

64 Simões, L.R., Maciel, G.M., Brandão, G.C. et al. (2011). Antiviral activity of Distictella elongate (Vahl) Urb. (Bignoniaceae), a potentially useful source of anti-dengue drugs from the state of Minas Gerais, Brazil. *Letters in Applied Microbiology* 53: 602–607.

65 Zandi, K., Teoh, B.T., Sam, S.S. et al. (2012). Novel antiviral activity of baicalein against dengue virus. *BMC Complementary and Alternative Medicine* 12: 1–9.
66 Zandi, K., Lim, T.H., Rahim, N.A. et al. (2013). Extract of *Scutellaria baicalensis* inhibits dengue virus replication. *BMC Complementary and Alternative Medicine* 13: 91.
67 Sarangi, M.K. and Padhi, S. (2014). Dengue and its phytotherapy: a review. *International Journal of Pharmaceutical and Phytopharmacological Research (eIJPPR)* 4 (1): 37–46.
68 Udom, C., Wej, C., Kittichai, K. et al. (2006). Larvicidal effect of pepper plants on *Aedes aegypti* (1) (diptera: culicidae). *Journal of Vector Ecology* 31 (1): 138–144.
69 Patil, M.B. and Khan, P.A. (2017). Primary phytochemical studies of *Catunaregam spinosa* (THUNB.) TIRVEN for secondary metabolites. *International Journal of Pharma and Bio Sciences* 8 (2): 320–323.
70 Rajalakshmi, K., Christian, G.J., Shanmugapriya, P., and Jeeva, G.R. (2016). A review on siddha formulation adathodai manappaagu in the management of dengue fever. *World Journal of Pharmaceutical Research* 5 (5): 414–423.

18

Medicinal Potentials of Phytochemicals for the Management of Rabies

D. Chandan[1], Shiva Prasad Kollur[2], Shashanka K. Prasad[1], Chandrashekar Srinivasa[3], Prabhurajeshwar[3], S. M. Gopinath[3], Chukwuebuka Egbuna[4,5], and Chandan Shivamallu[1]

[1] *Department of Biotechnology and Bioinformatics, School of Life Sciences, JSS Academy of Higher Education and Research, Mysuru, India*
[2] *Department of Sciences, Amrita School of Arts and Sciences, Mysuru, India*
[3] *Department of Biotechnology, Davangere University, Davangere, India*
[4] *Department of Biochemistry, Faculty of Natural Sciences, Chukwuemeka Odumegwu Ojukwu University, Uli, Nigeria*
[5] *Nutritional Biochemistry and Toxicology Unit, World Bank Africa Centre of Excellence, Centre for Public Health and Toxicological Research (PUTOR), University of Port Harcourt, Port Harcourt, Nigeria*

List of Abbreviations

CNS	Central Nervous System
RABV	Rabies Lyssavirus
RNP	Ribonucleoprotein Complex
RNA	Ribonucleic acid
WHO	World Health Organization

18.1 Introduction

Rabies is a fatal zoonotic viral disease that causes encephalitis in both warm-blooded animals as well as humans. *Rabies lyssavirus* (RABV), the virus responsible for rabies, belongs to the family *Rhabdoviridae* and order *Mononegavirales* [1]. Globally, it is estimated that around 55 000 peoples die due to rabies virus each year, out of which the highest number of death is found in Asia and African countries, due to the presence of endemic canine rabies and also dogs which remain as

Neglected Tropical Diseases and Phytochemicals in Drug Discovery, First Edition.
Edited by Chukwuebuka Egbuna, Muhammad Akram, and Jonathan Chinenye Ifemeje.

the important animal reservoirs in those regions [2]. Though dogs are the primary host for rabies virus which significantly contributes to the transmission of the virus to humans, the infection is mainly through physical contact (scratches and bites) [3]. The rabies virus exhibits a unique mode of transmission and is spread through the various nervous system [4]. After inoculation, these viruses spread to individual muscle cells and gains access to both motor and sensory nerve fiber endings. The main source of entering neurons is the neuromuscular junction. From the neuromuscular junction, the viruses spread to the central nervous system by retrograde axonal transport which multiplies within neuronal cell bodies and rapidly spread through CNS by axonal pathway that includes transsynaptic spread [5].

This viral disease is also known as hydrophobia (fear of water) which is a highly fatal acute disease of the central nervous system [6]. The disease is one of the oldest diseases known to mankind. The viruses are commonly caused by carnivores such as dogs, cats, jackals, and wolves. Dogs get infected through a bite by a rabid wild animal or fellow canine, in turn, bite from the dogs which carry the virus in their saliva is the common method of human infection [7]. Due to the increase in population density of dogs with very poor management and also limited vaccinations, especially in urban areas, has caused a high prevalence of the disease. Where most of the animal bites in India are particularly by dogs (91.5%), 60% are by stray dogs (Figure 18.1) while the remaining are pets. In all, it have been

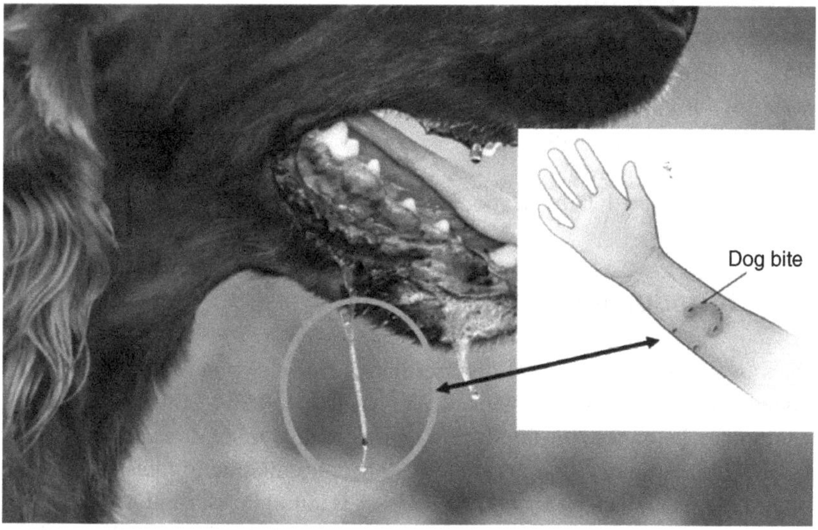

Figure 18.1 Stray dog bite is the major concern of human rabies. *Source:* reddogs/123 RF.

reported that the annual number of persons that die due to animal bites is around 38 million while the cost for post-bite treatment is approximately $25 million [8].

In South-Eastern Asia, rabies is common and accounts for 45% of the total cases of rabies in the world [9]. In India, about 18 000–20 000 cases are reported every year out of which 36% results in death. Rabies has an incubation period of one week to two years which may vary in children. Rabies is highly endemic in almost all states of India excluding the islands of Andaman, Nicobar, and Lakshadweep [10]. The incidence of rabies is prevalent in India, which is because of the lack of general awareness, underestimated incidences, lack of professional expertise, unchecked canine populations, and insufficient and irregular supply of both dog vaccines and anti-rabies vaccines.

The availability of vaccines is less for the majority of viruses causing encephalitis [11]. Although there are few antiviral drugs for rabies viruses, but once infection gets to the CNS, the vaccine may not be effective against the virus, which then leads to death. Antiviral drugs are used for the treatment of viral encephalitis but there are only a few or less potent antiviral drugs [12]. For example, Ribavirin (Figure 18.2) is a commonly used antiviral drug against rabies and many viruses such the Flavivirus, hepatitis C virus, human respiratory syncytial virus (RSV) infection, and viral hemorrhagic among others. Similarly, Acyclovir (Figure 18.2) and its derivatives are used against rabies and other viruses such as herpes simplex virus infections, chickenpox, and shingles. Again, interferon-alpha, a broad-spectrum antiviral drug is used against rabies virus. However, it has been reported that so far only one case of a female treated with the combination of antiviral drugs whose rabies progressed to the CNS survived. Therefore, it is important to concentrate on the discovery of novel antiviral agents against viral infection of the CNS [13].

In the past few decades, the therapeutics against viral infection has increased significantly. But unfortunately, many viruses are now resistant to vaccines. Moreover, most antiviral drug causes side effects and are expensive; thus, it is been important to develop a novel antiviral drug against these viruses [14]. Based

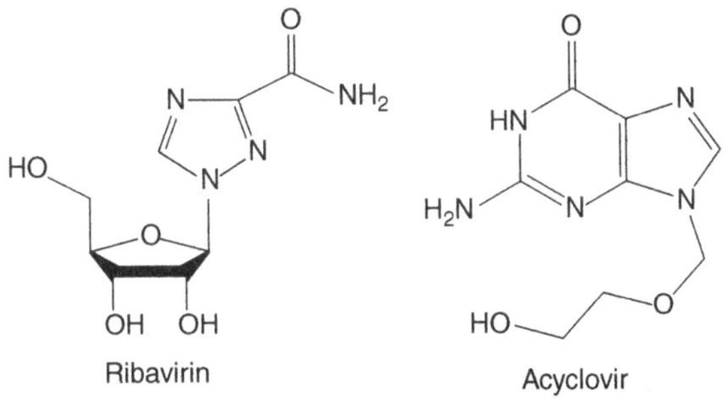

Figure 18.2 Chemical structure of ribavirin and acyclovir.

on Ayurveda, Unani, Siddha, Chinese herbal medicine, and traditional herbal medicine point of view, it appears that there is some prospects of the usefulness of natural compounds from plants for the treatment of rabies [15].

18.2 Epidemiology of Rabies

Rabies has been recognized for over 4000 years. Today, it has been prevalent in almost all parts of the world and is considered a threat to public health (Figure 18.3). Some countries like Hongkong, Singapore, Australia, New Zealand, Japan, and few small islands are reported to be free from the rabies virus. Rabies in animals appear in three different epidemiological forms which includes sylvatic (wide), Urban, and bat rabies. The Sylvatic life cycle is preserved by mongoose, Jackal, Hyena, fox, and a few other wild animals, whereas in Urban rabies, dogs are the main reservoir that carries the virus. The bat rabies is restricted to few Latin American countries and also in other parts of the USA. The saliva which comes from rabid animals is the primary source of infection (Figure 18.1) [16]. After infection, symptoms emerge three to four days or longer till death occur. Both animals and humans are susceptible to rabies. The transmission occurs by bite or scratches made by infected animals. In the United States and other European countries, the domestic animal's rabies was controlled during the 1940–1950s and now it has been reduced to less than 10% [17–19].

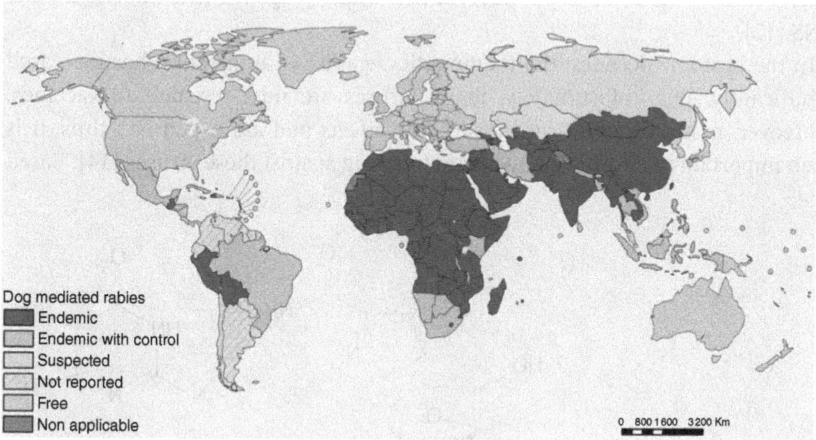

Figure 18.3 Distribution of rabies in the world. *Source:* Occurrence of rabies, World Health organization. Retrieved from https://www.who-rabies-bulletin.org/site-page/occurrence-rabies (accessed 19 September 2020).

18.3 Morphology of Rabies Virus

Rabies virus appears bullet shaped with approximately 180 nm in length and 75 nm in diameter (Figure 18.4). The genetic information of these viruses is organized in the form of helical ribonucleoprotein complex (RNP), in which the linear RNA is firmly connected with viral nucleoprotein N, polymerase protein L, and its cofactor P protein. The ribonucleocapsid complex ensures genome transcription and replication in the cytoplasm [21].

The RNA genome of rabies virus consists of 12 kilobases long, single-strand RNA with negative-sense RNA which is encoded with 5 viral proteins (31–51); Glycoprotein G, Phosphoprotein P, Polymerase L, Matrix protein M and a Nucleoprotein N (Figure 18.4). The envelop protein is coated with spikelike projection, corresponding to glycoprotein G trimers, which identify and binds to the cell receptor [22]. The Glycoprotein G consists of approximately 400 trimeric spikes, which are bound tightly on the surface of the virion. These glycoproteins play a vital role in the initiation of immune response and are also required for RABV pathogenicity, by binding to specific receptors, which then enter the nervous system via the endosomal transport pathway with low-pH-caused membrane union process [23]. The various points on this virus can form a means of new drug targeting especially as it concerns the receptor binding sites.

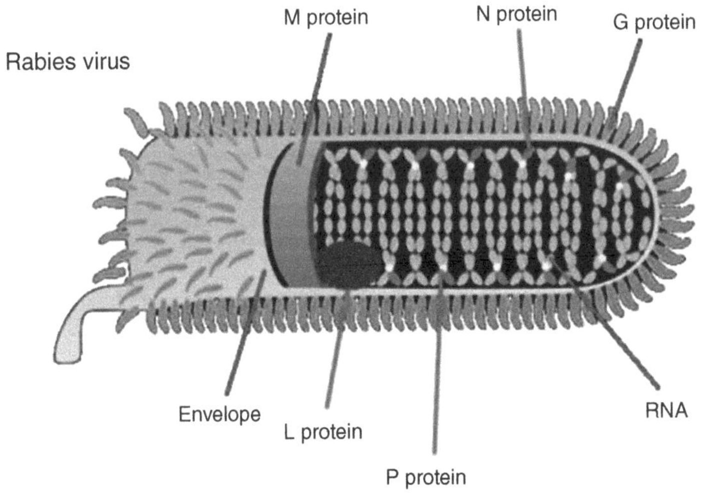

Figure 18.4 Morphology of rabies virus. *Source:* Barecha et al. [20]. Freely taken from: https://www.pimr.org.in/Barecha.PDF (accessed 19 September 2020).

18.4 Medicinal Plants a Rabies Virus

For many decades now, many studies have explored the medicinal potentials of phytochemicals against different kinds of diseases [24–29]. There are over 400 000 species of plants on earth out of which only about 15% has been studied for their medicinal benefits [27–29] According to WHO, about 80% of the world population in developing countries largely rely on medicinal plants and treatment. Recent studies shows that the use of plant-based materials is increasing in developing countries [30]. Notable plant families with antiviral activities include Euphorbiaceae, Poaceae, Apocynaceae, Lamiaceae among others [31]. Table 18.1 presents medicinal plants which have been reported to possess anti-RABV activities.

Table 18.1 Medicinal plants with anti-RABV activities.

S/no	Scientific name – family	Part used	Mode of use	Reference
1)	*Croton macrostachyus* – Euphorbiaceae	Root and stem	Extract	
2)	*Antidesma diandrum* – Euphorbiaceae	Leaves	Decoction	
3)	*Croton oblongifolius* Roxb – Euphorbiaceae	Root, stem and leaves	Powder and decoction	
4)	*Jatropha gossypifolia* L. – Euphorbiaceae	Bark	Blend	
5)	*Tragia involucrata* L. – Euphorbiaceae	Root and leaves	Extract	[1, 32]
6)	*Alchornea davidii* – Euphorbiaceae	Twig and leaves	Extract	
7)	*Phytolaca dodecandra* – Phytolaccaceae	Leaves and root	Extract	
8)	*Justicia schimperiana – Acanthaceae*	Leaves	Extract	
9)	*Ricinus communis* – Euphorbiaceae	Seed	Extract	
10)	*Brucea antidysenterica* – Simaroubaceae	Leaves and fruits	Crushed and baked	
11)	*Croton macrostachyus* – Euphorbiaceae	Bark, root, and leaves	Powder	
12)	*Cucumis cifolius* – Acanthaceae	Root	Powder	

Table 18.1 (Continued)

S/no	Scientific name – family	Part used	Mode of use	Reference
13)	*Salix subserrata* – Phytolaccaceae	Leaves	Decoction	
14)	*Euphorbia abyssinica* – Euphorbiaceae	Root, latex, and leaves	Powder and decoction	
15)	*Stephania abyssinica* – Minispermaceae	Root and leaves	Powder and extract	
16)	*Calpurnia aurea* – Fabaceae	Fruit, seed, and leaves	Crushed	
17)	*Datura stramonium* – Solanaceae	Leaves	Crushed	
18)	*Dorstenia barnimiana* – Acanthaceae	Root	Powder	
19)	*Solanum gigantum* – Solanaceae	Root	Crushed	
20)	*Lagenaria siceraria* – Cucurbitaceae	Root	Crushed	
21)	*Artemisia abyssinica* – Asteraceae	Leaves	Grinded	
22)	*Clucia lanceolata* – Euphorbiaceae	Leaves	To be chewed	
23)	*Dracaena steudneri* – Dracaenaceae	Leaves and fruits	Crushed and decoction	
24)	*Momordica foetida* – Cucurbitaceae	Root and stem	Decoction	
25)	*Blumea lacera* (Roxb.) – Asteraceae	Leaves	Powder	
26)	*Chamaedorea elegans* Mart. – Arecaceae	Seeds	Extract	
27)	*Glycosmis arborea* (Roxb.) – Rutaceae	Root, leaves, and fruits	Decoction and extract	
28)	*Artocarpus heterophyllus* Lam. – Moraceae	Stem	Latex	
29)	*Pedilanthus tithymaloides* (L.) Poit. – Euphorbiaceae	Leaves	Decoction	
30)	*Nothapodytes foetida* (Wight) Sleumer – Icacinaceae	Root	Suspension	
31)	*Elsholtzia winitiana* Craib – Lamiaceae	Leaves	Compressed	[1, 32, 33]
32)	*Ipomoea fistulosa* Mart. – Convulvulaceae	Leaves	Crushed	

(Continued)

Table 18.1 (Continued)

S/no	Scientific name – family	Part used	Mode of use	Reference
33)	*Azadirachta indica* A. Juss. – Meliaceae	Leaves	Crushed	———
34)	*Ricinus communis* L. – Euphorbiaceae	Leaves and seeds	Powder and infusion	———
35)	*Melothria maderaspatana* (L.) – *Cucurbitaceae*	Leaves	Extract	———
36)	*Centella asiatica* (L.) Urb. – Apiaceae	Seed and whole plant	Extract	———
37)	*Terminalia citrina* (Gaertn) Roxb. – Combretaceae	Fruits	Tonic or decoction	———
38)	*Acalypha indica* L. – Euphorbiaceae	Leaves	Powder	———
39)	*Enydra fluctuans* Lour. – Asteraceae	Whole plant	Decoction	———
40)	*Urtica dioica* L. – Urticaceae	Leaves	Paste	———
41)	*Alangium chinense* (Lour) – Alangiaceae	Root	Extract	———
42)	*Pupalia lappacea* (L.) A. Juss. – Amaranthaceae	Whole plant	Paste	———
43)	*Dioscorea pentaphylla* Linn. – Dioscoreaceae	Tuber	Paste	———
44)	*Ophiorrhiza mungos* L. – Rubiaceae	Root	Extract	———
45)	*Croton macrostachyus* Hochst. – Euphorbiaceae	Root, stem bark	Extract	———
46)	*Murraya koenigii* (L.) Spreng. – Rutaceae	Leaves	Extract	———
47)	*Solanum surattense* Burm. f. – Solanaceae	Seed	Paste	———
48)	*Ocimum gratissimum* L. – Lamiaceae	Leaves	Juice	———
49)	*Scoparia dulcis* L. – Scrophulariaceae	Root	Paste	———
50)	*Amaranthus spinosus* L. – Amaranthaceae	Seed	Powder	———
51)	*Justicia adhatoda* L. – Acanthaceae	Leaves	Powder	———
52)	*Alstonia scholaris* (L.) –Apocynaceae	Bark	Paste	———

Table 18.1 (Continued)

S/no	Scientific name – family	Part used	Mode of use	Reference
53)	*Ensete superbum* (Roxb.) – Musaceae	Rhizome	Extract	
54)	*Nigella sativa* L. – Ranunculaceae	Seed	Powder	
55)	*Streblus asper* Lour. – Moraceae	Root	Powder	
56)	*Allium sativum* L. – Alliaceae	Root	Juice	
57)	*Cassia fistula* L. – Ceasalpiniaceae	Leaves and root	Extract	
58)	*Cissus repens* Lam. – Ceasalpiniaceae	Leaves and root	Extract	
59)	*Rauvolfia serpentine* (L.) – Apocynaceae	Root	Powder	
60)	*Piper betle* L. – Piperaceae	Leaves	Juice	

Source: Based on Tsegaye and Elias [31].

18.5 Conclusion

This chapter justified that various medicinal plants can be utilized for the management of rabies viruses. Further research is required to evaluate the efficacy of medicinal plants and to discover potent anti-rabies bioactive compounds. Thus, modern knowledge, technical skills, and state-of-the-art tools are required for this investigation. It is also recommended that traditional medicine documentations be explored in order to identify native traditional plants used against rabies which could lead to the discovery of more potent antiviral drugs.

References

1 Wen, Y., Wang, H., Wu, H. et al. (2011). Rabies virus expressing dendritic cell-activating molecules enhances the innate and adaptive immune response to vaccination. *J. Virol.* 85 (4): 1634–1644.

2 Hendekli, C.M. (2005). Current therapies in rabies: brief review. *Arch. Virol.* 150 (6): 1047–1057.

3 Meresa, A., Degu, S., Tadele, A. et al. (2017). Medicinal plants used for the management of rabies in Ethiopia – a review. *Med. Chem. (Los Angeles).* 7 (2): 795–806.

4 Charlton, K.M. (1994). The pathogenesis of rabies and other lyssaviral infections: recent studies. *Curr. Top. Microbiol. Immunol.* 187: 95–119.

5 Knobel, D.L., Cleaveland, S., Coleman, P.G. et al. (2005). Re-evaluating the burden of rabies in Africa and Asia. *Bull. World Health Organ.* 83 (5): 360–368.

6 Jackson, A.C. (2013). Current and future approaches to the therapy of human rabies. *Antiviral Res.* [Internet] 99 (1): 61–67. http://dx.doi.org/10.1016/j.antiviral.2013.01.003.

7 Bögel, K. (2003). Control of dog rabies. In: *Rabies* (eds. A.C. Jackson and W.H. Wunner). US: Academic Press.

8 Sudarshan, M. (2005). Assessing burden of rabies in India: WHO sponsored national multicentric rabies survey, 2003. *Indian J. Community Med.* 30 (3): 100–101.

9 Gongal, G. and Wright, A.E. (2011). Human rabies in the WHO Southeast Asia region: forward steps for elimination. *Adv. Prev. Med.* 2011: 1–5.

10 Menezes, R. (2008). Rabies in India. *CMAJ* 178 (5): 564–566. https://doi.org/10.1503/cmaj.071488.

11 Fooks, A.R., Banyard, A.C., and Ertl, H.C.J. (2019). New human rabies vaccines in the pipeline. *Vaccine* 37 (Suppl 1): A140–A145. https://doi.org/10.1016/j.vaccine.2018.08.039.

12 Lafon, M. (2011). Evasive strategies in rabies virus infection [Internet]. In: *Advances in Virus Research*, 1e, vol. 79, 33–53. Elsevier Inc. http://dx.doi.org/10.1016/B978-0-12-387040-7.00003-2.

13 Vijayan, P., Raghu, C., Ashok, G. et al. (2004). Antiviral activity of medicinal plants of Nilgiris. *Indian J. Med. Res.* 120 (1): 24–29.

14 Folb, P.I., Bernatowska, E., Chen, R. et al. (2004). A global perspective on vaccine safety and public health: the Global Advisory Committee on Vaccine Safety. *Am. J. Public Health* 94 (11): 1926–1931. https://doi.org/10.2105/ajph.94.11.1926.

15 Schmitt, A.C., Ravazzolo, A.P., and Von Poser, G.L. (2001). Investigation of some Hypericum species native to Southern of Brazil for antiviral activity. *J. Ethnopharmacol.* 77 (2-3): 239–245. https://doi.org/10.1016/s0378-8741(01)00314-2.

16 Malerczyk, C., DeTora, L., and Gniel, D. (2011). Imported human rabies cases in Europe, the United States, and Japan, 1990 to 2010. *J. Travel Med.* 18 (6): 402–407. https://doi.org/10.1111/j.1708-8305.2011.00557.x.

17 Coleman, P.G., Fèvre, E.M., and Cleaveland, S. (2004). Estimating the public health impact of rabies. *Emerg. Infect. Dis.* 10 (1): 140–142. https://doi.org/10.3201/eid1001.020774.

18 Taylor, L.H. and Nel, L.H. (2015). Global epidemiology of canine rabies: past, present, and future prospects. *Vet. Med. (Auckl)* 6: 361–371. https://doi.org/10.2147/VMRR.S51147.

19 Fooks, A.R., Banyard, A.C., Horton, D.L. et al. (2014). Current status of rabies and prospects for elimination. *Lancet* 384 (9951): 1389–1399. https://doi.org/10.1016/S0140-6736(13)62707-5.

20 Barecha, C.B., Girzaw, F., Kandi, V., and Pal, M. (2017). Epidemiology and public health significance of rabies. *Perspect. Med. Res.* 5 (1): 55–67.

21 Albertini, A.A.V., Ruigrok, R.W.H., and Blondel, D. (2011). Rabies virus transcription and replication. *Adv. Virus Res.* 79: 1–22. https://doi.org/10.1016/B978-0-12-387040-7.00001-9.

22 Gaudin, Y. (2000). Rabies virus-induced membrane fusion pathway. *J. Cell Biol.* 150 (3): 601–612. https://doi.org/10.1083/jcb.150.3.601.

23 Osakada, F. and Callaway, E.M. (2013). Design and generation of recombinant rabies virus vectors. *Nat. Protoc.* 8 (8): 1583–1601. https://doi.org/10.1038/nprot.2013.094.

24 Egbuna, C., Kumar, S., Ezzat, S.M. et al. (eds.) (2019). *Phytochemicals as Lead Compounds for New Drug Discovery*, 1e. Cambridge, USA: Elsevier.

25 Egbuna, C. and Ifemeje, J.C. (2015). Biological functions and anti-nutritional effects of phytochemicals in living system. *IOSR J. Pharm. Biol. Sci. (IOSR-JPBS)* 10 (2): 10–19.

26 Shashank, K. and Egbuna, C. (eds.) (2019). *Phytochemistry: in vitro and In Silico Updates*, 1e. Singapore: Springer Nature.

27 Egbuna, C., Ifemeje, J.C., Udedi, S.C., and Kumar, S. (eds.) (2018). *Phytochemistry, Volume 1: Fundamentals, Methods, and Applications*, 1e. New York: Apple Academic Press/CRC Taylor & Francis.

28 Egbuna, C., Kumar, S., Ifemeje, J.C., and Kurhekar, J.V. (eds.) (2018). *Phytochemistry, Volume 2: Pharmacognosy, Nanomedicine, and Contemporary Issues*, 1e. New York: Apple Academic Press/CRC Taylor & Francis.

29 Egbuna, C., Ifemeje, J.C., Kumar, S., and Sharif, N. (eds.) (2018). *Phytochemistry, Volume 3: Marine, Industrial, and Advances*, 1e. New York: Apple Academic Press/CRC Taylor & Francis.

30 World Health Organization (2019). WHO global report on traditional and complementary medicine [Internet]. 228 p. https://apps.who.int/iris/bitstream/handle/10665/312342/9789241515436-eng.pdf?ua=1 (accessed 3 Septempter 2020).

31 Tsegaye, B.P.V. and Elias, T.K.N. (2015). Significance of traditional medicinal plants used for treatment of rabies at Ambo Town. *Med. Aromat. Plants* 4 (4): 1–6.

32 Tadesse, B., Abera, D., and Eyasu, E. (2015). Ethnobotanical Study of Medicinal Plants in Selected Horro Gudurru Woredas, Western Ethiopia. *J. Biol. Agric. Healthc.* 5 (1): 83–94.

33 Paudel, S., Rana, R.M., Mani Giri, P. et al. (2018). Review on medicinal plants used for treatment of dog bite Saru Paudel review on medicinal plants used for treatment of dog bite. *World J. Pharm. Pharm. Sci.* 7 (7): 498–510.

Part VI

Fungal and Ectoparasitic Infections

19

Roles of Medicinal Plants in the Diagnosis and Treatment of Eumycetoma

Chinaza Godswill Awuchi[1], *Muhammad Akram*[2], *and Chibueze Gospel Awuchi*[3]

[1] *Department of Physical Sciences, Kampala International University, Kampala, Uganda*
[2] *Department of Eastern Medicine and Surgery, Government College University Faisalabad, Faisalabad, Pakistan*
[3] *Department of Environmental Technology, Federal University of Technology Owerri, Owerri, Nigeria*

19.1 Introduction

Medicinal plants have been used for ages in traditional and modern medicine, and as a therapeutic resource for both humans and animals. Knowledge of some medicinal plants with palliative, therapeutic, and curative effects on many health complications and diseases, has been passed from one generation to the other, usually by herbalists and elders; scientists and researchers who have also contributed to the documentation of information on medicinal plants. This chapter focuses on the use of medicinal plants in the diagnosis and treatment of mycetoma. Here, mycetoma is often used in place of eumycetoma, except where clearly stated otherwise. Other applications of medicinal plants can be found in the other chapters. Medicinal plants are used in the treatment of many diseases, including fungal diseases [1]. Historically, many cultures have used plants/herbs as panacea for medical problems and for treatment of specific diseases and health complications. Initially, the specific use of medicinal plants was believed to be due to their color and/or shape. With this, leaves that are heart-shaped were used for the treatment of heart complications, while plants that have red leaves/flowers were used for the treatment of bleeding disorders and blood replenishment, and so on [2, 3]. Till today, many rural dwellers still hold this belief. The medicinal properties of plants are mostly due to their phytochemical compositions; alkaloids, terpenes, phenols (polyphenols), glycosides, etc. [4]. In developing countries today, plant

Neglected Tropical Diseases and Phytochemicals in Drug Discovery, First Edition.
Edited by Chukwuebuka Egbuna, Muhammad Akram, and Jonathan Chinenye Ifemeje.

Figure 19.1 Man with a swollen foot caused by mycetoma. *Source:* Fungal Diseases Mycetoma, Centers for Disease Control and Prevention, September 20, 2020).

remedies are common treatments for many health conditions, with the plant formula inherited from one generation to another, making them available [2, 5].

Mycetoma, often manifest as a chronic subcutaneous infectious disease (Figure 19.1) caused by fungi (eumycetoma) or aerobic actinomycetic bacterial infection (actinomycetoma) [6]. Although almost all cases of mycetoma are reported in Sudan, Mexico, India, and Venezuela, its true prevalence and incidence are not well-known [6, 7]. It has been frequently reported in those residing in rural regions, especially shepherds and farmers. It is in the WHO list of neglected tropical diseases (NTD) [7]. Eumycetoma is a fungal disease of humans characterized by chronic granulomatous features and affects mostly the limbs, chest walls, abdominal walls, and/or head [8, 9]. The most commonly reported mycetoma, mycetoma of the foot (*Mycetoma pedis*), is widely called the Madura foot (see Figure 19.1). Mycetoma (eumycetoma and actinomycetoma) is endemic disease in Africa, South America, Central America, and India. Mycetoma can also be seen in other regions of the world, however as endemic disease [10, 11]. Infection with mycetoma arises when the causative microorganism gains entrance into the body via penetrating injury (usually thorn pricks) or minor trauma [12]. A clear relationship exists between mycetoma people that walk barefoot and at the same time work manually.

Although mycetoma usually affects the foot, it can also affect any body part. The infection is mostly acquired through traumatic inoculations of some bacteria (actinomycetoma) or fungi (eumycetoma) into subcutaneous tissues [12]. Young adults are mostly affected, especially males between 15 and 30 years of age in developing nations. Persons with low socioeconomic status and those who work manually including herdsmen, laborers, and agriculturalists are mostly affected. Mycetoma has several adverse health, socioeconomic, and medical effects on affected people, health authorities, and communities [12]. There is no available accurate data on the prevalence and incidence of mycetoma. Early identification and treatments are vital to lessen injury and enhance the outcome of treatment.

Eumycetoma has been a common health and medical condition in subtropical and tropical regions [13]. It is a granulomatous subcutaneous disabling and disfiguring inflammatory disease [14]. *Madurella mycetomatis* is the major fungal agent causing eumycetoma in several parts of the globe [14–16]. Three factors including sero-purulent (or purulent) release which contains grains, sinuses formation, and painless subcutaneous mass are mycetoma pathognomonic [14, 17]. Mycetoma often spread to deep structures and the skin, causing deformities, destructions, and loss of functions, and may even be fatal [18]. As stated earlier, mycetoma has several adverse health, socioeconomic, and medical effects on affected people, health authorities, and communities [12, 14, 15].

Currently, the standard for eumycetoma treatment involves combining antifungal medications with surgical excision [14]. Antifungal drugs currently in use are itraconazole or ketoconazole, which are capital intensive with numerous adverse reactions; connected with high rate of recurrence. Surgical excision of eumycetoma include extensive local excisions, repeated debridements, and amputation [14]. Generally, eumycetoma treatment has several side effects, and has been associated with high rates of follow up dropout, recurrence, disabilities, as well as disfigurement [19]. Treating patients with eumycetoma remains difficult and challenging; the present treatment measures have some side effects, are capital-intensive, and have rate of high recurrence; as a result, poor patients' compliance has been reported in many cases [14]. Most patients of eumycetoma have many financial constraints and low socioeconomic status; many rely on medicinal plants as alternative treatment for their disease.

In recent years, the WHO acknowledged herbal and traditional medicines as therapeutic measures which have been in use for centuries, prior to the development of modern medicines, and still used currently [20]. Herbal treatments mostly use preparations of medicinal plants, usually in combinations, for therapeutic and treatment measures. In many countries such as Nigeria, Sudan, India, etc., medicinal plant treatments for numerous medical conditions are commonly in place. Several studies have been carried out to determine efficacy of medicinal plants for the treatment of eumycetoma [14, 21, 22]. A study involving 311 patients showed 42.4% of eumycetoma patients used medicinal plants for treatment at several stages of the disease. The study reported that the medicinal plants commonly used include *Cuminum cyminum, Citrullus colocynthis, Acacia nilotica,* and *Moringa oleifera* [22]. Table 19.1 presents some medicinal plants evaluated for anti-eumycetoma activities. Although most patients reported no significant benefits gained, several studies have shown

Table 19.1 Some medicinal plants evaluated for anti-eumycetoma activities and showed promising.

Common name	Botanical name	Mostly used parts
Clove	*Eugenia caryophyllus*	Flower bud [21, 22]
Cinnamon	*Cinnamum verum*	Stem bark [14, 22]
Black pepper	*Piper nigrum*	Fruit
Olipanum/Frankinscense	*Boswellia papyrifera*	Gum resin [22]
Acacia nubica	*Acacia nubica*	Root bark [14, 22]
Ginger	*Zingiber officinalis*	Rhizome [22]
Black cumin	*Nigella sativa*	Fruit [2, 21]
Moringa	*Moringa oleifera*	Seeds, leaves [14, 21]

Source: Based on Elfadil et al. [22].

the efficacies of some medicinal plants and plant extracts in the treatment of eumycetoma, even at low minimal inhibitory concentration (MIC) [21, 22]. Prevalence of treatment with medicinal plants can be associated with the patients' discontent with modalities of medical therapy [14] and the high cost of medications.

19.2 Brief History

The first report of mycetoma was in Madurai, India, in the middle of nineteenth century, and thereafter was first referred to as Madura foot [12]. Madura foot, also called maduramycosis or maduromycosis, is designated in Indian ancient writing as *Padavalmika*, meaning *Foot anthill* [12, 23]. In 1842, initial modern account of Madura foot was given from Madurai, India, by Gill; the disease was named *Madura* mycosis after the city Madurai [23]. The fungi causing the disease was made known by Carter in 1860 [12, 23]. Since then, mycetoma has been an endemic plague in most parts of tropics and subtropics, and currently listed among the neglected tropical diseases [7].

19.3 Signs and Symptoms of Mycetoma (Eumycetoma)

First lesion appears as slight subcutaneous swelling following small trauma. Subsequently, sinuses which release seropurulent and purulent exudates which have grains that are colonies of fungi are formed [23–25]. Loss of function, deformity, and deeper tissues destruction of the limbs affected might take place in subsequent stages [11]. Death due to eumycetoma is not common.

19.4 Clinical Features of Mycetoma

Mycetoma has been characterized by discharge that have grains, multiple sinuses, and painless subcutaneous mass. Mycetoma often spreads to bone, deep structures, and the skin, leading to deformity, loss of function, and destruction, and may lead to death [12]. The disease usually involves extremities, gluteal and back area; however, any body part may be affected. With its painlessness, lack of awareness, delayed progression, and paucity of health/medical facilities in endemic regions, some patients reports late when the infection is in advanced stage, in which amputation might be the only option. Secondary infection with bacteria (actinomycetoma) is also commonly encountered and may result in fatal septicaemia (severe infection of the whole human body system), disability, and increased pain, if untreated [12]. Mycetoma cannot transmit from humans to humans.

A notorious sign of mycetoma is the painless subcutaneous tissue swelling resembling a tumour, also known as tumefaction. The formation of grains is distinctive for actinomycetoma (bacterial) and eumycetoma (fungal). Grains signify microscopic colonies of causative microorganisms *in vivo* in vital tissues [24]. The fungal components are organized in subcutaneous fatty tissues in form of dense, circumscribed radial hyphae conglomerations encircled within amorphous matrix material. The grains can be easily visualized under light microscopy, by overview magnification; they can also be seen with the naked eyes in certain circumstances [25]. They have diameter of 0.3–1 mm and often released from inner tissues to the mycetoma surface through pressure. The discharge secretions occur via numerous sinuses; supposed draining sinus tracts. Depending on the microbial species causing the disease, the grains color can be pale (white), red, yellow, or black (Table 19.2) [30].

In Sudan, one of the most affected countries, a new causative fungus of eumycetoma was recently profiled. *Pleurostomophora ochracea* belongs to the *Calosphaeriales* order; *P. ochracea* is the first yellow grained fungus which is found in actinomycetoma [21].

19.5 Causes of Mycetoma

Mycetoma can result from fungal infection (eumycetes) in which it is known as eumycetoma, or bacterial infection of phylum actinomycetes where it is known as actinomycetoma [21, 23–25, 28]. Globally, eumycetoma is mostly caused by *M. mycetomatis* [21, 26, 27]. In most countries such as Sudan, where mycetoma is

Table 19.2 Stain of eumycetoma grains and distinct causative mold.

Grains	Causative agents	References
White and pale or unstained	*Scedosporium apiospermum* (white), *Scedosporium boydii* (previously called *Pseudallescheria boydii*), *Aspergillus nidulans*, *Neotestudina rosatii*, *Fusarium* spp., *Acremonium* spp., *Microsporum audouinii*, *Aspergillus flavus*, and *Microsporum ferrugineum*	[26, 27]
Dark and black	*Madurella mycetomatis*, *Trematosphaeria grisea* (previously called *Madurella grisea*), *Curvularia lunata*, *Falciformispora thompkinsii*, *Medicopsis romeroi* (previously called *Pyrenochaeta romeroi*), *Exophiala jeanselmei*, and *Falciformispora senegalensis* (previously called *Leptosphaeria senegalensis*)	[21, 24, 25, 28]
Yellow-brown	*Streptomyces somaliensis*, *Actinomadura madurae*, *Nocardia otitidiscaviarum* (previously called *Nocardia caviae*), and *Nocardia brasiliensis*	[2, 21, 29]
Yellow	*Pleurostomophora ochracea*	[24, 25]

Source: Aubock et al. [30].

Table 19.3 Characteristic discharge colors from mycetoma wounds caused by fungi and bacteria.

Discharge color	Fungi (eumycetoma)/bacteria (actinomycetoma)	Reference
Black discharge	*Pyrenochaeta romeroi*, *Madurella mycetomatis*, *Madurella grisea*, *Leptosphaeria tompkinsii*, *Leptosphaeria senegalensis*, *Exophiala jeanselmei*, *Cladophialophora bantiana*, *Curvularia lunata*, *Aspergillus terreus*	[21, 25, 28, 33–35]
White or yellow discharge	*Phaeoacremonium krajdenii*, *Pseudallescheria boydii*, *Noetestudina rosatii*, *Aspergillus nidulans*, *Actinomadura madurae* (bacterium), *Acremonium strictum*, *Nocardia brasiliensis* (bacterium), *Streptomyces somaliensis* (bacterium), *Nocardia asteroides* (bacterium)	[21, 25, 28]
Red discharge	*Actinomadura pelletieri* (bacterium), *Streptomyces somaliensis* (bacterium)	[21]

most common, majority of the patients are infected with *M. mycetomatis* [21]. Meanwhile, the exact natural habitation of *M. mycetomatis* has not been conclusively established [31]. However, the fungal DNA is easily seen in samples of plant and soil in eumycetoma endemic regions [32]. Table 19.3 presents the various characteristic discharge colors from mycetoma and the causative organisms.

19.6 Epidemiology of Mycetoma

Mycetoma is an endemic disease in the tropics and subtropics [12, 23]. Precise geographical distributions and occurrences of mycetoma worldwide is unknown mostly because the health condition is often slowly progressive, painless, and presented in late stages to health centers by most patients. The disease is unevenly distributed worldwide. The microorganisms (fungi and bacteria) that cause mycetoma are distributed globally, however, they are endemic in the tropics and subtropics regions which are the so-called "Mycetoma belt" (Figure 19.2), including the Sudan, Yemen, Thailand, Mexico, Mauritania, India, Ethiopia, Somalia, Senegal, Chad, and Venezuela [12, 21].

19.7 Pathogenesis of Mycetoma (Eumycetoma)

Mycetoma often affects field workers such as farmers, especially men of 20–40 years of age. It is contracted through inoculation of fungal grains in soil via skin breach caused by slight trauma such as thorn prick. Afterward, eumycetomas spread to deeper tissue, forming sinus tracts resulting in skin surfaces [23]. Mature eumycetomas present discharge of grain from the sinuses, which are ineffective

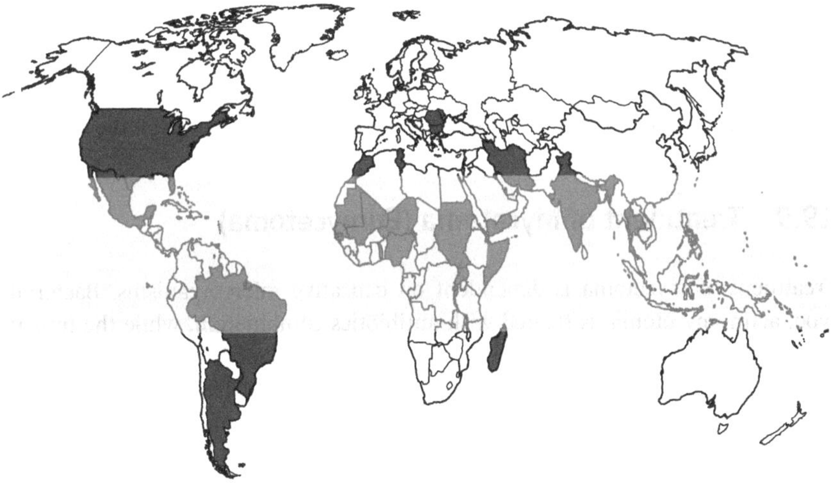

Figure 19.2 Global mycetoma distribution indicating mycetoma belt region (in shaded rectangular block) while shaded regions outside of it indicates some areas where cases of mycetoma have been reported in the medical literature (Adapted from Centers for Disease Control and Prevention). Freely taken from https://www.cdc.gov/fungal/diseases/mycetoma/index.html (accessed 20 September 2020).

and are made of fungal colonies. Internal spread of infections via lymph or blood is not common. Infections with black discharges often spread into the subcutaneous parts. Deeper spreads occur early in yellow and red varieties, penetrating bones and muscles, however, sparing tendons and nerves, which are extremely resistant to the invasive disease. Bacterial pseudomycosis, also called botryomycosis, has similar clinical pictures and is often caused by *Staphylococcus aureus*. There are a few bacteria which might also cause botryomycosis.

19.8 Mycetoma (Eumycetoma) Diagnosis

Mycetoma diagnoses are mostly done clinically in endemic regions such as Sudan, Ethiopia, Somalia, India, etc. Ultrasonography and X-rays can be used to evaluate level of the disease progression [12]. Results of X-rays are greatly variable. Eumycetoma is mostly detected in advanced stages which displays widespread destructions of all foot bones. One single lesion is rarely seen in tibia with identical pictures to chronic osteomyelitis. Tissue biopsy and the lesion pus or fine needle aspirate cytology might be done sometimes. A few reports reported a "dot in a circle sign" as a feature for MRI characteristics of the disease [11]. Same feature is described on ultrasound.

Before diagnosing mycetoma patient, clinical conditions that can be considered include Tuberculous ulcer, Tropical ulcer, Syphilis, Malignant neoplasm, Leprosy, Kaposi's sarcoma (vascular skin tumour often found in AIDS patients), and Botryomycosis (skin infections often caused *Staphylococcus aureus*) [23]. No available vaccine. Some hygienic measures such as washing feet and hands regularly and putting on sandals/shoes whilst working in the field can prevent the disease.

19.9 Treatment of Mycetoma (Eumycetoma)

Treatment of mycetoma is dependent on causative microorganisms. Bacterial type, actinomycetoma, is treated with antibiotics combination, while the fungal type, eumycetoma, is treated with combined antifungal medications and surgical excision. The treatment is often not satisfactory, expensive, has many side effects, and is unavailable in endemic regions [12]. Patients in affected regions often resort to treatment with medicinal plants, especially in the early stages of the disease [21]. Treatment varies according to the causative microorganisms, but may often include antifungal medications or antibiotics. The bacterial form (actinomycetoma) are cured using antibiotics [6]. The fungal form, eumycetoma, can be treated with antifungal medications, including antifungal medicinal plants [21, 36]. Surgical operation as bone resection might be required if cases are presented

Table 19.4 Medicinal plant compounds with antifungal activities as candidates for eumycetoma treatment.

Phytochemicals	Plant source	Plant parts	References
Alkaloid	*Datura metel*	Whole plant	[1, 4]
Alkaloid	*Haloxylon salicornium*	Aerial part	[1, 38]
Diterpenes	*Alpinia galangal*	Seeds	[1, 4]
Emodin, physcion and rhein	*Cassia tora*	Whole plant	[1, 38]
Essential oil	*Juniperus communis*	Leaves	[21]
Essential oil	*Ocimum gratissimum*	Bark	[21]
Essential oils	*Aegle marmelos*	Leaves	[1]
Flavonoid	*Camptotheca acuminate*	Leaves	[4]
Flavonoid luteolin	*Blumea balsamifera*	Leaves	[21]
Phenolic compounds	*Lycium chinense*	Root bark	[4]
Pinosylvin	*Pinus pinaster*	Leaves	[1]
Protein	*Ananas comosus*	Leaves	[1, 21]
Protein	*Euonymus europaeus*	Leaves	[1, 4]
Protein	*Musa acuminate*	Banana	[1]
Saponins	*Smilax medica*	Root	[1, 4]
Sesquiterpene	*Polygonum punctatum*	Whole plant	[1, 4, 21]
Triterpene saponin	*Capsicum frutescens*	Whole plant	[4]
Triterpenes	*Khaya ivorensis*	Stem bark	[1, 4, 21]

late or to improve effects of treatments with medications [11]. Amputation may be required in some cases [10]. Both eumycetoma and actinomycetoma may need extended treatment [6]. In some animals, such as cats, mycetoma may be treated by surgical excision. Antifungal medications have been reported to be rarely effective [37]. Some antifungal medicinal plants have been demonstrated to be alternative treatment for eumycetoma (Tables 19.1 and 19.4). The medicinal plants that may be used in the treatment of eumycetoma are discussed in other sections of this chapter; some of the phytochemicals used are also discussed.

Generally, drugs such as itraconazole, voriconazole, and ketoconazole are used for the treatment of the infection [39, 40]. Actinomycetoma often respond to medical treatments very well, while eumycetomas are resistant to treatment and might often require surgical treatments such as amputation or bone resection [10, 11, 23].

19.10 Experimental Evidence of Plants Tested for the Treatment of Eumycetoma

Results of an experiment conducted by Ezaldeen et al. [14] in 2013 on mycetoma herbal treatment and published in *PLoS Neglected Tropical Diseases*, present the significance of treating eumycetoma with medicinal plants. The study examined 311 eumycetoma patients; 22.5% (70) females and 77.5% (241) males. Their ages were within 18–72 years; 78.5% (244) were within 18–37 years, 18% (56) were within 38–57 years, while 3.5% (11) 58–77 years. In the research, 42.4% (132 patients) made use of medicinal plant treatment in the course of their disease, whereas 57.6% (179 patients) reported not to have used medicinal plants [14]. About 57.6% used different medicinal plants, as single herbal remedy or in combination of many medicinal plants. The study reported that the medicinal plants commonly used include *C. colocynthis* (4.5%), *C. cyminum* (5.3%), *A. nilotica* (9.8%), and *M. oleifera* (18.9%), among others plants [14]. Among the patients, 17.4% (23) made use of mixture of medicinal plants; 13.6% (18 patients) reported to have used *M. oleifera* in combination with other medicinal plants; 3.8% (5 patients) made use of *C. cyminum* also in combination with other medicinal plants [14]. In the study, *M. oleifera* use was 18.9%. While 50% (66 patients) could not recall the type of medicinal plants used, 13.6% (18 patients) made use of wide range of numerous medicinal plants [14]. Although, 29.3% (99 patients) reported of having had suspected complications due to treatment with medicinal plants (manifested as skin necrosis, burn, and infection), many patients reported that *M. oleifera,* which was the most commonly used plant, had medicinal properties and was helpful in the treatment of eumycetoma [14].

The result of the studies by Abdelgadir [21] in 2013 who screened seven medicinal plants for the treatment of eumycetoma found that the extracts of the medicinal plants inhibited the growth of *M. mycetomatis* at several concentrations; these medicinal properties were the first of their kind to be made known [21]. *B. papyrifera* was reported to be the most promising medicinal plant for treatment of eumycetoma [21]. Among the medicinal plants, two plants, *Nigella sativa* and *Acacia nubica* were reported to have similar healing effects on eumycetoma, but *B. papyrifera* had higher efficacy. The GC–MS hybrid analyses of the *B. papyrifera* active fractions identified phytochemicals, with nature of triterpenoids: Stigmatriene, Sitosterol, beta-amyrone, and beta-amyrin [21]. All the triterpenes extracted from *B. papyrifera* were shown to have significant anti-eumycetoma activities at varying levels of efficacy [21]. Stigmatriene showed the highest anti-eumycetoma activities in comparison to the other four triterpenoids phytochemicals when evaluated biologically against several *M. mycetomatis* strains. With GC–MS hybrid analysis, Stigmatriene datasheet was reported as the most constituent of *B. papyrifera* active fractions, making up 72.41% of the area percentage [21]. *B. papyrifera* active fractions were shown to have activity equal to 8 μg/ml at low MIC-50, validating the synergistic activities demonstrated by the other four triterpenoids in the extracts. Stigmatriene inhibited three strains of mycetoma at 0.125 μg/ml (low concentration). At 0.125 μg/ml concentration, Sitosterol killed 1/12 strains of *M. mycetomatis*. In medicine, Sitosterol is known to destabilize cell membranes of fungi [21, 22].

In an *in vitro* study by [2], the anti-mycetoma activities of the methanol and chloroform extracts from seed of *N. sativa* were evaluated against five fungal isolates (*M. mycetomatis* and *P. boydii*) and three bacterial isolates (*S. samaliensis, N. brasiliensis,* and *N. asteroides*) of mycetoma [2]. They reported that the chloroform and methanolic extracts from *N. sativa* seed had anti-mycetoma activities against *N. asteroids* and *N. brasiliensis* with a MICs of 8 and 4%, respectively. The least anti-mycetoma activities were reported against *Streptomyces somaliensis* using methanol extracts and MIC of 0.5% using chloroform extracts [2]. Meanwhile, chloroform extracts were reported to show activities against *M. mycetomatis* at 2% MIC, whereas the MIC was 8% for *P. boydii*, while methanolic extracts with 0.15% MIC showed activities against all fungi [2]. The methanol extracts were also screened for their antioxidant activities. Iron chelates were reported to have high activities than DPPH assays [2]. Qualitative screening of phytochemicals of the methanolic extracts showed the presence of coumarins, flavonoid, essential oil, cardiac glycosides, terpenoids, fatty acids, reducing sugars, carbohydrates, saponins, oxidized amines, unsaturated sterols, quaternary bases, alkaloids, and tannins [2]. Similarly, qualitative screening of phytochemicals of the chloroform extracts indicated the presence of terpenoids, flavonoid, alkaloids, quarternary bases, coumarins, unsaturated sterols, cardiac glycosides, saponins, and oxidized amines [2].

19.11 Medicinal Plants as Sources of Novel Drugs for the Treatment of Eumycetoma

Plants are the basis of traditional medicines in many communities for centuries. The history of modern medicine cannot be complete without giving some credits to plants. Thousands of years ago, fewer or no synthetic medicine existed; roughly 300 000 higher plant species were major sources of drugs and treatment of diseases [21]. Within that period, as the developed countries started shifting significantly from nature to synthetics, population and life expenditures increased by nearly twofold, people started arguing that natural substances are inherently healthier and safer than synthetic substances. The WHO estimates showed that about 80% of global populations mostly depend on traditional medicines which make use of medicinal plants for primary healthcare, in which plant-based medicines play a significant role [12, 41, 42]. In developed world, drugs derived from medicinal plants are very important as well: presently, no less than 119 phytochemicals obtained from plants are in use as important drugs, including those with antifungal activities [1, 21, 43]. Stigmatriene, beta-sitosterol, beta-amyrone, artemisinin, taxol, lapachol, quinine, morphine, digoxin, etc. are examples of medicinal plant phytochemicals that are currently in use in both modern and traditional medicines [4, 21, 22, 38, 44]. Additionally, aromatic and medicinal plants are regarded as main sources of valued extracts, including essential oils that are used globally in cosmetics, flavor, foods, pharmaceuticals, perfumery, and fragrance industries.

19.11.1 Phytochemicals with Antimicrobiological Activities Against Eumycetoma

Several active secondary metabolites from medicinal plants have excellent antimicrobial activities, including phenolic compounds as well as their derivatives. Essential oil, flavonoids, tannins, and/or the active components of several medicinal plants have been shown to have high antimicrobial activities [1, 4, 45]. Triterpenes which occur widely in nature in addition to their various pharmacological and medicinal properties have been demonstrated to have excellent antimicrobial effects [21, 46, 47].

Triterpenes are natural products which in general have common origin. The structures of triterpenes are known to be derived from structures of acyclic precursor squalene – C30 – triterpene. Triterpenes can be categorized into two major groups; pentacyclic and tetracyclic compounds: the former are commonly found in plants, while the latter are steroidal types which occur in microorganisms (mycosterols), animals (zoosterols), and plants (phytosterols) [21, 48]. Over 2500 triterpenes have so far received proper studies, some of them have been studied for their antifungal and antibacterial importance, including against fungal diseases such as eumycetoma [21, 22]. Many triterpenoids have shown several pharmacological properties including antimutagenic, antibacterial, and antifungal activities [21, 22, 47, 49, 50].

Boswellia papyrifera has been studied and reported to have four active phytochemicals of triterpenes against eumycetoma and the causative fungi [21, 22]. The phytochemicals including stigmatriene, beta-sitosterol, beta-amyrone, and beta-amyrin exert antifungal activities synergistically; stigmatriene, the most triterpene in the mixture, was reported to have greatest anti-eumycetoma activity among the four [21, 22]. The four triterpenoids show important line of novel medications with potentials for clinical applications in preventing and treating eumycetoma as well as other fungal diseases [21, 22].

It has been believed that triterpenes antimicrobial action mechanism may be because of change in cellular membrane permeability of microorganisms due to alteration of membrane lipids resulting in disruptions; these biochemical properties have been proved with phytosterols and triterpenes [4, 21, 51] to prevent growth many bacterial and fungal species, including the ones that cause eumycetoma. C24-Sterol methyl transferase enzyme (24-SMT) is responsible for the C24-Sterol acceptor conversion to C24-alkyl phytosterol. Animals do not synthesize this enzyme, which makes new target for designing rational drugs [52]. Though, fusidic acid that was reported to be first isolated from *Fusidium coccineum* (a fungus) in 1960s and cephalosporin P1 (which is chemically related to fusidic acid, but not cepholosporins which prevent the synthesis of peptidoglycan) are types of steroidal triterpene antibiotics called lanosta, and are shown to have activities against microorganisms through inhibiting synthesis of protein by preventing translocations on ribosome [21].

19.12 Specific Medicinal Plants with Promising Activities Against Eumycetoma

19.12.1 *Boswellia papyrifera*

Boswellia papyrifera (Del.) Hochst is a medicinal plant which belongs to the Burseraceae family. It is commonly found in abundance in East Africa, Arabia, and India [21]. In Sudan, a country in East Africa and one of the countries mostly affected by eumycetoma, *B. papyrifera* is common to Nuba and Ingasana mountains. *B. papyrifera* has been in use for medicinal purposes for thousands of years. It has essential oils and gum resin (frankincense/olibanum). Oleogum resin contains triterpenoids in rich amounts. Triterpenes, the main *B. papyrifera* active ingredients, have gained more relevance as antimicrobial, antitumor, and against chronic inflammatory disease [21, 22, 53]. Oleogum resin from *B. papyrifera* is fragrant, with yellowish-brown color and popularly used by the native communities as incense [53]. *B. papyrifera* is the most potential anti-eumycetoma agent among seven medicinal plants studied by Abdelgadir [21]. The GC–MS hybrid analyses of the *B. papyrifera* active fractions identified phytochemicals, with nature of triterpenoids: stigmatriene, sitosterol, beta-amyrone, and beta-amyrin [21]. All the triterpenes extracted from *B. papyrifera* were shown to have significant anti-eumycetoma activities at varying levels of efficacy [21].

19.12.2 *Nigella sativa* L. (Black Cumin)

Black cumin (*Nigella sativa*) is a plant that belongs to Ranunculaceae family, and erect annual medicinal plant. *N. sativa* is commonly cultivated in many regions worldwide, particularly in East Mediterranean nations. Black cumin has been used as spice and native to Mediterranean nations [2]. *N. sativa* seeds are used in local medicine in several countries in Asia, Middle East, and Far East for the treatment of rheumatism, asthma, diarrhea, abdominal pain, coughs, headache, etc. [2]. It has also been reported to be used in treatment of eumycetoma [2]. Oil extracts from *N. sativa* seeds have been demonstrated to have antimicrobial, analgesic, anticancer, anti-inflammatory, and antioxidant properties [2]. Black cumin seeds are used globally as spice and condiment in culinary activities. The seeds are composed of complex substances of many natural biocompounds which are mostly volatile. Organically, the benefits of the seeds include their usage as expectorant and carminative anti-asthmatic, as well as their potentials as immunomodulatory, anti-inflammatory, antimicrobial, antioxidant, and analgesic agents [54].

19.12.3 *Moringa oleifera*

Moringa oleifera, a tropical plant, has several medicinal properties, and has gained attention as medicinal plant for treating several health conditions [55]. *M. oleifera* is popular the tropical and subtropical regions. It has many medicinal properties. Many parts of *M. oleifera* including the bark, leaves, seed, roots, flowers, and fruit have been evaluated and showed properties as antibacterial, antifungal, circulatory and cardiac stimulants, antidiabetic, cholesterol lowering, hepatoprotective, antihypertensive, diuretic, antispasmodic, antitumor, anti-inflammatory, and antipyretic activities [56–58]. *M. oleifera* has been reported to be used in treatment of eumycetoma [14].

19.12.4 *Acacia nubica*

Acacia nubica is a plant that belongs to the Mimosaceae family, and among the most common in Africa and other parts of the world. Mimosine presence, as well as presence of related metabolites, in the plant limits its potential attributes as supplement in animal/human diet; Mimosine is among the non-protein amino acids [14, 21]. In a study done in 2013 by Ezaldeen et al., *Acacia nubica* was among the medicinal plants reported to be used for treatment of eumycetoma [14].

19.13 Antifungal Medicinal Plants That May Help Relieve Eumycetoma

The medicinal plants discussed in this section of the chapter are known to have antifungal activities. However, their efficacies against eumycetoma have not been sufficiently studied; this, of course, creates opportunities for further studies on these plants to establish their efficacies against eumycetoma, and also identify the active compounds which may have anti-eumycetoma activities.

***Aloe vera*:** This plant has many amino acids, enzymes, and elements that give antifungal, antiseptic, and astringent properties [59]. It is a natural treatment without any known side effects. Applying *Aloe vera* gel on fungal diseases, will at least prevent ability of fungi to spread. Eumycetoma may be managed with fresh parts of this plant

Garlic: Garlic has been used in treating infection caused by fungi [21] such as ringworm. Making paste with garlic cloves and small coconut oil has been shown to be effective against fungal diseases, through mixing the paste and applying thin layer on infected parts [59]. It should be covered and left for two hours and rinsed thereafter [59]. Stop using in case of any side effect. Garlic cloves are also consumed daily for the killing of fungi from inside.

Ginger: The gingerol and shogaols in ginger have antifungal activities [4, 21]. Ginger relieves yeast infection and inhibits the growth of *C. albicans* [59]. This root can treat Candida and many other fungal conditions. Drink ginger tea twice or thrice a day to prevent and cure fungal allergy.

Turmeric: The curcumin in turmeric has powerful antifungal activity and also prevents the growth of fungi as well as has protection against infection caused by yeast [4, 59]. It can be applied topically by mixing fresh powder of turmeric in water and apply on affected parts of the skins and the left to dry [59]. Once dried, gently wipe off using wet cloth. It can be done twice a day. Drinking turmeric tea two or three times daily can also give good results against fungal infections.

Coconut: Coconut has been shown to have antimicrobial activities [59]. Fatty acids in edible parts of coconut have antifungal activities [59]. Oil extracts from coconut may be topically applied on affected parts to treat infections caused by fungi.

Oregano oil: Oregano plant is among the most powerful antifungal medicinal plants. Oregano oil is obtained from its leaf by steaming and distilling to obtain its essential oil extracts. The active compounds are concentrated in the oil extracts [59]. Oregano oil consumption has been reported to be effective for Candida cleansing. Its antifungal effects may be effective against eumycetoma.

Cinnamon: Cinnamon is known to have excellent antibacterial and antifungal properties [4]. It can help destroy cells of *C. albicans* and inhibit its spread [59]. The plant can be an excellent candidate for the inhibition of the fungi that cause eumycetoma.

Onion: Allicin, a compound in onion, has shown to slow Candida growth and kill microbe in the body. Regular consumption of onion can inhibit and cure infection cause by fungi [59].

Apple cider vinegar: This plant has shown antibacterial and antifungal activities. It can be used on the skin parts infected with fungi [4, 59].

19.14 Novel Approach to Components and Extracts of Antifungal Plants

Novel systems of drug delivery have showed great potentials in delivering drugs made from medicinal plants such as implants, microcapsules, microparticles, and nanoparticles sustained tablets release or extended forms of release [1]. Many drugs made from medicinal plants with novel system of drug delivery have stamped mark in commercial settings; few are in stages of development in laboratories [60–66]. Active components of medicinal plants including curcumin, bromelain, atropine, and digoxin are promising for treatment of health conditions such as eumycetoma, inflammatory diseases, cancer, diabetes, viral and bacterial diseases among others [62–66]. Natural plant materials have many prospects based on documented scientific evidence explored by the authors and editors [67–76].

19.15 Future Prospects

Treatment with medicinal plants has been popular treatment right from the birth of human race. Going forward, the use of medicinal plant extracts and phytochemicals for the treatment of eumycetoma is quite promising. Worldwide, medicinal plants have been used for the treatment of the disease in countries such as Sudan, Ethiopia, and India, among others. It is high time we started looking toward the use of medicinal plants and plant extracts for the treatment of diseases, including NTDs such as eumycetoma and actinomycetoma. Government regulation, good policies, and control are needed to control the use of medicinal plants in treatment of eumycetoma, in order to maximize their benefits and minimize any side effects. Traditional herbalists should receive the proper education to help reduce any complications which may arise from treatment of eumycetoma with medicinal herbs. Additionally, more research should be done to identify antifungal components in many medicinal plants which are known to show high antifungal activities; these components can be tried on eumycetoma to determine their effectiveness.

19.16 Conclusion

Eumycetoma is fungal disease of humans usually caused *M. mycetomatis, T. grisea, C. lunata,* etc. and characterized by chronic granulomatous features and affects mostly the limbs, chest walls, abdominal walls, and/or head. It is endemic disease in Africa, South America, Central America, and India. Mycetoma can also be seen in other regions of the world, however as endemic disease. Madura foot is common disease of eumycetoma. The formation of grains is distinctive for both actinomycetoma (bacterial) and eumycetoma (fungal). Ultrasonography and X-rays can be used to evaluate the level of the disease progression. The treatment of eumycetoma with drugs such as itraconazole, voriconazole, and ketoconazole is often not satisfactory, expensive, has many side effects, and unavailable in endemic regions. Medicinal plants have been applied in treatment of the disease. Medicinal plants commonly used include *B. papyrifera, N. sativa, C. cyminum, C. colocynthis, A. nilotica, M. oleifera,* among others. Triterpenes (Stigmatriene, Sitosterol, Beta-amyrone, and Beta-amyrin) extracted from *B. papyrifera* were shown to have significant anti-eumycetoma activities at varying levels of efficacy. Many antifungal medicinal plants and their phytochemicals may also be effective in treating eumycetoma.

References

1 Mishra, K.K., Kaur, C.D., Sahu, A.K. et al. (2020). Medicinal plants having antifungal properties. In: *Medicinal Plants – Use in Prevention and Treatment of Diseases*, 1–14. IntechOpen http://dx.doi.org/10.5772/intechopen.90674.

2 Rania, M.A.O. and Yahia, M.A.E. (2015). Anti-mycetoma, anti-oxidant and phytochemical screening of *Nigella sativa* seeds. A thesis submitted to Department of Pharmacognosy, Faculty of Pharmacy, Ribat National University.

3 Cassileth, B.R. (1998). *The Alternative Medicine Handbook: The Complete Reference Guide to Alternative and Complementary Therapies*, 86–98. New York: W. W. Norton & Co.

4 Awuchi, C.G. (2019). Medicinal plants: the medical, food, and nutritional biochemistry and uses. *Int. J. Adv. Acad. Res.* 5 (11): 220–241.

5 Spencer, J.W. and Jacobs, J.J. (1999). *Complementary/Alternative Medicine: An Evidence Based Approach*, 143–148. Toronto: Mosby.

6 Welsh, O., Al-Abdely, H.M., Salinas-Carmona, M.C., and Fahal, A.H. (2014). Mycetoma medical therapy. *PLoS Negl. Trop. Dis.* 8 (10): e3218. https://doi.org/10.1371/journal.pntd.0003218.

7 van de Sande, W.W., Maghoub, E.S., Fahal, A.H. et al. (2014). The mycetoma knowledge gap: identification of research priorities. *PLoS Negl. Trop. Dis.* 8 (3): e2667. https://doi.org/10.1371/journal.pntd.0002667.

8 Motswaledi, H.M., Mathekga, K., Sein, P.P., and Nemutavhanani, D.L. (2009). *Paecilomyces lilacinus* eumycetoma. *Int. J. Dermatol.* 48 (8): 858–861. https://doi.org/10.1111/j.1365-4632.2008.04047.x.

9 Brownell, I., Pomeranz, M., and Ma, L. (2005). Eumycetoma. *Dermatol. Online J.* 11 (4): 10.

10 Efared, B., Tahiri, L., Boubacar, M.S. et al. (2017). Mycetoma in a non-endemic area: a diagnostic challenge. *BMC Clin. Pathol.* 17: 1. https://doi.org/10.1186/s12907-017-0040-5.

11 El-Sobky, T.A., Haleem, J.F., and Samir, S. (2015). Eumycetoma osteomyelitis of the calcaneus in a child: a radiologic-pathologic correlation following total calcanectomy. *Case Rep. Pathol.* 2015: 129020. https://doi.org/10.1155/2015/129020.

12 WHO (2019). Mycetoma. https://www.who.int/news-room/fact-sheets/detail/mycetoma (accessed 4 September 2020).

13 Fahal, A.H. and Hassan, M.A. (1992). Mycetoma. *Br. J. Surg.* 79 (11): 1138–1141.

14 Ezaldeen, E.A., Fahal, A.H., and Osman, A. (2013). Mycetoma herbal treatment: the Mycetoma Research Centre, Sudan experience. *PLoS Negl. Trop. Dis.* 7 (8): e2400. https://doi.org/10.1371/journal.pntd.0002400.

15 Fahal, A.H. (2004). Mycetoma thorn on the flesh. Review article. *Trans. R. Soc. Trop. Med. Hyg.* 98 (1): 3–11.

16 Fahal, A.H. (2011). Mycetoma. Review article. *Khartoum Med. J.* 4 (1): 514–523.

17 Fahal, A.H. (2010). Mycetoma. In: *Tropical Infectious Diseases: Principles, Pathogens and Practice*, 3e (eds. R.L. Guerrant, D.H. Walker and P.F. Weller), 234–239. Philadelphia: Saunders-Elsevier.

18 Fahal, A.H. (2013). Mycetoma. In: *Bailey and Love's Short Practice of Surgery*, 26e (eds. N.S. Williams, C.J.K. Bulstrode and P.R. O'Connell), 64–68. Oxford University Press.

19 Zein, H.A.M., Fahal, A.H., Mahgoub, E.S. et al. (2012). Predictors of cure, amputation and follow-up dropout among patients with mycetoma seen at the Mycetoma Research Centre, University of Khartoum, Sudan. *Trans. R. Soc. Trop. Med. Hyg.* 106 (11): 639–644.

20 Pal, S.K. and Shukla, Y. (2003). Herbal medicine: current status and the future. *Asian Pac. J. Cancer Prev.* 4 (4): 281–288.

21 Abdelgadir, H.E.H. (2013). Eumycetoma and plant-based pharmaco-therapy candidates. A thesis submitted in fulfillment of the requirements of the Degree of Doctor of Philosophy. Department of Pharmaceutics, Faculty of Pharmacy, University of Gezira

22 Elfadil, H., Fahal, A., Kloezen, W. et al. (2015). The in vitro antifungal activity of Sudanese medicinal plants against *Madurella mycetomatis*, the eumycetoma major causative agent. *PLoS Negl. Trop. Dis.* 9 (3): e0003488. https://doi.org/10.1371/journal.pntd.0003488.

23 Ananthanarayan, B.A., Jayaram, C.K., and Paniker, M.D. (2006). *Textbook of Microbiology*, 7e, 618. Orient Longman Private Ltd.

24 Nenoff, P., van de Sande, W.W.J., Fahal, A.H. et al. (2015). Eumycetoma and actinomycetoma – an update on causative agents, epidemiology, pathogenesis, diagnostics and therapy. *J. Eur. Acad. Dermatol. Venereol. (JEADV)* 29 (10): 1873–1883. https://doi.org/10.1111/jdv.13008.

25 Relhan, V., Mahajan, K., Agarwal, P., and Garg, V.K. (2017). Mycetoma: an update. *Indian J. Dermatol.* 62 (4): 332–340. https://doi.org/10.4103/ijd.IJD_476_16.

26 Hay, R.J. and Mackenzie, D.W. (1982). The histopathological features of pale grain eumycetoma. *Trans. R. Soc. Trop. Med. Hyg.* 76: 839–844.

27 Gupta, M.K., Banerjee, T., Kumar, D. et al. (2013). White grain mycetoma caused by *Scedosporium apiospermum* in North India: a case report. *Int. J. Low. Extrem. Wounds* 12: 286–288.

28 Ahmed, S.A., van den Ende, B.H., Fahal, A.H. et al. (2014). Rapid identification of black grain eumycetoma causative agents using rolling circle amplification. *PLoS Negl. Trop. Dis.* 8: e3368.

29 Shahapur, P.R., Peerapur, B.V., Shahapur, R.P. et al. (2014). Lymphocutaneous nocardiosis caused by *Nocardia otitidiscaviarum*: a case report and review of literature. *J. Nat. Sci. Biol. Med.* 5: 197–201.

30 Aubock, J., Pichler, E., and Fritsch, P. (1985). Myzetom durch *Petriellidium boydii*: Behandlung mit Ketoconazol. *Hautarzt* 36: 453–458.

31 van Belkum, A., Fahal, A., and van de Sande, W.W. (2013). Mycetoma caused by *Madurella mycetomatis*: a completely neglected medico-social dilemma. *Adv. Exp. Med. Biol.* 764: 179–189.

32 Ahmed, A., Adelmann, D., Fahal, A. et al. (2002). Environmental occurrence of *Madurella mycetomatis*, the major agent of human eumycetoma in Sudan. *J. Clin. Microbiol.* 40: 1031–1036.

33 Severo, L.C., Oliveira, F.M., Vettorato, G., and Londero, A.T. (1999). Mycetoma caused by *Exophiala jeanselmei*. Report of a case successfully treated with itraconazole and review of the literature. *Rev. Iberoam. Micol.* 16 (1): 57–59.

34 Vilela, R., Duarte, O.M., Rosa, C.A. et al. (2004). A case of eumycetoma due to *Madurella grisea* in northern Brazil. *Mycopathologia* 158 (4): 415–418. https://doi.org/10.1007/s11046-004-2844-y.

35 Ahmed, A.O., Desplaces, N., Leonard, P. et al. (2003). Molecular detection and identification of agents of eumycetoma: detailed report of two cases. *J. Clin. Microbiol.* 41 (12): 5813–5816. https://doi.org/10.1128/JCM.41.12.5813-5816.2003.

36 Karrakchou, B., Boubnane, I., Senouci, K., and Hassam, B. (2020). *Madurella mycetomatis* infection of the foot: a case report of a neglected tropical disease in a non-endemic region. *BMC Dermatol.* 20 (1): 1. https://doi.org/10.1186/s12895-019-0097-1.

37 Eldredge, D.M., Carlson, D.G., Carlson, L.D., and Giffin, J.M. (2008). *Cat Owner's Home Veterinary Handbook*, 160. Howell Book House.

38 Awuchi, C.G., Chinelo, K.E., Obinna, C.A. et al. (2020). Medicinal plant phytochemicals: the biochemistry and uses of the pharmacologically active alkaloids, terpenes, polyphenols, and glycosides. *Agro and Food Processing for Wealth Creation – The Nigerian Experience. Proceedings of the Nigerian Institute of Food Science and Technology.*

39 Capoor, M.R., Khanna, G., Nair, D. et al. (2007). Eumycetoma pedis due to *Exophiala jeanselmei*. *Indian J. Med. Microbiol.* 25 (2): 155–157. https://doi.org/10.4103/0255-0857.32726.

40 Loulergue, P., Hot, A., Dannaoui, E. et al. (2006). Successful treatment of black-grain mycetoma with voriconazole. *Am. J. Trop. Med. Hyg.* 75 (6): 1106–1107. https://doi.org/10.4269/ajtmh.2006.75.1106. PMID 17172376.

41 World Health Organization (2000). *General Guidelines for methodologies on research and evaluation of traditional medicine*. Geneva: WHO/EDM/TRM.

42 WHO (2005). *National Policy on Traditional Medicine and Complementary/Alternative Medicine*. Report of a WHO global survey. Geneva: WHO.

43 Mulholland, D.A. (2000). Medicinal plants, a source for new drugs. *African Conference on Medicinal Plants Research*, Alfateh University, Faculty of Pharmacy, Tripoli, Libya.

44 Evans, W.C. (2003). *Textbook of Pharmacognosy*, 15e. Arabic Center for Arabization, translation, authorship and publication.

45 Filamini, G., Cioni, P.L., Puleio, R. et al. (1999). Antimicrobial activity of the essential oil *Calamintha nepeta* and its constituent pulegone against bacteria and fungi. *Phytother. Res.* 13 (14): 349–351.

46 Hifzur, R.S. and Mohammed, S. (2011). Benefacial health effects of lupeol triterpene: a review of pre-clinical studies. *Life Sci.* 88: 285–293.

47 Raquel, C.J., Machael, K., Rayan, M.V. et al. (2011). New antimicrobial triterpenes from *Rhus taitensis*. *Planta Med.* 77 (14): 1651–1654.

48 Buckingham, J. (1996). *Dictionary of Natural Products*, vol. 11. England: CRC Press.

49 Molnar, J., Gyemant, N., Tanaka, M. et al. (2006). Inhibition of multidrug resistance of cancer cells by natural diterpenes, triterpenes, and carotenoids. *Curr. Pham. Des.* 12: 287–311.

50 Lin, S.B., Li, C.H., Lee, S.S., and Kan, L.S. (2003). Triterpene-enriched extracts from *Ganoderma lucidum* inhibit growth of hepatoma cells via suppressing protein kinase C, activating mitogen-activated protein kinases and G2-phase cell cycle arrest. *Life Sci.* 72: 2381–2390.

51 Tan, M.J., Ye, J.M., Turner, N. et al. (2008). Antidiabetic activities of triterpenoid isolated from better melon associated with activation of the AMPK pathway. *Chem. Biol.* 15: 263–273.

52 Jialin, L. and William, D.N. (2009). Steroid triterpenes: design of substrate-based inhibitors of Ergosterol and Sitosterol synthesis. *Molecules* 14: 4690–4706.

53 Sharma, A., Chhikra, S., Ghodekar, S.N. et al. (2009). Phytochemical and pharmacological investigations on *Boswellia serrata*. *Pharmacogn. Rev.* 3 (5): 206–215.

54 Ahmed, E.M. (2005). Phytochemical, antimalarial, molluscicidal and antimicrobial activity of selected Sudanese medicinal plants with emphasis on *Nigella sativa* seed L. Thesis of PhD, University of Gezira

55 Egbuna, C. (2015). *Moringa oleifera* "the mothers best friend". *Int. J. Nutr. Food Sci.* 4 (6): 624–630. https://doi.org/10.11648/j.ijnfs.20150406.14.

56 Mazumder, U.K., Gupta, M., Chakrabarti, S., and Pal, D. (1999). Evaluation of hematological and hepatorenal functions of methanolic extract of *Moringa oleifera* Lam. root treated mice. *Indian J. Exp. Biol.* 37 (6): 612–614.

57 Guevara, A.P., Vargas, C., Sakurai, H. et al. (1999). An antitumor promoter from *Moringa oleifera* Lam. *Mutat. Res.* 440 (2): 181–188.

58 Mehta, K., Balaraman, R., Amin, A.H. et al. (2003). Effect of fruits of *Moringa oleifera* on the lipid profile of normal and hypercholesterolaemic rabbits. *J. Ethnopharmacol.* 86 (2–3): 191–195.

59 Deepa, G. (2020). Anti-fungal foods to cure allergies and infections in natural way. https://www.thestatesman.com/lifestyle/health/anti-fungal-foods-cure-allergies-infections-natural-way-1502854858.html (accessed 5 September 2020).

60 Devi, V.K., Jain, N., and Valli, S.K. (2010). Importance of novel drug delivery systems in herbal medicines. *Pharmacogn. Rev.* 4 (7): 27–31.

61 Chakraborty, K., Shivakumar, A., and Ramachandran, S. (2016). Nanotechnology in herbal medicine. *Int. J. Herb. Med.* 4 (3): 21–27.

62 Egbuna, C., Kumar, S., Ezzat, S.M. et al. (eds.) (2019). *Phytochemicals as Lead Compounds for New Drug Discovery*, 1e. Cambridge, USA: Elsevier.

63 Shashank, K. and Egbuna, C. (eds.) (2019). *Phytochemistry: In vitro and In Silico Updates*, 1e. Singapore: *Springer Nature*.

64 Egbuna, C., Ifemeje, J.C., Udedi, S.C., and Kumar, S. (eds.) (2018). *Phytochemistry, Volume 1: Fundamentals, Methods, and Applications*, 1e. New York: Apple Academic Press/CRC Taylor & Francis.

65 Egbuna, C., Kumar, S., Ifemeje, J.C., and Kurhekar, J.V. (eds.) (2018). *Phytochemistry, Volume 2: Pharmacognosy, Nanomedicine, and Contemporary Issues*, 1ste. New York: Apple Academic Press/CRC Taylor & Francis.

66 Egbuna, C., Ifemeje, J.C., Kumar, S., and Sharif, N. (eds.) (2018). *Phytochemistry, Volume 3: Marine, Industrial, and Advances*, 1e. New York: Apple Academic Press/CRC Taylor & Francis.

67 Nwosu, O., Keskin, M., Lohani, H. et al. (2019). Bioactive lead compounds and molecular targets for the development of anti-inflammatory drugs. In: *Phytochemicals as Lead Compounds for New Drug Discovery* (eds. C. Egbuna, S. Kumar, S.M. Ezzat, et al.), 317–329. Cambridge, USA: Elsevier.

68 Tupas, G.D., Otero, M.C.B., Ebhohimen, E.I. et al. (2019). Anti-diabetic lead compounds and targets for drug development. In: *Phytochemicals as Lead Compounds for New Drug Discovery* (eds. C. Egbuna, S. Kumar, S.M. Ezzat, et al.), 127–138. Cambridge, USA: Elsevier.

69 Egbuna, C., Ezzat, S.M., Tijjani, H., and Srivastav, V.K. (2019). Synthetic Analogs of Phytochemicals. In: *Phytochemistry: In vitro and In Silico Updates* (eds. S. Kumar and C. Egbuna), 23–56. Singapore: Springer Nature.

70 Painuli, S., Semwal, P., and Egbuna, C. (2020). Mushroom: Nutraceutical, Mineral, Proximate Constituents and Bioactive Component. In: *Functional Foods and Nutraceuticals: Bioactive Components, Formulations and Innovations* (eds. C. Egbuna and G.-D. Tupas), 307–336. Switzerland AG: Springer Nature.

71 Egbuna, C., Gupta, E., Ezzat, S.M. et al. (2020). Aloe species as valuable sources of functional bioactives. In: *Functional Foods and Nutraceuticals: Bioactive Components, Formulations and Innovations* (eds. C. Egbuna and G.-D. Tupas), 337–387. Switzerland AG: Springer Nature.

72 Tijjani, H., Zangoma, M.H., Mohammed, Z.S. et al. (2020). Polyphenols: classifications, biosynthesis and bioactivities. In: *Functional Foods and Nutraceuticals: Bioactive Components, Formulations and Innovations* (eds. C. Egbuna and G.-D. Tupas), 389–414. Switzerland AG: Springer Nature.

73 Olatunde, A., Tijjani, H., Ishola, A.A. et al. (2020). Carotenoids as functional bioactive compounds. In: *Functional Foods and Nutraceuticals: Bioactive Components, Formulations and Innovations* (eds. C. Egbuna and G.-D. Tupas), 415–444. Switzerland AG: Springer Nature.

74 Walag, A.M.P., Ahmed, O., Jeevanandam, J. et al. (2020). Health benefits of organosulfur compounds. In: *Functional Foods and Nutraceuticals: Bioactive Components, Formulations and Innovations* (eds. C. Egbuna and G.-D. Tupas), 445–472. Switzerland AG: Springer Nature.

75 Das, S., Sharangi, A.B., Egbuna, C. et al. (2020). Health benefits of isoflavones found exclusively of plants of the Fabaceae family. In: *Functional Foods and Nutraceuticals: Bioactive Components, Formulations and Innovations* (eds. C. Egbuna and G.-D. Tupas), 473–508. Switzerland AG: Springer Nature.

76 Tijjani, H. and Egbuna, C. (2018). Medicinal plants with anti-venom activities. In: *Phytochemistry, Volume 2: Pharmacognosy, Nanomedicine, and Contemporary Issues* (eds. C. Egbuna, S. Kumar, J.C. Ifemeje and J.V. Kurhekar), 207–256. New York: Apple Academic Press.

20

Role of Phytochemicals in the Treatment of Ectoparasitic Infections

Scabies and Myiasis

Santwana Palai[1], *Shyam Sundar Kesh*[2], *Chinaza Godswill Awuchi*[3], *Surajudeen Abiola Abdulrahman*[4], *and Chukwuebuka Egbuna*[5,6]

[1] *Department of Veterinary Pharmacology and Toxicology, College of Veterinary Science and Animal Husbandry, Bhubaneswar, Odisha, India*
[2] *Department of Veterinary Clinical Complex (Veterinary Biochemistry), Faculty of Veterinary and Animal Sciences, West Bengal University of Animal and Fishery Sciences, Kolkata, West Bengal, India*
[3] *Department of Physical Sciences, Kampala International University, Kampala, Uganda*
[4] *Health Education England, Leicester, United Kingdom*
[5] *Department of Biochemistry, Faculty of Natural Sciences, Chukwuemeka Odumegwu Ojukwu University, Uli, Nigeria*
[6] *Nutritional Biochemistry and Toxicology Unit, World Bank Africa Centre of Excellence, Centre for Public Health and Toxicological Research (PUTOR), University of Port-Harcourt, Port Harcourt, Nigeria*

20.1 Introduction

Due to globalization of trade and commerce, ectoparasites are moving with human and animal host throughout the world by road, air, or waterways. Movement of population from rural to urban areas, from developing to developed countries; increasing resistance to the existing treatment regimen for head lice, scabies mite, myiasis flies; and increase in number of susceptible immunocompromised animals are few of the reasons for increasing endemicity and re-emergence of these ectoparasitic diseases [1]. These parasitic diseases are caused by scabies-causing mites (*Sarcoptes scabiei*), fleas (*Tunga penetrans, Pulex irritans*), myiasis-producing diptera flies (*Cochliomyia, Gasterophilus, Oestrus, Wohlfahrtia, Chrysomya, Sarcophaga, Dermatobia, Hypoderma, Cuterebra*), lice (*Phthirus pubis, Pediculus capitis*), and bed bugs (*Cimex*

Neglected Tropical Diseases and Phytochemicals in Drug Discovery, First Edition.
Edited by Chukwuebuka Egbuna, Muhammad Akram, and Jonathan Chinenye Ifemeje.

hemipterus, Cimex lectularius). These ectoparasites affect human health causing significant morbidity and socioeconomic losses.

In the last decade, WHO has recognized scabies and myiasis as "Neglected Tropical Diseases" (NTDs) through unification of global diagnostic criteria and implementation of integrated disease management programmes and preventive chemotherapy and transmission control in mass at the field level [2], aimed at reducing and/or eliminating these diseases from individuals or population groups around the world.

20.2 Scabies

Scabies is an infectious skin disease caused by obligate human mite, *S. scabiei* [3], which is one of the three most common disorders of the skin in children, together with bacterial skin infections and ringworm [4]. According to the 2015 estimate of the global burden of disease and injury incidence and prevalence, scabies affects at least 2.8% of the global population (>204 million individuals) [5]. It has been reported to be evenly common in both males and females [6]. However, the old and young are most commonly affected. Scabies is more common in tropical climates and developing countries, particularly among low socioeconomic groups. Infestation in animals is caused by different but similar mites, called Sarcoptic mange [7]. In general, the mite makes burrow in the stratum corneum of skin and finishes its life cycle there (Figure 20.1). It can be diagnosed clinically based on symptoms and signs like itching orpruritus and lesions as well as identification of mites, its eggs, or feces microscopically.

20.3 Myiasis

Myiasis is a parasitic condition caused by the invasion and infestation of living or dead organism tissue by flies which lay eggs and develop into maggots within the tissues of host animals or humans (Figure 20.2). In case of obligate myiasis, the maggots eat and grow in living tissues and in case of facultative myiasis, the flies incubate their larvae in degenerative necrotic conditions of wounds. Generally, obligate myiasis of humans occurs in tropical areas and facultative myiasis occurs all over the world [8]. Since many animals, especially domesticated animals, are not able to effectively react as humans do to the effect of myiasis, such infestation has continuing and severe implications for global and national livestock industries, resulting in huge economic losses wherever they cannot be addressed by actions of humans [9]. Though myiasis

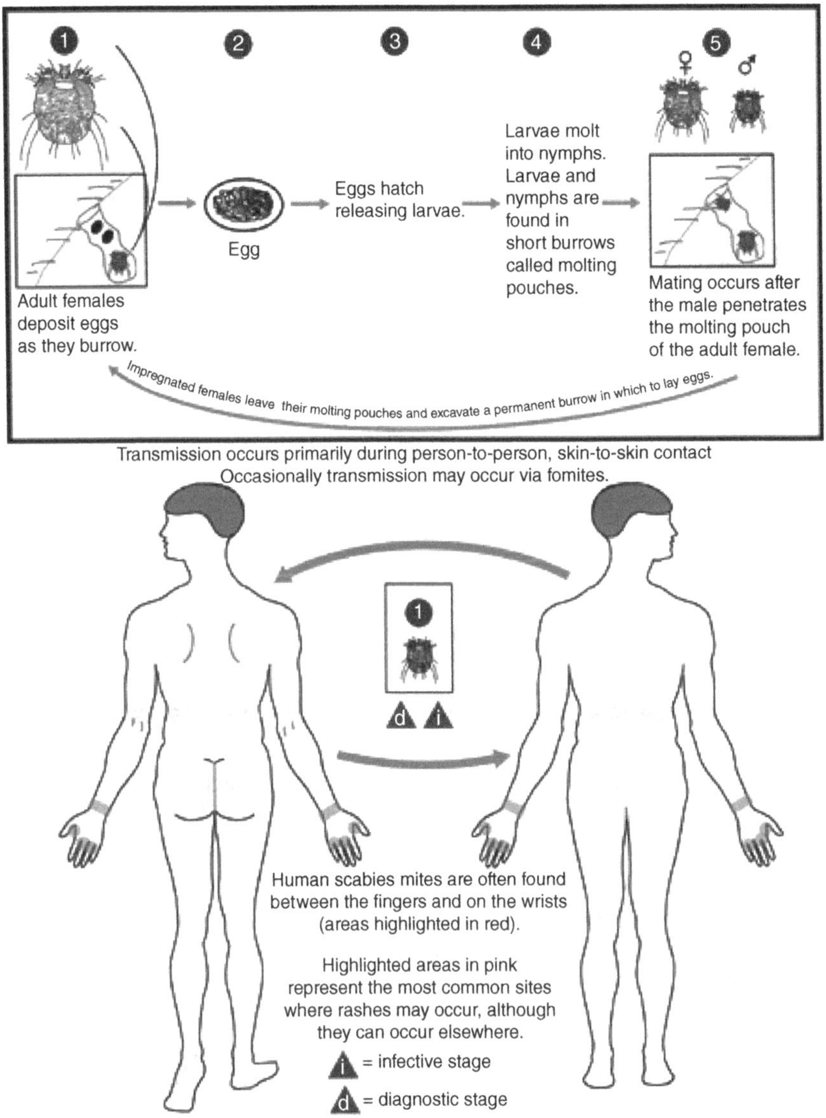

Figure 20.1 Scabies life cycle and transmission process. *Source:* Parasites – Scabies Biology, Centers for Disease Control and Prevention, 17 September 2020.

is typically a great concern for animal species, the high prevalence of flies that cause human myiasis in rural tropics means human affliction is inevitable, and may frequently require biomedical treatment or surgical removal of the parasites/maggots [10].

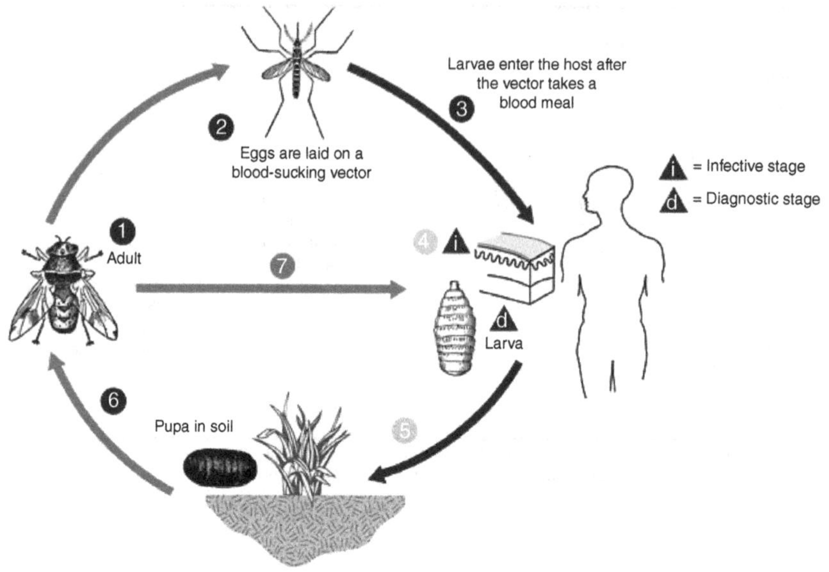

Figure 20.2 Myiasis life cycle and human transmission process. *Source:* Myiasis, Centers for Disease Control and Prevention, 17 September 2020.

20.4 Treatment and Control of Scabies and Myiasis

Both human scabies and human myiasis are wide-spreading conditions causing intense sufferings. Accurate diagnostic tests, convincing and acceptable treatment, and improvement of surveillance and control measures are important priorities for reducing the impact of scabies and myiasis on human populations.

Many medicinal treatments, including the use of phytochemicals and other plant materials, are available for the treatment of people infected with scabies. Scabies is effectively treated with antiscabietic topical drugs like permethrin, crotamiton, lindane, benzyl benzoate, malathion, monosulfiram, acaricide, and antiscabietic oral drugs like Ivermectin along with symptomatic treatment with antihistamines for itching and antibiotics for secondary bacterial infection [11, 12]. Treatment failure occurs due to improper application, reinfestation, and resistance to drugs such as permethrin, lindane, crotamiton, etc. Treatment of scabies must be extended to the household contacts of infested person, and also people who may have recently had close contact with infected person [4].

Among medicinal plants, tea tree oil has been known to be traditional and natural treatment against scabies [13, 14]. When used in a single-dose combination regime, plant extracts are believed to potentiate the antiscabietic effect of oral ivermectin in eliminating scabies, particularly among adults.

Permethrin, a synthetic chemical that acts like *Chrysanthemum* flower extract, is one of the treatments mostly effective against scabies, and is among the preferred treatments [4, 15]. Benzyl benzoate, which has been proven as an efficacious topical treatment for scabies [16, 17], can be isolated from some plant species such as the Polyathia. Cayenne has a component known as capsaicin, a phytochemical which is believed to reduce pain when applied to skin [13]. Studies have shown that scabies mites can be eliminated using *Capsicum annuum* on the skin; and capsaicin is believed to be responsible for this action [13]. Other plant extracts and phytochemicals such as *Azadirachta indica* extracts, clove oil, rosemary oil, *Eupatorium perfoliatum* extracts, turmeric (*Curcuma longa*), Lavandula oil, *Pimpinella anisum* seed oil, *Heliotropium indicum* have shown effectiveness against scabies and/or the mites that cause scabies [13].

Treatment of myiasis may require surgical removal of the larvae from furuncular lesion that contains embedded maggot [18] or simple suffocation of the larva to death by application of occlusive substances like petroleum jelly, nail polish, and bacon fat in the skin opening.

Prophylactic use of neurotoxic insecticides like diazinon, cypermethrin, deltamethrin, and insect growth regulators like cyromazine, dicyclanil, diflubenzuron, and so on, can control the mites and flies' infestation. Unfortunately, these otherwise efficacious treatments have become less desirable because of their toxicity, residual effect on environment, and developing resistance in the twentieth century.

Essential oils from some plants such as clove, basil, etc., act as insecticides which also inhibit myiasis directly or indirectly. The insecticidal/myiatic action is mostly associated with the eugenol which works through obstructing the octopamine receptors of insects and/or through possibly working via cascades of tyramine receptor, just like the thymol in thyme essential oil [19]. In *Cinnamomum camphora* L. Presl, linalool, a terpene, is a significant factor in insecticidal and repelling actions for myiasis control [20–22].

The hydrocarbon skeleton and structural characteristics of terpenoid [21, 22] provide hydrophobicity that is connected with enzyme inhibitions and deactivation of protein. Simas et al. [23] stated that myiatic larvicidal actions can be mainly associated with β-pinene, antethole, safrole, (*E*)-nerolidol, eugenol, and (*E*,*E*)-farnesol [23]. β-pinene and α-pinene (an enantiomer of β-pinene) in many plants have been associated with insecticidal and larvicidal actions for the control of myiasis [21].

The painful suffering from ectoparasitic diseases and effect of insecticides on environment have paved way for the development of new treatment strategies and acceptance of herbal anti-ectoparasitic drugs. Medicinal plants have a reputation from historically efficacious use by humans for the past thousands of years to treat various illness and disorders. Many plants are also recognized to be having antiparasitic properties (Table 20.1). The plant product or secondary metabolites stored in various parts of the plant can help to eradicate the external parasites.

Table 20.1 Medicinal plants and phytochemicals with projected scabicidal and anti-myiasis potentials.

SL. no.	Name of the plant CN: Common name F: Family	Geographical distribution	Phytochemistry	Other medicinal uses	References
1)	***Achyranthes aspera*** CN: Chaff-flower, devil's horsewhip, prickly chaff flower. F: *Amaranthaceae*	India, tropical Asia, and other parts of the world.	Seed, root, shoot Oleanolic acid Phenol, flavonoids, tannins, alkaloids, saponins, terpinoids, and reducing sugars	Cancer, leprosy, asthma, fistula, piles, arthritis, wound, snake bite, dandruff, hepatitis, renal disorders, dermatological disorders, malaria, fever, cough, gonorrhea, diabetes, pyorrhea, dysentery, ophthalmia, rabies, hysteria, toothache. Antimicrobial, antihypertensive, anti-anasarca, antinociceptive, prothyrodic, antispasmodic hypoglycemic, diuretic, hypolipidemic, antioxidant, cardiac stimulant, analgesic, anti-inflammatory, larvicidal, antifertility, immunostimulant, antipyretic and hepatoprotective properties	Dey [24]
2)	***Cannabis sativa*** CN: Marijuana, Ganja F: *Cannabaceae*	Origin: Central and Eastern Asia	Seeds, leaves, flower, fruit Cannabidiol, cannabichromene, and cannabigerol Alkaloids, flavonoids, cardiac glycosides, resins, terpins and steroids	Alzheimer's disease, glaucoma, multiple sclerosis, neuralgia depression, cancer AIDS. Analgesic, antiemetic, laxative anticonvulsive, anti-inflammatory, sedative actions	Audu et al. [25]
3)	***Ficus carica*** CN: Fig F: *Moraceae*	Origin: Eastern Mediterranean Southwest Asia	Dried fruits, root, and leaves Proanthocyanidins Fiber and polyphenols	Loss of appetite, gastric problems, diarrhea, sore throats, bronchial problems, cardiovascular disorders Anticancer, antimicrobial, antispasmodic, hypoglycemic, hypolipidemic, anti-inflammatory, hepatoprotective activities.	Mawa et al. [26]

4)	***Ficus racemosa*** CN: Gular F: *Moraceae*	Southeast Asia, India, Australia, and throughout the world.	Fruit, leaves, bark root, latex Gluanol acetate, racemosic acid Carbohydrates, glycosides, fixed oils and fats, proteins and fats, phenolic compounds, tannins, phytosterols, alkaloids, flavonoids, saponins, gums, and mucilages. Triterpenoids alkaloids, and tannins.	Nerve weakness, dysentery, diabetes, mumps, hydrophobia Antibacterial, hepatoprotective, antidiabetic, hypolipidemic, antifilarial, antipyretic, anti-inflammatory, analgesic, muscle relaxant, antidiabetic, antioxidant, anti-diarrheal, anti-inflammatory, antifungal activities.	Sachan et al. [27] and Yadav et al. [28]
5)	***Ficus benghalensis*** CN: Banyan tree F: *Moraceae*	Origin: Asia (India, Burma, Thailand, Malaysia, Southern China)	Stem bark, aerial roots, vegetative buds, leaves, fruits, and latex Phytosterols, anthocyanidin, leucoanthocyanidins, flavonoids, triterpenoids	Rheumatic inflammations, pimples, dysentery, wounds, ulcers, abscesses, sores, cracked soles, diabetes, erysipelas, nervous disorders, menorrhagia, leucorrhoea, seminal weakness. Antihyperlipidemic, hypocholesterolemic, anti-inflammatory, antioxidant, antibacterial, antifungal, larvicidal, antimutagenic, hepatoprotective activity	Khaliq [29]
6)	***Mangifera indica*** CN: Mango F: *Anacardiaceae*	Originated from India to West Africa	Stem barks and leaves Mangiferin, kaempferol, quercitin Tannins and flavonoids, anthocyanidins, leucoanthocyanins, tannins	Gastric disorder, toothache, sore gums, sore throat, malaria, diarrhea and dysentery, breast cancer, cystic ovaries, endometriosis, coronary arteriosclerosis, respiratory, cardiovascular, endocrine, ischemia, carcinogenesis. Antioxidant, anti-inflammatory and immunomodulatory functions.	Okwu and Ezenagu [30]

(Continued)

Table 20.1 (Continued)

SL. no.	Name of the plant CN: Common name F: Family	Geographical distribution	Phytochemistry	Other medicinal uses	References
7)	***Sarcococca saligna*** CN: Piruli, Geru F: *Buxaceae*	India, Nepal, Myanmar, Afghanistan, Sri Lanka	Leaves, stem Pachyaximine-A Sarcosine, saracodine, salignenamide-A, B, salignarine-C	Gastrointestinal, respiratory disorders, liver diseases, fever, inflammation, rheumatism, syphilis, Alzheimer's disease. Antibacterial, antitumor, antiulcer, acetylcholine esterase inhibitory, hypotensive, ganglion-blocking, antileishmanial, antiplasmodial properties.	Sahu and Arya [31]
8)	***Eucalyptus globules*** CN: southern blue gum F: *Myrtaceae*	Origin: Tasmania, Southeastern Australia	Barks, roots, fruits, buds, leaves Quinones, saponins, phenols, carbohydrates, tannins, flavonoids, terpenoids, reducing sugars	Pain, pyrexia, hypertension, kidney stone, and sunburn diseases. Antimalarial, antiseptic, astringent, balsamic, hypoglycemic, rubefacient, antispasmodic, antiviral, decongestant, deodorant, expectorant, febrifuge, stimulant action.	Mishra et al. [32]
9)	***Allium cepa*** CN: Onion F: *Liliaceae*	Origin: Central Asia (between Turkmenistan and Afghanistan) Found all over the world	Bulb Quercetin, alkaloids, tannins, saponins, carbohydrates, cardiac glycosides, proteins	Stomach upset, gastrointestinal disorders, cataracts, cardiovascular diseases, cancer, appetizer. Anti-inflammatory, anthelmintic, hypoglycemic, antiseptic, antispasmodic, expectorant, diuretic, carminative, febrifuge, hypotensive properties.	Arora et al. [33]
10)	***Aloe zebrina*** CN: Zebra leaf aloe, spotted aloe F: *Asphodelaceae*	Northern South Africa, Namibia, Zimbabwe.	Stem, leaf bases, bitter juice Glycosides, anthrones, lectins	Infections, internal parasites, digestive ailments, wound, psoriasis Skin healing, disinfectant for wounds, worms expelling properties	Grace et al. [34]

11)	***Anethum graveolens*** CN: Dill F: *Apiaceae*	Asia, southern Europe, Italy	Seeds, leaves, and roots α-phellandrene, limonene, dill ether, myristicin	Carminative, stomachic, and diuretic	Jana and Shekhawat [35]
12)	***Apium graveolens*** CN: celery F: *Apiaceae*	Europe and parts of Asia	Stems, leaves, and root Flavonoids, steroids, glycosides, alkaloids, phenols, furocoumarins, volatile oils, fatty acids	Gastrointestinal, cardiovascular, cytotoxic, antimicrobial, anthelmintic, hypolipidemic, anti-inflammatory, central nervous effects	Al-Snafi [36]
13)	***Azadirachta indica*** CN: Neem F: *Meliaceae*	India, Pakistan, Nepal, Sri Lanka, Bangladesh, Maldives, Southern Iran	All parts of the tree Nimbidin Glycosides, steroids, phenols, terpenoids, alkaloids, flavonoids, tannins.	Eczema, arsenical dermatitis, burn ulcers, furunculosis, scabies and seborrheic dermatitis, sore throats, diarrhea, dysentery, hemorrhaging, fatigue, skin ulcers, gangrenous wounds. Cytotoxic effect, analgesic effects, anti-inflammatory effects, anti-bacterial, anti-diarrheal, hemostatic, anti-hemorrhoidal, anti-inflammatory, astringent	Emran et al. [37]
14)	***Bambusa. multiplex*** CN: Hedge Bamboo F: *Poaceae or Gramineae*	Origin: China Nepal, Bhutan, Assam, Sri Lanka, Indo-China, southern China, Africa, Asia, Central & South America	Leaves, young shoots Phenols and flavonoids which have an antimicrobial activity. Flavonoids found in bamboo extracts are flavonides, such as orientin, homoorientin, vitexin	Anti-free-radicals, antioxidant, antiaging, antifungal, insecticide, lipid-regulating and cardio-cerebro-vascular protection.	Ogita and Sasamoto [38] and Valentino et al. [39]

(*Continued*)

Table 20.1 (Continued)

SL. no.	Name of the plant CN: Common name F: Family	Geographical distribution	Phytochemistry	Other medicinal uses	References
15)	***Brassica compestris*** CN: Mustard F: *Brassicaceae*	Mediterranean region Cosmopolitan distribution	Root, stem, leaves, flowers leaves, florets, and seeds Phenolics Caffeic, ferulic, sophoroside-glucosides, sinapic acids, kaempferol, ketoglutaric, malic, shikimic acids.	Antioxidant, radical scavenging capacity, cytotoxic activity	Jing et al. [40]
16)	***Cassia alata*** ***Senna alata*** CN: winged senna/ emperor's candlesticks/ ringworm bush F: *Caesalpinioideae*	Mexico, West Indies, Paraguay, Southeast Asia	Leaves, flowers, roots, stem Yellow chrysophanic acid, anthraquinone	Ringworm, skin problems Laxative or purgative, fungicidal properties.	Singh et al. [41]
17)	***Cinnamon zeylanicum*** CN: Cinnamon F: *Laureaceae*	Origin: Indian subcontinent, India, Nepal, Bhutan, Pakistan, Sri Lanka	Bark Cinnamaldehyde and eugenol	Dyspeptic conditions, flatulence, loss of appetite, abdominal pain with diarrhea, inflammation of the eye, leukorrhea, vaginitis, rheumatism, neuralgia, wounds, and toothache	Manosi et al. [42]
18)	***Clausena anisata*** CN: Horsewood F: *Rutaceae or Citrus family*	India, Nepal, Sri Lanka, and Africa, Western Ghats, Eastern Cape, South Africa	Roots, Crushed leaves Phenolics, alkaloids, flavonoids, saponins, Proanthocyanidins and tannin	Indigestion, mouth infections, otitis abscesses, burns, hemorrhoids, rheumatism, general body pains. Antiseptic, analgesic, open wounds, chronic cough, tuberculosis.	Lawal et al. [43]

19)	***Conyza dioscoridis*** CN: Burnoof F: *Asteraceae*	Egypt	Leaf, flower, and root Isorhamnetin, sorhamnetin, rutinoside, rhamanetin, epicatechin	Rheumatic pain, diarrhea. Antimicrobial, antioxidant, and anti-inflammatory activities	Aziz Balah [44] and Zalabani et al. [45]
20)	***Erythrina lysistemon*** CN: common coral tree, lucky bean tree F: *Fabaceae*	North West Province, Mpumalanga, Limpopo, Swaziland, KwaZulu-Natal, Eastern Cape.	Leaves, flowers, stem, bark Alkaloids Benzenoids, phenylpropanoids	Radical scavenging ability, antifungal activity, antibacterial activity.	Juma and Majinda [46]
21)	***Eugenia aromatica*** CN: Clove F: Myrtaceae	The Moluccas. Myanmar.	Undeveloped flowers bud Volatile oil, gallotannic acid; oleanolic acid lactone, β-sitosterol, nigricin, flavaellagic acid, caryophyllin	Nausea, emesis, flatulence, languid indigestion, dyspepsia Carminative, germicide, astringent, antiseptic, local anesthetic, expectorant	Begum et al. [47]
22)	***Lactuca sativa*** CN: Lettuce F: *Asteraceae*	Eurasia, North America. Throughout most of the world.	Seed oil Oleic, stearic, palmitic, myristic, cis-palmitoleic, behenic and lignoseric acids. β-Sitosterol and β-amyrin	Sedative effect, hypnotic effect, analgesic effect, anticonvulsant activity.	Said et al. [48]
23)	***Lavender augustifolia*** CN: England lavender F: *Lamiaceae*	Origin: France Italy and Spain, Europe, Africa, southwest Asia to southeast India.	Flower spikes 1,8-Cineole, borneol, and camphor, linalool, lavandulyl acetate, and linalyl acetate	Anticonvulsant, anxiolytic, antioxidant, anti-inflammatory, antimicrobial activities.	Kirimer et al. [49] and Cardia et al. [50]
24)	***Matricaria chamomilla*** CN: Chamomile F: *Asteraceae*	Europe, Asia, North America, South America, Australia.	Flowers Herniarin, umbelliferone Sesquiterpenes, flavonoids, coumarins, and polyacetylenes	Sleeplessness, anxiety, flatulence, intermittent fever, skin condition, mouth ulcers resulting from cancer treatment Anti-inflammatory properties antibacterial and fungicidal.	Zhao et al. [51]

(*Continued*)

Table 20.1 (Continued)

SL. no.	Name of the plant CN: Common name F: Family	Geographical distribution	Phytochemistry	Other medicinal uses	References
25)	***Nigella sativa*** CN: Nigella black cumin, kalojeere F: *Ranunculaceae*	Northern India, Bangladesh, East Africa, Europe, North America, South-East Asia	Seeds Thymoquinone Flavonoids, anthocyanins, tannins, alkaloids, linoleic, oleic acid.	Amenorrhea, anorexia, dyspepsia, flatulence, jaundice, dysentery, diarrhea, hydrophobia, ascites, cough, fever, paralysis. Appetizer, stimulant, diuretic, anodyne, deodorant emmenagogue, galactagogue, anthelmintic, thermogenic, carminative, digestive, constipating, sudorific, febrifuge, expectorant, purgative, immunomodulatory, anti-inflammatory, spasmolytic, abortifacient, local anesthetic, analgesic, antimicrobial, antidiabetic, anticancer properties	Desai et al. [52]
26)	***Pimpinella anisum*** CN: anise anis, Aniseed F: Umbelliferae/ Apiaceae	Eastern Mediterranean Region, Southwest and West Asia, Egypt, Spain	Seeds Petroselinic acid Naringin, chlorogenic acid and rosmarinic acid Phenolic, flavonoid, and tannin	Flatulence, dry cough, infant catarrh food additive natural antioxidant, antiseptic	Bettaieb Rebey et al. [53]
27)	***Piper betle*** CN: Betel leaf F: *Piperaceae*	Origin: Malaysia South and Southwest China. India, Sri Lanka, Malaysia, Thailand, Taiwan, Southeast Asian countries	Leaves Chavicol, cyneole, alkaloid, sugar, tannins, sesquiterpene	Indigestion, bronchitis, constipation, congestion, gum hardener, conservation of teeth. Carminative, digestive, appetizer, expectorant, stimulant Antioxidative, antiprotozoan, antifungal, cytotoxic, antibacterial, aphrodisiac activities	Shah et al. [54]

28)	***Podocarpus nivalis*** CN: New Zealand gymnosperms F: *Podocarpaceae*	North Island of New Zealand, Australia, China, Japan	Fruit Terpenoids, norditerpenoid dilactones, biflavonoids	Fevers, asthma, coughs, cholera Anti-inflammatory, antioxidant, antitumor, antimicrobial, larvicidal, plant and insect growth regulation functions.	Abdillahi et al. [55]
29)	***Punica gratum*** CN: Pomegranate F: *Lythraceae*	Asia, particularly to Iran, Afghanistan and the Himalayas. Throughout the tropics and subtropics.	Seed, flower, leaves Ellagitannins, gallotannins, punicacortin D, punicalin	Kidney stone, bleeding of kidney, irritable condition of bladder inflammation, painful urination, burning sensation, problem in urine discharge. diarrhea, dysentery, hyperacidity, piles, sterility, cough. Antibacterial, antioxidant, anti-arthrogenic, anti-inflammatory, antitumor, hepatoprotective, antidiarrheal properties.	Jayaprakash and Sangeetha [56]
30)	***Raphanus sativus*** CN: Radish F: *Brassicaceae*	Eeastern Mediterranean and the Caspian Sea	Mature radish taproot Isothiocyanates, flavonoids, phenolics, anthocyanidins, glucosinolates, brassinosteroids	Obesity, diabetes, constipation, chronic cough, tracheitis, hypertension, constipation. Antimicrobial, antiviral properties	Sham et al. [57]
31)	***Rosmarinus officinalis*** CN: Rosemary F: *Lamiaceae*	Origin: Mediterranean region, North Africa and Europe, Java and Philippines South America, global distribution	Leaves, roots, stems, flowers Rosmarinic acid, carnosic acid, apigenin, diosmin, luteolin, genkwanina	Dyspepsia, circulation problems, minor wounds, rashes, headache, colds, expectorant, diuretic and antispasmodic in renal colic, rheumatism, muscles and joints pain. Antibacterial, antidiabetic, anti-inflammatory, antitumor, antioxidant properties	Andrade et al. [58]

(Continued)

Table 20.1 (Continued)

SL. no.	Name of the plant CN: Common name F: Family	Geographical distribution	Phytochemistry	Other medicinal uses	References
32)	***Sesamum indicum*** CN: Sesame F: *Pedaliaceae*	Africa, India. Tropical regions around the world	Edible seeds, which grow in pods, roots Anthrasesamones Polyphenols, alkaloids, flavanoids, terpenoids, and glycosides.	Body pains, neuralgia, old-age disease, paralysis, gout, rheumatism, piles, dysentery, urinary diseases Tonic, diuretic, lactagogue, poultice	Shah [59]
33)	***Spirostachys africana*** CN: Tamboti, African mahogany tree F: *Euphorbiaceae*.	South Africa, North Africa.	Fresh latex, bark, whole plant Beyerene derivatives, diosphenols, ketols, stachynones	Painkiller in toothache, stomach pains, rhinitis, painful eyes	Maroyi [60]
34)	***Tadehagi triquetrum*** CN: Trefle Gros F: *Fabaceae*	South Asian, East Asian, and Southeast Asian countries. tropical, subtropical, and Pacific regions of the world	Root, leaves Isoflavones, triquetrumones A B, C, biisoflavanone, (*R*)-triquetrumone	Chronic coughs and tuberculosis. Hypoglycemic activity, immunoprotection effects, antioxygenic properties, anti-inflammatory activity	Aye et al. [61]
35)	***Thevetia peruviana*** CN: Yellow oleander, Pila Kanher. F: *Apocynaceae*	Origin: Central & South America	Flower, leaf, seed bark, kernals Thevetin A and B (Cerebroside), peruvoside, neriifolin, thevetoxin, ruvoside	Astringent to bowels, urethral discharge, worms, skin diseases, leucoderma, wounds, piles, eye trouble, itching, fever, bronchitis, rheumatism, dropsy, heart diseases. Anti-tumor, abortifacient, purgative functions.	Ghavri and Adhav [62]

36)	***Trigonella foenum-graecum*** CN: Fenugreek F: *Fabaceae*	Eastern Mediterranean, western Asia, India	Seeds Trigonellin, cholin Mucilage, sugars, alkaloid	Allergies, ulcers, uterine problems, loss of appetite, bronchial catarrh, bronchitis, cancer, swollen eyes, high cholesterol, body odor, fevers, gas, gastric disorders, lung infections, mucus excessive, sore throat, abscesses, anemia, asthma, boils Antidiabetic, anticancer, antifungal, antipyretic, antibacterial, antioxidant activities	Yadav and Kaushik [63]
37)	***Uraria crinita*** CN: Blue lady orchid F: *Leguminosae*	India and southern China throughout Southeast Asia	Root, whole plant, fruit, seed, leaf Genistein Triterpene glycosides, 24-deoxyoxytrogenin 3-*O*-α-L-rhamnopyranosyl (1 → 2) [β-D-glucopyranosyl]-β-D-galactopyranosyl	Treat bone dysplasia in children, dysentery, diarrhea, enlarged spleen and liver, tumors or stopping bleeding, fever, coughs, intestinal worms and other parasites, pustules and fistulae Carminative for children. Antioxidant activity	Tropical Plants Database, Ken Fern [64]
38)	***Chrysopogon zizanioides/ Vetiveria zizanioides*** CN: vetiver grass or Khus grass F: *Poaceae*	Northern India, Bangladesh, Burma (Myanmar), South-East Asia.	Roots Khusimol, α-vetivone, β-vetivone Sesquiterpenols, sesquiterpenones.	Ringworm, indigestion and loss of appetite Insecticidal, antibacterial, antifungal, antioxidant, herbicidal, antinociceptive and anti-inflammatory activities	Chahal et al. [65]

(*Continued*)

Table 20.1 (Continued)

SL. no.	Name of the plant CN: Common name F: Family	Geographical distribution	Phytochemistry	Other medicinal uses	References
39)	***Chenopodium ambrosiodes/ Dysphania ambrosioides*** CN: American wormseed F: *Chenopodiaceae*	Origin: Central and South America Europe, Africa, Asia	Leaf, bud seed, whole plant Chenopodium oil Saponins, fatty acids, glycosides	Pain-killers, pulmonary troubles, laxatives Vermifuges, anus, hemorrhoids, fabrifuges, dropsy, swellings, edema, indigestion, coughs, tuberculosis, fever, sores, abscesses, fevers, headache, stomach pains. Antibiotic, bacteriostatic, febrifuge, fungistatic, anthelmintic, vermifuge, dysenteric amoeba	Sousa et al. [66]
40)	***Thymus vulgaris*** CN: Thyme, garden thyme F: Lamiaceae, mint family	Origin: southern Europe Western Mediterranean, Southern Italy.	Alkaloids, steroids, tannins, and phenol compounds Geraniol, terpineol, linalool, carvacrol, thymol	Rheumatism, insect bites, pains, muscle swelling. Anti-inflammatory, immunomodulatory, antioxidant, antibacterial, and antifungal properties.	Hosseinzadeh et al. [67]

20.5 Conclusion

This chapter summarizes the role of phytochemicals in the management of scabies and myiasis at individual and population levels. The presence of alkaloids, saponins, glycoside, and flavonoids in plants elicit toxic effect to scabitic mites and dipterian flies for control of scabies and myiasis, respectively. These phytochemicals generally distort and kill the larvae and adult stages of these ectoparasites when used at high concentrations. They also prevent reinfestation. These effects are species-specific, and the plants (or extracts) are easily degradable and less susceptible to resistance when compared with current synthetic chemicals/drugs in practice. Whereas the scope for combination of these phytochemicals and current synthetic drugs as a regimen for treatment of ectoparasitic diseases is continuing to gain acceptance worldwide, it is believed that these potential plant compounds will also facilitate the development of newer drugs for control of insects, and can be a valuable part of insect management programmes for scabies and myiasis.

References

1 Diaz, J.H. (2015). Introduction to ectoparasitic diseases. In: *Mandell, Douglas, and Bennett's Principles and Practice of Infectious Diseases* (eds. J.E. Bennett, R. Dolin and M.J. Blaser), 3243–3245. WB Saunders.

2 Chandler, D.J. and Fuller, L.C. (2019). A review of scabies: an infestation more than skin deep. *Dermatology* 235 (2): 79–90.

3 Gates, R.H. (2003). *Infectious Disease Secrets*, 2e, 355. Philadelphia: Elsevier, Hanley Belfus. ISBN: 978-1-56053-543-0.

4 Andrews, R.M., McCarthy, J., Carapetis, J.R., and Currie, B.J. (2009). Skin disorders, including pyoderma, scabies, and tinea infections. *Pediatric Clinics* 56 (6): 1421–1440.

5 Vos, T., Allen, C., Arora, M. et al. (2016). Global, regional, and national incidence, prevalence, and years lived with disability for 310 diseases and injuries, 1990–2015: a systematic analysis for the Global Burden of Disease Study 2015. *The Lancet* 388 (10053): 1545–1602.

6 Vos, T., Flaxman, A.D., Naghavi, M. et al. (2012). Years lived with disability (YLDs) for 1160 sequelae of 289 diseases and injuries 1990–2010: a systematic analysis for the Global Burden of Disease Study 2010. *The Lancet* 380 (9859): 2163–2196.

7 Georgis, B.D.D. (2014). *Parasitology for Veterinarians*, 10e, 68. Elsevier Health Sciences. ISBN: 9781455739882.

8 Burgess, I.F. (2003). Myiasis: maggot infestation. *Journal of Bone and Joint Surgery* 13: 438–475.

9 Otranto, D. (2001). The immunology of myiasis: parasite survival and host defense strategies. *TRENDS in Parasitology* 17 (4): 176–182.

10 John, D.T. and Petri, W.A. Jr. (2006). *Markell and Voge's Medical Parasitology*, vol. 9, 178–181. Missouri: St. Louis, Saunders Elsevier.

11 Karthikeyan, K. (2005). Treatment of scabies: newer perspectives. *Postgraduate Medical Journal* 81 (951): 7–11.

12 Vañó-Galván, S.E. and Moreno-Martin, P. (2008). Generalized pruritus after a beach vacation. Diagnosis: scabies. *Cleveland Clinic Journal of Medicine* 75 (7): 474–478.

13 Akram, M., Riaz, M., Noreen, S. et al. (2020). Therapeutic potential of medicinal plants for the management of scabies. *Dermatologic Therapy* 33 (1): e13186. https://doi.org/10.1111/dth.13186.

14 Kumalasari ML, Marwing A, Kusumawati E, Mustika I, Lusiana N, Widayanti LP, Andiarna F, Andyarini EN, Hidayati S. Development of extract of herbal plants as anti-scabies in Pesantren. Conference Paper. AICIS 2019, 1–4 October, Jakarta, Indonesia. doi:10.4108/eai.1-10-2019.2291663.

15 Walker, G.J. and Johnstone, P. (2000). Interventions for treating scabies. *Cochrane Database of Systematic Reviews* (3): CD000320. https://doi.org/10.1002/14651858.CD000320.pub2. PMC 6532717. PMID 17636630.

16 Hay, R.J. (2009). Scabies and pyodermas-diagnosis and treatment. *Dermatologic Therapy* 22 (6): 466–474.

17 Chinaza, G.A., Chinelo, K.E., Obinna, C.A., et al. (2020). Medicinal plant phytochemicals: the biochemistry and uses of the pharmacologically active alkaloids, terpenes, polyphenols, and glycosides. Agro and food processing for wealth creation – the Nigerian experience Conference Paper Presented at the 44th NIFST Conference Themed "Agro and Food Processing for Wealth Creation – The Nigerian Experience", October 2020. Abstract No.: AB 013.

18 Smith, S.M. (2015). Treating infestations of the human botfly, *Dermatobia hominis*. *The Lancet Infectious Diseases* 15 (5): 512.

19 Rattan, R.S. (2010). Mechanism of action of insecticidal secondary metabolites of plant origin. *Crop Protection* 29 (9): 913–920.

20 Jiang, H., Wang, J., Song, L. et al. (2016). GC× GC-TOFMS analysis of essential oils composition from leaves, twigs and seeds of *Cinnamomum camphora* L. Presl and their insecticidal and repellent activities. *Molecules* 21 (4): 423.

21 Chaaban, A., Gomes, E.N., Santos, V.M. et al. (2017). Essential oils for myiasis control: potentialities for ecofriendly insecticides. *European Journal of Medicinal Plants*: 1–25.

22 Awuchi, C.G. (2019). Medicinal plants: the medical, food, and nutritional biochemistry and uses. *International Journal of Advanced Academic Research* 5 (11): 220–241.

23 Simas, N.K., Lima, E.D., Conceição, S.D. et al. (2004). Produtos naturais para o controle da transmissão da dengue: atividade larvicida de *Myroxylon balsamum* (óleo vermelho) e de terpenóides e fenilpropanóides. *Química Nova* 27 (1): 46–49.
24 Dey, A. (2011). *Achyranthes aspera* L: phytochemical and pharmacological aspects. *International Journal of Pharmaceutical Sciences Review and Research* 9 (2): 72–82.
25 Audu, B.S., Ofojekwu, P.C., Ujah, A., and Ajima, M.N. (2014). Phytochemical, proximate composition, amino acid profile and characterization of Marijuana (*Cannabis sativa* L.). *The Journal of Phytopharmacology* 3 (1): 35–43.
26 Mawa, S., Husain, K., and Jantan, I. (2013). *Ficus carica* L.(*Moraceae*): phytochemistry, traditional uses and biological activities. *Evidence-Based Complementary and Alternative Medicine* 2013, Article ID 974256, 8 pages. doi:https://doi.org/10.1155/2013/974256.
27 Sachan, N.K. and Kumar, Y. (2010). Phytochemical investigation of *Ficus racemosa* bark – an ethanomedicinal plant. *Annals of Pharmacology and Pharmaceutical Sciences* 2: 80–84.
28 Yadav, R.K., Nandy, B.C., Maity, S. et al. (2015). Phytochemistry, pharmacology, toxicology, and clinical trial of *Ficus racemosa*. *Pharmacognosy Reviews* 9 (17): 73.
29 Khaliq, H.A. (2017). A review of pharmacognostic, physicochemical, phytochemical and pharmacological studies on *Ficus bengalensis* L. *Journal of Scientific and Innovative Research* 6 (4): 151–163.
30 Okwu, D.E. and Ezenagu, V. (2008). Evaluation of the phytochemical composition of mango (*Mangifera indica* Linn) stem bark and leaves. *International Journal of Chemical Sciences* 6 (2): 705–716.
31 Sahu, N. and Arya, K.R. (2017). Ethnobotanical and ethnopharmacological activities of *Artemisia nilagirica, Lyonia ovalifolia, Sarcococca saligna and Taraxacum officinale. International Journal of Pharmaceutical Sciences and Research* 8 (11): 4818–4825.
32 Mishra, A.K., Sahu, N., Mishra, A. et al. (2010). Phytochemical screening and antioxidant activity of essential oil of Eucalyptus leaf. *Pharmacognosy Journal* 2 (16): 25–28.
33 Arora, E.K., Sharma, V., Khurana, A. et al. (2017). Phytochemical analysis and evaluation of antioxidant potential of ethanol extract of *Allium cepa* and ultra-high homoeopathic dilutions available in the market: a comparative study. *Indian Journal of Research in Homoeopathy* 11: 88–96.
34 Grace, O.M., Simmonds, M.S., Smith, G.F., and Van Wyk, A.E. (2008). Therapeutic uses of Aloe L. (*Asphodelaceae*) in southern Africa. *Journal of Ethnopharmacology* 119 (3): 604–614.
35 Jana, S. and Shekhawat, G.S. (2010). *Anethum graveolens*: an Indian traditional medicinal herb and spice. *Pharmacognosy Reviews* 4 (8): 179.

36 Al-Snafi, A.E. (2014). The pharmacology of *Apium graveolens* – a review. *International Journal for Pharmaceutical Research Scholars* 3 (1): 671–677.

37 Emran, T.B., Nasir Uddin, M.M., Rahman, A. et al. (2015). Phytochemical, antimicrobial, cytotoxic, analgesic and anti-inflammatory properties of *Azadirachta indica*: a therapeutic study. *Journal of Bioanalysis and Biomedicine* 12: 2.

38 Ogita, S. and Sasamoto, H. (2017). *in vitro* bioassay of allelopathy in four bamboo species; *Bambusa multiplex, Phyllostachys bambusoides, P. nigra, Sasa kurilensis,* using sandwich method and protoplast co-culture method with digital image analysis. *American Journal of Plant Sciences* 8 (7): 1699–1710.

39 Valentino, M.J., Ganado, L.S., Gando, M.R., and Undam, J.R. (2015). Phytochemical screening and bioassay of the anti-microbial activity of three species of bamboo in Nueva Ecija, Philippines. *Advances in Environmental Biology* 9 (24): 389–396.

40 Jing, W.G., Wang, Z.M., Zhao, Y. et al. (2014). Chemical constituents from seeds of *Brassica campestris*. Zhongguo Zhong yao za zhi=Zhongguo zhongyao zazhi=. *China Journal of Chinese Materia Medica* 39 (13): 2521–2525.

41 Singh, S., Singh, S.K., and Yadav, A. (2013). A review on *Cassia* species: pharmacological, traditional and medicinal aspects in various countries. *American Journal of Phytomedicine and Clinical Therapeutics* 1 (3): 291–312.

42 Manosi, D., Suvra, M., Budhimanta, M., and Jayram, H. (2013). Ethnobotany, phytochemical and pharmacological aspects of *Cinnamomum zeylanicum* blume. *International Research Journal of Pharmacy* 4 (4): 58–63.

43 Lawal, I.O., Grierson, D.S., and Afolayan, A.J. (2015). Phytochemical and antioxidant investigations of a *Clausena anisata* hook, a South African medicinal plant. *African Journal of Traditional, Complementary and Alternative Medicines* 12 (1): 28–37.

44 Aziz Balah, M.A. (2016). Chemical and biological characterization of *Conyza dioscoridis* (L.) desf. family (*Compositae*) in some perennial weeds control. *South African Journal of Botany* 103: 268–274.

45 Zalabani, S.M., Hetta, M.H., and Ismail, A.S. (2013). Anti-inflammatory and Antimicrobial activity of the different *Conyza dioscoridis* L. Desf. Organs. *Biosafety* 2 (1): 1–3. https://doi.org/10.4172/2167-0331.1000106.

46 Juma, B.F. and Majinda, R.R. (2006). Constituents of *Erythrina lysistemon*: their brine shrimp lethality, antimicrobial and radical scavenging activities. *Natural Product Communications* 1 (2): 106–107.

47 Begum, S., Siddiqui, B.S., Khatoon, R., and Aftab, F. (2014). Phytochemical studies on *Syzygium aromaticum* Linn. *Journal of the Chemical Society of Pakisthan* 36 (3): 512.

48 Said, S.A., Kashef, H.E., Mazar, M.E., and Salama, O. Phytochemical and pharmacological studies on *Lactuca sativa* seed oil. *Fitoterapia* 67 (3): 215–219.

49 Kirimer, N., Mokhtarzadeh, S., Demirci, B. et al. (2017). Phytochemical profiling of volatile components of *Lavandula angustifolia* Miller propagated under *in vitro* conditions. *Industrial Crops and Products* 96: 120–125.

50 Cardia, G.F., Silva-Filho, S.E., Silva, E.L. et al. (2018). Effect of lavender (*Lavandula angustifolia*) essential oil on acute inflammatory response. *Evidence-Based Complementary and Alternative Medicine* 2018, Article ID 1413940, 10 pages. doi:https://doi.org/10.1155/2018/1413940.

51 Zhao, J., Avonto, C., Wang, M. et al. (2012). Phytochemical investigation of chamomile. *Planta Medica* 78 (05): P_50.

52 Desai, S.D., Saheb, S.H., Das, K.K., and Haseena, S. (2015). Phytochemical analysis of *Nigella sativa* and it's antidiabetic effect. *Journal of Pharmaceutical Sciences and Research* 7 (8): 527.

53 Bettaieb Rebey, I., Bourgou, S., Aidi Wannes, W. et al. (2018). Comparative assessment of phytochemical profiles and antioxidant properties of Tunisian and Egyptian anise (*Pimpinella anisum* L.) seeds. *Plant Biosystems – An International Journal Dealing with All Aspects of Plant Biology* 152 (5): 971–978.

54 Shah, N.C. (2016). *Sesamum indicum* (sesame or til): seeds and oil – a historical and scientific evaluation from Indian perspective. *Asian Agri-History* 20 (1): 3–19.

55 Abdillahi, H.S., Stafford, G.I., Finnie, J.F., and Van Staden, J. (2010). Ethnobotany, phytochemistry and pharmacology of *Podocarpus sensu latissimo* (sl). *South African Journal of Botany* 76 (1): 1–24.

56 Jayaprakash, A. and Sangeetha, R. (2015). Phytochemical screening of *Punica granatum* Linn. peel extracts. *Journal of Academia and Industrial Rresearch* 4 (5): 160–162.

57 Sham, T.T., Yuen, A.C., Ng, Y.F. et al. (2013). A review of the phytochemistry and pharmacological activities of *raphani semen*. *Evidence-based Complementary and Alternative Medicine* 2013, Article ID 636194, 16 pages. doi:https://doi.org/10.1155/2013/636194.

58 Andrade, J.M., Faustino, C., Garcia, C. et al. (2018). *Rosmarinus officinalis* L.: an update review of its phytochemistry and biological activity. *Future Science OA* 4 (4): FSO283.

59 Shah, S.K., Garg, G., Jhade, D., and Patel, N. (2016). *Piper betle*: phytochemical, pharmacological and nutritional value in health management. *International Journal of Pharmaceutical Sciences Review and Research* 38: 181–189.

60 Maroyi, A. (2013). Traditional use of medicinal plants in south-central Zimbabwe: review and perspectives. *Journal of Ethnobiology and Ethnomedicine* 9 (1): 31.

61 Aye, M.M., Aung, H.T., Sein, M.M., and Armijos, C. (2019). A review on the phytochemistry, medicinal properties and pharmacological activities of 15 selected Myanmar medicinal plants. *Molecules* 24 (2): 293.

62 Ghavri, K. and Adhav, M. (2018). Comparative phytochemical screening of morphotypes of *Thevetia peruviana* (Pers.). *Journal of Pharmacognosy and Phytochemistry* 7 (5): 539–541.

63 Rashmi, Y. and Rahul, K. (2011). A study of phytochemical constituents and pharmacological actions of *Trigonella foenum-graecum*: a review. *International Journal of Pharmacy and Technology* 3 (2): 1022–1028.

64 Tropical Plants Database (2019). Ken Fern. tropical.theferns.info. 2019-10-31. tropical.theferns.info/viewtropical.php?id=Uraria+crinita (accessed 9 September 2020).

65 Chahal, K.K., Kaushal, S., and Sandhu, A.K. (2015). Chemical composition and biological properties of *Chrysopogon zizanioides* (L.) *Roberty syn. Vetiveria zizanioides* (L.) Nash—a review. *Indian Journal of Natural Products and Resources (IJNPR) [Formerly Natural Product Radiance (NPR)]* 6 (4): 251–260.

66 Sousa, Z.L., de Oliveira, F.F., da Conceição, A.O. et al. (2012). Biological activities of extracts from *Chenopodium ambrosioides Lineu* and *Kielmeyera neglecta Saddi*. *Annals of Clinical Microbiology and Antimicrobials* 11 (1): 20.

67 Hosseinzadeh, S., Kukhdan, A.J., Hosseini, A., and Armand, R. (2015). The application of *Thymus vulgaris* in traditional and modern medicine: a review. *Global Journal of Pharmacology* 9: 260–266.

Part VII

Non Classified NTDs: Malaria and Human Tuberculosis

21

Malaria and Recent Developments in Antimalarial Drugs

Mithun Rudrapal[1] *and Dipak Chetia*[2]

[1] *Department of Pharmaceutical Chemistry, Rasiklal M. Dhariwal Institute of Pharmaceutical Education and Research, Chinchwad, Pune, Maharashtra, India*
[2] *Department of Pharmaceutical Sciences, Dibrugarh University, Dibrugarh, Assam, India*

List of Abbreviations

ACTs	Artemisinin-based combination therapies
AM	Artemether
AQ	Amodiaquine
ART	Artemisinin
AS	Artesunate
ATQ	Atovaquone
BSS	Blood-stage schizonticide
CDCP	Centers for Disease Control and Prevention
CQ	Chloroquine
CQ-R Pf	Chloroquine-resistant *P. falciparum*
DHA	Dihydroartemisinin
DOX	Doxycycline
DV/FV	Digestive vacuole/food vacuole
Hb	Hemoglobin
HL	Halofantrine
LUM	Lumfantrine
MDR	Multidrug resistant
MQ	Mefloquine
Pf	*Plasmodium falciparum*
PG	Proguanil
PIP	Piperaquine

Neglected Tropical Diseases and Phytochemicals in Drug Discovery, First Edition.
Edited by Chukwuebuka Egbuna, Muhammad Akram, and Jonathan Chinenye Ifemeje.

PQ	Primaquine
PYN	Pyronaridine
QN	Quinine
SBDD	Structure-based drug design
SP	Sulfadoxine-pyrimethamine
TET	Tetracycline
TSS	Tissue-stage schizonticide
WHO	World Health Organization

21.1 Introduction

Malaria remains to be one of the most common infectious diseases and a serious public-health problem across the globe. It is a growing endemic disease that is widespread, particularly in tropical and subtropical parts of the world. In spite of all efforts to eradicate and control malaria by World Health Organization (WHO), it is still prevalent and also a leading cause of morbidity and mortality worldwide.

Historically, malaria is one of the oldest diseases known to mankind. History of malaria is as ancient as the history of civilization and the history of mankind itself. Mentions of malaria disease can be found in ancient Chinese, Indian, and Egyptian manuscripts. Historians and archaeologists found evidence of existence of malaria fever in the Xia Dynasty and Medieval Europe. "Malaria-like febrile illness" (with names like "ague" or "paludism") was described since Hippocrates, the Greek physician, in the fifth century BC, as unique periodic fevers associated with marshes and swamps. He was the first person who describes the manifestations of malaria disease. The linguistic origin of the word "malaria" stems from Medieval Italian: "mala aria" meaning "bad air," since it was believed that the disease was manifested as intermittent fever caused by putrid marsh air associated with swamps. The word was introduced to English by Horace Walpole, who wrote about a "horrid thing" called "malaria," in 1740, which comes to Rome every year and kills one. The term "malaria," without the apostrophe, evolved into the name of the disease in the twentieth century. Scientific studies on malaria were made in 1880, when a French army doctor working in Algeria named Charles Louis Alphonse Laveran observed parasites inside the red blood cells (RBCs) of patients suffering from malaria. Laveran also first proposed the notion of protozoan (single-cell parasite) as the cause of malaria disease. He was awarded the Nobel Prize in Physiology or Medicine for his pioneering discovery, in 1907. Thereafter, the protozoan was called *Plasmodium* by the Italian scientists, Ettore Marchiafava and Angelo Celli. Later, Carlos Finlay, a Cuban doctor treating patients with yellow fever in Havana, first proposed that the disease was transmitted to humans by mosquitoes. In 1898, Sir Ronald Ross, a physician in the British army working in India, first demonstrated that the *Anopheles* mosquito was the vector for the disease. His pioneering work on establishing the main features of the parasitic life cycle earned him the Nobel Prize in Medicine in 1902 [1, 2].

21.2 Epidemiology of Malaria

Malaria along with six other diseases viz. diarrhea, HIV/AIDS, tuberculosis, measles, hepatitis B, and pneumonia accounts for about 85% of global infectious disease burden. Malaria afflicts approximately 36% of the worlds' population annually. It infects between 200 and 300 million people, causing between one and three million deaths annually around the world. In 2018, nearly half of the world's population was at risk of malaria, in 91 countries and areas situated mostly in the tropical and/or subtropical regions. Most malaria cases and deaths occur in Sub-Saharan Africa. The WHO regions of Southeast Asia, Eastern Mediterranean, Western Pacific, and the Americas are also at high risk (Figure 21.1). Some population groups are at considerably higher risk of malaria infection and developing severe disease than others. These include infants, children below five years of age, pregnant women, patients with HIV/AIDS, nonimmune migrants, and travelers [2–5].

According to the latest World Malaria Report [5], released in November 2018, it was estimated that there were 216 million cases of malaria and 445 000 deaths in 2016. In 2018, 93% of malaria cases were observed in the WHO African region. Highest malarial deaths occurred in the African region (94%) which is followed by the Southeast Asian region (7%) and the Eastern Mediterranean region (3%).

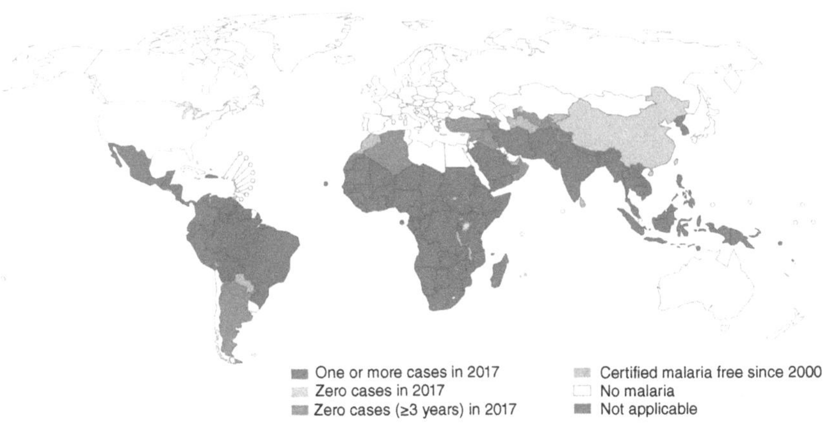

Figure 21.1 Malaria: Global Health Observatory (GHO) data. *Source:* The Global Health Observatory, Malaria, World Health Organization. Retrieved from https://www.who.int/gho/malaria/en/.

About more than two-thirds (67%) of all malaria deaths occur in children below 5 years of age, mostly in Sub-Saharan Africa. Most of these deaths were due to *Plasmodium falciparum*.

In the Southeast Asian region of WHO, out of about 1.4 billion people living in 11 countries, 1.2 billion (85.7%) are exposed to the risk of malaria and most of them live in India. Out of 2.5 million reported cases in the Southeast Asia, India alone contributes about 76% of the total cases. Currently, 80.5% of the 1.4 billion population of India lives in malaria risk areas [6].

21.3 Disease Transmission

In most cases, malaria is transmitted to people through the bites of infected female *Anopheles* mosquito, called mosquito vector. This is the only insect vector that spreads infection of the disease (Figure 21.2). Among approximately 480 known species of *Anopheles*, only about 50 species have been implicated in the transmission of malaria. *Anopheles gambiae* complex is the chief vector in Africa, *Anopheles freeborni* in North America, and *Anopheles culicifacies, Anopheles fluviatilis, Anopheles minimus, Anopheles philippinensis, Anopheles stephensi*, and *Anopheles sundaicus* in the Indian subcontinent [2].

During nighttime, when the female *Anopheles* mosquito bites humans for a blood meal, it inoculates the sporozoites of *Plasmodium* into human blood stream, thus spreading the infection. The blood meal from human host is indispensable for mosquitoes to nourish eggs [2].

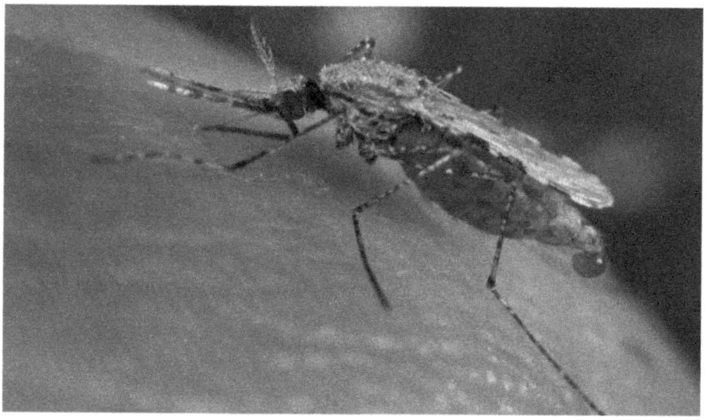

Figure 21.2 An *Anopheles stephensi* mosquito (vector) obtaining a blood meal from a human host through its pointed proboscis. Note: Through the abdomen, blood droplet is expelled after engorging on host blood. *Source:* Adapted from Centers for Disease Control and Prevention. PHIL identification number #5814. Freely taken from: https://phil.cdc.gov/Details.aspx?pid=5814. Retrieved: 20/09/2020.

Malaria can also be transmitted congenitally by blood transfusion (transfusion malaria), from mother to growing fetus (congenital malaria), and through sharing of contaminated needles and transplant of infected organs. Transfusion malaria is fairly common in malaria endemic areas. In the second case, transmission of infection from mother to child (intrauterine) occurs during pregnancy, and also placenta becomes heavily infected with the parasites. Accidental transmission through needle stick injury can occur among drug addicts who share syringes and needles [2, 3].

21.3.1 Malaria Parasites

In humans, malaria is caused by protozoan parasites of the genus *Plasmodium*, belonging to the parasitic phylum *Apicomplexa*. There are four species that infect humans, namely, *P. falciparum*, *Plasmodium vivax, Plasmodium Ovale,* and *Plasmodium malariae*. Recently, *Plasmodium knowlesi*, a primate (monkey) malaria parasite, has also been reported as the fifth causative organism of human malaria. This group of pathogenic *Plasmodium* species is usually referred to as human malaria parasites. The majority of human malaria infections are caused by *P. falciparum* and *P. vivax*, while, *P. ovale*, *P. malariae,* and *P. knowlesi* also infect humans, but relatively less frequently. *P. falciparum* and *P. vivax* are the most widespread among these five species, and cause severe forms of the disease and even deaths. *P. falciparum* and *P. vivax* cause 80% of total malaria cases and more than 90% of deaths worldwide. On the other hand, *P. malariae* and *P. ovale* are less prevalent and cause comparatively less severe, but persistent infections in humans. *P. knowlesi* has been reported to be responsible for severe disease in humans [2, 7–9].

P. falciparum is the most widespread species geographically, especially observed in tropical and subtropical parts of the world. It is found in Africa, South America, and Southeast Asia. It is the most predominant and virulent organism which causes a majority of malaria-related morbidity and mortality. It is the only species that can cause pernicious and potentially fatal malaria. It is accounted for about 90% of all malaria-related deaths due to *P. falciparum* every year worldwide. ***P. vivax*** has the widest distribution around the globe. It is the most widely found species in Latin America, Africa, and Southeast Asia. It is probably the most widespread due to the high population of Asia. Approximately 60% of infections in India are caused by *P. vivax*. It was previously considered benign, but it also causes life-threatening symptoms. ***P. ovale*** is much less widespread, which is mostly found in Ghana, Liberia, Nigeria, and the tropical West African region. It is biologically and morphologically very similar to *P. vivax*. *P. malariae* is found worldwide, but its geographic distribution is irregular. It is located in the tropical and subtropical parts of South and Central America,

Southeast Asia, and Africa. Although *P. malariae* is not lethal, if left untreated, it can cause a chronic infection that can even last a lifetime and also may cause severe nephrotic syndrome. ***P. knowlesi*** is a parasite of Old World monkeys (macaques, *Macaca fascicularis*). It is located in Southeast Asia, and causes severe infections in humans [2, 3].

21.4 Life Cycle of *Plasmodium*, Hemoglobin Digestion, and Drug Targets

21.4.1 Life Cycle of *Plasmodium*

The life cycle of *Plasmodium* parasite is a complex biological phenomenon that involves two hosts: the human and the female *Anopheles* mosquito [2, 10, 11]. The whole life cycle consists of asexual replicating phase in human host that involves human liver (Tissue schizogony or Pre-erythrocytic schizogony) stages, and human blood (Erythrocytic schizogony) stages, and the sexual reproductive phase in mosquito vector (Sporogony). The complete life cycle of Plasmodium and pathogenesis of malaria infection is depicted in Figure 21.3.

21.4.1.1 Human Liver Stage

While ingesting a blood meal, an infected mosquito injects *Plasmodium* sporozoites into a human host (1). Within minutes of inoculation, sporozoites (that are not killed by the immune system of human) reach the human liver (after entering into the blood stream) (2), where they invade hepatocytes, and either lie dormant (as **hypnozoites**) or develop to trophozoites that reproduce by asexual multiplication (or mitosis) over several days (5–16 days) to eventually form tissue schizonts (3) comprising tens of thousands of uninucleated merozoites **(Tissue Schizogony/ Exoerythrocytic schizogony, A)**, which rupture and release merozoites (4) that start the blood-stage infection.

21.4.1.2 Human Blood Stage

Following the initial replication in the liver, after bursting from hepatocytes, the merozoites are released into the blood stream and each merozoite invades erythrocytes (5), where they undergo again asexual multiplication **(Erythrocytic schizogony, B)** to form new merozoites. Infected erythrocytes burst and release merozoites (6) that infect a naïve erythrocyte and begin the erythrocytic cycle again. This repetitive cycle of invasion–multiplication–release–invasion continues and is referred to as the intraerythrocytic cycle of *Plasmodium*. The intraerythrocytic cycle takes about 48 hours in infections with *P. falciparum, P. vivax,* and *P. ovale* and 72 hours in case of *P. malariae* infection. A fraction of the merozoites

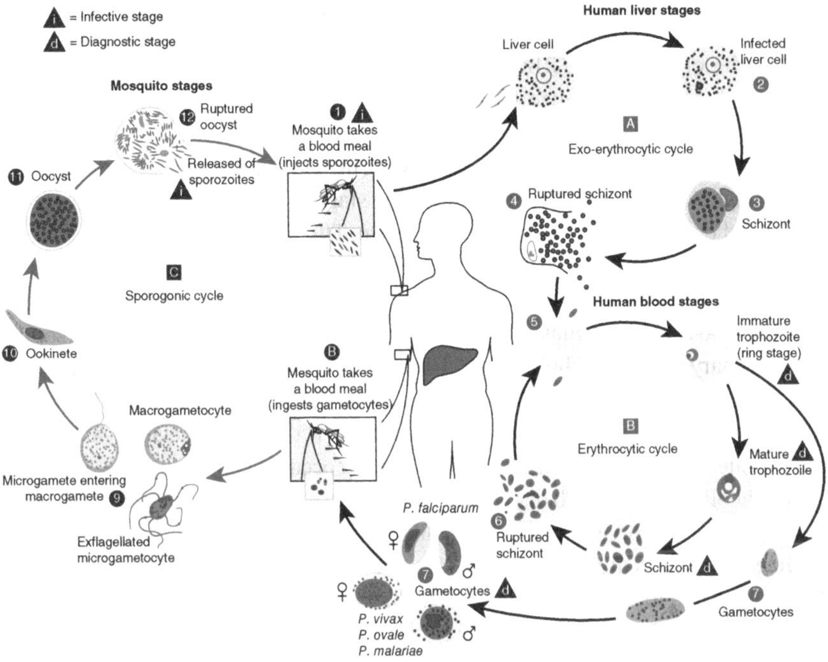

Figure 21.3 Life cycle of *Plasmodium* and pathogenesis of malaria. *Source:* Malaria. Centers for Disease Control and Prevention, 10 September 2020.

differentiate into gamete-producing cells (male and female gametocytes) (7) in the blood late in the erythrocytic (sexual blood stage) stage of infection. These gametocytes remain inactive in the blood stream of the human.

21.4.1.3 Mosquito Stage

The gametocytes, male (microgametocytes) and female (macrogametocytes), are usually ingested by the *Anopheles* mosquito during a blood meal from the human host (8). The gametocytes begin sexual reproduction in the digestive tract of the mosquito vector. The multiplication of parasites in the mosquito vector is known as the sporogenic cycle **(Sporogony)**, the sexual stages of parasites in the mosquito. While in the mosquito's stomach, the male gametes (microgametes) fertilize the female gametes (macrogametes) generating zygotes (9). The zygotes in turn become ookinetes (10) which invade mosquito's midgut wall where they develop into oocysts (11). The oocysts grow, rupture, and release sporozoites (12), which goes from the midgut of mosquito to the mosquito's salivary glands. Thus, inoculation of the sporozoites (1) into a new human host occurs and the cycle begins again.

21.4.2 Blood Stages of *Plasmodium* and Hemoglobin Digestion

Within the erythrocytes (RBCs), the parasite develops through several asexual stages such as ring, trophozoite, mature trophozoite, early schizont, and mature schizont (Figure 21.4).

When viewed under light microscopy, the ring-stage parasite resembles a ring, called "signet ring stage." The immature ring stage continues to feed on host cytoplasm and develops into a mature trophozoite. The nucleus then undergoes multiple fissions followed by a process of cell division and produces thousands of merozoites as mature schizont, depending on the species of malaria parasite.

At the blood stage, the malaria parasite feeds on host hemoglobin (Hb) as a source of its primary food. The digestion of Hb results in the production of amino acids which are utilized for protein synthesis in parasites. In *P. falciparum* infection, about 60–70% of host Hb is degraded by the parasites. In the more metabolically active trophozoite stage, 75% of host cell Hb degradation occurs in approximately 12 hours. This catabolic process of Hb digestion is believed to occur in a specialized acidic (pH 5.0–5.4) metabolic organelle (lysosome-like), called digestive vacuole (DV) or food vacuole (FV) within the parasitized red blood cell (PRBC). It is the primary site of Hb digestion. After ingestion, Hb is broken down into smaller peptides, and finally hydrolyzed to amino acids by the successive action of a series of various proteolytic enzymes in the FV/DV of parasites. Free amino acids are thereby transported across the vacuolar membrane into the cytoplasm, where they are utilized for the biosynthesis of parasite proteins leaving the heme behind, as residue [12–14]. The degradation process of Hb in the FV of parasite is depicted in Figure 21.5.

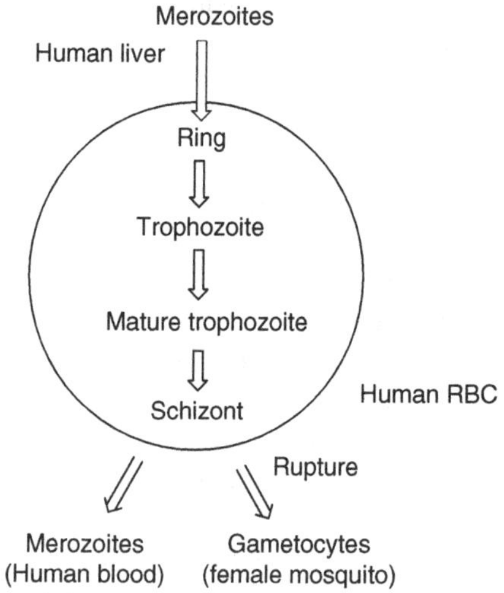

Figure 21.4 Blood stages of malaria parasites.

Two aspartic proteinases, plasmepsin I and plasmepsin II (analogous to cathepsin D), and two cysteine proteinases, falcipain 2 and falcipain 3 (analogous to cathepsin L), are involved in the proteolysis of Hb. The first aspartic proteinase (plasmepsin I) makes a single cleavage in the Hb, unraveling the molecule for further proteolysis. The second aspartic proteinase (plasmepsin II) has preferential activity on acid denatured globin. These two aspartic proteinases are found in the DV of the parasite at different time in the parasitic cycle.

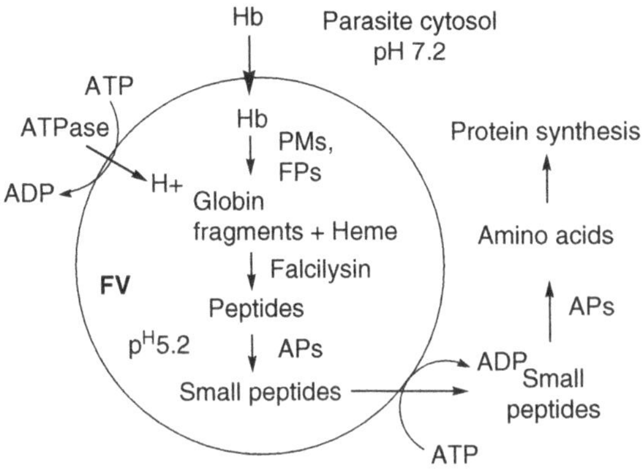

Figure 21.5 Hemoglobin digestion in the FV of parasites. APs, aminopeptidases; FV, food vacuole; FPs, falcipain 2 and 3; PMs, plasmepsin I and II.

Plasmepsin I plays an important role at the ring stage, whereas plasmepsin II acts on trophozoites and young schizonts. The plasmepsin generates Hb fragments that are further catabolized into globin fragments by the action of cysteine protease (falcipain 2 and falcipain 3). Another group of proteanases, called metalloproteinase, aminopeptidases (analogous to cathepsin L) degrades peptides into smaller peptides in the FV and also to amino acids in the parasite cytosol [12, 15].

Falcipains are a group of proteolytic enzymes belonging to the cysteine protease family of *P. falciparum*. Falcipains include falcipain-1 (FP1), falcipain-2 (FP2), and falcipain-3 (FP3). Falcipain-2 is best studied among these three proteases. They are involved in Hb degradation in the erythrocytic stage of malaria parasites. It has been reported that inhibitors of these enzymes block the degradation of Hb and subsequent lysis of RBCs, suggesting the importance of these enzymes as promising drug targets in the development of new antimalarial drug candidates. The falcipain proteases, namely, FP2 and FP3, have already been defined as potential alternative targets for newer generation antimalarial drugs [16–18].

21.4.3 Hemoglobin Degradation and Antimalarial Drug Targets

Hb degradation is considered essential for the survival of malaria parasites, because *Plasmodium* has a limited capacity for *de novo* synthesis of amino acids. Amino acids derived from proteolysis of Hb are converted into parasite proteins and also used for energy metabolism. In the acidic condition of FV, the degradation of Hb also produces heme, which is not metabolized or recycled, but instead stored upon further conversion into an inert insoluble polymeric compound, known as the malaria pigment (dark-colored heme crystals) or hemozoin (HZ). During this process, an amount of four equivalent heme [Ferroprotoporphyrin IX, (Fe(II)PPIX)] is produced which eventually gets oxidized to α-hematin

[Ferriprotoporphyrin IX, (Fe(III)PPIX)] by molecular oxygen (with the transfer of an electron to oxygen) and also generates free oxygen radicals (superoxide anion), called reactive oxygen species (ROS). Free hematin (Ferriprotoporphyrin, FP) is toxic to the parasite, since it acts as pro-oxidant to parasite proteins, and superoxide anion is also toxic to the parasite because it induces disruption of parasite cell membranes or damages parasite cell structures probably by the lipid peroxidation process. The superoxide ion is spontaneously scavenged to H_2O_2 by the action of oxidant defense enzymes like SOD and CAT available in the FV of parasites. As a natural phenomenon, the toxic heme, i.e. hematin, is detoxified by the action of heme polymerase (HP) enzymes (also known as histidine-rich proteins, HRPs) and is converted to nontoxic HZ. This biochemical conversion of FP (soluble, toxic) to HZ (insoluble, nontoxic) is considered a major pathway of heme detoxification which is considered important for the survival of parasites. FP may also be destroyed by reaction with H_2O_2 in the FV through the formation of an activated intermediate, FP-Fe(IV)*=O, that further degrades to the degraded hematin, FP_{deg}. Alternatively, FP may diffuse into the parasite cytosol where it is destroyed by an oxidation reaction with GSH through the formation of FP–GSH complex [13, 19, 20].

It is now understood that the whole biochemical process that occurs in the FV of parasites primarily involves Hb digestion (by proteolytic enzymes) and heme detoxification, which are indispensable for the survival of parasites. The potential therapeutic targets in the parasite might be either targeting enzymes (proteolytic enzymes or proteinases) involved in the Hb degradation process or the mechanism of heme detoxification. Inhibition of plasmepsins is considered to block Hb degradation and kill the parasite in culture systems. Inhibition of falcipains also results in impairment of vacuolar functions with the death of parasites. The selectivity of antimalarial drug action for quinoline-based drugs appears to derive from the fact that they inhibit directly specific enzymes or substrate molecules (heme) in the Hb digestion process. The schematic of generation and detoxification of heme in the FV of parasites, and chemotherapeutic targets of major antimalarial drugs (chloroquine/artemisinin [ART]) is represented in Figure 21.6.

HP enzyme or HRPs, Hb degrading proteases enzymes (plasmepsins and falcipains) and heme (hematin/FP) are some potential targets involved in the antimalarial drug action of major drugs like chloroquine (CQ), quinine (QN), and ART.

Quinolines bind to hematin molecule (FP) interfering with the heme detoxification process (causes disruption of pathway), thereby inhibiting the polymerization of toxic FP to nontoxic HZ, which leads to accumulation of toxic heme that causes the death of parasites. It has been suggested that hematin [Fe(III)], upon reduction to heme [Fe(II)], provides ferrous iron that is claimed to be responsible for the cleavage of endoperoxide bridge of ART-based drugs to cytotoxic carbon

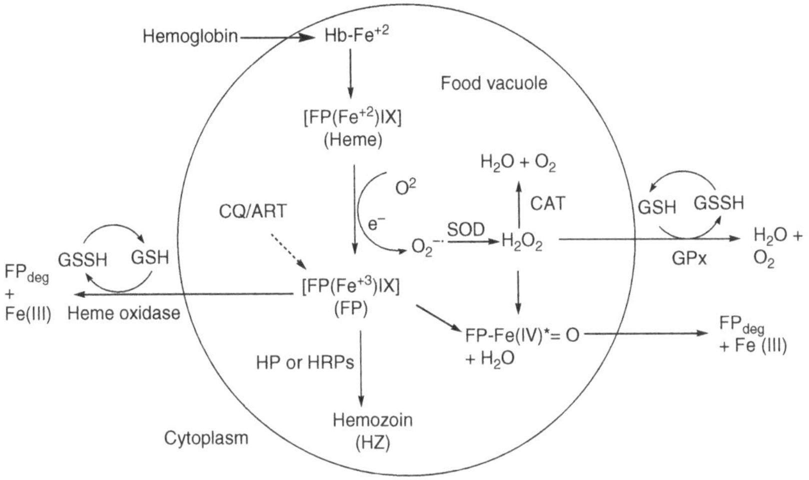

Figure 21.6 Mechanism of heme detoxification and targets of antimalarial drugs [21]. ART, artemisinin; CAT, catalase; CQ, chloroquine; GSH, redued glutathione; GSSH, oxidized glutathione; GPx, glutathione peroxidase; FP, ferriprotoporphyrin (Hematin); HRPs, histine-rich proteins; HP, heme polymerase; HZ, hemozoin; SOD, superoxide dismutase. *Source:* Meshnick and Dobson [21].

radical species which destroy parasites. Table 21.1 depicts the site(s), pathway(s), and target protein(s) involved in antimalarial drug action.

It is therefore noteworthy that the formation of heme during the Hb digestion process is the basis of discovery and development of antimalarial drugs. The Hb degrading enzymes are also considered the novel drug targets in antimalarial drug discovery.

However, a majority of antimalarial drugs are weak bases which accumulate in the acidic FV of parasites by pH trapping (or ion trapping) mechanism. Therefore, basicity (pK_a) of drug molecules and vacuolar pH are other important criteria in antimalarial drug action. Apart from enzymes involved in Hb digestion and heme detoxification mechanism, enzymes involved in macromolecular and metabolite synthesis (DNA and RNA synthesis) and those responsible for membrane processes and signaling are also considered to be new drug targets. Some new categories of drug targets have been reported which include ion channels, proteases, kinases, and metabolic pathways (purine pathways, pyrimidine metabolism, folate pathways, phospholipid metabolism, glycolysis) and miscellaneous targets like DNA topoisomerases, tubulin assembly [21–25]. These targets can be used as the basis for screening compounds in structure-based drug design (SBDD) approach to identify new antimalarial lead molecules.

Table 21.1 Site(s), pathway(s), and drug target(s) involved in antimalarial drug action.

Site(s) of action	Drug(s)	Mechanism/ pathway(s)	Target(s)
Lysosomal food vacuole (FV)	Quinolines (QN, CQ, AQ, MQ, PQ)	Hemoglobin digestion, Heme polymerization	Heme polymerase (HRPs), heme protein (Hematin)
	Endoperoxides (ART and its semisynthetic derivatives, newer endoperoxides (OZ439))	Free radical generation, Heme alkylation	Plasmepsins, Falcipains, Heme polymerase (HRPs), Proteases, Glucose transporter, $Na^{+/}H^{+}$ antiport
Mitochondria (MT)	ATO	Electron transport	Cytochrome b reductase Dihydroorotate dehydrogenase (DHOD)
Cytosol	Antifolates (Sulfadoxine, Dapsone, Pyrimethamine, Proguanil)	Folate metabolism Glycolysis Nucleotide biosynthesis	DHPS DHFR-TS Lactate dehydrogenase (LD)

ART, artemisinin; ATO, atovaquone; AQ, amodiaquine; CQ, chloroquine; DHPS, dihydropteroate synthase; DHFR-TS, dihydrofolate reductase-thymidylate synthase; HRPs, histidine-rich proteins; MQ, mefloquine; PQ, primaquine; QN, quinine.

21.5 Malaria Infections, Symptoms, and Pathogenesis

Immediately after the infective bite by the *Anopheles* mosquito, the parasite starts spreading infection (by inoculation of sporozoites) into human host, but the first symptoms of malaria (fevers, chills, and profuse sweating) appear after a period of time (usually 9–14 days), called "incubation period." After a variable incubation period, 5–7 days for *P. falciparum*, 6–8 days for *P. vivax*, 9 days for *P. ovale,* and 12–16 days for *P. malariae*, extraerythrocytic merozoites (~5 000–40 000 per sporozoite) are released after rupture of tissue schizonts from the liver cells into the blood stream. The time interval for completion of this pre-erythrocytic tissue phase is called as prepatent period. During development and multiplication in the tissue phase, the host is asymptomatic. The first symptoms of malaria after the prepatent period (time between inoculation of sporozoites and appearance of symptoms, i.e. the time period when the sporozoites undergo schizogony in the liver) are called the primary attack. In infections with *P. falciparum* and

P. malariae (non-relapsing malarias), all the tissue schizonts rupture at about the same time and none of them persist in the liver cells; the entire infection moves rapidly into the blood stage. In case of *P. vivax* and *P. ovale* (relapsing malarias), some sporozoites may go into hibernation, providing a reservoir of parasites in the liver and this nondividing (dormant) tissue stage is called as "hypnozoites" (the cryptobiotic phase). In this stage, parasites can lie dormant for months or years and on reactivation, they may cause clinical relapse. In malaria endemic areas, people may develop partial immunity that allows asymptomatic infections to occur. In patients at risk, symptoms appear in about 7 days or more (usually 10–15 days) after the infective mosquito bite. The time required for the appearance of first symptoms (fever, chills, and profuse sweating) of malaria is determined by the complete time span (24–72 hours) of intraerythrocytic life cycle of parasites. The type and intensity of primary attack is therefore dependent on the time required for the rupture of infected erythrocytes and also the type of invading *Plasmodium* species.

Classically (but infrequently observed), the attacks of malaria occur every second day with the "tertian" parasites (*P. falciparum*, *P. vivax*, and *P. ovale*) and every third day with the "quartan" parasite (*P. malariae*) depending on the intraerythrocytic cycle of *Plasmodium* parasites. Therefore, variations in biologic and pathologic features of clinical importance appear depending on the type of infecting parasite species. During life cycle, the erythrocytic stage of parasite is an important determinant to define the level of parasitemia which in turn determines the types of symptoms to be appeared. *P. vivax* and *P. ovale* tend to invade only reticulocytes, while *P. malariae* invades senescent erythrocytes. In these three cases, parasitemia densities are found low. By contrast, *P. falciparum* can invade all RBC stages, with considerably high levels of parasitemia. *P. falciparum* malaria is therefore the cause of all the mortality and most of the morbidity in malaria, although, but rarely, *P. vivax* or *P. ovale* also produce serious complications, debilitating relapses, and even death. Deaths from malaria are normally caused by cerebral, renal, or pulmonary fever [2, 3, 10, 26]. Different types of malaria infections are described as follows:

***P. falciparum* malaria (or malignant tertian malaria)** is characterized by an erythrocytic cycle of 48 h. This parasite species is recognized as the most lethal parasite that causes most severe infections and deaths related to malaria. The individual infected by *P. falciparum* experiences fever, headache, fatigue, dizziness, abdominal pain, aching muscles, sore back, joint pain, vomiting, nausea, enlarged spleen, convulsions, and some neurological symptoms as well. Neurological dysfunction occurs because this infection has an adverse effect on brain and the central nervous system. Neurologic defects (ataxia, palsies, speech difficulties, deafness, and blindness) may occasionally persist following cerebral malaria, especially in children. Recurrent infections *cause* severe malaria, and can cause severe blood loss (or anemia) that is frequently observed in children and pregnant women. Malaria in pregnancy caused by *P. falciparum* leads to severe disease in mother, and may even lead to premature delivery of a low-birth-weight baby. *P. falciparum* may lead to serious and life-threatening conditions if left untreated. Untreated cases can progress to pulmonary edema, cerebral dysfunction, hypoglycemia, convulsions, liver failure, renal failure, sometimes paralysis, coma, and even death [2, 3, 10].

In contrast, ***P. vivax*** **malaria (benign tertian malaria)** presents acute symptoms with erythrocytic cycle of 41–45 hours. It is relatively nonlethal, but is often accompanied by debilitating conditions. Severe anemia can also be seen with *P. vivax* infections. This strain has a dormant liver stage that can activate and invade the immature RBCs after months or years, causing many patients to relapse up to five years after apparent clinical cure. Although it seldom causes death or other serious problems like rupture of spleen, it can still cause major illness. Some of the common symptoms of this infection are flu-like symptoms, chills, fatigue, and diarrhea [2, 3].

P. ovale **malaria (Ovale tertian malaria)** is rarely the cause of morbidity even in children, its features are similar to that of *P. vivax* with an erythrocytic cycle of 49–50 hours. *P. vivax* and *P. ovale* infect only young RBCs. In some cases, *P. ovale* reoccurs because the parasite can stay in the dormant stage in the liver for a few months to four to five years after getting infected initially by the mosquito-carrying malaria. These parasites are most likely to relapse and invade RBCs and render the patient fall sick again. This is the rarest of all malaria types [2, 3].

P. malariae **malaria (quartan malaria)** presents persistent reappearance of symptoms like fever, which recurs every third day with an erythrocytic cycle of 72 hours. It is milder than other types of malaria and is mostly found in old RBCs and may not be clinically noticeable. Red cells are destroyed in the peripheral capillaries and anemia results. Nephrotic syndrome can result from repeated infections with *P. malariae* [2, 3].

P. knowlesi **malaria (tertian malaria)** has a 24 hours erythrocytic cycle and can, therefore, multiply rapidly in the blood, causing an uncomplicated case to become serious very quickly. Fatal cases of infection with this parasite species have also been reported [2, 3].

In general, infection with malaria parasites may result in a wide range of symptoms, ranging from absent or mild symptoms (asymptomatic infections or no apparent illness) through the classic symptoms to severe disease complications, and even death. The first symptoms include fever, chills, sweating, headache, muscle pains, vomiting, and diarrhea that may be mild and difficult to recognize as malaria. If not treated immediately, infection with *P. falciparum* malaria can progress to severe illness which often leads to death. Children with severe

P. falciparum malaria frequently develop one or more of the symptoms like severe anemia, respiratory distress, metabolic acidosis, or cerebral malaria. In adults, multi-organ failure is also frequent. Cerebral malaria is a dangerous complication with several manifestations of cerebral dysfunction such as high fever, headache, drowsiness, delirium, confusion, abnormal neurological signs (including focal and generalized convulsions, or seizures), and impaired consciousness or coma. However, the severity of the disease illness depends on several factors, such as the species of infecting parasite and the human's acquired immunity and genetic background. According to the Centers for Disease Control and Prevention (CDCP), malaria can be categorized as uncomplicated or severe/complicated malaria depending on the type of attack, and subsequent occurrence of signs and symptoms [2, 3, 10, 26, 27].

21.5.1 Uncomplicated Malaria

It is a symptomatic infection with low parasitemia without signs of severity and/or evidence of vital organ dysfunction. Uncomplicated malaria is diagnosed when symptoms are present, but there are no clinical or laboratory signs to indicate a severe infection or the dysfunction of vital organs. Individuals suffering from this form of the disease can eventually develop severe malaria if the disease is left untreated, or if they have poor or no immunity to the disease. The clinical manifestations involve typical and atypical features. The typical/classical (but rarely observed) malaria attack lasts 6–10 hours. The uncomplicated malaria is defined as acute febrile illness and typically occurs in cycles with every second day, although some parasite species can cause a longer cycle or mixed symptoms. It consists of:

- a cold stage (shaking chills, shivering), lasts for 15 minutes to 1 hour, usually at mid-day between 11 a.m. to 12 noon),
- a hot stage (high grade fever even reaching above 106 °F, headaches, vomiting), lasts for two to six hours, and
- a sweating stage (profuse sweating, fever return to normal temperature, tiredness), persist for two to four hours.

The febrile paroxysms describe above are usually accompanied by atypical symptoms. These include atypical fever, headache, perspirations, abdominal pain, slight cough, enlarged liver and spleen, jaundice, rigors, tiredness, dehydration, muscle ache, back pain, joint pains, chest pain, malaise, nausea, sometimes vomiting, cough, delirium, anorexis, breathlessness, weakness, anxiety, restlessness, seizures in young children, altered behavior, acute psychosis and sometimes puffiness of lids [2, 3, 10].

21.5.2 Complicated or Severe Malaria

It is characterized by signs of severity and/or clinical or laboratory evidence of vital organ dysfunction. This occurs when *P. falciparum* infections are complicated by serious organ failures or abnormalities in the patient's blood or metabolism. This form has the capacity to be fatal if left untreated. The manifestations of severe malaria includes high fever accompanied with headache and shivering and chills, abnormal behavior, prostration, multiple convulsions or repeated generalized convulsions, hyperpyrexia, hyperparasitemia (≥5%), delirium, severe anemia, impairment of consciousness or coma and other neurologic abnormalities, hemoglobinuria due to hemolysis, abnormalities in blood coagulation, thrombocytopenia, cardiovascular collapse and shock, jaundice, pulmonary edema or acute respiratory distress syndrome (ARDS) and acute renal failure (kidney failure). Symptoms like hypotension, hypoglycemia, abnormal bleeding, severe anemia and fetal and maternal complications are also seen in pregnant women with sever malaria [2, 3, 10].

All the clinical symptoms and severe pathological features associated with malaria are caused by the asexual erythrocytic or blood stages of parasites. The clinical manifestations and pathogenesis of malaria is depicted in Figure 21.3. When the parasite develops in the erythrocyte (undergoes nuclear divisions), numerous known and unknown waste substances such as red cell membrane products, HZ pigment and other toxic factors like glycosylphosphatidylinositol (GPI) accumulate in the infected red blood cells (iRBCs). These are dumped into the blood stream when the infected erythrocytes lyse and release invasive merozoites along with cell debris in the host cytoplasm. These products, particularly the GPI, activate macrophages and endothelial cells to secrete cytokines and inflammatory mediators such as tumor necrosis factor (TNF), interferon-γ, interleukin-1 (IL-1), IL-6, IL-8, macrophage colony-stimulating factor, and lymphotoxin, as well as superoxide and nitric oxide (NO). The systemic manifestations of malaria have been attributed to the various cytokines in response to the parasites and red cell membrane products. The inflammatory mediators act to produce fever and rigors and probably influence other severe pathophysiology associated with malaria. The presence of the cell debris is also the cause of the episodes of febrile illness (or paroxysms), i.e. fever and chills, associated with malaria. HZ is also been linked to the induction of apoptosis in developing erythroid cells in the bone marrow, thereby causing anemia. The merozoites released by the red cell rupture go on to infect more *P. falciparum*-infected erythrocytes, particularly those with mature trophozoites, adhere to the vascular endothelium of venular blood vessel walls and do not freely circulate in the blood. When this sequestration of infected erythrocytes occurs in the vessels of the brain, it is believed to be a factor in causing the severe disease syndrome known as cerebral malaria, which is associated particularly with *P. falciparum* malaria with high mortality [2, 11].

21.6 Malaria Control and Prevention

During the last few decades, significant advances were made in malaria eradication and control programme due to the improved diagnosis of disease, introduction of effective vector control measures, and the use of combination therapy (CT). WHO recommends protection of all people at risk of malaria with effective vector control. Two forms of vector control, insecticide-treated mosquito nets (ITNs) and indoor residual spraying with insecticides (IRS), are effective and widely used. However, there are a number of approaches to the control and prevention of malaria as summarized below, and the choice of intervention in a country/region are usually most dependent on cost-effectiveness [2, 28, 29].

- **Early diagnosis and prompt treatment** with antimalarial drugs is an essential approach in controlling the spread of the disease. Chemotherapy still remains the mainstay in malaria control strategy. WHO recommended multidrug-based CT such as artemisinin-based combination therapies (ACTs) that are widely used for the treatment and prevention of malaria.
- **Prophylaxis** can also be achieved by adopting personal protection measures such as the use of insecticide-treated mosquito bed nets (ITNs), screening doors and windows, wearing protective clothing and IRS. All these ways are to protect people from being bitten by an infected mosquito. The other way of personal control is the use of preventative or prophylactic drugs. Drug regimens started before and after finishing traveling to a malaria endemic region or the use of chemoprophylactic drugs (primaquine [PQ], mefloquine [MQ]) to prevent the risk of attack (or the development of the disease, but not the initial infection) to nonimmune travelers in endemic region.
- **Vector or mosquito control** remains a traditional approach in malaria control programme. It is the main way to prevent transmission of malaria. Spraying of insecticides (DDT) and lervicides (Fenthion) to kill the adult or larvae mosquitoes. The CDCP recommends the application of insect repellents to the skin when anticipating the exposure to mosquitoes. The specific repellents recommended are DEET (*N,N*-Diethyl-*meta*-toluamide), permethrin, pyrethroids, picaridin, and PMD (*para*-Menthane-3,8-diol). The use of IRS, and mosquito repellants or coils and vaporizers also remains to be effective in vector control programme. Vector control was previously quite effective in controlling malaria. However, it is now largely ineffective due to the emergence of insecticide-resistant mosquitoes.
- Ideally, a protective vaccine would be the most effective approach to control malaria.

21.6.1 Malaria Vaccine

As an alternative to antimalarial drugs, a malaria vaccine has the potential of inducing long-term immunity and accordingly might be given only once or twice to provide life-long protection against malaria infection. RTS, S/AS01 (RTS, S), also known as Mosquirix is an injectable vaccine that provides partial protection against malaria in young children. It is the world's first malaria vaccine. The vaccine is being evaluated in Sub-Saharan Africa as a complementary malaria control tool that potentially could be added to the core package of WHO-recommended preventive and treatment measures. Attempts to develop a vaccine, however, have been hindered by the great genetic diversity of the parasite, its multistage life cycle, as well as the complex and inefficient human immune response [5].

With the concerted efforts of chemotherapy/chemoprevention and the use of ITNs and IRS, the control of malaria has been geared toward reducing overall malaria burden over the years. However, several problems have impeded the effectiveness of antimalarial interventions. Several barriers to malaria control [2, 5, 28, 29] are mentioned below:

- **Drug resistance:** It is a major problem since the emergence of drug-resistant malaria cases slowed the prevention efforts because they made available drugs less effective and potentially unsafe medications for the management.
- **Vector resistance:** Vector resistance to commonly used insecticides makes interventions like ITNs and IRS less efficient.
- **Poor health conditions:** Weak, overwhelmed health systems in poor endemic countries makes the implementation of interventions ineffective.
- **Poverty:** Malaria thrives in poor regions aided by lack of knowledge and education. Uneducated and poor peo.ple do not know how to prevent malaria, or cannot afford drugs and ITNs and other preventive measures.

Malaria is not just a disease commonly associated with poverty, but is also a cause of poverty and a major hindrance to economic development. Malaria disease has been associated with negative economic effects in places where the disease is widespread or in countries with heavy malaria burden. In malaria endemic countries, the disease accounts for about 40% of public health expenditure. The total funding for control and elimination of malaria has been estimated US$2.7bn in 2016. The economic impact includes costs of health care, working days lost due to sickness, days lost in education, decreased productivity due to cerebral malaria, and loss of investment and loss of earning from tourism [5].

21.7 Treatment of Malaria

21.7.1 Traditional Treatment of Malaria

The oldest antimalarial known to man is the bark of the cinchona tree (*Cinchona officinalis* L.). The first recorded treatment of malaria dates back to 1600 when the bark of the Cinchona tree was first used by the native Peruvian Indians to treat the intermittent fevers associated with such illnesses because of its antifebrile properties. The powdered cinchona bark administered in various forms (tincture/infusion) and through various means remained the main antimalarial agent until 1820, when two French Pharmacists, Pierre Joseph Pelletier and Joseph Bienaimé Caventou, succeeded in separating its active ingredients, which they named QN and cinchonine. This was the basis for the subsequent development of QN and its derivatives, which are still in wide spread use today. A Chinese medicinal herb, *Qinghaosu* (*Artemisia annua* L.), was also known to be used for curing "ague," as malaria was then known, and the decoction of leaves was used in Chinese Traditional Medicine (CTM) to treat malaria for over 2000 years. Later, ART isolated (1972) from this source and its semisynthetic derivatives have also been developed as antimalarial agents [2, 21, 24, 30]. Thus, herbal medicine served as a starting point for the development of novel antimalarial drug leads.

21.7.2 Current Treatment of Malaria

Traditional medicines form the foundation of antimalarial drugs in current use. Today, a myriad of drugs is available for the treatment and prophylaxis of malaria. However, the growing incidence of resistant cases, particularly the multidrug-resistant (MDR) *P. falciparum* infections have made currently available drugs redundant in the treatment or control of malaria.

21.7.2.1 Currently Available Antimalarial Drugs

The current antimalarial drugs used for the treatment and prevention of malaria infection belong to four broad classes, namely, the quinolines, the antifolates, the artemisinins, and the antibiotics. The most widely used antimalarial drugs are QN, CQ, amodiaquine (AQ), PQ (QN and its derivatives); MQ, halofantrine (HL), lumefantrine (LUM, arylaminoalcohol compounds); sulfadoxine (SX), pyrimethamine (PYR), proguanil (PG) (antifolate combination drugs); artesunate (AS), artemether (AM) (ART derivatives); and atovaquone (ATO) and doxycycline (DOX) (antibiotic).

21.7.2.1.1 Quinoline-based Drugs

The first and most commonly used are the quinoline-based antimalarials, which include QN, the parent compound and its derivatives CQ, AQ, and MQ. QN, **a**, a cinchona alkaloid isolated from cinchona bark, was the only known antimalarial drug for more than three centuries, and until the 1930s was the only effective agent for the treatment of malaria. It was the first medicine against malaria. Later, the structure of QN served as a platform for the development of several potent synthetic drugs such as CQ, **b**, a 4-aminoquinoline derivative of QN. With the introduction of clinically effective, less toxic, and cheaper drug, CQ in the mid-twentieth century (1940), ceased the wide spread use of QN. Soon after its introduction, CQ became the mainstay of therapy and prevention, since this drug was effective against all strains of the parasite. Another synthetic antimalarial, PQ, **c** (1950), was also developed thereafter as based on the structure of lead QN molecule. PQ is 8-aminoquinoline analogue of QN. MQ, **d**, a potent synthetic 4-aminoquinolinemethanol (or quinolinemethanol derivative of QN), structurally related to QN, was developed (1975) after CQ to treat resistant malaria. Malaria parasite resistance to CQ and MQ began to appear within a few decades of introduction. AQ, **e**, another 4-aminoquinoline synthetic antimalarial, was found effective against CQ-resistant strains of *P. falciparum*, but its use in the treatment of malaria has been restricted because of its severe hepatotoxicity and agranulocytosis (also myelotoxicity). The structures of some clinically useful quinoline-based antimalarial drugs are represented in Figure 21.7.

However, among all these drugs, CQ, MQ, and PQ are the most widely used antimalarial drugs. To avoid the toxicity of AQ, a new drug was developed named

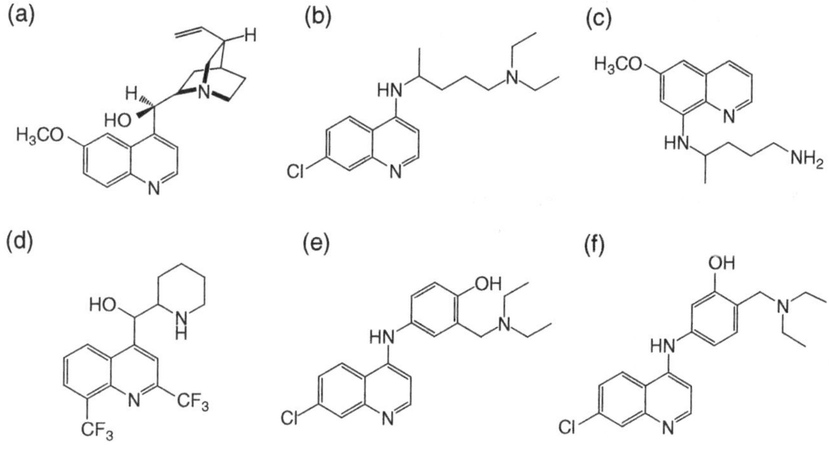

Figure 21.7 Structures of some quinoline-based antimalarial drugs.

isoquine (IQ), **f**, the modified form of AQ. IQ does not possess the toxicity associated with AQ and the antiparasitic activity is retained.

Related to MQ is HL, **a**, a phenanthrene methanol with activity against the erythrocytic stages of the malaria parasite. Its use has been recommended in CT with semi-artemisinin derivatives of ART (AS) in cases with multiple drug-resistant *P. falciparum*. Cardiotoxicity is a potential problem associated with this antimalarial drug. Two drugs were originally synthesized in China that was also found effective in MDR *P. falciparum* infections, but when used in combinations with ART drugs. These are LUM, **b**, another phenanthrene methanol derivative, and pyronaridine (PYN), **c**, a benzonaphthyridine derivative. LUM is used as a fixed-dose combination with AM (as Co-Artem). PYN was reportedly effective in clinical trial in CT with AS. It possesses toxicities like hepatotoxicity. Piperaquine (PIP), **d**, a bisquinoline analogue, was also developed to overcome the problem of drug-resistance with *P. falciparum* malaria. PIP is combined with dihydroartemisinin (DHA) (as Artekin) to treat CQ-resistant *P. falciparum* infection. Ferroquine, **e**, was developed by incorporation of ferrocene into the basic scaffold of CQ, which is under clinical trial. This drug is being developed for the treatment of resistant malaria. A new 8-aminoquinoline drug, tafenoquine, **f**, is currently under clinical trial. It has been proposed to be indicated for the treatment as well as for the prevention of malaria. The structures of these newer quinoline-based antimalarials are presented in Figure 21.8.

Tafenoquine can be used for treating malaria relapses (against hypnozoites) caused by *P. vivax* and *P. ovale* malaria. It has longer half-life than PQ, but it causes

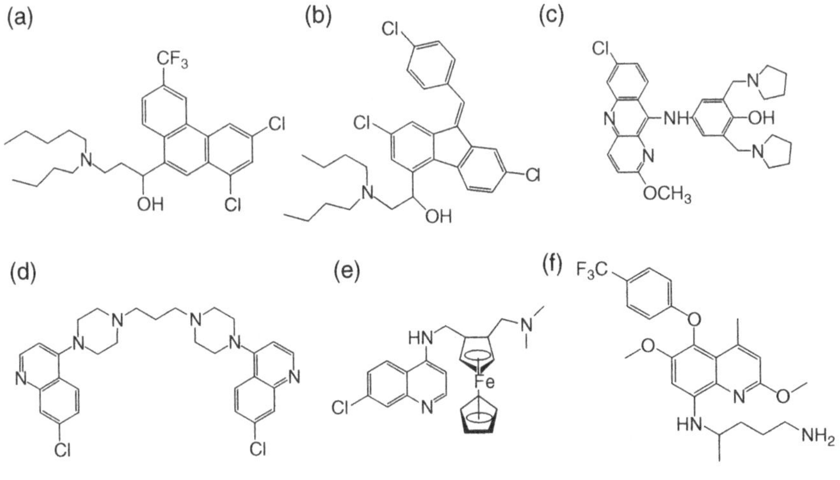

Figure 21.8 Structures of some newer quinoline-based antimalarials.

hemolysis in patients with glucose-6-phosphate dehydrogenase (G6PD) deficiency like PQ [21, 24, 31].

Quinoline-based antimalarials are used widely in the treatment and prophylaxis of malaria. QN still remains an important antimalarial drug due to the emergence of CQ-resistant and MDR strains of malaria parasites, especially *P. falciparum*. Due to its undesirable side effects, it is now only used as an intravenous injection (as sulfate salt) to treat severe malaria. CQ has long been the drug of choice for treating malaria. It still remains the most widely used drug in the treatment of malaria, despite increasing parasite resistance for its availability, low cost, and good tolerability. CQ is also preferred as prophylactic agent (in combination with PG) in malarious regions without resistant *P. falciparum* malaria. CQ formulated as the phosphate salt for oral use, and it is the drug of choice in the treatment of non-*P. falciparum* and sensitive *P. falciparum* malaria. It remains the first-line treatment in uncomplicated CQ-sensitive *P. falciparum* infections. It rapidly terminates fever (within 24–48 hours) and clears parasitemia (within 48–72 hours) caused by sensitive parasites. It also remains the drug of choice for treating *P. vivax* malaria. In CQ-resistant malaria, the next drug of choice is MQ, followed by QN in combination with tetracycline (TET), DOX, or sulfadoxine-pyrimethamine (SP). MQ and AQ are widely available and are used to treat cases of uncomplicated malaria in areas where CQ resistance is prevalent. MQ is administered orally and it possesses a long half-life which has resulted in the emergence of drug resistance. However, MQ is recommended to be used in long-term chemoprophylaxis. AQ is useful in CQ-resistant *P. falciparum* infections. It has been an effective agent against drug-resistant *P. falciparum*, but it is not recommended for prophylaxis, since it causes agranulocytosis and hepatitis. PQ is the only drug currently available for the prevention of *P. vivax* and *P. ovale* relapses. Terminal PQ prophylaxis is recommended against any dormant infection (to eliminate latent liver-stage parasites) to ensure a complete cure. MQ or PQ is not recommended in patients with G6PD deficiency, to avoid severe drug-induced hemolysis [31–35].

Quinoline-based drugs are blood-stage schizonticidal. CQ/MQ is selectively active against the intraerythrocytic mature forms (trophozoites) and also younger ring forms of malaria parasites, without any activity against gametocytes. The quinolines inhibit the HP enzyme resulting in specific toxicity during the parasite development. CQ accumulates by a weak base mechanism in the acidic FV of trophozoite-infected cells and act by forming a complex with heme in the parasite FV, which prevents heme polymerization and consequently, HZ formation. Simply, they these drugs block the polymerization of heme to hemozoin (malaria pigment). As a result, the heme which is released during Hb degradation builds up toxic accumulation of heme (hematin), thereby kills the parasite with its own toxic effects. The mode of action of QN is similar to CQ. QN binds strongly to heme protein and forms complexes that are toxic to the malarial parasite. MQ also acts by inhibiting the HP, similar to CQ [19, 20, 22, 31].

However, rapid development of resistance has severely limited the use of quinoline-based drugs alone, and therefore, they are used in combination with others in the treatment of resistant malaria. The increasing prevalence of MDR strains of malaria parasites, particularly *P. falciparum* in most malaria endemic areas (Southeast Asia including Myanmar, Thailand, Vietnam, and India, African continent, and Eastern Mediterranean region), has significantly reduced the efficacy of CQ and other potent quinoline antimalarials in the treatment of malaria.

21.7.2.1.2 Antifolate Drugs

Resistance to CQ led to the extensive use of antifolate drugs in the treatment of malaria as combination antimalarial drugs. Antifolates are of two types, namely, the dihydropteroate synthase (DHPS) inhibitors which include sulfa-drugs (sulphonamides like sulfadoxine (SX), **a**, sulfalene) and sulfones [dapsone (DDS), **b**], and the dihydrofolate reductase (DHFR) inhibitors that are diaminopyrimidines [PYR **c**, trimethoprim] and biguanides [PG, **d** and its derivatives, chlorproguanil (CHLORPROG), **e**, cycloguanil, **f** (a metabolite of PG)]. Although these drugs have antimalarial activity (relatively weak) when used alone, resistance can develop rapidly. Therefore, a drug of the first group is usually used in combination with a drug of the second one. When used in combination, they produce a marked synergistic effect and can be effective even in the presence of resistance to the individual components. Some typical combinations include SP (Fansidar), sulfalene-pyrimethamine (metakelfin), and sulfamethoxazole-trimethoprim (co-trimoxazole). Sulphadoxine-pyrimethamine [1970, SP] is the most widely used in the treatment of CQ-resistant malaria. A new antifolate combination of chlorproguanil and DDS (2003, Lapdap) was developed as an effective and cheap antimalarial, which replaced SP because of its more potency against resistant malaria than that of SP. Later, chlorproguanil-dapsone-artesunate (CDA) combination was also developed [32–35]. The structures of some antifolate antimalarials are depicted in Figure 21.9.

SP is one of the frequently used antifolate combination drugs used in the treatment of CQ-resistant malaria. Antifolate drugs prevent the nuclear division of malaria parasites at both the erythrocytic and exoerythrocytic stages. Antifolate combination drugs, such as SP, act through sequential and synergistic blockade of two key enzymes involved in folate synthesis. Sulfonamides and sulfones inhibit the DHPS, while PYR and related compounds inhibit dihydrofolate reductase-thymidylate synthase (DHFR-TS). These compounds inhibit the synthesis of folate cofactors that are required for parasitic nucleotide synthesis and thus parasitic DNA and amino acid metabolism. Repeated administration of SP can develop severe skin reactions like Stevens-Johnson syndrome. Another combination, pyrimethamine/dapsone (Maloprim) may cause toxic effects like agranulocytosis, if increased doses are given over time [19, 20, 22].

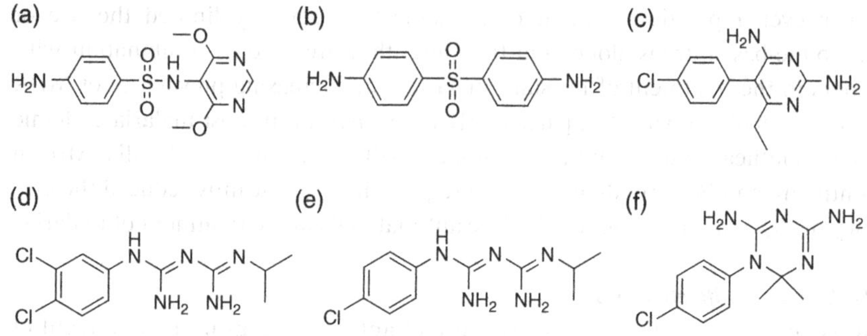

Figure 21.9 Structures of some antifolate drugs.

21.7.2.1.3 Artemisinin and Its Derivatives

Artemisinin (ART, **a**), an active constituent of *A. annua* L. (Sweet wormwood) and its semisynthetic derivatives are considered as the most important, effective, and new antimalarial agents. Chemically, ART is a sesquiterpene lactone endoperoxide bearing 1,2,4-trioxane ring as the peroxide functional moiety. ART is not soluble in water or oil and hence has poor solubility, oral absorption, and bioavailability. However, since the peroxide bridge of the compound is stable under various chemical reactions, several oil/water-soluble derivatives of ART have been synthesized. So, reduction of ART produced DHA, **b**, a sesquiterpene lactol, which served later as a template for the synthesis of a series of semisynthetic analogues such as AM, **c**, arteether (AE), **d**, AS, **e**, and artelininic acid, **f**. They are collectively known as the first-generation derivatives of ART (Figure 21.10). First-generation ART derivatives can be grouped into oil-soluble C(10) β-alkyl ethers (AM and AE) and water-soluble C(10) β-(substituted) esters (sodium artesunate and sodium artelinate) (Figure 21.10).

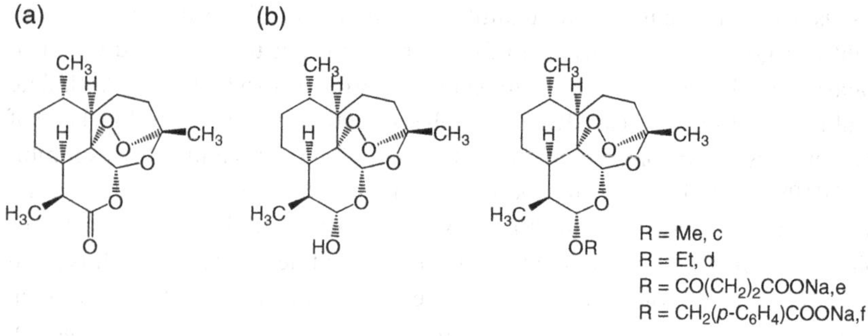

Figure 21.10 ART and its semisynthetic derivatives.

These drugs possess better oil/water solubility, and therefore, has antimalarial efficacy than the parent compound, ART. These are all cyclic endoperoxides or sesquiterpene lactones. The semisynthetic derivatives of ART have improved pharmacokinetic properties and are also of current clinical use. Because of having excellent clinical efficacy against resistant parasites, they mostly replaced the use of quinolines and antifolate-based drugs [35–41].

ART and its derivatives are fast-acting antimalarials effective against MDR strains of *P. falciparum* and used for the treatment of severe malaria. They are also active against *P. vivax* malaria. ART-based drugs showed very rapid parasite clearance times and faster fever resolution than QN. They have high recurrence rates when administered alone. In some areas of Southeast Asia, combinations of ART-based drug and MQ in combination offer the only reliable treatment for uncomplicated MDR *P. falciparum* malaria [42, 43].

ART and its analogues are active against all blood stages, particularly against younger ring forms and gametocytes. They have no activity on hepatic stages of parasites. ART derivatives reduce parasitemia very rapidly and are well tolerated in both adults and children. ART and related compounds are concentrated in parasite-infected erythrocytes and exert their parasiticidal activity subsequent to reductive activation by heme in an irreversible redox reaction, which produces toxic carbon-centered free radicals (through cleavage of endoperoxide bridge). The toxic free radicals may lead to alkylation of heme or bring about oxidative damage to parasite proteins/lipids. The bridging endoperoxide group therefore appears to be crucial for the antimalarial activity. The antimalarial activity of artemisinins may also result from the inhibition of a parasite calcium ATPase [44–46].

21.7.2.1.4 Atovaquone

ATO (1992) is a 2-hydroxynapthoquinone derivative that is currently the most widely used drug in the treatment of malaria infections, particularly in immunosuppressed patients. It is effective against CQ-resistant *P. falciparum*, but because, when used alone, resistance develops rapidly, ATO is often given in combination with PG. A new fixed-dose antimalarial combination of ATO and PG (Malarone, 1998) is available now in the market worldwide. ATO represents a novel class of expensive antimalarial drugs [34, 47]. The structure of ATO is given in Figure 21.11.

ATO is used in combination with PG or TET for the prevention and treatment of CQ-resistant malaria, including cerebral malaria caused by *P. falciparum*. It is as effective as MQ or DOX. Malarone shows good tolerability with minimal side effects. ATO acts through inhibition of electron transport at the cytochrome bc1 complex in plasmodial mitochondria and depolarizes the membranes of plasmodial mitochondria [32, 34].

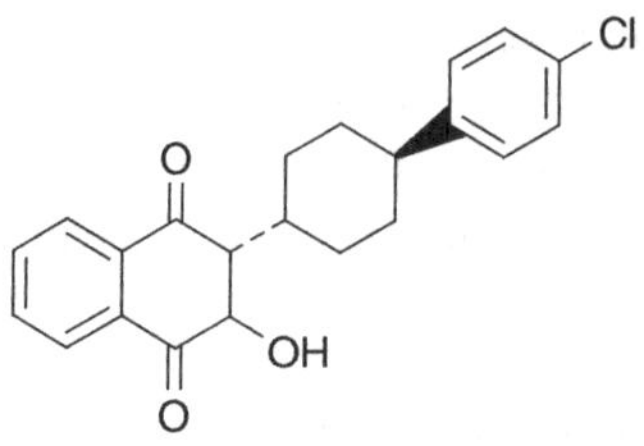

Figure 21.11 Structure of ATO.

21.7.2.1.5 Antibiotics

TET, **a**, and its derivatives are very potent antimalarials and are used for both treatment and prophylaxis. In areas where QN has developed resistance, tetracyclines are often used in combination with QN to improve cure rates. DOX, **b**, is an effective agent when used prophylactically against CQ-resistant or MDR-resistant *P. falciparum*. Clindamycin (CL), **c**, has been used but offers only limited advantage as compared with other available antibiotics. Response of parasites to CL is slow and recrudescence rates are high. Azithromycin (AZ) is also used in the treatment of malaria, which is as effective as MQ or DOX as prophylactic agent. Chloramphenicol is also used in the treatment of malaria (Figure 21.12) [2].

Antibiotics play an important role in the treatment and prophylaxis (short-term chemoprophylaxis) against MDR *P. falciparum* infection. They are used as CT with quinolines (for example, CQ-DOX) to combat the resistant malaria. Antibiotics concentrate in RBCs/tissues and exert the antimalarial activity against the erythrocytic/exoerythrocytic stage of malaria parasite. The antimalarial activity is due to inhibition of mitochondrial protein synthesis (by binding reversibly to the 30S subunit of ribosome) in parasites [32, 34]. The antimalarial drugs of current use and their mode(s) of action are enlisted in Table 21.2.

21.7.2.2 Approaches to Current Antimalarial Treatment

In general, malaria is a curable disease if diagnosed promptly and treated correctly. The treatment of malaria, therefore, depends on the proper diagnosis of the infective parasite species, the stage of infection, the level of parasitemia, the type of drug resistance, and finally, the availability of effective antimalarial agents. The diagnosis of malaria involves the demonstration of parasites in the blood, usually by microscopy. Additional laboratory findings may include anemia, thrombocytopenia

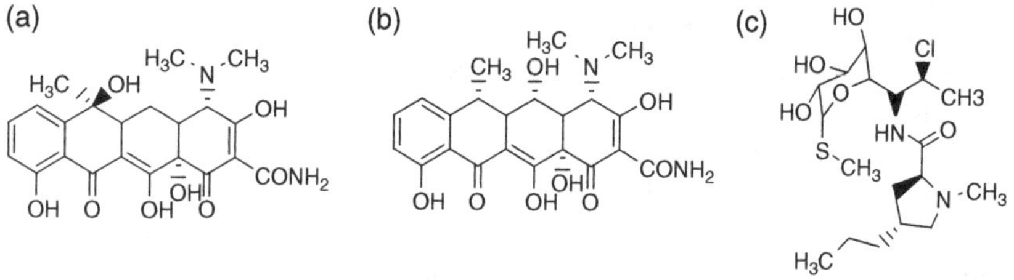

Figure 21.12 Structures of some antibiotics. *Source:* Based on Malaria Site. Available from http://www.malariasite.com/. Accessed 10 November 2019.

Table 21.2 Currently available antimalarial drugs.

Drug	Target of action	Mode(s) of action	Uses
CQ, AQ	BSS	Inhibit heme polymerase Alkylate heme	Treatment and prophylaxis of sensitive parasites
QN	BSS	Same as CQ	Treatment of CQ-R *P. falciparum*
MQ	BSS	Same as CQ	Treatment and prophylaxis of CQ-R *P. falciparum*
PQ	TSS and Gametocytocide	Inhibit heme polymerase	Radical cure and terminal prophylaxis of *P. ovale* and *P. vivax*
HL, LUM, PYN	BSS	Inhibit heme polymerase	Treatment of CQ-R *P. falciparum*
ATO	BSS	Inhibit mitochondrial electron transport	Treatment and prophylaxis of CQ-R *P. falciparum*, in combination with PG
PG	BSS	Inhibit DHFR and stops pyrimidine biosynthesis in DNA synthesis	Treatment and prophylaxis of CQ-R *P. falciparum* (with ATO) Prophylaxis of CQ-R *P. falciparum* (with CQ)
SP	BSS	Inhibit DHPS/DHFR-TS and hinders parasitic DNA	Treatment of CQ-R *P. falciparum*
ARTs (AS, AM, DHA)	BSS and Gametocytocide	Formation of iron-catalyzed free radical and alkylate heme	Treatment of MDR *P. falciparum* (ACTs)
TET, DOX	BSS	Inhibit mitochondrial protein synthesis and blocks nucleic acid synthesis	Treatment and prophylaxis of CQ-R *P. falciparum*

ACTs, artemisinin-based combination therapies; AM, artemether; AS, artesunate; ARTs, artemisinins; ATO, atovaquone; AQ, amodiaquine; BSS, blood-stage schizonticide; CQ, chloroquine; CQ-R, chloroquine-resistant; DHA, dihydroartemisinin; DOX, doxycycline; HL, halofantrine; LUM, lumfantrine; MDR, multidrug resistant; MQ, mefloquine; PG, proguanil; PQ, primaquine; PYN, pyronaridine; QN, quinine; SP, sulfadoxine-pyrimethamine; TET, tetracycline; TSS, tissue-stage schizonticide.

(decrease in blood platelets count), and jaundice (elevation of bilirubin and aminotransferases). If infection is not treated in time, sudden deterioration in the patient health may happen, leading to severe and life-threatening conditions since the parasites replicate rapidly in the blood. Severe malaria with *P. falciparum* infection may lead to death, if not treated promptly and adequately [2, 3, 5, 47]. Severe malaria is, therefore, a medical emergency and should be treated appropriately and urgently.

A sound knowledge of the life cycle of the malaria parasite is fundamental to understanding the various approaches of malaria prevention and treatment. The understanding of the biochemistry of malaria parasites led to identification of many potential targets and possible mode(s) of action of the drugs already in use. In *Plasmodium* life cycle, the asexual blood stage is responsible for the clinical signs and symptoms of malaria. The periodic fever and chills (appeared as first symptoms of malaria infection) occur when RBCs rupture and release the merozoites. Blood stage is therefore the primary target of chemotherapy/chemoprophylaxis. Most curative drugs act at this stage only. The sexual blood stage (gametocytes) do not cause any known pathology and is not a primary target for therapy [10, 20].

There are generally three classes of drugs used for the treatment and prevention of malaria. Drugs are classified according to their therapeutic action against different stages of life cycle of the parasite in human hosts: (i) the trophozoites and schizonts in the RBCs (erythrocytic stage of parasites); (ii) schizonts in the liver (exoerythrocytic stage of parasites), and (iii) gametocytes in RBCs (sexual erythrocytic stage). The drugs that are used for treating infection with blood stages of parasites (trophozoite/schizont) are called as **erythrocytic schizonticides (or Blood schizonticides)**. Examples of drugs in this category include CQ, QN, mepacrine, AQ, MQ, PG, HL, ARTs, ATO, SX, PYR, DDS, and tetracyclines. Drugs used for curing infection with liver schizont parasites are known as **exoerythrocytic schizonticides (or Tissue schizonticides)**. Examples of drugs include PQ, PG, LUM, PYR, and antibiotics. Drugs that are used to treat infection with the blood-stage gametocytes are known as **gametocytocides**. Examples of drugs are CQ, AQ, QN, and ART. These drugs may also act against liver schizonts. PQ is effective against both gametocytes and hypnozoites.

Antimalarial drugs are used for prophylactic and therapeutic purposes or both in malaria infection. Drugs are used for prophylactic purposes to prevent the spread of infection after inoculation. Drugs that are used to prevent infection from occurring attack the pre-erythrocytic stages of parasites, and are therefore termed as **liver schizonticidal**. The term schizonticidal is referred to as those interfere with the growth and multiplication of merozoites in blood or liver stages of parasites. In **causal prophylaxis**, drugs are used not to prevent sporozoites from reaching the liver of the human host, but, rather, to attack the early stages of the infection where parasites are still confined to the hepatocytes (liver parenchymal cells) and to prevent them (inhibition of growth and multiplication) from releasing merozoites into the bloodstream. For example, PG has poor schizonticidal action on intraerythrocytic stages **(blood schizonts)** of the parasite, while, it has high schizonticidal action on exoerythrocytic stages **(tissue schizonts)** of the parasite. PQ has greater effect on tissue schizonts than blood schizonts. Another type of prophylaxis exists where schizogony is not prevented, but parasites are kept at a sub-patent level so that no clinical symptoms are observed. This is known as **suppressive or clinical prophylaxis**. Blood schizonticides are effective as clinical prophylactic drugs. Antimalarial drugs are also used for curative purposes to act on existing infections. Drugs that are used for **therapeutic purposes** primarily act on erythrocytic stages (blood schizonts) of the parasites and thus stops the parasite life cycle at the erythrocytic stage. Dormant liver-stage parasites (hypnozoites) of *P. vivax* and *P. ovale* may reactivate weeks or months after the initial infection, producing relapses. The treatment of such clinical relapses or recrudescences is usually followed by anti-relapse drugs like PQ or PYR.

Chemoprophylaxis is generally meant for travelers, migrant laborers, and military personnel who travel to highly malaria endemic areas. It is generally recommended that chemoprophylaxis should be initiated two weeks prior to travel to malaria endemic areas. However, if DOX was prescribed, it is normally taken one to two days before departure. Prophylaxis should be continued for four weeks after leaving an endemic area to ensure success. MQ remains a drug of choice for prophylaxis before traveling to malaria areas where CQ-resistant *P. falciparum* exists [2, 3, 32–34].

21.8 Antimalarial Drug Resistance

The emergence of resistance to antimalarial drugs has become a challenging task in malaria control and prevention worldwide. This presents an acute clinical problem, which severely limited the effective use of available drugs in the treatment or prevention of malaria. It has increased the global incidence of severe malaria and malaria-related mortality.

In 1967, WHO defined drug resistance as the "ability of a parasite strain to survive and/or multiply despite the administration and absorption of a drug given in doses equal to or higher than those usually recommended but within the tolerance of the subject." This definition was later modified to include the sentence, "the form of the drug active against the parasite must be able to gain access to the parasite or the infected red blood cells for the duration of the time necessary for its normal action" [48, 49].

21.8.1 Development and Spread of Resistance

Various factors relating to drug, parasite, and human host interactions contribute to the development and spread of drug resistance (Figure 21.13). In addition, characteristics of drugs such as drugs with a long terminal elimination half-life enhance the

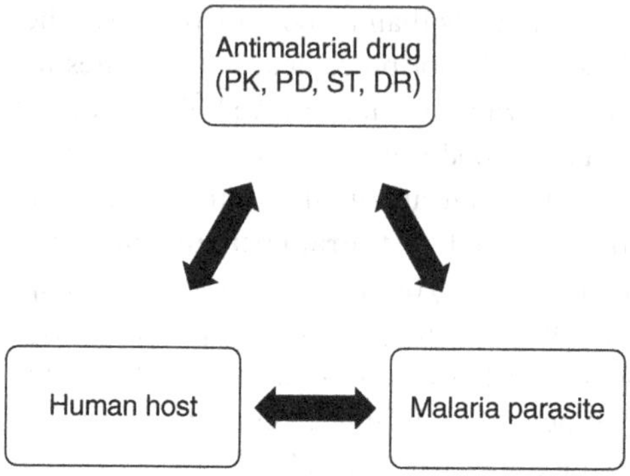

Figure 21.13 Interactions among the three components of malaria chemotherapy. PK, pharmacokinetics; PD, pharmacodynamics; ST, selective toxicity; DR, drug resistance.

development of resistance, particularly in areas of high transmission. Human host factors include the widespread incorrect use or irrational use of antimalarial drugs and possibly the level of host immunity. Parasite factors associated with the development of resistance include the biology of *Plasmodium* species concerned and the intensity of transmission. Vector and environmental factors also contribute to the spread of resistance [2].

The molecular mechanism of drug action is a critical element that defines the fundamental of development of resistance. However, genetic mutations of drug targets confer resistance to the parasites in nature. The malaria parasite is well known for its frequent, *de novo* mutations, mostly single, and sometimes multiple. In the presence of heavy infection and inadequate drug levels, the resistant mutants survive and propagate. Development of resistance requires a high level of parasitemia, coupled with low or inadequate drug levels. Most cases of resistance have emerged out of Southeast Asian region. This region is known for low immunity, high level of parasitemia, and a long history of indiscriminate use of different antimalarial drugs. One study also suggested that *P. falciparum* in Southeast Asia has an inherent propensity to develop drug resistance through genetic mutation.

Increased drug pressure is a significant contributor to drug resistance. As increased amounts of a drug are used, the likelihood that exposure of parasites to inadequate drug levels raises and resistant mutants are more readily developed. Drug pressure leads to higher release of gametocyte and this facilitates the propagation of the resistant mutants that have escaped the drugs. The use of PQ as a gametocytocidal agent in *P. falciparum* aids such spread of resistance. Administration of drugs with long elimination half-life also facilitates the spread of resistant malaria parasites. The residual antimalarial activity that is present during the post-treatment period serves as a "selective filter," which prevents infection by sensitive parasites but allows infection by resistant parasites. Drugs such as CQ, MQ, and PIP, which persist in the blood for months, provide a selective filter long after their administration has ceased. The development of drug resistance is more intense with *P. falciparum* compared with *P. vivax* and the higher incidence of resistance in *P. falciparum* is observed. Drug resistance results in a delay in or failure to clear asexual parasites from the blood, which allows production of the gametocytes of parasite that are responsible for transmission of the resistant genotype [2, 49, 50].

21.8.2 Current Status of Antimalarial Drug Resistance

Over the past three decades, the emergence of drug resistance, particularly against *P. falciparum*, has been increasing and is considered a major contributor to the global resurgence of malaria. Till date, drug resistance has been documented in *P. falciparum*, *P. vivax*, and *P. malariae* malaria. *P. falciparum* has developed resistance to nearly all antimalarial drugs currently in use. *P. vivax* has been reported to be resistant to CQ and PQ, and *P. malariae* has been found to be resistant to CQ and PYR in some endemic areas. *P. falciparum* has also become resistant to multiple drug therapy in many areas where the parasite is endemic. Multidrug resistance of *P. falciparum* is observed when the parasite is resistant to more than two operational drugs of different chemical classes and modes of action. Moreover, cross-resistance between CQ and AQ is common and development of resistance to MQ may lead to resistance to HL and QN.

Two classes of drugs that have been reported first to develop parasitic resistance are the 4-aminoquinolines (especially CQ) and the antifolate drugs (PYR, sulfonamides). CQ resistance is now common in most malaria-affected areas where *P. falciparum* occurs. Owing to widespread drug resistance, these drugs are becoming increasingly ineffective in malaria endemic areas of the world. Replacement of CQ with SP provided a temporary respite, until resistance to this combination has been emerged in Southeast Asia, South America, and Africa. Resistance to SP is also widespread and develops much more rapidly. Resistance to MQ and ATO is confined only to those areas where it has been widely used (Thailand, Cambodia, and Vietnam). Recently, evidence of resistance for ART-based drugs against *P. falciparum* has also been reported from the Southeast Asian region [49–51].

21.8.3 Mechanism of Drug Resistance

The biochemical mechanisms involved in the antimalarial drug resistance are a consequence of decreased accumulation of the drug at the target site as a result of enhanced efflux, reduced uptake of the drug, or a combination of both. Resistance to antimalarial drugs arises as a result of spontaneously occurring mutations that affect the structure and activity of the drug target (at the molecular level) in the malaria parasite or affect the access of the drug to that target.

21.8.3.1 CQ and Related Compounds

The generally accepted explanation for resistance against CQ is that resistant parasites accumulate less CQ than sensitive parasites. Thus, lethal concentrations of the drug are hindered from reaching the parasitic FV. The decrease in CQ accumulation can be attributed to a higher rate of CQ efflux, a lower rate of CQ uptake, or varying combinations of both these processes. It is believed that resistance of *P. falciparum* to CQ is related to an increased capacity for the parasite to expel CQ at a rate that does not allow CQ to reach levels required for inhibition of haem polymerization in the FV of parasites. Different genes/proteins that are involved in antimalarial drug resistance are depicted in Table 21.3.

However, several other mechanisms such as altered pH and decreased drug uptake into the parasite FV were also reported for the development of CQ resistance. It is still unclear whether parasite resistance to other quinoline-based antimalarials (AQ, MQ, HL, and QN) occurs via similar mechanisms. The resistance of CQ in *P. falciparum* may be multigenic and is initially conferred by mutations in a gene encoding a transporter called *P. falciparum* chloroquine-resistant transporter (*pfcrt*). Further, molecular studies suggested that point mutation of the *P. falciparum* multi-drug resistant (*pfmdr-1*) gene might be the cause of reduced susceptibility of QN, MQ, HL (and possibly CQ) [10, 14, 30].

21.8.3.2 Antifolate Drugs

There is also evidence of increasing antifolate resistance in malaria parasites. Antifolate resistance is generally observed due to a combination of mutations in

Table 21.3 Genes/proteins involved in antimalarial drug resistance.

Gene/protein	Function	Principal drugs affected
Pfcrt	Transporter	CQ, AQ, MQ, PQ, AQ, HL, LUM, ART derivatives, QN
Pfmdr-1	Transporter	CQ, AQ, MQ, HL, LUM, ART derivatives, QN
Pfdhps	Folate pathway enzyme (DHPS)	SX, DDS
pfdhfr-ts	Folate pathway enzyme (DHFR-TS)	PYR, PG, CHLORPROG
Cytb	Subunit of complex III (Cyt. bc1 complex) in electron transport chain	ATO
pfatp6	Membrane bound Ca^{2+}-transporting ATPase	ART derivatives
pfkelch13	Kelch protein gene on chromosome 13	ART derivatives

Pfcrt, *P. falciparum* chloroquine-resistant transporter; *pfmdr-1*, *P. falciparum* multidrug resistant gene 1; *pfdhps*, *P. falciparum* dihydropteroate synthase; *pfdhfr-ts*, *P. falciparum* dihydrofolate reductase-thymidylate synthase; *pfatp6*, *P. falciparum* sarco/endoplasmic reticulum calcium dependent ATPase (SERCA); *pfkelch13*, *P. falciparum* kelch protein gene on chromosome 13.

the target enzymes and the use of an alternative pathway to recover folate. The molecular basis for *P. falciparum* resistant to antifolates such as SP has been identified. Successive point mutations in *dhps* and *dhfr-ts* genes that encode *P. falciparum* DHPS and DHFR-TS are associated with the development of resistance to SX and PYR, respectively [2, 10].

21.8.3.3 Atovaquone and Artemisinins

Resistance is conferred by specific single-point mutations in the cytochrome-b (*cytb*) gene. Resistance to ART derivatives has also been reported recently. *P. falciparum* kelch protein gene on chromosome 13 (*pfkelch13*) is attributed to be responsible for the development of resistance in ART derivatives [2, 10].

21.8.4 Prevention of Drug Resistance

Interventions aimed at preventing drug resistance generally focus on reducing overall drug pressure through more selective use of drugs, improving the way the drugs are used through modified prescribing follow-up practices and patient compliance, or using drugs or drug combinations which are inherently less likely to foster resistance or have properties that do not facilitate development or spread of resistant parasites [2, 48–50]. The following approaches can be adopted to reduce the emergence and spread of drug resistance:

- Selection of drugs, use of conventional drugs first in uncomplicated cases.
- Avoiding of drugs with longer half-life, if possible.
- Avoiding of basic antimalarial drugs for non-malarial indications, for example, the use of CQ for rheumatoid arthritis in a malaria endemic area.
- Ensuring high degree of patient compliance by direct observation therapy (use of drugs with single-dose regimens such as SP and MQ).
- Monitoring for resistance and early treatment of resistant cases to prevent their spread.
- Early, accurate diagnosis, re-treatment, and closely follow-up patients.
- Clear policy of using newer antimalarial agents.
- Use of vaccines.
- Use of WHO-recommended multidrug CT such as SP, ATO-PG, MQ-AS, and LUM-AM.

21.8.4.1 Antimalarial CT

The concept of CT is based on the synergistic or additive activity of two or more drugs, to improve therapeutic efficacy and also to delay the development of resistance to the individual drugs of the combination. In antimalarial CT, two or more drugs are used together that act with independent mode of action probably at different biochemical targets in the life cycle of *Plasmodium* parasite. The criteria for selection of combinations of antimalarial drugs are:

1) Therapeutic efficacy of the combination, irrespective of the efficacy of the individual components.
2) Safety of the drugs in combination, especially among high risk groups.
3) Potential for widespread use of the combination at all levels of the health care, system, including its use for home management.
4) Potential for consumer compliance.
5) Cost effectiveness.
6) Potential to delay or prevent development of resistance.
7) Other factors including drug availability, and potential for widespread use.

Efforts to malaria control have been seriously impaired because of the emergence and spread of resistance to quinoline-based drug, mainly CQ. SP was, until recently, used to treat resistant cases of CQ. However, resistance to SP has also been developed, which reduces the clinical usefulness of this combination drug. Multidrug resistance to both CQ and SP occurs frequently in African and Southeast Asian regions. QN, either alone or in combination with TET, or MQ are used for the treatment of such MDR malaria. Decline in QN efficacy and high rates of MQ resistance have also been reported in the Southeast Asian region. The combination of AS with MQ has therefore been the drug of choice in MDR *P. falciparum* cases, with reduced recurrence rate and improved cure rate (particularly, in Southeast Asia). ART-based (for example, AS) monotherapy is not recommended as it increases the disease recurrence due to the rapid development of degree resistance. In order to prevent recurrence of the disease, ART derivatives are therefore given along with long-acting antimalarial drug(s) (drugs with longer half-life) in combined drug regimens. WHO recommended combining the rapid schizonticidal ART derivative (AS, AM, or DHA) with one or more partner drugs (from different antimalarial class having longer half-life) for the treatment of resistant *P. falciparum* malaria. Such combination antimalarial drug regimens are known as ACTs. According to WHO, ACTs are currently considered as the front-line treatment of MDR resistant *P. falciparum* malaria [2, 41, 52, 53]. It has been reported that such antimalarial combination drugs inhibit drug resistance and also decrease significantly the level of malaria transmission. The rationale for ACTs is that the ART derivative rapidly clears 95% of the parasites and the remaining 5% are cleared by the longer acting partner drug [2]. However, serious CNS and neurotoxicity are important concerns with ART-based therapy. In recent days, the therapeutic efficacy of ACTs has become limited due to the development of MDR strains of *P. falciparum*, particularly in Southeast Asia.

The antimalarial CT of current use and combinations in the pipeline are summarized in Table 21.4.

Table 21.4 Antimalarial combination therapy.

Type	Combination therapy
Combination in use	CQ + SP
Non-ART-based	CQ + SP, AQ + SP, ATO + PG, MQ + SP, QN + TCN/DOX, AQ + CHLORPROG
ART-based	AM + LUM (Co-Artem, fixed dose, AL), AS + CQ, AS + AQ. AS + MQ (AM), AS + SP, AS + DOX, AS + DOX + CQ
Combinations in the pipeline (new combinations)	PIP-DHA-Trimethoprim (Artecom), PIP-DHA-Trimethoprim (Artecom plus), Naphthoquine + DHA, PYN + AS, DHA + PIP (Artekin), Chloroguanil-Dapsone + AS (Lapdap plus), Artemisone (BAY449585) + MQ (withdrawn), Methylene blue + PQ (Phase I), Arterolane (OZ277) + PIP (Phase II–III), Artefenomel (OZ439) + PIP (Phase II-III), DHA + PIP + MQ (under TRAC-II, TACTs)

AM, artemether; AS, artesunate; ATO, atovaquone; AQ, amodiaquine; CHLORPROG, chloroproguanil; CQ, chloroquine; DHA, dihydroartemisinin; DOX, doxycycline; LUM, lumfantrine; MQ, mefloquine; PG, proguanil; PIP, piperaquine; PYN, pyronaridine; QN, quinine; SP, sulfadoxine-pyrimethamine; TCN, tetracycline; TACTs, triple artemisinin-based combination therapies; TRAC, tracking resistance toward artemisinin resistance.

21.9 Need for New Antimalarial Drugs

The emergence and spread of resistance of malaria parasites to available antimalarial drugs necessitates the discovery and development of novel antimalarials agents. To achieve this goal, antimalarial drug research needs to identify new drug candidates based on novel drug targets.

21.10 Antimalarial Drug Discovery

The aim of antimalarial drug discovery programme is to develop safe and affordable new drugs that would be reasonably cheap to produce, and effective against drug-resistant malaria with any toxic effects on human body. The antimalarial efficacy and toxicity are essential parameters for the selection of new drugs. Suitable *in vitro* and *in vivo* screening methods are therefore necessary for the evaluation of their efficacy and toxicity toward the development of novel antimalarial drugs. The antimalarial drug development, therefore, follows several strategies, ranging from traditional evaluation of potent agents from natural products to minor modifications of existing agents on a rational basis to develop novel drug molecules that act against new targets.

Table 21.5 General approaches to the discovery and development of new antimalarial drugs.

Approach	
Natural products and their derivatives	QN and ART, and their derivatives
Analogues of existing drugs	CQ, AQ, MQ, PQ, LUM, PIP, tafenoquine, tebuquine, new endoperoxides
Optimized therapy with available drugs including combination therapy	ACTs, folate drug combinations (SP), ATO/PG, CHLORPROG/DDS
Compounds active against other diseases	Antifolate drugs, ATO, antibiotics
Drug resistance reversers (chemosensitizers)	Chlorpheniramine and antidepressants
Compounds active against new targets	Protease inhibitors, phospholipid inhibitors

ACTs, artemisinin-based combination therapies; ATO, atovaquone; AQ, amodiaquine; CHLORPROG, chloroproguanil; CQ, chloroquine; DDS, dapsone; LUM, lumfantrine; MQ, mefloquine; PG, proguanil; PIP, piperaquine; PQ, piperaquine; QN, quinine; SP, sulfadoxine-pyrimethamine.
Source: Based on [54].

Table 21.5 depicts multiple general approaches that are pursued toward the development of new antimalarial drugs with novel modes of action, effective against MDR malaria.

21.10.1 Traditional Approaches

21.10.1.1 Development of New Antimalarial Drugs from Natural Sources

The investigation of medicinal plants having folkloric uses as antimalarial medicine may be potential sources of novel natural products that can be used for further development to potent antimalarial drugs or lead molecules. Natural products like QN and ART obtained from plant sources served as lead molecules in the development of many potent antimalarial drugs of current use. However, extensive phytochemical and pharmacological/toxicological evaluations are required for the discovery and development of antimalarial drugs/lead molecules from plant-based traditional remedies or bioactive natural products.

21.10.1.2 Development of New Antimalarial Agents from Synthetic Sources

Synthetic quinolines like CQ, AQ (4-aminoquinolines), PQ (8-aminoquinolines), MQ, HL, LUM (quinoline aminoalcohols), PIP (bisquinoline analogue), and PYN (benzonaphthyridine derivative) were developed based upon the structural template of QN. Several chemical strategies were involved in structural modification of QN/existing drugs to improve the therapeutic efficacy as well as toxicity. Tebuquine (4-aminoquinoline derivative, a CQ analogue) and tafenoquine (8-aminoquinoline derivative, a derivative of PQ) are two newer drugs added to the list. Ferroquine (4-aminoquinoline derivative, a CQ analogue, Phase II terminated) and AQ-13 (4-aminoquinoline analogue, Phase II) are presently under development. Synthetic endoperoxides related to ARTs such as trioxanes and trioxolane analogues are also being developed as newer antimalarial agents. Some newer peroxide-based antimalarials such as aterolane (OZ277), artefenomel (OZ439), and artemisone (BAY449585) are under clinical phase of development. Some newer drugs belonging to different classes that are under development are DSM265 [*Pf* dihydrooroate dehydrogenase (DHOH) inhibitor, a triazolopyrimidine-based drug, Phase II], MMV390048 [*Pf* phosphatidylisositol-4-kinase (PI4K) inhibitor, Phase I], and KAE609 or cipargamin (Na^+-ATPase 4 inhibitor). An antibiotic, fosmidomycin, is also under clinical phase of development (Phase IIa).

21.10.2 Modern Approaches

Traditionally, drugs are discovered by testing naturally derived or synthetically obtained compounds in time-consuming multistep processes against a battery of *in vitro* and *in vivo* screening methods. Compounds having promising therapeutic potential are further investigated for their development as drug candidates after pharmacokinetic, metabolism, and toxicity studies. Today's modern drug discovery process involves rational design and development of novel drug molecules based on a particular disease target using modern tools and techniques of virtual and experimental screening techniques (Figure 21.14). In virtual screening, computational methods screen large chemical libraries targeted toward a specific biological receptor, using advanced high-performance computing environments, data management software, and Internet. It delivers new drug candidates quickly and at lower costs [55]. Virtual screening is an approach of SBDD that uses computer-based (*in silico*) methods to discover and develop new drug molecules on the basis of biological structures of particular disease of interest.

SBDD methods mainly focus on the design of molecules for a disease target with known 3D structures followed by the determination of their binding affinity for the target by molecular docking along with other *in silico* screening methods (ADMET and toxicity screening) for optimization of molecules during development. Antimalarial drug discovery based on the SBDD approach involves the application of modern tools of molecular modeling and other *in silico* techniques in the development of novel antimalarial drug candidates (Figure 21.15).

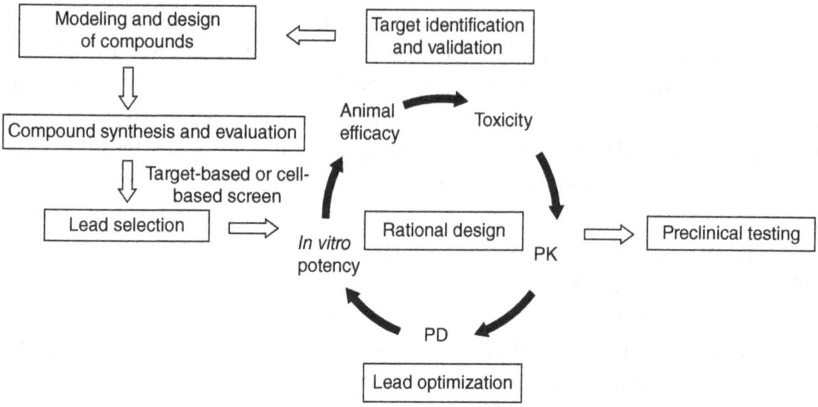

Figure 21.14 Drug discovery and development based on the SBDD approach.

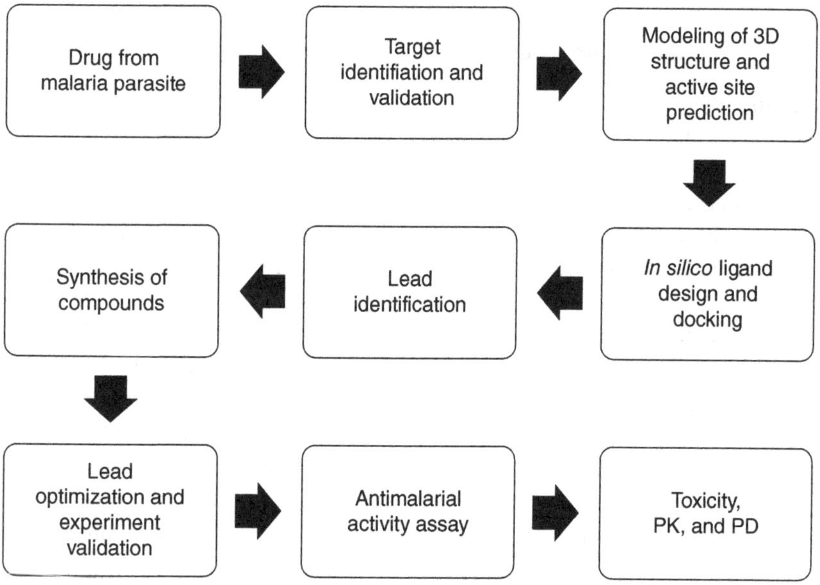

Figure 21.15 Antimalarial drug development based on the SBDD approach.

21.11 Conclusion

Chemotherapy is considered to be the mainstay in malaria control and elimination programme worldwide. Increasing resistance of malaria parasites to currently available drugs has significantly limited the success of malaria chemotherapy leading to the resurgence of the disease with increased severity and deaths. In recent years, the emergence of resistant *P. falciparum* strains is increasing at an alarming rate, which seriously threatens the clinical effectiveness of available antimalarial drugs and/or therapies. The number of effective drugs available to treat such resistant cases of malaria is gradually declining. Keeping it in view, there is an urgent need for the discovery and development new and affordable antimalarial drugs candidates from natural or synthetic sources with proven clinical efficacy and novel modes of antimalarial action.

References

1 Cox, F.E.G. (2010). History of the discovery of the malaria parasites and their vectors. *Parasit. Vectors* 3: 5.

2 Malaria Site (2019). http://www.malariasite.com/ (accessed 10 November 2019).

3 WHO Global Health (2020). Observatory (GHO) data: malaria. https://www.who.int/gho/malaria/en/ (accessed 20 September 2020).

4 Autino, B., Noris, A., Russo, R., and Castelli, F. (2012). Epidemiology of malaria in endemic areas. *Mediterr. J. Hematol. Infect. Dis.* 4 (1): e2012060.

5 WHO (2018). *World Malaria Report*. Geneva: World Health Organization. http://www.who.int/ (accessed 2 December 2019).

6 Dash, A.P., Valecha, N., Anvikar, A.R., and Kumar, A. (2008). Malaria in India: challenges and opportunities. *J. Biosci.* 33 (4): 583–592.

7 Rudrapal, M. and Chetia, D. (2016). Endoperoxide antimalarials: development, structural diversity and pharmacodynamic aspects with reference to 1,2,4-trioxane-based structural scaffold. *Drug Des. Dev. Ther.* 10: 3575–3590.

8 White, W.J. (2008). *Plasmodium knowlesi*: the fifth human malaria parasite. *Clin. Infect. Dis.* 46: 172–173.

9 Hay, S.I., Guerra, C.A., Tatem, A.J. et al. (2004). The global distribution and population at risk of malaria: past, present, and future. *Lancet Infect. Dis.* 4: 328–336.

10 Griffith, K.S., Lewis, L.S., Mali, S., and Parise, M.E. (2007). Treatment of Malaria in the United States (a systematic review). *Clin. Rev.* 297 (20): 2265.

11 Malaria (2020). Centers for Disease Control and Prevention. http://www.cdc.gov/malaria/about/biology (accessed 10 September 2020).

12 Banerjee, R. and Goldberg, D.E. (2001). The *Plasmodium* food vacuole. In: *Antimalarial Chemotherapy: Mechanisms of Action, Resistance, and New Directions in Drug Discovery* (ed. P.J. Rosenthal), 43–56. New Jersey: Humana Press.

13 Egan, T.J. (2003). Haemozoin (malaria pigment): a unique crystalline drug target. *Targets* 2: 115–124.

14 Foley, M. and Quinoline Antimalarials, T.L. (1998). Mechanism of action and resistance and prospects for new agents. *Pharmacol. Ther.* 79: 60–67.

15 Teixeira, C., Gomes, J.R.B., and Gomes, P. (2011). Falcipains, *Plasmodium falciparum* cysteine proteases as key drug targets against malaria. *Curr. Med. Chem.* 18: 1555–1572.

16 Singh, N., Sijwali, P.S., Pandey, K.C., and Rosenthal, P.J. (2006). *Plasmodium falciparum*: biochemical characterization of the cysteine protease falcipain-2. *Exp. Parasitol.* 112: 187–192.

17 Rosenthal, P.J., Sijwali, P.S., Singh, A., and Shenai, B.R. (2002). Cysteine proteases of malaria parasites: targets for chemotherapy. *Curr. Pharm. Des.* 8: 1659–1672.

18 Oliveira, M.E.D., Cenzi, G., Nunes, R.R. et al. (2013). Antimalarial activity of 4-metoxychalcones: docking studies as falcipain/plasmepsin inhibitors, ADMET and lipophilic efficiency analysis to identify a putative oral lead candidate. *Molecules* 18: 15276–15287.

19 Pandey, A.V., Bisht, H., Babbarwal, V.K. et al. (2001). Mechanism of malarial Haem detoxification inhibition by chloroquine. *Biochem. J.* 355: 333–338.

20 Egan, T.J. and Marques, H.M. (1999). The role of haem in the activity of chloroquine and related antimalarial drugs. *Coord. Chem. Rev.* 190–192: 493–517.

21 Meshnick, S.R. and Dobson, G.M.J. (2001). The history of antimalarial drugs. In: *Antimalarial Chemotherapy: Mechanisms of Action, Resistance, and New Directions in Drug Discovery* (ed. P.J. Rosenthal), 2–25. New Jersey: Humana Press.

22 Dorn, A., Stoffel, R., Matile, H. et al. (1995). Malarial hemozoin/β-hematin supports haem polymerization in the absence of protein. *Nature* 374: 269–271.

23 Sullivan, D.J., Gluzman, I.Y., Russell, D.G., and Goldberg, D.E. (1996). On the molecular mechanism of chloroquine's antimalarial action. *Proc. Natl. Acad. Sci.* 93: 11865–11870.

24 Casteel, D.A. and Agents, A. (2003). Burger's medicinal chemistry and drug discovery. In: *Chemotherapeutic Agents*, vol. 5 (ed. P.J. Abraham), 920–931. New York: Wiley.

25 O'Neill, P.M. and PM, P.G.H. (2004). A medicinal chemistry perspective on artemisinin and related endoperoxides. *J. Med. Chem.* 47: 2945–2964.

26 Rudrapal, M. (2011). A brief review on malaria and current antimalarial drugs. *Curr. Pharma. Res.* 1 (3): 286–292.

27 Rudrapal, M. and Chetia, D. (2014). *Malaria: Chemotherapy and Development of Novel Antimalarial Agents*. Germany: LAP LAMBERT Academic Publishing.

28 Bhagavathula, A.S., Elnour, A.H., and Shehab, A. (2016). Alternatives to currently used antimalarial drugs: in search of a magic bullet. *Infect. Dis. Poverty.* 5: 103.

29 (2019). *Malaria Control Programmes*. Geneva: World Health Organization. http://www.who.int/ (accessed 16 October 2019).

30 Robert, A., Benoit-Vical, F., Dechy-Cabaret, O., and Meunier, B. (2001). From classical antimalarial drugs to new compounds based on the mechanism of action of artemisinin. *Pure Appl. Chem.* 73: 1173–1180.
31 O'Neill, P.M., Bray, P.G., Hawley, S.R. et al. (1998). 4-Aminoquinolines: past, present and future: a chemical perspective. *Pharmacol. Ther.* 77: 29–58.
32 Chiang, P.K., Bujnicki, J.M., Su, S., and Lanar, D.E. (2006). Malaria: therapy, genes and vaccines. *Curr. Mol. Med.* 6: 309–314.
33 Warrell, D.A. (1992). Cerebral malaria. *Schweiz Med. Wschr.* 122: 879–886.
34 Ridley, R.J. (2002). Medical need, scientific opportunity and the drive for antimalarial drugs. *Nature* 415: 686–693.
35 Bray, P.G., Ward, S.A., and O'Neill, P.M. (2005). Quinolines and artemisinin: chemistry, biology and history. *Curr. Top. Microbiol. Immunol.* 295: 3–38.
36 Dinio, T., Gorka, A.P., McGinniss, A. et al. (2012). Investigating the activity of quinine analogues versus chloroquine resistant *Plasmodium falciparum*. *Bioorg. Med. Chem.* 20: 3292–3297.
37 Luo, X.D. and Shen, C.C. (1987). The chemistry, pharmacology, and clinical applications of Qinghaosu (artemisinin) and its derivatives. *Med. Res. Rev.* 7: 29–52.
38 Balint, G.A. (2001). Artemisinin and its derivatives: an important new class of antimalarial agents. *Pharmacol. Therap.* 90: 261–265.
39 Liu, Y., Cui, K., Lu, W. et al. (2011). Synthesis and antimalarial activity of novel dihydro-artemisinin derivatives. *Molecules* 16: 4527–4538.
40 Kyle, D.E., Teja-Isavadharm, P., Li, Q. et al. (1998). Pharmacokinetics and pharmacodynamics of qinghaosu derivatives: how do they impact on the choice of drug and the dosage regimens? *Med. Trop.* 58: 38–44.
41 Garner, P. and Graves, P.M. (2005). The benefits of artemisinin combination therapy for malaria extend beyond the individual patient. *PLoS Med.* 2: e105.
42 Park, V.K., O'Neill, P.M., Maggs, J.L. et al. (1998). Safety assessment of peroxide antimalarials: clinical and chemical perspectives. *Br. J. Clin. Pharmacol.* 46: 521–529.
43 Krungkrai, S.R. and Yuthavong, Y. (1987). The antimalarial action on *Plasmodium falciparum*, *Qinghaosu* and artesunate in combination with agents that modulate oxidative stress. *Trans. R. Soc. Trop. Med. Hyg.* 81 (5): 710–714.
44 White, N.J. (1994). Clinical pharmacokinetics and pharmacodynamics of artemisinin and derivatives. *Trans. R. Soc. Trop. Med. Hyg.* 88: S41–S43.
45 White, N.J. (1997). Assessment of the pharmacodynamic properties of antimalarial drugs *in vivo*. *Antimicrob. Agents Chemother.* 41: 1413–1422.
46 Meshnick, S.R. (2002). Artemisinin: mechanisms of action, resistance and toxicity. *Int. J. Parasitol.* 32: 1655–1660.
47 Geleta, G. and Ketema, T. (2016). Severe malaria associated with *P. falciparum* and *P. vivax* among children in Pawe Hospital, Northwest Ethiopia. *Malaria Res. Treat.* 6: 1–7.

48 Farooq, U. and Mahajan, R.C. (2004). Drug resistance in malaria. *J. Vect. Borne Dis.* 41: 45.

49 Bloland, P.B. (2001). *Drug Resistance in Malaria*. Geneva: World Health Organization.

50 Winstanley, P.A., Ward, S.A., and Snow, R.W. (2002). Clinical status and implications of antimalarial drug resistance. *Microb. Infect.* 4: 157–164.

51 Bethel, D., Se, Y., Lon, C. et al. (2011). Artesunate dose escalation for the treatment of uncomplicated malaria in a region of reported artemisinin resistance: a randomized clinical trial. *PLoS ONE* 6 (5): e19283.

52 NIH, U.S. National Library of Medicine (2015). *A Study by the Tracking Resistance to Artemisinin Collaboration (TRAC) (TRACII)*. London: University of Oxford. (updated 2 December 2015). https://clinicaltrials.gov/ct2/show/NCT02453308 (accessed 8 September 2019).

53 Golenser, J., Waknine, J.H., Krugliak, M. et al. (2006). Current perspectives on the mechanism of action of artemisinins. *Int. J. Parasitol.* 36 (14): 1427–1441.

54 Aguiar, A.C.C., da Rocha, E.M.M., de Souza, N.B. et al. (2012). New approaches in antimalarial drug discovery and development – a review. *Mem. Inst. Oswaldo Cruz* 107: 831–845.

55 Biamonte, M.A., Wanner, J., and Le Roch, K.G. (2013). Recent advances in malaria drug discovery. *Bioorg. Med. Chem. Lett.* 23: 2829–2843.

22

Therapeutic Potentials of Phytochemicals Against *Mycobacterium tuberculosis*, the Causative Agent of Tuberculosis

Alloysius Chibuike Ogodo[1], Verwiyeh Silas Tatah[2], Francis Ushie Ebuara[1], and Chidera Favour Ogodo[3]

[1] *Department of Microbiology, Faculty of Pure and Applied Sciences, Federal University Wukari, Wukari, Nigeria*
[2] *Department of Biochemistry, Faculty of Pure and Applied Sciences, Federal University Wukari, Wukari, Nigeria*
[3] *Department of Medical Services, Federal University Wukari, Wukari, Nigeria*

List of Abbreviations

AMK	Amikacin
Cap-NMR	Capillary nuclear magnetic resonance
CS	Cycloserine
DNA	Deoxyribonucleic acid
EMB	Ethambutol
ETH	Ethionamide
FLD's	First-line drugs
FQs	Fluoroquinolones
HIV	Human immunodeficiency virus
INH	Isoniazid
KAN	Kanamycin
MDR-TB	Multidrug resistance tuberculosis
MIC	Minimum inhibitory concentration
NAD	Nicotinamide adenine dinucleotide
NTD	Neglected tropical disease
PAS	*p*-Amino salicyclic acid
PTH	Prothionamide
PZA	Pyrazinamide
RIF	Rifampin

Neglected Tropical Diseases and Phytochemicals in Drug Discovery, First Edition.
Edited by Chukwuebuka Egbuna, Muhammad Akram, and Jonathan Chinenye Ifemeje.

RNA	Ribonucleic Acid
rRNA	Ribosomal ribonucleic acid
SLD's	Second-line drugs
SM	Streptomycin
SPE	Solid-Phase Extraction
TB	Tuberculosis
TDR	Totally drug-resistant
TMPs	Traditional medicine practitioners
WHO	World Health Organization
XDR-TB	Extensively multidrug resistance tuberculosis

22.1 Introduction

Mycobacterium tuberculosis, also known as *Tubercle bacilli* is the etiologic agent of tuberculosis (TB). The organism is characterized by its unusual cell wall structure with complex lipid called mycolic acid that is responsible for its acid fastness. Mycolic acid, which forms a waxy layer on the cell, protects the organism against antibacterial effects of some antibiotics [1]. *Mycobacterium tuberculosis* is contained in airborne aerosol (droplet nuclei) which are generated when people infected with pulmonary or laryngeal tuberculosis disease shout, cough, or sneeze. The infectious particle can remain suspended in the air for a long time through which they can be transmitted by inhalation of the droplet nuclei containing *M. tuberculosis,* which then enters the mouth or passes through the nasal passages to the upper respiratory tract and the bronchi thereby reaching the alveoli of the lungs, leading to infection (tuberculosis) (Figure 22.1). The infection can also occur through oral route [2].

Tuberculosis is an age-long disease that has existed for millennia yet remains a major global health problem. It is among the top 10 causes of death worldwide, ranking above HIV/AIDS. Although TB is not currently classified among the neglected tropical disease (NTD) by the WHO because tuberculosis receives greater funding and treatment alongside malaria and HIV/AIDS. However, there is a comparison of the effect of NTD to tuberculosis and malaria in sub-Saharan Africa. Moreover, the co-infection of neglected tropical diseases with HIV/AIDS and tuberculosis results in a high mortality rate [3–6].

The introduction of effective drug therapy regimen using isoniazid (INH), rifampicin, pyrazinamide (PZA), and ethambutol (EMB) did not stop the global spread of tuberculosis, therefore, TB remains a global public health problem and emergency [5]. There is a need for efforts to be intensified towards the eradication of TB through the use of new drugs with higher activity and low side effects. Plants used as food or medicine with phytochemicals could be a potential

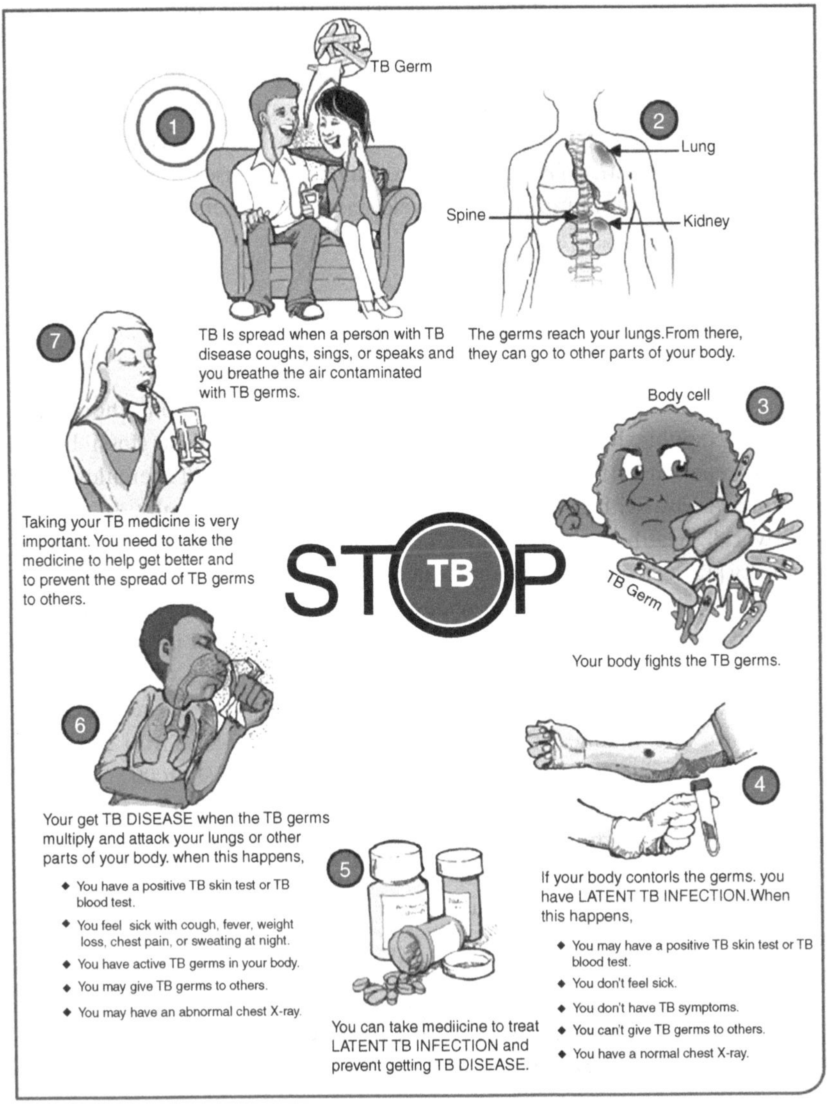

Figure 22.1 How TB is spread and the difference between latent TB infection and TB disease. *Source:* Stop TB, Centers for Disease Control and Prevention, 17 September 2020. https://npin.cdc.gov/publication/stop-tb-2 (accessed 17 September 2020).

antimycobacterial or antituberculosis agents. Natural products derived from plants or phytochemicals such as alkaloids, terpenoids, flavonoids, etc. have been reported to be a potential antituberculosis agent. Natural antimicrobials are now considered to be more promising in combating the many increasing concerns of antibiotic resistance and multidrug resistance TB than combinatorial chemistry and other synthetic procedures [5, 7, 8]. Spices have been used in India with natural herbs in food or as medicine to maintain health, hygiene, and proper sanitation to increase longevity [9]. The use of plants or plant natural products as food or medicine reduces the chance of developing antibiotic-resistant strains of the infectious agent (*M. tuberculosis*) as a result of drug abuse, decreases the cost of treatment, and minimizes the development of adverse drug reactions (ADRs) [5].

22.2 Disease Burden of Tuberculosis

Tuberculosis has been known since prehistoric times to be among the world's most dreaded public health diseases. Moreover, spinal tuberculosis (fossil bones) has been reported to be evidenced around 800 BC [10, 11]. Tuberculosis is a pandemic disease with a high mortality rate in humans globally. In fact, in 1993 world health organization (WHO) declared tuberculosis a global emergency [12]. According to the report of Christopher et al. [13], one-third of the population of the world have *M. tuberculosis* infection (TB). In 2012, approximately 8.6 million cases of tuberculosis were reported globally, most of these cases were found in Asia (58%) and Africa (27%), Eastern Mediterranean (8%), European (4%), and America (3%). India has the highest incidence of TB, followed by China, South Africa, and Indonesia. In Indonesia, the incidence of TB in 2012 was estimated at 0.5 million and the prevalence is estimated at 297/100 000 population and the mortality rate of 27/100 000 population. WHO reported 9.6 million cases of TB in 2014 with 1.5 million deaths. More so, an estimated 1 million children were infected with 0.14 million deaths [14, 15]. In 2016, 6.3 million new cases of Tb were reported with 1.3 million deaths in addition to 374 000 cases reported among HIV patients [16].

22.3 Drugs for Tuberculosis (First- and Second-Line Treatment)

Tuberculosis has continued to be an enigma to tackle, ravaging the health of many people globally. TB is still considered a pandemic disease and a copious infectious agent plaguing human health with a high mortality rate worldwide [11]. The existence of *M. tuberculosis* strains with multidrug resistance has inspired efforts made so far in drug discovery and development to tackle this menace. However, more extensive research is still ongoing by scientists to developed antimycobacterial drugs with improved modes of action and presents more tolerable side effects [11].

Chemotherapy has been the most commonly employed treatment method used to control the spread of *M. tuberculosis*, provided some conditions are ensured, this include;

i) Combination of antituberculosis drugs to curtail the emergence of new resistance strains,
ii) Strict adherence to medication during the treatment period
iii) Prevention of disease degeneration.

Research has shown that most antibiotics that can be used in treating TB, work more effectively during the exponential growth phase of bacteria [17]. In this phase of TB treatment, the antibiotics destroy rapidly growing bacteria that are responsible for sputum formation, and the removal of clinical symptoms [18]. However, the slow-growing strains bacterial cannot be easily treated by most antibiotics, since the easily gain resistance during treatment. Therefore, in order to kill the persistent or slow-growing strains of *M. tuberculosis*, there is a need for the continuation phase of the treatment.

There has been a lot of research to develop more potent TB drugs that can combat the propagation of the disease, following the emergence of resistant strains. The painstaking research by scientist over the years has led to the discovery of effective first-line antituberculosis drugs (FLD's) between 1950s and 1970s, this discovery drastically had an immediate impact in decreasing the number of cases of TB, especially in developed countries that had access to the drugs [18]. For effective treatment of tuberculosis, a course chemotherapy is normally judiciously followed.

First-line drugs (FLD) such as EMB, INH, PZA, streptomycin (SM), and rifampin (RIF) (Table 22.1), can effectively be used in the management of TB. However, there are cases where treatment with the FLD fails to cure tuberculosis due to poor adherence to treatment instructions, which contributes in creating strains of the bacteria (*Mycobacterium tuberculosis*) that are resistant to antibiotics [19]. The advent of multidrug-resistant tuberculosis (MDR-TB), TB which is resistant to at least INH and RIF [17], became a source of great concern, hence leading to the development of second-line drugs (SLD) that are scarce, more toxic, and expensive than FLDs [20]. Therefore, the detection and treatment of drug-susceptible or single drug-resistant TB is a smart technique that has been used in preventing the spread of MDR-TB [21].

22.3.1 First-Line Antituberculosis Drugs

The FLD used against tuberculosis are INH, rifampicin, EMB, PZA, and SM. These are discussed below.

Table 22.1 Some first- and second-line drugs used against tuberculosis.

Drugs	Mechanism of action	Resistant gene	MIC (μg/ml)	Drug line	References
Pyrazinamide	It is involved in cellular energy depletion and interference with transmembrane activities	*pncA*	20.0–100.0	First	[28]
Streptomycin	It inhibits the ribosomal protein biosynthesis in cell	*rpsL*	2.0–8.0	First	[32]
Isoniazid	It acts by inhibiting cell wall mycolic acid, lipids, carbohydrates, and NAD metabolism	*katG*	0.02–0.2	First	[23]
Ethambutol	Bactericidal inhibition of cell wall arabinogalactan synthesis	*embCAB*	1.0–5.0	First	[28]
Rifampicin	It acts by inhibiting the synthesis of RNA	*rpoP*	0.5–1.0	First	[24]
Fluoroquinolones	Interferes with protein biosynthesis by inhibiting DNA gyrase	*gyrA/gyrB*	0.5–2.0	Second	[39]
Kanamycin	Inhibits biosynthesis of proteins	*rpsL*	1.0–8.0	Second	[40]
Amikacin	Inhibits biosynthesis of proteins	*rrs*	2–4 (mg/l)	Second	[40]
Cycloserine	Involved in the inhibition of peptidoglycan synthesis	*alrA*	5.0–20.0	Second	[40]
p-Aminosalicylic acid	Inhibition of the biosynthesis of thymine	*thyA*	0.5	Second	[41]
Ethionamide	—	*inhA*	2.5–10	Second	[23]

22.3.1.1 Isoniazid

INH was developed in 1952 and has been used as one of the most effective and specific first-line antituberculosis drug that has played a vital role in the management of TB. *M. tuberculosis* has been reported to be highly susceptible to INH at a minimum inhibitory concentration (MIC) range of 0.02–0.2 μg/ml [22]. INH under anaerobic conditions acts only against developing tubercle bacilli, rather than non-replicating bacilli [22]. The drug gain entrance into the mycobacterial cell through passive diffusion with its most adverse reactions being hepatotoxicity and neurotoxicity [23].

22.3.1.2 Rifampicin

Following the development of INH in 1952, more research led to the discovery of Rifampicin in 1972 a potent antituberculosis drug that served as an excellent alternative. Its mechanism of action involve binding to the β-subunit of RNA polymerase (*rpoB*), an enzyme that is involved in the transcription and expression of mycobacterial genes, leading to inhibition of the bacterial transcription activity and thereby destroying the organism [24]. Rifampicin unlike INH has a unique characteristic in that it is active against actively growing and slowly metabolizing bacilli [25] and it produces lesser side effects than INH following administration. The MICs of rifampicin ranges between 0.05 and 1 μg/ml on solid or liquid media, but the MIC is higher in egg media (MIC = 2.5–10 μg/ml) [26].

22.3.1.3 Ethambutol

EMB [dextro-2,2′-(ethylenediimino) di-1-butanol], is an important FLD in the treatment of TB and drug-resistant TB. It is a bacteriostatic agent that supports the activity of other antibiotics such as quinolones and rifamycins as well as aminoglycosides in the treatment of mycobacterial infections. However, EMB has been reported to have some side effects which include loss of appetite, dizziness, nausea, blurred vision, vomiting, stomach pain, color blindness, headache, rash, itching, face swelling, and fingers/toes tingling [27]. EMB inhibits *M. tuberculosis* with MICs between 0.5 and 2 μg/ml. Its mechanism of action is by affecting the cell wall integrity through the inhibition of arabinogalactan biosynthesis as well as the biosynthesis of lipoarabinomannan which causes D-arabinofuranosyl-P-decaprenol (intermediate in the biosynthesis of arabinan) to accumulate [27, 28].

22.3.1.4 Pyrazinamide

The discovery of PZA was a milestone in the treatment of MDR-TB. PZA is an important first-line antituberculosis (anti-TB) drug that has been utilized in succinct chemotherapy and is one of the cornerstone drugs in MDR-TB treatment [25]. Its unique feature is based on its ability to inhibit semi-dormant bacilli growing in acidic environments [25]. PZA is a pro-drug which is analogous in structural to nicotinamide. The enzymes pyrazinamidase or nicotinamidase (PZase) converts PZA into pyrazinoic acid, the active form of the drug [29], hence PZA can only be active against *M. tuberculosis* at an acid pH with MICs of 6.25–50 μg/ml [28]. Reported side effects of PZA include stomach upset and hypersensitivity reactions [30].

22.3.1.5 Streptomycin

SM, an aminocyclitol glycoside antibiotic was first used to treat TB in 1948. SM is effective against *Mycobacterium tuberculosis* that is in its active growth stage (MICs = 2–4 μg/ml) and do not show activity against intracellular tubercle bacilli that is not actively growing [31]. SM interferes with the proofreading of translation by binding to 16S ribosomal ribonucleic acid (rRNA), leading to the inhibition of protein biosynthesis [32]. Although SM administrations are reported to have side effects such as nephrotoxicity and ototoxicity with vestibular dysfunction, its renal toxicity is less than that of either capreomycin or kanamycin. There have also been reports of potential hearing and renal dysfunction. Hence, patients administered with SM should be monitored properly [32].

22.3.2 Second-Line Antituberculosis Drugs

The SLD for the treatment of tuberculosis which are highlighted below are *p*-Aminosalicylic acid, fluoroquinolones, cycloserine, aminoglycosides, and ethionamide.

22.3.2.1 *p*-Amino Salicylic Acid

The combination of *p*-Aminosalicylic acid (PAS) with SM and INH have been reported to be effective in treating TB [33]. *p*-Aminosalicylic acid is one of the early first-line antituberculosis drug, however, its use as a first-line drug was discontinued following the discovery of potentially more effective drugs such as rifampicin. Moreover, PAS is still among the drugs used to treat extensively drug resistance tuberculosis (XDR-TB), although a high level of toxicity has been associated with its administration in addition to other side effects such as gastrointestinal disturbances [34].

22.3.2.2 Fluoroquinolones

The fluoroquinolones (FQs), second-line antituberculosis drug have a broad-spectrum antimicrobial function and so are broadly used in treating a variety of infections due to bacteria including respiratory tract infection, gastrointestinal tract infection and urinary tracts infections, as well as sexually transmitted diseases (STD), and chronic osteomyelitis [35]. Report indicated that FQs have exceptional anti-*M. tuberculosis* (both *in vitro* and *in vivo*) activity [36]. Other drugs in the same class with FQs include moxifloxacin, ofloxacin, and ciprofloxacin in addition to levofloxacin. Currently, FQs are used as SLD in TB treatment with little or no side, although gastrointestinal intolerance, dizziness, headache, and rashes have been reported on short-term administration for bacterial infections [17].

22.3.2.3 Cycloserine

Cycloserine (CS) is an antibiotic used for tuberculosis treatment. Its mechanism of action is still unclear; however, studies show that it can prevent tubercle bacilli from synthesizing peptidoglycans, which are a major component of the bacteria cell wall. This leads to a vulnerable bacteria cell wall, resulting in the killing of the bacteria. Cycloserine possesses very high gastric tolerance and has no cross-resistance to other compounds. But reports indicate that it can cause adverse psychiatric effects [35, 36] which is a major side effect. Hence, it is of paramount importance to carry out a psychiatric interrogation before prescribing cycloserine drug. Cycloserine has proven to be one of the backbones of treatment for MDR and XDR tuberculosis [17].

22.3.2.4 Aminoglycosides (Kanamycin, Amikacin, and Capreomycin)

Amikacin (AMK), kanamycin (KAN), and capreomycin (CAP) drugs are used against MDR-TB and are in injectable forms. They exert their metabolic effect by interfering with the translation of protein. There have been reports of renal toxicity following administration of the drugs. Hence there is a need for constant monitoring of hearing and renal function to avoid any form of dysfunctions [37].

22.3.2.5 Ethionamide/Prothionamide

Ethionamide (ETH, 2-ethylisonicotinamide), an anti-tuberculosis drug, is derived from isonicotinic acid and is one of the effective drugs in the treatment of tuberculosis as early as 1956. ETH shows *in vitro* anti-*Mycobacterium tuberculosis* activity with MICs ranging from 0.5 to 2 μg/ml (liquid medium), from 2.5 to 10 μg/ml (7H11 agar), and from 5 to 20 μg/ml (LJ medium) [33]. Ethionamide and prothionamide (pro-drugs) activated by a mono-oxygenase [37] and inhibits *InhA* during the biosynthesis of mycolic acid [38]. Its mechanism of action involves the oxidation of sulfur groups by flavin monooxygenase, and the drug is then converted to 2-ethyl-4-aminopyridine. The intermediate products formed before 2-ethyl-4-aminopyridine have been reported to be toxic to mycobacteria, though they are not known yet possibly due to their instability [38].

22.4 The Role of Plants as Antimycobacterial Agents

Studies have shown that multi drugs resistance-TB (MDR-TB) can be treated with FLDs in combination with SLD like *p*-aminosalicylic acid, amikacin, ofloxacin, ethionamide, kanamycin, and scapreomycin [11]. However, there are strain termed XDR-TB (extensively drug-resistant TB) [42] that are resistant to either INH or rifampicin and any of injectable SLD like amikacin, capreomycin, and kanamycin. Despite the effectiveness of the regiment active TB treatment, it is associated with so many ADRs, making chemotherapy of tuberculosis a challenge [43].

Recently, there have also been reports of the emergence of totally drug-resistant (TDR) tuberculosis strains [44]. Studies show that these new strains are intolerant to all the FLD and most SLD used for TB treatment [42]. This has led to the quest and search for alternative antituberculosis drugs with high potency for TB treatment base on plant secondary metabolites.

Medicines from plant sources are known to be easily absorbed by the body with less adverse side effects to the body [45]. Unlike MICs of antimicrobial agents derived from microorganisms, plant-based MICs are much higher with much more inhibitory and resistance altering capacity [45]. The presence of multiple functional groups as shown by GC-MS analysis of many plant extracts has been reported by Venkitanarayana [46] to be responsible for the antimicrobial action in a plant as a result of several mechanisms of action, hence altering the bacterial capacity to develop resistance against secondary metabolites. Therefore, plants could serve as a potent complementary alternative therapy in combating the spread of MDR strains of *Mycobacterium tuberculosis.*

22.4.1 The Metabolic Role of Plant Extract in Combating *Mycobacterium tuberculosis*

There are different ways in which plant secondary metabolites alter the cells of microbes amongst which includes; changing membrane structure and function, interruption of DNA replication, inhibition of RNA transcription and protein synthesis, coagulation of cytoplasmic components, interference with quorum sensing, metabolic processes, gene regulation, and inhibition of the synthesis of cell enzymes, etc. [47]. Literature reports show that the primary antimicrobial mechanism of action of plant secondary metabolites is by the alteration of cell membrane integrity of bacteria [48]. They affect membrane structural integrity, permeability, or function. Studies show that as lipophilicity increases, the action of plant metabolites as antibacterial agents also increases due to an increase in their interaction with the cell membrane [49].

Inhibition of quorum sensing by plants secondary metabolites is one of the basic mechanisms used in combating microbial infections since pathogens employ it during the production of toxins, biofilm, and other chemical substances that increase virulence [50, 51]. Reports have shown that most extracts from plants help increase antibiotic potency [52, 53]. Hence, apart from their antimicrobial activity, many secondary metabolites from plants are also evaluated for their potentiating role as antibiotics and their capacity to reverse antibiotic resistance [54]. Most medicinal plants possess broad-spectrum bioactive compounds [55] with various phytochemicals having a different antimicrobial mode of action [56]. The major action of phytochemicals includes causing changes in the structure and function of proteins, interruption nucleic acid synthesis, alteration of the structure and function of the cell membrane, and inhibition of cytochrome P450 thereby enhancing the bioavailability of active substances in secondary metabolites from plants [55, 57].

22.4.2 Global Reports of Plants with Antimycobacterial Potential

The use of ethnomedicine as a way of combating mycobacterial diseases has been commonly practiced in different parts of the world over the years. Currently, a lot of progress has been made to promote the ethnomedicinal knowledge of plant-based therapeutics used to treat mycobacterial diseases with plant species globally. A lot of Surveys have been carried out to determine the antimycobacterial potential of various plant parts from different parts of the world to serve as suitable substitute or complement for chemotherapy to combat multidrug-resistant TB and some other side effects recorded with the use of conventional anti-TB drugs. Ethno-medicine surveys from different parts of the world have identified various plant parts used to treat mycobacterial diseases in various countries. The phytochemical analysis of these plant extracts showed that the extracts predominantly contains flavonoids, alkaloids, saponins, tannins, terpenes/terpenoids, steroids, flavonoids, and glycosides. This extract has shown high MICs against mycobacterial compared to conventional drugs.

An ethnobotanical survey carried out by Lawal et al. [58] in Eastern Cape province, South Africa reported 30 different plants used in the treatment of tuberculosis by traditional medicine practitioners (TMPs). In this survey, *Clausena anisata*, *Artemisia afra* and *Haemanthus albiflos* were reported to be the most commonly used plants for the treatment of tuberculosis (Table 22.2).

22.4.3 Plant Metabolites with Anti-Mycobacterial Activity

Secondary metabolites from plant sources have continued to be of great interest due to their potential to provide novel chemical structures that are fundamental in drug discovery with high efficacy as an antimicrobial agent. The high biodiversity of plants, have made them the primary repository of phytochemicals owing to their rich secondary metabolite content. These secondary metabolites contain active biomolecules with antimycobacterial properties. Phytochemical like alkaloids, flavonoids, phytosterols, saponins, tannins, etc. have been extracted from different plants and their GC–MS analysis has identified active compounds with antimycobacterial activity.

Mativandlela [63], reported that the active components of *Galenia africana* leaves were (2S)-5,7,2′-trihydroxy flavanone (E)-3,2′,4′- trihydroychalcone, (E)-2′,4′-dihydroychalcone and (E)-3,2′,4′-trihydroy-3-methoxychalcone. *Withania somnifera* leaves and roots were reported by Adaikkappan et al. [82] to have holine and ashwag with a ferin A, Pterocarpan-hydroxy-tuberosin, hydroxyl-isoflavone, tuberostan, puerarone as active ingredients with anti-mycobacterial potentials. *Zanthoxylum capense* roots have benzophenanthridine, decarine, 6-acetonyldyhydronitidine, *N*-isobutyl-(2E,4E)-2,4-tetradecadienamide as

Table 22.2 Reports of plants with antimycobacterial potentials.

S/N	Name of plant	Plant parts	Country/region	References
1)	*Clausena anisate, Haemanthus albiflos, Artemisia afra*	—	South Africa	[58]
2)	*Zanthoxylum leprieurii, Piptadeniastrum africanum, Albizia coriaria,* and *Rubia cordifoli*	Leaves, roots and stem	Mpigi and Butambala regions of Uganda	[59]
3)	*Buddleja salviifolia, Clematis brachiate, Dianthus basuticus, Dicoma anomala* subsp. *anomala, Drimia depressa, Elephantorrhiza elephantina, Eragrostis plana, Gazania krebsiana* subsp. *krebsiana, Helichrysum caespititium, Helichrysum odoratissimum, Helichrysum pallidum, Hypoxis hemerocallidea, Leobordea lanceolata, Malva parviflora* var. *parviflora, Metalasia muricate, Morella serrata, Othonna natalensis, Pennisetum glaucum, Pentanisia prunelloides, Senecio asperulus, Trifolium burchellianum* subsp. *burchellianum, Withania somnifera, Xysmalobium undulatum, Zantedeschia albomaculata* subsp. *Albomaculata*	Leaves	Lesotho (South Africa)	[60]
4)	*Zanthoxylum capense*	Root	Mozambique	[61]
5)	*Sterculia setigera*	Leaves	Nigeria	[62]
6)	*Galenia africana*	Leaves	South Africa	[63]
7)	*Dodonaea angustifolia*	Leaves	South Africa and Ethiopia	[63]
8)	*Acalypha indica* and *Adhatoda vasica*	Leaves	India	[64]
9)	*Allium cepa* Bulb	Bulb	India	[64]
10)	*Aloe vera*	Leaves	Punjab (India)	[64]
11)	*Calpurnia aurea*	Roots	—	[65]
12)	*Ocimum basilicum*	Seeds	Ethiopia	[66]

Table 22.2 (Continued)

S/N	Name of plant	Plant parts	Country/region	References
13)	*Artemisia abyssinica, Croton macrostachyus, Eucalyptus camaldulensis*	Leaves	Ethiopia	[66]
14)	*Excoecaria agallocha, Aegiceras corniculatum, Avicennia officinalis,* and *Achyrocline alata*	Leaves and stems	Colombia Bueno	[67]
15)	*Swinglea glutinosa*	Fruit	Colombia	[68]
16)	*Piper nigram*	Seeds	Maharashtra (India)	[68]
17)	*Andrographis paniculata*	Leaves	Madhya Pradesh (India)	[69]
18)	*Uvaria rufa*	Leaves	Philippines	[70]
19)	*Globularia alypum*	Leaves	Tunisia	[71]
20)	*Calotropis gigantean*		India	[72]
21)	*Vetiveria zizanioides*	Roots	India	[73]
22)	*Alstonia scholaris*	Bark	Tamil Nadu (India)	[74]
23)	*Pavetta corymbosa*	Leaves	Nigeria	[75]
24)	*Euphorbia scarlatica*	Stem	Kenya	[76]
25)	*Mallotus philippensis*	Leaves	Uttaranchal (India)	[64]
26)	*Alpinia galanga*	Rhizomes	Uttaranchal (India)	[77]
27)	*Syzygium aromaticum*	Flower	Tamil Nadu (India)	[77]
28)	*Cinnamomum verum*	Bark	Tamil Nadu (India)	[78]
29)	*Allium ursinum*		Colombia	[78]
30)	*Calophyllum brasilience, Vismia baccifera, Vismia maxicana*	Bark and leaf	Mexico	[79]
31)	*Dodonaea angustifolia, Petrolobium stellatum*		Ethiopia	[78]
32)	*Crinum jagus*	Bulb	Nigeria	[80]
33)	*Scadoxus multiflorus, Acacia nilotica*		Kenya	[76]
34)	*Vitex negundo*	Leaf, seeds	Punjab (India)	[81]

active ingredients [61]. *Bocconia arborea* leaves and roots contain 6-methoxydihydrochelerytrin, Pinostrobin, 1-hydroxy-benzoisochromanquinone, 23-hydroxy-5a-lanosta-7,9 and 24-triene-3-one as active components [83]. *Achyrocline alata* stem and leaves contain Carvacrol as active component while *Swinglea glutinosa* fruit peel contains Thymol, *p*-cymene, *p*-cymene, 1,8-cineole, limonene and β-pinene as active components [67]. *Aloe vera* leaves have barbaloin as an active ingredient according to Pandey et al. [84]. *Andrographis paniculata* leaves according to Shrivastava and Garg [69], contain andrographolide. *Cirtullus colosnthis* aerial parts and deseeded fruits contains ursolic acid, cucurbitacine E, and cucurbitacin I according to Mehta et al. [85]. *Adhatoda vasica* leaves contains vasicine acetate, 2-acetyl benzylamine as reported by Gupta et al. [64]. *Trichosanthes dioica* stem and leaves have been reported by Arya [81] to contain, berberine, columbin, chasmanthin, palmarin, tinosporon, tinosporic acid, and tinosporol as active components. *Caesalpinia pulcherrima* leaves and flowers contain Myricitroside while *Ocimum sanctum* flowers, leaves and seeds contain ursolic acid, apigenin, orientin, luteolin, apigenin-7-Oglucuronide, and luteolin-7-Oglucuronide as active components according to Birdi et al. [68].

According to Lall et al. [86] *Morinda citrifolia*, Leaves, roots, and fruits contains anthraquinone salizarin, nordamnacanthol, ursolic acid; β-sitosterol, asperuloside, and caproic acid while *Myrtus communis* fruits contain Myricetin, kaempferol, α-pinene, cineole, myrtenol, nerol, geraniol, and dipentene. *Vitex trifolia* leaves, roots, and fruits containing Artemeti, luteolin, orientin, casticin, iridoid, aucubin, agnuside, and vitricin. *Mimosa pudica* Leaves, roots Mimosine, and turgorin while *Piper nigrum* Fruits Piperine as active component. Lall et al. [86] also reported that *Helichrysum melanacme* shoots contains 2,4,6-trihydroxy-3-prenylc halcone, 4,6,5-trihydroxy-6 and dimethydihydroxyoyrano [2,3-2,3] chalcone as active ingredients with antimycobacterial potential. *Garcinia polyantha* Stem bark according Kuete et al. [87] contain 1,5-dihydroxyxanthone, bangang xanthone, 1,3,5-trihydroxyxanthone and tetrahydroxyxanthone as active components. *Garcinia livingstonei* contains the active substance amentoflavone and 4′-methoxyamentoflavone as reported by Kaikabo and Eloff [88]. *Terminalia avicennioides* root bark contains arjunolic acid, friedelin, and friedelin-3β-ol as active substances [89] with anti-mycobacterial properties. *Combretum imberbe* leaves contain the active substance hydroxyimberbic acid, glycosides of hydroxyimberbic acid according to Katerere et al. [90]. Tangmouo et al. [91], reported the presence of active substances like Crassiflorone, diospyrone, plumbagin in *Diospyros crassiflora* stem bark. *Euclea natalensis* roots was reported by Van der Kooy et al. [92] to contain diospyrin, isodiospyrin, neodiosspyrin 7-methyljuglone, mamegakinone and Shinanolone while *Fichus cordata* stem bark contains β-sitosterol-3-β-glucopyranoside, catechin, and Epiafzelechin. Kokusaginine was reported to be the active substances in *Balsamocritus camerunensis* stem bark [93]. *Oricia suaveolens* stem bark contain chemical components such as Evoxanthine and 1-hydroxy-2,3-dimethoxy-10-methylacridone. *Ajuga remota* Whole plant contain ergosterol-5-8-endoperoxide [94].

22.5 Phytochemicals Against Multidrug-Resistant Tuberculosis

Different classes of plants exist, with angiosperms being one of the largest (about 45 000 species so far identified). Out of the 45 000 species of angiosperms, only 5–15% of them have been extensively investigated for their biological activity according to Cragg et al. [95]. A number of these plants, either as a whole or as one of their components, serve as prototypes for modern medicine [96–101]. Plants contain a variety of chemical substances that are of high benefits to ensure the healthy living of man [102].

Several investigations have been conducted into the *in vitro* antimycobacterial activity of plants especially those reported to have medicinal properties. About 25% of the drugs dispensed in developed countries like the United States contain at least one active ingredient that is derived from plant materials while others are synthesized to mimic a plant metabolite [103]. About 13 antimycobacterial compounds including, humulone, lupulone, geraniol, allicin, nevadensin, berberine, tuberosin, alantolactone, helenine, berberastine, ailanthone, andrographolide, and bakuchiol were reported from various plants [104]. Rifampicin is a semi-synthetic drug derived from rifamycin which is produced by *Amycolatopsis mediterranei* [105].

M. tuberculosis is fast becoming a major disease of concern because it is rapidly acquiring resistance against conventional drugs, thus frightening the global health community [106]. MDR-*M. tuberculosis* requires much longer treatment regimen which are effective than those for *M. tuberculosis* strains which have not acquired resistance to antibiotics. *Andrographis paniculata, Annona muricata, Centella asiatica,* and certain Indonesian plants have demonstrated excellent potentials in the containment of MDR-*M. tuberculosis*. Their phytochemicals extracted using water, methylene chloride, ethanol, *n*-hexane, and ethyl acetate significantly inhibited MDR-*M. tuberculosis* tested *in vitro* [5]. Similarly, phytochemicals from *Allium sativum* [107], *Humulus lupulus* [108], *Hypericum* sp. [109], *Diospyros anisandra* [110], *Croton tonkinensis* [111], and a host of other plants have been reported to show inhibitory activity against many drug resistant strains of *M. tuberculosis*.

22.6 Phenolic Compounds as Potential Drugs for Tuberculosis

Phenolic compound is a diverse and widely distributed class of phytochemical among the plant kingdom with about 8000 structure known and reported. These phenolic compounds is made up of aromatic phenyl ring attached with hydroxyl substituents. C6–C3 phenylpropanoid subunit is the most abundant carbon skeleton/building block for phenolic compounds. Phenolic compounds are all aromatic and pigmented which can be visualized. Phenolics mostly combine with sugars like glycosides and this contributes to their water solubility. Furthermore, phenolic compounds can broadly be divided into flavonoids and non-flavonoids on the basis of their structure [112–115]. Some of these phenolic compounds target major biosynthetic pathways in mycobacteria thereby inhibiting their growth. For instance, pinocembrin isolated from *Alpinia katsumadai* inhibits *Mycobacterium tuberculosis* with MIC of 128 μg/ml [116]. Other examples of phenolic compounds that are isolated from plants with anti-*Mycobacterium tuberculosis* properties which are believed to have the potential for the treatment of TB include flavonoid 7-demethylartonol E (MIC = 50 μg/ml), chromone artorigidusin (MIC = 12.5 μg/ml), artonol (MIC = 100 μg/ml), cycloartobixanthone (MIC = 25 μg/ml) isolated from *Artocarpus rigidus*, monospermoside (MIC = 25 μg/ml), formononetin (MIC = 50 μg/ml) derived from *Butea monosperma* L, acetin isolated from *Chromolaena odorata* L (MIC = 175 μg/ml), kanzanol C from *Dorstenia barteri* (MIC = 2.44 μg/ml) and naringenin (MIC = 30 μg/ml) derived from *Ficus nervosa* [115, 117–120].

22.7 Mechanism of Action of Phenolic Compounds Against Mycobacterium

In recent years, MDR-TB and XDR-TB have posed a great challenge to public health in dealing with tuberculosis. MDR-TB results when a strain of *Mycobacterium tuberculosis* is resistant to FLD (INH and rifampicin) of TB while XDR-TB is as a result of *M. tuberculosis* being resistant to FLD, at least one of three SLD (kanamycin, capreomycin, and amikacin) that are injectable and any fluoroquinolones. MDR and XDR in TB could be acquired or intrinsic. Acquired resistance is associated with chromosomal mutations in the genes that codes for drug activating enzymes or targets [115, 116, 121]. This leads to altered target protein configuration thereby reducing the susceptibility of the organism to an administered drug. It should be noted that the intrinsic TB resistance is mostly due to human actions such as poor physician prescription, poor patient adherence, poor supply or quality of drugs, and difference of metabolism and nutrition. Moreover, intrinsic TB resistance is due to the permeability of cell wall and efflux pumps systems that allows only selected particles (hydrophilic or hydrophobic), to enter into the bacteria cell while expelling foreign materials out from the cell [115, 116, 122–124]. This type of resistance reduces the effectiveness of current antibiotics in use by preventing them from entering into the mycobacteria cell (their protein targets) and expelling them before reaching their targets [125]. Phenolic compounds have been reported to serve as efflux system inhibitor, proteasome (provides intrinsic resistance) inhibitor, and mycolic acid biosynthesis inhibitor as well as inhibition of nitric oxides by macrophages [115].

22.8 Future Projections

It is safe to assert that plant resources have not yet been exploited to the extent required to ensure the control of MDR-tuberculosis and other pathogens notwithstanding the fact that phytochemicals may not be as effective as some drugs of synthetic chemical/microbial origin. But even at that, the diversity of plant products should allow synergistic effects in combinations for the control of pathogens. In plants particularly the angiosperms, only less than 10% have been screened for natural products discovery [126]. Therefore, there is need for more approaches in harnessing the plant natural bioactive compounds for the development of new drugs to treat tuberculosis especially MDR-TB. To ascertain the active ingredient in these plants, the use of techniques such as high-performance liquid chromatography coupled with mass spectrometry, liquid chromatography–mass spectrometry, liquid chromatography–nuclear magnetic resonance (NMR) – mass spectrometry, capillary NMR spectroscopy and LC–solid-phase extraction (SPE)–NMR along with bioassay-guided fractionation as well as high-throughput bioassays should be encouraged [127]. If well utilized, natural products, phenolic compounds, and other secondary metabolites derived from plants are potential reservoirs for development of new drugs to combat TB, especially MDR-TB.

22.9 Conclusion

Plants will continue to be excellent source of secondary metabolites with active chemical molecules against MDR and XDR strains of mycobacteria. Several studies have reported active molecules from plant extracts with high MIC against MDR strains. Literature review of both traditional knowledge and analysis of plants phytochemicals for antimycobacterial activity propose plants as potential source of antimycobacterial drugs that are cost effective, with less adverse side effect than conventional drugs. Despite the high MIC of plant extract against antimycobacterial they still have resistance modifying properties that enable them to combat drug resistant-TB.

References

1 Hogg, S. (2005). *Essential Microbiology*, 15–422. England: John Wiley & Sons, Ltd.

2 Giftillda, S.E.T., Poonkuzhali, S., Sathiya, H.X., and Velpandian, V. (2018). Anti-tuberculosis potential of siddha herbs – a review. *European Journal of Pharmaceutical and Medical Research* 5 (12): 551–555.

3 Hotez, P.J. (2013). NTDs V.2.0: "blue marble health" – neglected tropical disease control and elimination in a shifting health policy landscape. *PLoS Neglected Tropical Diseases* 7 (11): e2570.

4 Hotez, P.J. and Kamath, A. (2009). Neglected tropical diseases in sub-saharan Africa: review of their prevalence, distribution and disease burden. *PLoS Neglected Tropical Diseases* 3 (8): e412.

5 Chaudhary, N., Mehra, V., Mago, P., and Khatri, M. (2017). Antimycobacterial potential of Indian spices: a review. *International Journal of Pharmacognosy and Phytochemical Research* 9 (8): 1127–1134.

6 Hotez, P.J., Aksoy, S., Brindle, P.J., and Kamhawi, S. (2020). What constitutes a neglected tropical disease? *PLoS Neglected Tropical Diseases* 14 (1): e0008001.

7 Newman, D.J., Cragg, G., and Snader, K.M. (2000). The influence of natural products upon drug discovery. *Natural Product Reports* 17: 215–234.

8 Ngwoke, K.G., Odimegwu, D.C., and Esimone, C.O. (2011). Antimicrobial natural products. In: *Science Against Microbial Pathogens: Communicating Current Research and Technology Advances* (ed. A. Mendez-Vilas). Badajoz, Spain: FORMATEX.

9 De, A.K. (2004). *Spices: Traditional Uses and Medicinal Properties*, vii–xvii. Daryaganj: Asian Books Pvt Ltd.

10 Ayvazian, L.F. (1993). History of tuberculosis. In: *Tuberculosis: A Comprehensive International Approach* (eds. L.B. Reichman and E. Hershfield), 1–20. New York: Marcel Dekker Inc.

11 Muniyan, R. and Gurunathan, J. (2017). Antimycobacterial activity of potential plant metabolites with emphasis on management of drug resistant Mycobacterium tuberculosis strains. *Research Journal of Biotechnology* 12 (12): 75–86.

12 WHO (2004). The World Health report, changing history. http://www.who.int/whr/2004/en/report04_en.pdf (accessed 28 October 2019).

13 Christopher, D., Suzanne, S., Paul, D. et al. (1999). Global burden of tuberculosis: estimated incidence, prevalence and mortality by country. *JAMA* 282: 677–686.

14 WHO (2013). Global tuberculosis report, 1–145. http://www.who.int/tb/publications/global_report/en (accessed 28 December 2019).

15 Radji, M., Kurniati, M., and Kiranasari, A. (2015). Comparative antimycobacterial activity of some Indonesian medicinal plants against multi-drug resistant *Mycobacterium tuberculosis*. *Journal of Applied Pharmaceutical Science* 5: 19–22.

16 WHO (2017). Global tuberculosis report 2017. Geneva, Switzerland: WHO. http://apps.who.int/iris/bitstream/10665/259366/1/9789241565516-eng.pdf (accessed 22 November 2019).

17 Wang, J.Y., Hsueh, P.R., Jan, I.S. et al. (2006). Empirical treatment with a fluoroquinolone delays the treatment for tuberculosis and is associated with a poor prognosis in endemic areas. *Thorax* 61 (10): 903–908.

18 Chan, E.D. and Iseman, M.D. (2002). Current medical treatment for tuberculosis. *British Medical Journal* 325: 1282–1286.

19 Ma, Z., Lienhardt, C., McIlleron, H. et al. (2010). Global tuberculosis drug development pipeline: the need and the reality. *Lancet* 375: 2100–2109.

20 Espinal, M.A., Laszlo, A., Simonsen, L. et al. (2001). Global trends in resistance to antituberculosis drugs. World health organization-international union against tuberculosis and lung disease working group on anti-tuberculosis drug resistance surveillance. *New England Journal of Medicine* 344: 1294–1303.

21 Masjedi, M.R., Farnia, P., Sorooh, S. et al. (2006). Extensively, drug-resistant tuberculosis: 2 years of surveillance in Iran. *Clinical Infectious Diseases* 43 (7): 841–847.

22 Bernstein, J., Lott, W.A., Steinberg, B.A., and Yale, H.L. (1952). Chemotherapy of experimental tuberculosis. V. Isonicotinic acid hydrazide (nydrazid) and related compounds. *American Review of Tuberculosis* 65 (4): 357–364.

23 Bardou, F., Raynaud, C., Ramos, C. et al. (1998). Mechanism of isoniaziduptake in *Mycobacterium tuberculosis*. *Microbiology* 144 (Pt 9): 2539–2544.

24 Ramaswamy, S. and Musser, J.M. (1998). Molecular genetic basis of antimicrobial agent resistance in *Mycobacterium tuberculosis*: 1998 update. *Tubercle and Lung Disease* 79 (1): 3–29.

25 Mitchison, D.A. (1979). Basic mechanisms of chemotherapy. *Chest* 76 (6 Suppl): 771–781.

26 Telenti, A., Imboden, P., Marchesi, F. et al. (1993). Direct, automated detection of rifampin-resistant *Mycobacterium tuberculosis* by polymerase chain reaction and single-strand conformation polymorphism analysis. *Antimicrobial Agents and Chemotherapy* 37 (10): 2054–2058.

27 Takayama, K. and Kilburn, J. (1989). Inhibition of synthesis of arabinogalactan by ethambutol in *Mycobacterium smegmatis*. *Antimicrobial Agents and Chemotherapy* 33 (9): 1493–1499.

28 Zhang, Y. and Yew, W.W. (2009). Mechanisms of drug resistance in *Mycobacterium tuberculosis*. *The International Journal of Tuberculosis and Lung Disease* 13 (11): 1320–1330.

29 Konno, K., Feldmann, F.M., and Mcdermott, W. (1967). Pyrazinamide susceptibility and amidase activity of tubercle bacilli. *The American Review of Respiratory Disease* 95 (3): 461–469.

30 Mcdermott, W. and Tompsett, R. (1954). Activation of pyrazinamide and nicotinamide in acidic environment in vitro. *American Review of Tuberculosis* 70 (4): 748–754.

31 Mitchison, D.A. (1985). The action of antituberculosis drugs in short-course chemotherapy. *Tubercle* 66 (3): 219–225.

32 Gale, E.F., Cundliffe, E., Reynolds, P.E. et al. (1981). *The Molecular Basis of Antibiotic Action*. New York: John Wiley and Sons, Inc.

33 Zhang, Y., Heym, B., Allen, B. et al. (1992). The catalase/peroxidase gene and isoniazid resistance of *Mycobacterium tuberculosis*. *Nature* 358 (6387): 591–593.

34 Douglass, J. and Steyn, L.M. (1993). A ribosomal gene mutation in streptomycin-resistant *Mycobacterium tuberculosis* isolates. *The Journal of Infectious Diseases* 167 (6): 1505–1506.

35 Spies, F.S., da Silva, P.E., Ribeiro, M.O. et al. (2008). Identification of mutations related to streptomycin resistance in clinical isolates of *Mycobacterium tuberculosis* and possible involvement of efflux mechanism. *Antimicrobial Agents and Chemotherapy* 52 (8): 2947–2949.

36 Bartlett, J.G., Dowell, S.F., Mandell, L.A. et al. (2000). Practice guidelines for the management of community-acquired pneumonia in adults. Infectious Diseases Society of America. *Clinical Infectious Diseases* 31 (2): 347–382.

37 De Barber, A.E., Mdluli, K., Bosman, M. et al. (2000). Ethionamide activation and sensitivity in multi-drug resistance *Mycobacterium tuberculosis*. *Proceedings of the National Academy of Sciences* 97: 9677–9682.

38 Banerjee, A., Dubnau, E., Quemard, A. et al. (1994). inhA, a gene encoding a target for isoniazid and ethionamide in *Mycobacterium tuberculosis*. *Science* 263: 227–230.

39 Kocagöz, T., Hackbarth, C.J., Unsal, I. et al. (1996). Gyrase mutations in laboratory selected, fluoroquinolone resistant mutants of *Mycobacterium tuberculosis H37Ra*. *Antimicrobial Agents and Chemotherapy* 40 (8): 1768–1774.

40 Jugheli, L., Bzekalava, N., DeRijk, P. et al. (2009). High level of cross-resistance between kanamycin, amikacin, and capreomycin among *Mycobacterium tuberculosis* isolates from Georgia and a close relation with mutations in the *rrs* gene. *Antimicrobial Agents and Chemotherapy* 53 (12): 5064–5068.

41 Rengarajan, J., Sassetti, C.M., Naroditskaya, V. et al. (2004). The folate pathway is a target for resistance to the drug para-aminosalicylic acid (PAS) in mycobacteria. *Molecular Microbiology* 53 (1): 275–282.

42 Klopper, M., Warren, R.M., Hayes, C. et al. (2013). Emergence and spread of extensively and totally drug-resistant tuberculosis, South Africa. *Emerging Infectious Diseases* 19: 449–455.

43 Marra, F., Marra, C.A., Bruchet, N. et al. (2007). Adverse drug reactions associated with firstline anti-tuberculosis drug regimens. *The International Journal of Tuberculosis and Lung Disease* 11: 868–875.

44 WHO (2012). Drug resistant tuberculosis. http://www.who.int/tb/challenges/mdr/tdrfaqs/en (accessed 28 October 2019).

45 Sibanda, T. and Okoh, A.I. (2007). The challenges of overcoming antibiotic resistance: plant extracts as potential sources of antimicrobial and resistance modifying agents. *African Journal of Biotechnology* 6: 2886–2896.

46 Venkitanarayana, K. (2014). The potential for plant derived antimicrobials for controlling zoonotic and food-borne diseases. In: *Proceedings of the 2nd International Conference on Animal and Dairy Sciences, 15–17 September 2014.* Hyderabad International Convention Centre: India.

47 Radulovic, N.S., Blagojevic, P.D., Stojanovic-Radic, Z.Z., and Stojanovic, N.M. (2013). Antimicrobial plant metabolites: structural diversity and mechanism of action. *Current Medicinal Chemistry* 20: 932–952.

48 Ahmad, A., Khan, A., Akhtar, F. et al. (2011). Fungicidal activity of thymol and carvacrol by disrupting ergosterol biosynthesis and membrane integrity against *Candida. European Journal of Clinical Microbiology and Infectious Diseases* 30: 41–50.

49 Sikkema, J., de Bont, J.A., and Poolman, B. (1995). Mechanisms of membrane toxicity of hydrocarbons. *Microbiology and Molecular Biology Reviews* 59: 201–222.

50 Koh, K.H. and Tham, F.Y. (2011). Screening of traditional Chinese medicinal plants for quorum-sensing inhibitors activity. *Journal of Microbiology, Immunology and Infection* 44: 144–148.

51 Pan, J. and Ren, D. (2009). Quorum sensing inhibitors: a patent overview. *Expert Opinion on Therapeutic Patents* 19: 1581–1601.

52 Adwan, G. and Mhanna, M. (2008). Synergistic effects of plant extracts and antibiotics on *Staphylococcus aureus* strains isolated from clinical specimens. *Middle-East Journal of Scientific Research* 3: 134–139.

53 Braga, L.C., Leite, A.A.M., Xavier, K.G.S. et al. (2005). Synergic interaction between pomegranate extract and antibiotics against *Staphylococcus aureus. Canadian Journal of Microbiology* 51: 541–547.

54 Ulrich-Merzenich, G., Panek, D., Zeitler, H. et al. (2009). New perspectives for synergy research with the omic-technologies. *Phytomedicine* 16: 495–508.

55 Efferth, T. and Koch, E. (2011). Complex interactions between phytochemicals. The multi-target therapeutic concept of phytotherapy. *Current Drug Targets* 12: 122–132.

56 Simoes, M., Bennett, R.N., and Rosa, E.A.S. (2009). Understanding antimicrobial activities of phytochemicals against multidrug resistant bacteria and biofilms. *Natural Product Reports* 26: 746–757.

57 Wink, M. (2008). Evolutionary advantage and molecular modes of action of multi-component mixtures used in phytomedicine. *Current Drug Metabolism* 9: 996–1009.

58 Lawal, I.O., Grierson, D.S., and Afolayan, A.J. (2014). Phytotherapeutic information on plants used for the treatment of tuberculosis in Eastern Cape province, South Africa. *Evidence-based Complementary and Alternative Medicine* https://doi.org/10.1155/2014/735423.

59 Bunalema, L., Obakiro, S., Tabuti, J.R.S., and Waako, P. (2014). Knowledge on plants used traditionally in the treatment of tuberculosis in Uganda. *Journal of Ethnopharmacology* 151: 999–1004.

60 Kose, L.S., Moteetee, A., and van Vuuren, S. (2015). Medicinal plants used for the treatment of tuberculosis in Lesotho: an ethnobotanical survey. *South African Journal of Botany* 98: 183–183.

61 Luo, X., Pires, D., Ainsa, J.A. et al. (2013). *Zanthoxylum capense* constituents with antimycobacterial activity against *Mycobacterium tuberculosis in vitro* and *ex vivo* within human macrophages. *Journal of Ethnopharmacology* 146: 417–422.

62 Babalola, I.T., Adelakun, E.A., Wang, Y., and Shode, F.O. (2012). Anti-TB activity of *Sterculia setigera* Del., leaves (Sterculiaceae). *Journal of Pharmacognosy and Phytochemistry* 1: 17–23.

63 Mativandlela, S.P.N. (2009). Antituberculosis activity of flavonoids from *Galenia africana* L. var. Africana. PhD thesis, Faculty of Natural and Agricultural Sciences, University of Pretoria, Pretoria.

64 Gupta, R., Thakur, B., Singh, P. et al. (2010). Anti-tuberculosis activity of selected medicinal plants against multi-drug resistant *Mycobacterium tuberculosis* isolates. *Indian Journal of Medical Research* 131: 809–813.

65 Gemechu, A., Giday, M., Worku, A., and Ameni, G. (2013). *In vitro* anti-mycobacterial activity of selected medicinal plants against *Mycobacterium tuberculosis* and *Mycobacterium bovis* Strains. *BMC Complementary and Alternative Medicine*: 13. https://doi.org/10.1186/1472-6882-13-291.

66 Muthuswamy, M., Asan, P., and Kumar, V. (2013). Screening of antitubercular activity of some medicinal plants from Western Ghats, India. *International Journal of Pharma and Bio Sciences* 4: 328–334.

67 Bueno-Sanchez, J.G., Martinez-Morales, J.R., Stashenko, E.E., and Ribon, W. (2009). Anti-tubercular activity of eleven aromatic and medicinal plants occurring in Colombia. *Biomédica* 29: 51–60.

68 Birdi, T., D'Souza, D., Tolani, M. et al. (2012). Assessment of the activity of selected Indian medicinal plants against *Mycobacterium tuberculosis*: a preliminary screening using the microplate alamar blue assay. *European Journal of Medicinal Plants* 2: 308–323.

69 Shrivastava, A. and Garg, H.K. (2014). Effect of andrographolide on proliferation of *Mycobacterium canetti*. *International Journal for Pharmaceutical Research Scholars* 1: 410–413.

70 Macabeo, A.P.G., Tudla, F.A., Krohn, K., and Franzblau, S.G. (2012). Antitubercular activity of the semi-polar extractives of *Uvaria rufa*. *Asian Pacific Journal of Tropical Medicine* 5: 777–780.

71 Khlifi, D., Hamdi, M., El Hayouni, A. et al. (2011). Global chemical composition and antioxidant and anti-tuberculosis activities of various extracts of *Globularia alypum* L. (Globulariaceae) leaves. *Molecules* 16: 10592–10603.

72 Kumar, J.K., Prasad, A.G.D., and Chaturvedi, V. (2015). Phytochemical screening of five medicinal legumes and their evaluation for *in vitro* anti-tubercular activity. *AYU* 35: 98–101.

73 Saikia, D., Parveen, S., Gupta, V.K., and Luqman, S. (2012). Anti-tuberculosis activity of Indian grass KHUS (*Vetiveria zizanioides* L. Nash). *Complementary Therapies in Medicine* 20: 434–436.

74 Antony, M., James, J., Misra, C.S. et al. (2012). Antimycobacterial activity of the plant extracts of *Alstonia scholaris*. *International Journal of Current Pharmaceutical Research* 4: 40–42.

75 Nvau, J.B., Oladosu, P.O., and Orishadipe, A.T. (2011). Antimycobacterial evaluation of some medicinal plants used in plateau state of Nigeria for the treatment of tuberculosis. *Agriculture and Biology Journal of North America* 2: 1270–1272.

76 Mariita, R.M., Ogol, C.K.P.O., Oguge, N.O., and Okemo, P.O. (2010). Antitubercular and phytochemical investigation of methanol extracts of medicinal plants used by the samburu community in Kenya. *Tropical Journal of Pharmaceutical Research* 9: 379–385.

77 Shivakumar, A. and Jayaraman, G. (2011). Anti-tuberculosis activity of commonly used medicinal plants of south India. *Journal of Medicinal Plant Research: Planta Medica* 5 (31): 6881–6884.

78 Balcha, E., Mengiste, B., Gebrelibanos, M. et al. (2014). Evaluation of *in-vitro* anti-mycobacterial activity of selected medicinal plants in Mekelle, Ethiopia. *World Applied Sciences Journal* 31: 1217–1220.

79 Gomez-Cansino, R., Espitia-Pinzon, C.I., Campos-Lara, M.G. et al. (2015). Antimycobacterial and HIV-1 reverse transcriptase activity of julianaceae and clusiaceae plant species from Mexico. *Evidence-Based Complementary and Alternative Medicine* 2015: 1–8. https://doi.org/10.1155/2015/183036.

80 Akintola, A.O., Kehinde, A.O., Adebiyi, O.E., and Ademowo, O.G. (2013). Anti-tuberculosis activities of the crude methanolic extract and purified fractions of the bulb of *Crinum jagus*. *Nigerian Journal of Physiological Sciences* 28: 135–140.

81 Arya, V. (2011). A review on anti-tubercular plants. *International Journal of PharmTech Research* 3: 872–880.

82 Adaikkappan, P., Kannapiran, M., and Anthonisamy, A. (2012). Anti-mycobacterial activity of *Withania somnifera* and *Pueraria tuberosa* against *Mycobacterium tuberculosis* H37Rv. *Journal of Academia and Industrial Research* 1: 153–156.

83 Camacho-Corona, M.R., Favela-Hernandez, J.M.J., Gonzalez-Santiago, O. et al. (2009). Evaluation of some plant-derived secondary metabolites against sensitive

and multidrug-resistant *Mycobacterium tuberculosis*. *Journal of the Mexican Chemical Society* 53: 71–75.

84 Pandey, D.K., Tabarak, M., and Banik, R.M. (2012). Quantitative estimation of barbaloin in *Aloe vera* and its commercial formulations by using HPTLC. *International Journal of Medicinal and Aromatic Plants* 2: 420–427.

85 Mehta, A., Srivastva, G., Kachhwaha, S. et al. (2013). Antimycobacterial activity of *Citrullus colocynthis* (L.) Schrad. against drug sensitive and drug resistant *Mycobacterium tuberculosis* and MOTT clinical isolates. *Journal of Ethnopharmacology* 149: 195–200.

86 Lall, N., Hussein, A.A., and Meyer, J.J.M. (2006). Antiviral and antituberculous activity of *Helichrysum melanacme* constituents. *Fitoterapia* 77: 230–232.

87 Kuete, V., Poumale, H.M.P., Guedem, A.N. et al. (2010). Antimycobacterial, antibacterial and antifungal activities of the methanol extract and compounds from *Thecacoris annobonae* (Euphorbiaceae). *South African Journal of Botany* 76: 536–542.

88 Kaikabo, A.A. and Eloff, J.N. (2011). Antibacterial activity of two biflavonoids from *Garcinia livingstonei* leaves against *Mycobacterium smegmatis*. *Journal of Ethnopharmacology* 138: 253–255.

89 Mann, A., Ibrahim, K., Oyewale, A.O. et al. (2011). Antimycobacterial friedelane-terpenoid from the root bark of *Terminalia avicennioides*. *American Journal of Chemistry* 1: 52–55.

90 Katerere, D.R., Gray, A.I., Nash, R.J., and Waigh, R.D. (2003). Antimicrobial activity of pentacyclic triterpenes isolated from African Combretaceae. *Phytochemistry* 63: 81–89.

91 Tangmouo, J.G., Meli, A.L., Komguem, J. et al. (2006). Crassiflorone, a new naphthoquinone from *Diospyros crassiflora* (Hien). *Tetrahedron Letters* 47: 3067–3070.

92 Van der Kooy, F., Meyer, J.J.M., and Lall, N. (2006). Antimycobacterial activity and possible mode of action of newly isolated neodiospyrin and other naphthoquinones from *Euclea natalensis*. *South African Journal of Botany* 72: 349–352.

93 Fouotsa, H., Mbaveng, A.T., Mbazoa, C.D. et al. (2013). Antibacterial constituents of three Cameroonian medicinal plants: *Garcinia nobilis*, *Oricia suaveolens* and *Balsamocitrus camerunensis*. *BMC Complementary and Alternative Medicine* 13 https://doi.org/10.1186/1472-6882-13-81.

94 Cantrell, C.L., Rajab, M.S., Franzblau, S.G. et al. (1999). Antimycobacterial ergosterol-5,8-endoperoxide from *Ajuga remota*. *Planta Medica* 65: 732–734.

95 Cragg, G.M., Newman, D.J., and Snader, K.M. (1997). Natural products in drug discovery and development. *Journal of Natural Products* 60: 52–60.

96 Egbuna, C., Mishra, A., and Goyal, M.R. (eds.) (2020). *Preparation of Phytopharmaceuticals for the Management of Disorders*, 1ste. Academic: Press (Elsevier).

97 Egbuna, C., Kumar, S., Ezzat, S.M. et al. (eds.) (2019). *Phytochemicals as Lead Compounds for New Drug Discovery*, 1ste. Cambridge, USA: Elsevier.

98 Shashank, K. and Egbuna, C. (eds.) (2019). *Phytochemistry: In vitro and In Silico Updates*, 1e. Singapore: *Springer Nature*.

99 Egbuna, C., Ifemeje, J.C., Udedi, S.C., and Kumar, S. (eds.) (2018). *Phytochemistry, Volume 1: Fundamentals, Methods, and Applications*, 1e. New York: Apple Academic Press/CRC Taylor & Francis.

100 Egbuna, C., Kumar, S., Ifemeje, J.C., and Kurhekar, J.V. (eds.) (2018). *Phytochemistry, Volume 2: Pharmacognosy, Nanomedicine, and Contemporary Issues*, 1e. New York: Apple Academic Press/CRC Taylor & Francis.

101 Egbuna, C., Ifemeje, J.C., Kumar, S., and Sharif, N. (eds.) (2018). *Phytochemistry, Volume 3: Marine, Industrial, and Advances*, 1e. New York: Apple Academic Press/CRC Taylor & Francis.

102 Egbuna, C. and Tupas, G.D. (eds.) (2020). *Functional Foods and Nutraceuticals – Bioactive Components, Formulations and Innovations*, 1e. Cham, Switzerland: Springer Nature.

103 Newman, D.J., Cragg, G.M., and Snader, K.M. (2003). Natural products as sources of new drugs over the period 1981–2002. *Journal of Natural Products* 66: 1022–1037.

104 Schauenberg, P. and Paris, F. (1990). *Guide to Medicinal Plants*, First paperback edition (Paris 1974). Lutterworth Press.

105 Tribuddharat, C. and Fennewald, M. (1999). Integron-mediated rifampin resistance in *Pseudomonas aeruginosa*. *Antimicrobial Agents and Chemotherapy* 43 (4): 960–962.

106 Zignol, M., Hosseini, M.S., Wright, A. et al. (2006). Global incidence of multidrug-resistant tuberculosis. *Journal of Infectious Diseases* 194: 479–485.

107 Hannan, A., Ikram Ullah, M., Usman, M. et al. (2011). Anti-mycobacterial activity of garlic (*Allium sativum*) against multi-drug resistant and non-multi-drug resistant *Mycobacterium tuberculosis*. *Pakistan Journal of Pharmaceutical Sciences* 24: 81–85.

108 Serkani, J.E., Isfahani, B.N., Safaei, H.G. et al. (2012). Evaluation of the effect of *Humulus lupulus* alcoholic extract on rifampin-sensitive and resistant isolates of *Mycobacterium tuberculosis*. *Research in Pharmaceutical Science* 7: 235–242.

109 Nogueira, T., Medeiros, M.A., Marcelo-Curto, M.J. et al. (2013). Profile of antimicrobial potential of fifteen *Hypericum* species from Portugal. *Industrial Crops and Products* 47: 126–131.

110 Uc-Cachon, A.H., Borges-Argaez, R., Said-Fernandez, S. et al. (2014). Naphthoquinones isolated from *Diospyros anisandra* exhibit potent activity against pan-resistant first-line drugs *Mycobacterium tuberculosis* strains. *Pulmonary Pharmacology & Therapeutics* 27: 114–120.

111 Jang, W.S., Jyoti, M.A., Kim, S. et al. (2016). In vitro antituberculosis activity of diterpenoids from the Vietnamese medicinal plant *Croton tonkinensis*. *Journal of Natural Medicines* 70: 127–132.

112 Irchhaiya, R., Kumar, A., Yadav, A. et al. (2015). Metabolites in plants and its classification. *World Journal of Pharmacy and Pharmaceutical Science* 4: 287–305.

113 Harbone, J.B. (1973). *Phytochemical Methods: A Guide to Modern Techniques of Plant Analysis*. London, UK: Chapman and Hall Ltd.

114 Pereira, D.M., Valentão, P., Pereira, J.A., and Andrade, P.B. (2009). Phenolics: from chemistry to biology. *Molecules* 14: 2202–2211.

115 Mazlun, M.H., Sabran, S.F., Mohamed, M. et al. (2019). Phenolic compounds as promising drug candidates in tuberculosis therapy. *Molecules* 24: 2449.

116 Groblacher, B., Kunert, O., and Bucar, F. (2012). Compounds of *Alpinia katsumadai* as potential efflux inhibitors in *Mycobacterium smegmatis*. *Bioorganic & Medicinal Chemistry* 20: 2701–2706.

117 Namdaung, U., Aroonrerk, N., Suksamrarn, S. et al. (2006). Bioactive constituents of the root bark of *Artocarpus rigidus* subsp. rigidus. *Chemical & Pharmaceutical Bulletin* 54: 1433–1436.

118 Chokchaisiri, R., Suaisom, C., Sriphota, S. et al. (2009). Bioactive flavonoids of the flowers of *Butea monosperma*. *Chemical & Pharmaceutical Bulletin* 57: 428–432.

119 Suksamrarn, A., Chotipong, A., Suavansri, T. et al. (2004). Antimycobacterial activity and cytotoxicity of flavonoids from the flowers of *Chromolaena odorata*. *Archives of Pharmacal Research* 27: 507–511.

120 Chen, L.W., Cheng, M.J., Peng, C.F., and Chen, I.S. (2010). Secondary metabolites and antimycobacterial activities from the roots of *Ficus nervosa*. *Chemistry & Biodiversity* 7: 1814–1821.

121 Gandhi, N.R., Nunn, P., Dheda, K. et al. (2010). Multidrug-resistant and extensively drug-resistant tuberculosis: a threat to global control of tuberculosis. *Lancet* 375: 1830–1843.

122 Louw, G.E., Warren, R.M., Gey van Pittius, N.C. et al. (2009). A balancing act: efflux/influx in mycobacterial drug resistance. *Antimicrobial Agents and Chemotherapy* 5: 44–99.

123 Islam, M.M., Hameed, H.M.A., Mugweru, J. et al. (2017). Drug resistance mechanisms and novel drug targets for tuberculosis therapy. *Journal of Genetics and Genomics* 3: 13–34.

124 Rodrigues, L., Ramos, J., Couto, I. et al. (2011). Ethidiym bromide transport across *Mycobacterium smegmatis* cell wall: correlation with antibiotic resistance. *BMC Microbiology* 11: 35.

125 Darwin, K.H., Ehrt, S., Gutierrez-Ramos, J.-C. et al. (2003). The proteasome of *Mycobacterium tuberculosis* is required for resistance to nitric oxide. *Science* 302: 1963–1966.

126 Houghton, P.J. (2001). Old yet new – pharmaceuticals from plants. *Journal of Chemical Education* 78: 175–184.

127 Subramani, R., Narayanasamy, M., and Feussner, K.D. (2017). Plant-derived antimicrobials to fight against multi-drug-resistant human pathogens. *3 Biotech* 7 (3): 172. https://doi.org/10.1007/s13205-017-0848-9.

Index

a

Neglected Tropical Diseases and Phytochemicals in Drug Discovery, First Edition.
Edited by Chukwuebuka Egbuna, Muhammad Akram, and Jonathan Chinenye Ifemeje.

f

n

O

t

u

v

Printed in the USA/Agawam, MA
September 16, 2026

910661.016